Studies in Logic

Logic and Cognitive Systems

Volume 45

Errors of Reasoning

Naturalizing the Logic of Inference

Studies in Logic Series Editor
Dov Gabbay   dov.gabbay@kcl.ac.uk

Errors of Reasoning
Naturalizing the Logic of Inference

John Woods

University of British Columbia

and

University of Lethbridge

© Individual author and College Publications 2013.
All rights reserved.

ISBN 978-1-84890-114-8
Second corrected printing, 2014

College Publications
Scientific Director: Dov Gabbay
Managing Director: Jane Spurr
Department of Informatics
King's College London, Strand, London WC2R 2LS, UK

http://www.collegepublications.co.uk

Original cover design by Orchid Creative   www.orchidcreative.co.uk
Printed by Lightning Source, Milton Keynes, UK

I dedicate this book to the memory of

JONATHAN E. ADLER
1948-2012

Preface to the Second Printing

This second printing is an improved and corrected version of the first. We are grateful for the assistance in reformatting to Jane Spurr, and to all those who have helped out by spotting typos and the like, especially Frank Zenker and Woosuk Park.

Vancouver
20 July 2014

Preface to the First Printing

This book started out to be volume three of the omnibus work, *A Practical Logic of Cognitive Systems,* to be co-authored with my friend Dov Gabbay. The previous two volumes of that work are *Agenda Relevance: A Study in Formal Pragmatics* (2003) and *The Reach of Abduction: Insight and Trial* (2005). Each reflects a certain division of labour. I would write the philosophical chapters (collectively called "the conceptual model") and Dov would write the mathematical chapters (called there "the formal model"). Our aim was to bring our respective results into a serviceable harmony within the covers of a single book. This arrangement had some clear advantages. As one reviewer kindly observed, our joint authorship was to good logic as good song-writing teams are to good music. But there were disadvantages too. One was that the two models approach led to some rather large and very pricey books, with *Agenda Relevance* weighing in at 508 pages and *The Reach of Abduction* at a scarcely less hefty 476. A second difficulty was that the volumes' intended readership lacked the homogeneity we had tried to achieve in these books. In consequence, many would-be readers (more particularly, purchasers) would be faced with the decision to acquire a book, a significant chunk of which they were unlikely to read.

As planning started for volume three, we found ourselves drawn to the view that, whereas the division of labour is right, packaging its respective products in a single volume left more to say against it than for. We were aided in these reflections by a change of circumstances at our publisher, Elsevier. One of the highpoints of our professional lives was to have our work appear under the publisher's North-Holland imprint. But on the heels of a business-model restructuring, Elsevier decided that production of monographs, especially in philosophy, would be significantly curtailed. An early upshot of this turn-about was the death of the magnificent *Studies in Logic and the Foundations of Mathematics,* a very sad occasion in its own right and a powerful motivation for the subsequent founding of College Publications.

For all these reasons we have decided that, beginning with this one, any further volumes in the spirit of *A Practical Logic of Cognitive Systems* – a name to which we lay no further title – would be single-authored. The present volume, *Errors of Reasoning,* offers a philosophico-logical treatment of the logic of premiss-conclusion reasoning. A companion volume, whose unoficial title is *Formalizing the Logic of Error,* is in the works for Dov, and may appear in due course.

Still further volumes are contemplated, to be produced in this same new way. In preparing *Errors of Reasoning*, I have incurred many debts. Dov Gabbay and I have had a productive intellectual partnership for over a dozen years. It is a colleagueship graced by an affable and aggressive enthusiasm overlain by a nourishing friendship. In Dov's company I have never worked so hard or had so much fun. *Errors of Reasoning* is no exception. A number of its chapters would be less good than I hope they now are, but for Dov's eagle eye.

Ralph Johnson, Lorenzo Magnani and my Research Assistant Alirio Rosales have done me the kindness of reading drafts of the book in its entirety, and have favoured me with their helpful criticisms. Others have read various pieces of the book and have been constructive and generous with their candour. They are Adam Morton, Louise Cummings, Harvey Siegel, Peter Bruza, Maurice Finocchiaro and the late Jonathan Adler. Also of great help was correspondence or conversation with a number of people whose influence on various of the book's particularities is more or less palpable. I especially wish to acknowledge Nicholas Rescher, Johan van Benthem, Jeff Pelletier, Jeff Horty, the late John Pollock, Hans Hansen, Jaakko Hintikka, Patrick Suppes, David Makinson, Ivor Gratton-Guiness, the late Peter Houtlosser, Douglas Niño, Bryson Brown, Ori Simchen, Roberta Ballarin, Paul Bartha, Ulrike Hahn, Mike Oaksford, Andrew Irvine, the late L.J. Cohen, Franz Berto, Graham Priest, Stephen Read, Gabriella Pigozzi, Woosuk Park, Shahid Rahman, Juan Redmond, Ahti-Vikko Pietarinen, Mathieu Marion, Jean-Yves Béziau, Dale Jacquette, Paul Thagard, Atosha Aliseda, Nick Griffin, David Hitchcock, Mark Weinstein, Cyrille Imbert, Fabio Paglieri and Amirouche Moktefi.

Two short courses based on this book were delivered to the Institute of Logic and Cognition at Sun Yat-sen University in November 2012. I thank all involved for their friendly hospitality and colleagueship: Director of the Institute Shier Ju, Vice-Director Minghui Xiong, Jing Zhu, Wei Xiong and Yun Xie, and students Xiaojing Wu, Zhicong Liao and Wei Bin. A similar welcome awaited me at the South China Normal University, kindly arranged by Yi Zhao and Ming Xiong.

During the book's lengthy gestation I was also ably assisted by undergraduate Research Assistants Kian Mintz-Woo, Emily Cooley and Frank Hong. Their work was done efficiently and well, and was supplemented by many hours of smart conversation generously offered, from which I have learned a lot, and for which I extend my warmest thanks.

Portions of the book have appeared in various places elsewhere and reappear here in refurbished or much-corrected form. For permission to re-use I extend my thanks to editors/publishers/copyright holders of *The Review of Symbolic Logic, Logic Journal of IGPL, Topoi, Informal Logic, Argumentation,* Springer, Elsevier and College Publications.

In the first forty years of my academic career, thirty-three of them included administrative duties, as Head, Dean, or President. These days university administrators seem not to enjoy the confidence or affection of their non-administrative colleagues. No doubt there are cases in which these enmities are entirely warranted. But, overall, hating the "suits" is a stupid and undeserved default position. Heads, Deans, and Presidents are there to foster scholarship and research. I have had the good luck to have been on the receiving end of this support in all stages of the writing of this book. I would like to make a point of saying how much I value and appreciate the involvement of department Heads Bryson Brown, Michael Stingl and Kent Peacock (in Lethbridge), Tom Maibaum, Andrew Jones and Michael Luck (in London) and Mohan Matthen, Margaret Schabas, Paul Bartha and Alan Richardson (in Vancouver); of Deans Bhagwan Dua and Christopher Nicol (in Lethbridge), and Nancy Gallini and Gage Averill (in Vancouver); and of Presidents William Cade and Michael Mahoney (in Lethbridge), and Martha Piper (in Vancouver).

Work on the book has been supported in whole or part by grants from the Engineering and Physical Sciences Research Council of the United Kingdom, the Social Science and Humanities Research Council of Canada, the Abductive Systems Group, and the Dean's Discretionary Fund, Faculty of Arts and Science, University of Lethbridge, with thanks to all.

In all matters technical and administrative, Carol Woods runs things in Vancouver. Jane Spurr does the same in London. Without Jane, Dov would be in real estate. Without Carol, John would simply be out of business.

Vancouver
20 June 2013

CONTENTS

CHAPTER ONE

ERRING

"The philosophy of reasoning, to be complete, ought to comprise the theory of bad as well as of good reasoning."   John Stuart Mill

"There is no such thing as a classification of the ways in which men may arrive at an error; it is much to be doubted whether ever *can be.*"   Augustus De Morgan

"But the old connection [of logic] with *philosophy* is closest to my heart right now .... I hope that logic will have another chance in its mother area."   Johan van Benthem

1.1. *Errors of logic*

This is a book about premiss-conclusion reasoning, with a special emphasis on factors in play when it falls into error, when bad reasoning is mistaken for good. My point of departure is that logic hasn't done very well at this and, in a number of respects, its provisions for error are seriously misguided. Doubtless many readers will be contrary-minded. After all, isn't it true, especially in the variations that have come down to us in the past 140 years, that logic has achieved an extraordinary success? Isn't right and wrong reasoning what logic is about? How could it have come about that this extraordinary success hasn't done well – and sometimes has done badly – in dealing with logic's very subject matter?

Part of what I will say here is that the reason why it hasn't done well with reasoning is that logic is not suitably equipped to handle it well, and that a successful logic of reasoning will have to be rejigged and retrofitted in requisite ways. A further objective is to consider how this restructuring of logic might go. My proposals in this regard will include two of particular importance, made so in no small measure by the complexities and uncertainties they occasion for the new logic. One is that a logic of human reasoning as it occurs in the circumstances of real life will not do well unless it opens itself to instruction from the relevant empirical sciences. The other is that the logic we seek can't be got unless we adjust its provisions to the cognitive natures of real life reasoning agents. The first requirement calls on the new logic to be "empirically

1

sensitive". The second asks it to be "epistemologically aware". These are tall orders, entertained here rather more with experimental intent than axiomatic assurance.

As traditionally understood, logic investigates the principles of right reasoning. Since error is what right reasoning is meant to avoid, it would appear that a theory of reasoning must work up an account of bad reasoning. Although error is certainly in logic's traditional ambit, its typical treatment serves to play down the connection. The error in which mainstream logicians have the largest stake is invalidity. But deductive logic is a theory of *validity*. Invalidity trails along as a byproduct, as a kind of privation. It is the absence of validity. In deductive logic it is validity that wears the trousers. Seen in this way, errors of reasoning are parasitic on the rules of right reasoning. There is a clear procedural suggestion in this for the would-be theorist: Get the rules of right reasoning sorted, and errors of reasoning will fall out more or less of their own weight.

De Morgan is one of the many logicians to have favoured this idea. De Morgan thought that since there are numberless ways in which reasoning can go wrong, a logic of wrong reasoning could only be a logic of right reasoning, the number of whose principles are comparatively few.[1] One might say with De Morgan – and with the rule-violation community generally – that there are no stand alone theories of deductive error in logic. What might such a theory be? In no small measure, it would have to satisfy the following:

Proposition 1.1a
THE CONCEALMENT-DETECTION PROBLEM: *A trouser-wearing theory of error would have something helpful to say, among other things, about how errors are both inapparent enough to escape initial detection and yet recognizable enough to permit subsequent correction, and about the factors implicated in this transition from concealment to detection.*

---

[1] De Morgan (1847), chapter 13: "As to mere inference, the main object of this work [*Formal Logic*], it is reducible to rules; these rules being all obeyed, an inference, as inference, is good; consequently a bad inference is a breach of one or more of these rules." See also Joseph (1916): "Truth may have its norms, but error is infinite in its aberrations, and they cannot be digested in any classification." (p. 569).

Traditionally minded logicians have long been accustomed to regard the rules of right reasoning as formal constraints, that is, as constraints that hold independently of propositional content. This gives rise to a further difficulty. James Oliver and Gerald Massey have argued – independently – that a formal theory of invalidity for natural languages is not possible.[2] Given that natural language arguments instantiating invalid forms can themselves be informally valid, these critics disavow the capacity of mathematical logic to generate a theory of erroneous reasoning in natural languages. Although a natural language argument instantiating a valid form is valid, there are valid such arguments that instantiate nothing but invalid forms.[3] This produces an asymmetry, in which a formal theory of validity is thought to be possible, but not one of invalidity. In so saying, we learn something important about a stand alone logic of deductive error. Not all of it can be squeezed into the formal logics of deduction.[4]

The old deontological idea that an error of reasoning is reasoning that violates the prescriptive rules for right reasoning also echoes in logic's traditional approach to inductive error, in which the property of inductive strength is the principal target. In the logician's technical sense, a piece of reasoning is inductively strong just in case it exhibits a conditional probability of requisitely high value linked to an appropriately conditioned data-set, typified by sample-to-population inferences in the context of experimental trial. Whereas deductively valid reasoning is truth preserving, inductively successful reasoning is probabilistically clinching. So reasoning is inductively defective when it is not inductively strong. Here, too, rightness rather than wrongness is the focal issue. Inductive strength wears the trousers in inductive logic. The mainstream treatment of inductive error is mainly thin.

---

[2] Oliver (1967), Massey, (1975a, 1975b, 1981).

[3] Consider, for example, the argument ⟨Harry's shirt is red, Harry's shirt is coloured⟩. Although semantically valid, it instantiates the invalid form ⟨α, β⟩. This is what some writers call "material validity". See, for example, Brandom (2000). In this book I use a standard convention for the simplified representation of arguments, in which $\ulcorner\langle\alpha_1, ..., \alpha_n, \beta\rangle\urcorner$ denotes an ordered sequence of statements α_i, β whose terminal, i.e. rightmost, element is the argument's conclusion, and the preceding elements its premisses.

[4] Also imperiled is the cognitivist programme in psychology and AI, according to which the human deductive reasoner is a concrete realization of a Turing machine which draws conclusions from premises by way of mechanical operations on syntactic properties of formulas. See Turing (1950) and Fodor and Pylyshyn (1988). The idea of reasoning as computational is at least as old as Hobbes (1651/1962). See below, section 1.6.

3

We have it, then, that a significant part of logic's traditional understanding of error can be set out in this biconditional:

Proposition 1.1b
RR-RULE VIOLATION: *E is an error of reasoning if and only if there exists a truth-preserving prescription R for right reasoning or a probabilistically clinching prescription R' for right reasoning, and E violates R or R'.*

It would be wrong to leave the impression that logicians have never been attracted by prospects of a stand alone theory of mistakes of reasoning. Nicholas Rescher's *Error* is a notable recent development in that direction.[5] The same was true of Peter Ramus in his day, of Richard Whately in his, and also of J.S. Mill.[6] Similarly, in saying that this or that is or has been true of "traditionally minded" of "mainstream" logicians, or that these things are or have been true of them "for the most part", the quoted qualifications are essential for accuracy. Logic is an ancient discipline with a vast literature. Even since taking the mathematical turn at the comparatively recent mid-point of the 19th century, logic has produced a sprawling pluralism and, in a number of respects, a rivalrous one. The mainstream in logic is not a precisely defined idea, and "by and large" connotes trend lines, not strict invariability. Even so, it is hard to miss the dominant influences – classical approaches to deduction, Bayesian approaches to induction, and comparatively slight regard for what we might call "third way reasoning".

1.2 The Gang of Eighteen

Notwithstanding its generally thin approach to errors of reasoning, there is a category of error coincident with the founding of logic itself, and off and on a part of its research programme ever since, which has attracted special and somewhat thicker attention than usual. These are the fallacies, whose traditional list is a loose assemblage of which the following is not untypical: *ad baculum, ad hominem, ad populum, ad verecundiam, ad ignorantiam, ad misericordiam,* affirming the consequent, and denying the antecedent, begging the question, many

---

[5] Rescher (2007). See also *Synthese*, 163 (2008), 299-442, a special dedicated to error and Mayo and Spanos (2010). Mayo and her colleagues take what has come to be called the *Severe Test* approach to inductive logic. See also Mayo (1996).
[6] Ramus (1543/1964), Whately (1836), Mill (1843/1961), bk. 5.

questions, hasty generalization, equivocation, biased statistics, the gambler's fallacy, *post hoc, ergo propter hoc,* composition and division (of which *secundum quid* is a special case), faulty analogy, and *ignoratio elenchi* (of which straw man is a special case). If we don't count *secundum quid* and straw man separately, this makes for a list of eighteen, known – with a certain light heartedness – as the Gang of Eighteen. Some of these, especially question-begging and many questions, are seen by some investigators as dialectical fallacies rather than errors of inference. But the traditional view is that most, if not all, of the items on this list instantiate the rule-violation conception of error as we have formulated it here.[7]

A further feature of the traditional conception is that fallacies are errors which people in general have a natural tendency to commit, and do commit with a notable frequency. They are also attended by a significant likelihood of postdiagnostic recurrence. They are like bad habits. They are hard to break. A further task for a thick theory of fallacious reasoning is to offer an account of this seductiveness. A related idea is that part of the appeal of fallacious reasoning is that it takes less time and effort to pull off than reasonings that meet a stricter and more demanding rectitude. This gives them a further attractiveness. They are shortcuts to conclusions as rightly reasoned and more easily produced than the stricter standards allow. It is customarily supposed that the gang of eighteen – the traditional list – falls into the extension of the traditional concept of fallacy. I will call this the "concept-list instantiation thesis."

In *Fallacies* (1970), the Australian logician Charles Hamblin handed down a harsh indictment of logic's failure to provide for the fallacies in theoretically deep ways. He admonished logicians to repair this omission. It was Hamblin's contention that "we have no *theory* of fallacy at all, in the sense in which we have theories of correct reasoning

---

[7] It is instructive to compare this list with another recently assembled one: slippery slope, appeal to precedent, many questions, vagueness, begging the question, *ad hominem, ad populum, ad verecundiam, ad ignorantiam, ad baculum, ad misericordian, ignoratio elenchi,* straw man, red herring, equivocation, affirming the consequent, denying the antecedent, biased statistics, hasty generalization, non-cause, *post hoc ergo propter hoc,* consequent, illicit major, illicit minor, *secundum quid,* two wrongs, gambler's and false analogy (Tindale, 2007). Tindale's is a Gang of Twenty-Eight. Other lists reflect similar differences, but all retain a substantial common core, of which the Gang of Eighteen is a reasonably representative assortment.

or inference."[8] Much in the spirit of Mill, Hamblin was calling upon logicians to produce a thick theory of mistakes of reasoning. Responses to this challenge weren't late in coming, beginning with what came to be known as the Woods-Walton Approach,[9] and followed in short order by developments in pragma-dialectics and informal logic.[10]

One of the questions triggered by the lament for the sorry state of the 1970s fallacies programme in logic is:

> HAMBLIN'S QUESTION: *Why was its state so sorry? Why was fallacy theory so difficult? Could it be that logicians, who were smart enough for, say, inaccessible cardinals and Bayesian nets, weren't smart enough for fallacies? If so, what would have made it so?*

In Woods (2004) I floated the idea that a suitably trouser-wearing logic of error might lend some support to one or all of the following three ideas. My purpose in announcing them now, so early in the proceedings, is not in hope of their acceptance off the bat, but rather of a willingness to give them some fair consideration. The burden of supporting them is distributed throughout the book. I introduce them here in the spirit of cards on the table. The first idea is the *rarity thesis*, which claims that so-called fallacies are rarely committed. The other, a corollary of the first, is the *negative thesis* that the so-called fallacies aren't really fallacies at all. Should these turn out to be supportable claims, we would have the means of suggesting an answer to Hamblin's Question:

> **Proposition 1.2a**
> THE CONCEPT-LIST MISALIGNMENT THESIS: *Contrary to the traditional concept-list instantiations thesis, the items on the traditional* list *are not to be found in the extension of the traditional* concept *of fallacy.*

---

[8] Hamblin (1970), p. 11; emphasis in the original. What Hamblin here intends by "theories of correct reasoning or inference" are the mainstream deductive logics.
[9] See, for example, Woods and Walton (1972, 1974, 1975, 1982), among others. These papers are included in Woods and Walton, (1989), reissued as Woods and Walton (2007), with a Foreword by Dale Jacquette. See also Woods (2004).
[10] Van Eemeren and Grootendorst (1992), Walton (1995), and Johnson and Blair (1980). See also Finocchiaro (2005) and (2013) and Blair (2012). A good collection of readings is Hansen and Pinto (1995).

Why then have logicians done so poorly with the fallacies? The answer might be that the fallacies they try to investigate aren't fallacies as they themselves conceive them to be.

The third idea is the one conveyed by:

Proposition 1.2b
THE COGNITIVE VIRTUE THESIS: *To a nontrivial extent, the items on the traditional list are cognitively virtuous ways of reasoning.*

If the concept-list misalignment thesis is true, it reflects badly on the mainstream. It convicts a long-lived tradition in logic of a costly carelessness, and invites a further question. In recognition of the colleague who first pressed it upon me, I'll call it

FINOCCHIARO'S QUESTION: *What led the logical mainstream into so a serious confusion? Why were traditionalists unable to see this failure of fit between concept and instantiation?*

We can readily agree that *circa* 1970 fallacy theory was in crisis. By and large, the logicians who were most centrally involved in advancing the modern subject's research programmes paid no attention to fallacies, and such work as managed to find its way into First Year logic texts was pretty thin gruel. Hamblin's exposé did not go unnoticed. But the attention it attracted was mainly from outside the logical mainstream, and even more so from respondents who weren't logicians at all.

Nearly a half-century has passed since Hamblin raised the alarm. It is only natural to wonder whether, were he still with us, Hamblin would be happy with the current state of things. I think that, on the whole, he would not. For one thing, mainstream logicians haven't revived the fallacies programme, and have not bothered with post-1970 treatments in other areas of enquiry – notably speech communication and discourse analysis. For another, the abundant literature generated by those enquiries have yet to generate a genuinely robust theoretical consensus. Were he with us today, this would give Hamblin yet another question to press. In his absence, I'll press it for him.

WOODS' QUESTION: What explains the thinness and theoretical unsettledness of present-day work on fallacies?

7

Securing an answer to Hamblin's Question will take most of the book. As we proceed, when the occasion presents itself to consider an item on the traditional list, I shall pause to consider whether it can be made to lend some confirmation to the misalignment thesis. In this way we will build up a "misalignment-thesis confirmation tally". The answer to Finocchiaro's Question and Woods' Question I defer until the end of chapter 15.

The misalignment and cognitive virtue theses make some very strong claims. In their raw and unqualified states, it can only be expected that they would run into some stiff opposition. It is necessary to ask what would it take to make good on these radical pronouncements? Should we demand a demonstration that not a single one of the eighteen is a fallacy, that none at all fits the traditional concept of fallacy? Should we require that each of them without exception be shown as a cognitively benign reasoning strategy? I myself would be happy with a somewhat less demanding burden of proof. What if it were possible to show that typically, if not invariably, what logicians traditionally classify as fallacies are not fallacies as traditionally conceived of? If this could be accomplished, we could say that it "craters" the traditional position on the fallacies, that it craters the concept-list alignment thesis. Similarly, suppose it could be shown that typically, if not invariably, what logicians traditionally classify as fallacies are not cognitively defective modes of reasoning, but rather the opposite? Then here too we could say that this craters the traditional approach. My view is that, while it might not kill it outright, the cratering of a view inflicts a mortal wound. So I will propose:

Proposition 1.2c
THE DAMAGE OF CRATERING: *Arguments that crater the received view of the fallacies provide a sufficient reason to accept the concept-list misalignment and cognitive virtue theses.*

To the best of my knowledge, Maurice Finocchiaro is the first logician to have suggested a rarity thesis, in a way that also suggests a version of the negative thesis.[11] Finocchiaro invites us to examine the documentary record – for example, the writings of Galileo – which are replete with efforts to arrive at the truth of things by reasoning from premisses. Where, Finocchiaro asks, do we find the eighteen? The

---

[11] Finocchiaro (1981/2005).

8

question is rhetorical; he thinks it answers itself. Galileo made errors of reasoning; but he didn't make *those* errors.[12]

These days there is more cutting-edge technical logic done outside departments of philosophy than within them; and psychologists, to whom mainline logicians still pay little heed,[13] have started to invest in fallacies research in rather influential ways. It is wide-ranging work, at times overlapping with the research programmes of AI, statistics and machine learning. It includes "predictive modeling theory, multiple regression formulae, neural networks, naïve Bayes's classifiers, Markov Chains, Monte Carlo algorithms, decision tree models and support vector machines, [as well as] the well-known heuristics and biases program ...."[14] The psychologists' list is also quite substantial, although with not much in common with the fallacies that have traditionally occupied logicians. Prominent are the base-rate neglect fallacy, the conjunction fallacy, covariation illusions, the interview effect, hindsight bias, the prosecutor's fallacy, the Lake Wobegon Effect, and the overconfidence bias.[15] Still, some psychologists do direct their attention to the logician's

---

[12] Finocchiaro (1980).

[13] An important exception is L.J. Cohen, who criticizes the heuristics and bias approach in psychology, especially as relates to the conjunction fallacy. See, for example, Cohen (1981, 1982). See also Gigerenzer and Selten (2001).

[14] Bishop and Trout (2005), p. 12.

[15] For a (small) sample of the psychological literature on error see Griliches (1974), Edwards (1968), Wason (1966), Kahneman *et al*, (1982), Kahneman and Tversky (1996), Cheng and Van Ness (1999), Kahneman and Tversky (2000), Nisbett and Ross (1980), Gigerenzer and Selten (2001), Gigerenzer (2005), Mellers, Hertwig and Kahneman (2001), Dunn (2004), Fillenbaum (1977), Sweetser (1990). Psychological studies to date have concentrated on deductive and probabilistic and inductive reasoning, with somewhat less attention given to decisional and causal reasoning and very little to argument as such. There is no simple paradigm at present; in fact, there are at least four main approaches that are currently in contention. These are the *mental models* account e.g., Johnson-Laird and Byrne (1991), *mental logics* e.g., Rips, (1994), *rational analysis and information gain* e.g., Chater and Oaksford (1999), Oaksford, Chater, Grainger and Larking (1997), *domain specific reasoning schemas* e.g., Evans and Over (1996), and *dual-process models*, e.g. Evans (2007). Nisbett and Ross (1980) calls upon philosophers to involve themselves in sorting out normative issues that arise in psychology (pp. 13-14). Bishop and Trout (2005) is an interesting response to this invitation. See also a special issue of the journal *Informal Logic*, 29 (2009) on psychology and argumentation, edited by Lance Rips, and Rips' introductory essay, "Argumentative thinking: An introduction to the special issue on psychology and argumentation" (pp. 327-336)

traditional fallacies, particularly to the conditions affecting people's ability to recognize them. The psychological approaches roughly subdivide into those that retain the traditional idea that the fallacies really are fallacious[16] and those that hold that whether or not an argument in the form of a traditional fallacy is actually fallacious will vary with context.[17]

These psychological researches greatly enlarge the fallacies component of any logic of error that aims to meet Hamblin's challenge. In some ways this is unfortunate. It presents the logician of error with a heavier than usual workload. It requires him to enter areas that traditionally lie beyond his own theoretical reach. It also makes for a significantly more complex theory of reasoning than anything logicians have yet to produce. It would be unrealistic to think that a book like this could pronounce definitively on the psychological literature. Even so, there are aspects of these scientific contributions that are of importance for my developing account, and I shall make some effort to give them their due. A point of interest is that psychologists do a better job overall in locating their work on the rationality of human performance within a network of assumptions about the psychological make-up of reasoning agents. This is an advantage (and another complication) that a logic of error should be ready to follow up on.

An especially egregious favouritism is the one lavished by logicians upon deduction:

> There was a long tradition in philosophy according to which good reasoning had to be deductively valid. However, that tradition began to be questioned in the 1960s, and is now thoroughly discredited (Pollock, 2008, p. 451).

If the empirical record is anything to go on, most human reasoning is third way reasoning. It is reasoning for which deductive validity and inductive strength are inappropriate standards for the assessment of performance. Imposing those standards on all of reasoning is a momentous interference. It makes us all inferential-misfits. It subjects human reasoning on the ground to a corrosive scepticism. So let us introduce a further condition:

---

[16] For example, Neuman (2003), Neuman, Weinstock and Glasser (2006), Neuman and Weitzman (2003), Ricco (2003).
[17] See for example Rips (2002), Oaksford and Hahn (2004), Hahn and Oaksford (2006a, 2006b, 2007a, 2007b), and Hahn, Oaksford and Bayindir (2005).

Proposition 1.2d
THE THIRD WAY MANDATE: *It falls to an epistemically and epistemologically sensitive logic of error to describe the character of third way reasoning and to propose performance standards appropriate to its nature.*

1.3 *Naturalizing logic*

In implementing the third-way mandate I will draw upon the resources of a *naturalized logic*. Naturalized logic stands to orthodox logic as naturalized epistemology stands to orthodox epistemology. In each case, there is a principled openness to empirical factors, especially the lawlike disclosures of the relevant branches of natural science. A naturalized logic is an empirically sensitive logic. It is an approach to reasoning that pays attention to what people are like, how they are put together and what they get up to when they reason.

The naturalization of logic is far from a new idea. In the modern era alone, it is the founding principle of Dewey's experimental logic.

> Logic is a social discipline ... [E]very inquiry grows out of background of culture and takes effect in greater or less modification of the conditions out of which it arises. Merely physical contacts with physical surroundings occur. But in every interaction that involves intelligent direction, the physical environment is part of a more inclusive social or cultural environment. (Dewey, LW 12, 27)

A similar theme is present in the early writings of Toulmin:

> Logic ... may have to become less an apriori subject than it has recently been .... Not only will logic have to become more empirical; it will inevitably tend to be more historical. (Toulmin, 1958, p. 257)[18]

Putnam also advances the idea with the question "Is logic empirical?"[19], and Finocchiaro responds affirmatively (and rhetorically):

---

[18] See also Johnson (1987).
[19] Putnam (1968).

[I]s logic an abstract science that studies entailment, truth functions, the calculus of propositions, predicates, relations, identity, etc.; or is it a social science that studies the mental activities of reasoning and argument? If the former is the case, how does logic relate to mathematics? Is it just a branch of mathematics? Or does it provide the foundations of mathematics? If logic is a social science, how does it relate to experimental cognitive psychology? .... so we may ask, what is or ought to be the relationship between formal or symbolic logic and reasoning or argument? (2005, pp. 6-7)[20]

The success of mathematical logic is tied in no small way to its freedom from the drag of context, and from the semantic distractions of sentences chock-a-block with propositional content. In the new paradigm, sentences don't say anything; and they are treated without regard to when they were uttered, by whom or in response to what. There are no *people* in the models of mainstream mathematical logic. Nearly everyone thinks that the discovery of people-free, context-independent symbolic languages was the linch-pin of the success of the mathematical turn. Whence the iconic importance of Frege's notational breakthrough in the *Begriffsschrift* of 1879 and of the subsequent improvements wrought by Peano.[21] It is an achievement that prompted Quine to remark in the preface to *Methods of Logic* that logic is an ancient discipline, but since 1879 it has been a great one.[22]

There are plenty of systems of logic, calling themselves in one way or another "agent-centred" and "goal-oriented", in which representations of agents appear; that is, symbols for agents are a dedicated part of syntax. But theories with notations for agents are not agent-centred in the sense that I intend, unless they also take into account the resources an agent has at his (or its) disposal for the discharge of that class of tasks – chiefly premiss-conclusion tasks – in which logic has historically shown a dominant interest. I will say that theories that fail this requirement are *name-droppers* about agency and goal-seeking. They are accounts of agency and goal-seeking *façon de parler.*[23]

---

[20] See also Krabbe *et al.* (1993) and Finocchiaro (2005).
[21] Frege (1879/1967), Peano (1908).
[22] Quine (1982).
[23] Here is an example. One might combine categorical grammar with relational algebras. This would enable us to construe natural language "as a sort of cognitive programming language for transforming information". One might then

Most logics take up the task of saying what reasoning is like without bothering to say what reasoners are *like*.[24] The common assumption appears to be that in getting the reasoning part right the reasoner part drops into place virtually automatically. For if good reasoning is reasoning in such-and-such ways, what could a good reasoner be if not a being (or device) that reasons in those ways? Against this, an empirically sensitive logic urges the converse dependency. Without independent consideration of what reasoners are like, an account of right reasoning is one that seeks leverage without a fulcrum. Accordingly,

Proposition 1.3a
REASONERS: *An empirically sensitive logic of* reasoning *requires independent engagement with an account of* reasoners.[25]

---

link this to "modal logic and process theories in computer science", arriving at "a dynamic logic of programs". With still further developments, notably in game theory, one might arrive at "A general theory of agents that produce, transform, and convey information – and in all this, their social interaction should be understood as much as their logical powers." (van Benthem, 2011, p. ix)

[24] An important exception is Adam Morton. See Morton (2006) and Morton (2013). The title of (2013) is *Bounded Thinking*. It has just now appeared, I regret that I haven't had an earlier acquaintance.

[25] Wooldridge and Jennings (1995) distinguish between a weak sense of agency, which is generally accepted in the computer science community, and a stronger sense that has attracted less of a consensus. According to the weak conception (p. 116), agents are *autonomous*; they exercise control of their inner states and actions. They are *social*; they interact linguistically; they are *reactive*; they read their environment and respond to changes in it. They are *proactive*; they initiate goal-directed actions. The stronger conception provides (p. 117) that agents are *mobile*; they move about physical and/or electronic networks. They are *honest*; they don't knowingly communicate misinformation. They are *cooperative*; they will do what is asked, and will avoid having conflicting goals. Finally, agents are *rational*; they act to achieve their goals in line with their beliefs. The Wooldridge-Jennings list is indifferent to the distinction between institutional and individual (concerning which, see below). As they themselves acknowledge, some elements of their characterization are more plausible than others. There is something right about the autonomy criterion, but, as we shall see in due course, it requires qualification. Similarly, we might better think of honesty and cooperation as default features of agency, rather than as necessary conditions of it. Perhaps this most interesting item on their list is rationality, which is nothing more or less than using one's "reason" in trying to achieve one's goals.

Proposition 1.3a is a special case of a more general methodological principle. Its main idea is easy to state. Let T be a theory about a certain kind of human behaviour – shopping, organizing a trip, negotiating a collective agreement, proving Fermat's Last Theorem, or whatever else. This is the behaviour which the theory strives to account for in certain ways. Although it seems an obvious point, hardly worth mentioning, I will mention it anyway:

Proposition 1.3b

RESPECT FOR DATA: *If T is an empirically sensitive theory of human behaviour of kind K as performed in the conditions of real life, then the theorist should take pains to assure himself that what he takes to be K-behaviour on the ground actually fits T's descriptive and assessment criteria.*

COROLLARY: *In particular, if it is found that behaviour on the ground disconforms to the assessment norms of T, the investigator must take care to check that this behaviour actually is the behaviour that T undertook to describe and whose performance its norms sought to constrain.*

Proposition 1.3b counsels the theorist to take care that phenomena his theory pronounces upon are the phenomena the theory set out to clarify in the first place. The advice it gives is as elementary as it is valuable: In matters of theory-building, make sure that you keep your eye on the ball.

It would not be wrong to say that "everyone knows" that data on the ground are often not what they seem. So why say it? I say it because, in the actual contact a theory seeks with its target goings-on on the ground, it is a truth that is often not properly *heeded*. Consider again the logician's commonplace that invalid reasoning is wrong reasoning. A lazy glimpse of our data of that actual reasoning behaviour discloses that it is hardly ever valid. A mindful inspection of the same data discloses that, given the contexts in which such reasoning actually occurs, it is typically the case that the validity standard is not appropriately in play. So, though it is not met, it is not violated either. The price of this disregard is the absurdity that reasoning on the ground is mainly wrong. Far from being too obvious to mention, the respect for data principle is not sufficiently respected not to be mentioned. So I am mentioning it here, and will do so again, on and off, as we go along.

1.4. *Practical reasoning*

An empirically sensitive naturalized logic is a logic centred on the practical. It is a logic that takes "the practical turn".[26] There are different interpretations of the word "practical", both in its application to reasoning and to the logics that study it. Mine is a resource-bound conception of practicality. It is not, perhaps, the name that philosophers are most familiar with. But despite its comparative novelty of the name, the *notion* is far from unfamiliar. So it is helpful to have a conception of practicality that locks deeply into the gears of agency.

What, after all, *is* an agent? An agent is a being (or device) that *does things*. We may take it, then, that the reasonings undertaken by an agent, no matter how abstract or *recherché*, are those that have some bearing on his conduct as an agent. His reasoning about breakfast may prompt him to reach for the Rice Krispies rather than the Coco-Puffs; his reasoning about the Continuum Hypothesis may prompt him to try to undermine the independence proofs or merely to reflect further on the cost to set theory of making do without it.

The short story about practical reasoning can be set out as follows: Some cognitive tasks require comparatively large numbers (quantities, amounts) of cognitive resources, such as information, time and computational capacity. Call these *big tasks*. Call the others *small tasks*. Bigness and smallness are comparative notions and, as used here, neither is intended evaluatively. 'Big' and 'small' are technical terms. To classify a task as big is not to praise it; and to classify a task as small is not to make light of it. There is also a critical distinction between *institutional* agents, such as NASA or Her Majesty's Government, and *individual* agents, such as any reasonably well-accoutered human being. One way of marking this distinction is by saying that it is typical of institutional agents to execute big tasks and of individual agents to execute small tasks.

Accordingly, it is *typical* of institutional agents that they transact their cognitive agendas with greater numbers and kinds of cognitive assets than typically available to individual agents in the transaction of their cognitive agendas. I will say that an agent is *reasoning practically* when he is reasoning in the discharge of a small task. I will say that a *practical logic* is a logic of practical reasoning.[27]

---

[26] Woods *et al.* (2002).

[27] For the present use of "practical" there is no natural non-negative antonym. But since the human individual is a paradigm of this sense of practicality, it will

15

Suppose now, that contrary to the position I am advancing here, cognitive success were reserved for outcomes that met the standards governing big tasks. If true, then individual knowers, in the performance of the undertakings that are typical of them, would be the subjects of a massive ignorance. Beings like us would hardly ever know anything. The consequences for a theory of error would be as catastrophic as they are direct. For we would have it that, with regard to the tasks that are typical of him as a practical agent, the reasoning that underpins its performance is nearly always in error. My complaint, so far, against the standard approaches logicians take to reasoning is that, if sound, the catastrophe thesis would be part of the correct account of the reasoning of human individuals. There are some logicians and theorists of reasoning who see in this implication a *reductio ad absurdum* of the received view. Perhaps they, no less than I, would be startled to learn (or recall) that the *reductio* view is nowhere close to the majority opinion in the philosophical tradition. Throughout its long history, philosophers have thronged to the idea – if only tacitly – that the catastrophe thesis might well be true, that its truth is a real possibility. No reasoning theorist who knew his onions would venture forth without bearing that fact in mind. One way or another, he must make up his mind about the catastrophe thesis. Here is what my mind makes of it: I am not prepared to be a big-box sceptic about right reasoning. *The catastrophe thesis should be resisted like the plague.*

1.5 *Psychologism*

An empirically sensitive practical logic of erroneous reasoning is agent-centred, goal-oriented and resource-bound. Investigators who make room for context and agency are drawn to a form of what used to be called the Laws of Thought approach[28] and, needless to say, are committed to an element of psychologism in logic.[29] This psychologism

---

be enough to regard institutional agents as supplying most of the desired contrast.

[28] See, for example, Ellis (1979), p.v.

[29] Virtually everyone is (respectfully) aware of Frege's hostility to psychologism. Many fewer appear to know the grounds of his opposition. Frege's case – a rather quaint one to modern ears – is directed against psychologism in mathematics. But since he thought that mathematics reduces to logic, it can be assumed that his anti-psychologicistic arguments apply equally to logic. Frege's case is to be found in *Die Grundlagen* (1953), and includes the following points

is not inadvertent. It is psychologism on purpose.[30] Since human agents come with psychologies as standard equipment, once you admit people to logic, you admit their psychological make-ups as well, warts and all. A naturalized logic is naturalization on the model of the naturalization of epistemology. The expression "epistemology naturalized" originated with Quine as the title of a celebrated paper.[31] Its original subtitle was "The case for psychologism."[32] Of course, Quine wasn't talking about logic. He wouldn't dream of psychologizing logic. Quine wants it to be true that nothing is logic if it is not first order classical logic. Given logic's actual development in the course of Quine's own life time, it is clear that his is the minority view. Quine aside, I say that psychologism is once again an open question in the research programme of logical theory. Its re-emergence should not be prejudged. Better to wait and see how a psychologically real, agent-based logic fares as a theory of reasoning once it is up and running. In this I cast my lot with John Macnamara: There is no need for a "border dispute" between logic and psychology.[33]

My openness to psychological inputs is not carelessly ventured. It is made in recognition of the fact that logic is an ancient subject and that psychology, in comparison, is a Johnny-come-lately. It is made in recognition that logic won its spurs before psychology was ever thought of as a stand alone science. And it is made in recognition of the fact that it is only fairly recently that psychology could claim a mature and stable scientific respectability. I myself am of the view that by any fair measure the psychology of perception is a mature and deep science, and that, although not yet as well-advanced, the psychology of cognition is on track for the same kind of success. Of course, cognitive psychology has

---

among others: We couldn't prove the Pythagorean theorem without taking into account the phosphorous content of our brains (p. vi). Since we all have different ideas of numbers, each of us would have his very own number two (p. 37). Since our minds are finite, there couldn't be infinitely many numbers (pp. 37-38). Sensory experience would be foundational for mathematics, thus costing mathematics the objectivity which reason alone can provide (p. 38). Psychology is too immature a science to support a highly advanced science such as mathematics. (p. 38).

[30] Grumpiness about psychologism in philosophy is ably assessed in Jacquette, (2003). See also Pelletier and Elio (2005), Gabbay and Woods (2003), chapter 2, and Gabbay and Woods and others in a special issue of *Studia Logica*, 88 (2008).

[31] Quine (1969). Quine's naturalism is an extreme version of it. I'll come back to this point in the chapter to follow.

[32] See Willard (1989).

[33] Macnamara (1986).

no lessons to pass on to pure first order quantification theory, which is as near to mined-out as a great theory gets. Naturalized logic is another matter. It is built for psychological inputs, and it is nowhere close to the stages of development that the better parts of psychology have now attained. So neither is my openness to psychology a patronizing one.

Proposition 1.5
EMPIRICAL SENSITIVITY: *A naturalized logic of erroneous reasoning should take note of, when they are available, the lawlike pronouncements of the relevant parts of descriptively adequate psychology. Whenever one's logic deviates from such provisions, it should explain the discrepancy and assess its significance.*

Logicians have something to learn from psychology, but not everything that psychology pitches is worth catching. For example, progress made by cognitive science these past few decades derives in no small measure from the abandonment of the idea that psychology, on its own and in its own sweet time, can do all the work of epistemology.

However drawn we might be to orthodox admonitions to forgo the psychological in logic, the present day determination to circle the wagons around mathematical logic and refuse admittance to the empirical outliers is compromised by an historical oddity. It is that mathematical logic has to a quite extraordinary degree *retained* the ancient presumption that logic is a theory of reasoning[34] – indeed on some tellings it is about reasoning at its most perfect.[35] Since reasoning is what reasoners do, even mathematical logic retains the pretence of a tie between its disclosures and the conditions under which reasoners behave when they reason well.

---

[34] Harman and Kulkarni (2007), p. 5, observe that "in the traditional view, a deductive logic is a theory of reasoning". (They themselves are non-believers.) See also Hahn and Oaksford (2007a), p. 705: "By extending a normative, probabilistic approach to at least some aspects of argumentation, we hope to show that such an approach can generalize beyond the narrow confines of deduction and induction as construed in the psychology of reasoning to the real human activity of which these modes of reasoning are but a small part."
[35] This is a view also commonly held by psychologists. Evans (2002) calls it the "deduction paradigm". (Evans, too, is a non-believer). Syllogisms "have been extensively studied by psychologists in search of 'the fundamental human deduction mechanism' ...." (Stenning, 2002, p. 8)

How could logic succeed as a theory of reasoning without somehow taking reasoners into account? The question is (in the logician's sense) a complex one. It admits of two answers. Hardliners such as Quine and Harman will say that logic isn't about reasoning. Others will allow, at least in principle, for *some* kind of connection between the rules of logic and the rules of right reasoning. In a variation already mentioned, a reasoner is a virtual entity implicit in the theory but without formal recognition there. A reasoner is any device, actual or counterfactual, implementing the logic's norms. Other variations are more forthcoming. They show a degree of readiness to admit agents and contexts into their logics as expressly load-bearing objects of theory. And so it has been, with greater or less vigour, for some forty years and more.

What is so striking about the large bulk of agent-admitting logics is the conservatism with which they seek to preserve the provisions of the mainstream. While they give at least notational recognition of agents, they are unable to shake off their addiction to deduction. Agent-admitting logics are extensions or adaptations of mathematical logics, fashioned in the spirit of Quine's famous maximum of minimum mutilation.[36] It provides that in making these transformations the methodology of mainstream logics be trifled with as little as possible. Agent-admitting logics would still be grammars of semi-interpreted formal languages, overlain by syntactic theories of proof and set theoretic theories of truth, each fashioned with an eye on the metatheoretical plums of consistency, completeness and soundness, or on the categoricity of axioms. I need a name for such things. Suppose we say that they are "mathematically virtuosic".

1.6 Competence psychologism

It is interesting to compare the forgoing remarks with the competence psychologism exhibited in contemporary semantic theories. Roughly and informally, it is the view that for human individuals to understand their own language they must be able to specify the semantic content of each of its sentences, an ability that requires that the link between expression and content be recursive. For this to be so, language users are assumed to be symbol-manipulating mechanisms, finite in

---

[36]"The price [of straying from the mainstream] is not quite prohibitive, but the returns had better be good." (Quine, 1970/1986, p. 86) Quine's maxim has an AI-counterpart in theories of belief revision (Gärdenfors, 1984, 1988, Gärdenfors and Makinson, 1988, Willard and Yuan, 1990, and Nebel, 1991).

nature but subject to an unlimited enlargement of capacity. The job of a theory is to investigate the symbolic-structure systems manipulated by these mechanisms.[37] Given the idealized nature of the theory's language users – they are, among other things, logically omniscient – this relation can be modelled by an ideal computer, hence recursively.

How tenable is this idealized approach to a psychologically valid semantics? There is an argument advanced by Richmond Thomason (1983/2011) asserting that it is not tenable at all. Here, adapting Montague (1963), is its gist: Imagine that our ideal language-user is able to suppose that truth is an attribute of the semantic contents of sentences. Given the rules for sentence-production in his language, it would be possible to formulate a sentence to the effect that its own semantic content is not true. So the hypothesis that truth is an attribute of the semantic contents of the sentences of this idealized language yields an inconsistency in the manner of Tarski (1936/1983). This has three discouraging implications. The relation between a sentence and its semantic content cannot be formulated arithmetically. It cannot be true that the relation is recursive. It cannot be modelled by an ideal computer.

What holds for truth in such contexts also holds for the other propositional attitudes. Indeed, "if certain assumptions about knowledge and belief are granted which are plausible in the context of cognitive psychology, the result concerning truth extends to these propositional attitudes as well." (Thomason, p. 6) If we add these assumptions to a semantical theory of the present kind – if, for example, the theory represents the language-user's beliefs and these beliefs "include arithmetic" – and if we then teach this theory to the ideal language-user, he can't but be landed in an inconsistency.

Why, then, would we suppose that the inconsistency-inducing theory could be taught to our idealized speaker? The answer is that given its psychological character, semantics is a recursive concept, representable in both formal and human languages. Accordingly we have

---

[37] Included in this class of theories are what Jerrold Katz calls "semantic theory" (1972), in interpretative (Jackendoff, 1976) and generative semantics (Lakoff, 1971), theories à la Carnap in which "meanings are equivalence classes of phrase structure trees modulo intensional isomorphism" (Thomason, 1983/2011), theories that explain meaning by translation into a system of logic or any sort of suitably ideal formalized language, and most accounts developed by A. I theorists. See also Davidson (1965), Evnine (1991), and Lepore and Ludwig (2007).

Proposition 1.6a

THOMASON'S PARADOX: *Competence psychologism in semantics is inconsistent.*

Cross (2001) has shown that the inconsistency holds even where belief is not closed under consequence.

Thomason is instructively diffident about the paradox. He allows that successful scientific theories are rarely up-ended by difficulties engendered by philosophical attempts to foundationalize them. Even so,

> I do not wish to deny the strength and plausibility of the instructions behind the idea that thought consists in computational operations on syntactic representations. But I believe that the problems I raise here call for a much more critical approach to theories based on this idea. And critical versions of these theories are [excluding, for example, Asher and Kamp, 1986, and Creswell, 1985] seldom if ever to be found. (p. 13)[38]

It is well to note that Thomason's target is not the idea that a philosophical theory of meaning might benefit from a sensitivity to the lawlike disclosures of empirical psychology. In other words, Thomason's case against psychologism in semantic theory is no case against my own psychologism in the logic of reasoning. It does, however, target the ideal agent component of competence psychologism in a way that carries over to any like idealization of human reasoning. Accordingly,

Proposition 1.6b

CONTRA SYNTACTIZED IDEAL AGENCY: *Thomason's Paradox inflicts serious damage on any approach on which human reasoning is modelled as an idealized mechanism for computational operations on syntactic representations.*

1.7 A methodological tension

---

[38] In (2011), Bas van Fraassen proposes a way out of Thomason's Paradox by complexifying the relationship between expression and content, but concludes that, under the plausible assumption that the semantic theorist considers himself as the subject of such representations, "Thomason's devastating conclusion returns." (p. 15)

In the province of logic, mathematically virtuosic theories have enormous appeal. They are case studies in technical finesse, and in some instances they provide deep and satisfying reconstructions of target concepts. In the period since 1879, systems of logic have proliferated. There are more mathematically virtuosic systems of logic than you can shake a stick at. As I have said, this has engineered a large and contentious pluralism in logic. Anyone who has paid any attention to the historical unfolding of these multiplicities will have no trouble in pointing to cases in which a system's mathematical virtuosity trumps its conceptual adequacy. There lies a tension of no small moment. Here are a pair of cases briefly to consider.

Almost no one believes that there are true contradictions. Nearly everyone thinks that the very idea of true contradictions is absurd, an insult to reason itself. Yet there are systems of dialethic logic – the logic of true contradictions – meeting all the conditions of mathematical success.[39] They have recursively specifiable symbolic languages, formally adequate proof rules and semantics, and they are sound and complete with respect to their semantics. Dialethic logicians publish their work in the best of the mainstream journals and with leading university presses.[40] While nearly everyone thinks they are duds on the score of conceptual or philosophical adequacy, they do well on the score of technical virtuosity. Similarly, the system of relevant logic known as first-degree entailment (FDE) is sound and complete with respect to a possible world semantics (the so-called "star" semantics) in which the negation operator has no intuitive interpretation. FDE recalls the words of Quine about another but related matter: "They think they are talking about negation, '~', 'not'; but surely the notation [has] ceased to be recognizable as negation …". Yet FDE is widely studied and well-regarded as a technical achievement.[41] It is virtually impossible for a

---

[39] Dialethism is the view that, in certain select cases, two (*di*) things are true (*aletheia*) – namely, a statement α and its negation ⌐~α⌐. The Australian coiners of the term, seeking to honour its Greek roots, spell it "dialetheism". It is a mistake. Nothing in Greek (or in English) mandates the second occurrence of "e". We should grant the term to its inventors, but not their spelling of it.

[40] For example, Graham Priest's highly counterintuitive system LP is developed in a paper that appeared in the *Journal of Philosophical Logic* (Priest, 1979) and his book, *Beyond the Limits of Thought* (Priest, 1995), is in its second edition with the Cambridge University Press.

[41] It even has its apologists on the conceptual side. See, for example, Restall (1999). It is worth mentioning, by the way, that notwithstanding the mutilation of negation occasioned by its star semantics, a good many sensible people are of the view that FDE gives

logician acquainted with developments in relevant logic to be unaware of FDE.

I want to be clear about what I mean by conceptual fidelity. No one seriously doubts the existence of conceptual revolutions in science, in which old concepts are radically re-thought or entirely new ones are thought up for the first time. It is reasonable to suppose that, when these re-thinkings or originating stipulations meet various conditions of the requisite sort, they are scientifically valuable undertakings – thus, in that sense, good, hence conceptually adequate forms of thinking. This is not the sense of conceptual adequacy I intend it. My sense is a generalization of Quine's notion as applied to the "not" of dialethism. Though no enemy of re-definition and stipulation, there are some notions with respect to which Quine is a stickler. Negation is one and consequence is another. In each case, Quine thinks that there are established pretheoretical uses of these concepts in the formalizations of which it is desirable that they have a discernible presence. Generalizing, a pretheoretical concept is handled by a theory in a conceptually faithful way to the extent that, it is recognizably present in the theory's formal representations. This gives to the would-be theorist some useful general instruction:

Proposition 1.7
CONCEPTUAL FIDELITY: *If the object of a theorist's attention is a concept he calls "K" – for example, the concept of following from – and if there exists a concept, also called "K", which is pre-theoretically present in the theorist's home language, then, in the absence of the requisite disavowals, the default assumption is that the concept targeted by the theory is the one already present in his home language, and the default requirement is that the theory should not unfold in ways indifferent to the factor of conceptual fidelity.*

COROLLARY: *This does not preclude a theorist originating new concepts under the old name "K". But when he does so, he should say so.*

---

a perfectly credible account of entailment. If this is right, we see that in order to get its target concepts right – entailment, logical truth, etc. – it is not necessary for a logic to get all its connectives right (that is, right according to their common meanings in natural speech).

1.8 *The consequence rule*

From its systematic beginnings, logic's dominating preoccupation has been with relations of logical consequence. "Relations" in the plural is the right word for it. Even in his earliest writings on logic, *Topics* and *On Sophistical Refutations*, Aristotle distinguishes between a propositional relation called necessitation (and which is treated as a primitive) and a premiss-conclusion relation of *syllogistic* necessitation. It would not be going too far to say that the fixation on consequence is, if anything is, a defining feature of logic both ancient and modern, and that this is something by which we ourselves should be guided – as far as we can – in our present enterprise. Accordingly, though subject to the possibility of subsequent recall, I am going to impose

Proposition 1.8a
THE CONSEQUENCE RULE: *In examining premiss-conclusion reasoning of distinctive type k – deductive, inductive or third-way – the logician's central task for logic is to determine whether there exist consequence relations $\vdash_k$ (one or more) peculiar to that type, and if so to expose their characteristic properties.*

EXAMPLES: If some demonstrative mathematical reasoning is underway, it is natural to suppose that deductive consequence is in play. If some statistico-experimental reasoning is underway, inductive or probabilistic consequence is involved. If explanatory reasoning is underway, abductive consequence is often in play. If defeasible reasoning is underway, then default-consequence is frequently involved.

COROLLARY: *Once $\vdash_k$ is specified, it is desirable to supply some guidance in determining the memberships of its domain and converse domain. The first is a matter of consequence recognition. The second pertains to premises-searches.*

Aristotle's logical writings reflect an important further contrast. Although it is not one he explicitly identifies, its influence has been felt from his day to ours. This is the distinction between the consequences a set of beliefs *has* and the consequences that a reasoner will, may, or should *draw*, hence a distinction between consequence recognition and consequence selection. There is a like distinction between the rules that generate the consequences that a set of beliefs has and the rules that regulate which of them it would be appropriate to draw. It is natural to ask whether these are the same rules. On the face of it, it would appear not. Whether something is a consequence of a set of premisses is wholly a matter of satisfying the requisite semantic conditions. Whether it is a consequence which it would be right (or necessary) to draw is partly a matter of semantics but also in large part tied to psychological factors. Accordingly,

Proposition 1.8b
HAVING AND DRAWING: *Consequence-having occurs in logical space. Consequence-drawing occurs in a reasoner's mind.*

This gives us an obvious question to reflect upon. What does right reasoning require of the consequence-drawer? Does it demand that a reasoner draw every consequence had by his premisses? The question divides logicians into the No camp and the Yes Camp. Those in the Yes camp tend to frame the requirements of right reasoning for the ideal, not the actual, reasoner. According to the No camp, a good reasoner will draw only a proper subset of consequences of anything he believes, not excluding the null set. There is, thus, a loose but discernible concurrence between the Yes camp and the logical mainstream, and between the No camp and more empirically sensitive logics, whereupon the consequence rule assigns to the reasoner a further task – the consequence-drawing task.

For the Yes side, as for mainstream inductivists, belief is closed under consequence. For the No side it is not. The trouble with the Yes answer is that the number of consequences that a set of beliefs has is always transfinitely large for anyone to draw. The Yes people offset this disadvantage by dressing up their affirmative answer as a normative ideal, as something for the logically omniscient reasoner. So understood, the more consequences you can manage to draw from anything you believe the better your reasoning, never mind that you can't manage to draw anything remotely, i.e. finitely, close to them all.[42]

---

[42] Note, however, that there are some particular limitations that the Yes camp will concede. Thanks to Gödel's incompleteness results, not even the most ideal of ideal reasoners can draw every consequence of first order Peano arithmetic formalized in the manner of *PM*.

25

The No camp displays a greater sense of realism about these things. It is reluctant to attach the knock of the subpar to what we never do and couldn't. According to the Yes camp we are notably and systematically deficient in our inference-making ways. According to the No camp, this is an indictment too heavy for serious countenance. Beyond that, there is a point which the No camp can use to its considerable advantage. It is that when a proposition α is a consequence of something we believe, α is not always a candidate for drawing. The No camp subscribes to the view that one always draws at most a proper subset of the consequences of one's beliefs. What we now see is that the No view extends to the empty set of those consequences. For forty years Harman and others have been telling us (rightly) that when α is a consequence of β sometimes the right consequence to draw is that β is false.[43] But $\ulcorner$not-$\beta\urcorner$ is not a consequence of β. The fact that the consequence which is sometimes right to draw when α is a consequence of β is neither α nor any other consequence of it leads Harman to reject the traditional idea that a deductive logic is a theory of reasoning.

There is something right about this, and something not. What is right is that a logic that tells us what the consequences of something are can't in the general case tell us what consequences to draw. What is wrong is the idea that the logic of deduction has nothing to do with deducing. This goes too far. Deducers have a natural interest in not playing fast and loose with inconsistency, hence with the consequences with which inconsistency is interdefinable. Logics that are good at inconsistency spotting have an obvious role to play. They help reasoners to consider what consequences they should *not* draw. Accordingly, as a supplement to the consequence rule, we must also have

Proposition 1.8c
THE DRAWING RULE: *If in the course of examining reasoning of type k in which a consequence relation $\vdash_k$ is implicated, a logic's further task is to set out conditions under which a k-consequence should be drawn.*

So far we have taken note of two kinds of consequence-mistake. There is the mistake of drawing consequences that aren't there, which is the ancient mistake of *non sequitur*. There is also the mistaking a drawing

---

[43] Proposed by Harman (1970) and, independently, Woods and Walton (1972), the latter reprinted as chapter 1 of their (1989/2007).

26

a conclusion which, though *sequitur*, is not an appropriate consequence to draw, given one's interests and one's resources. But there is a third way of consequence-mistakenness. It is the failure to draw a consequence that is there and appropriate to draw. It is the mistake of not *seeing* what follows from what. This, too, has been a preoccupation of logicians since ancient times. The instantiation of proof designed to overcome consequence-blindness in mathematical contexts. And the great dialecticians laid great store on the interrogative techniques by which such blindspots could be removed. There is a useful lesson in this.

Proposition 1.8d
ERRORS OF OMISSION: *Some errors of reasoning are consequences overlooked. In other words, the consequence-recognition task is sometimes not successfully performed.*

1.9 Premiss-searches

Proof-searches are but a special class of problem-solving tasks. At a certain level of abstraction a problem calls for the presentation of considerations that make it go away. In the case of a proof search, the problem is that there is some proposition β that stands in need of a certain kind of support. The problem, in effect, is a request for properly supported propositions α_1, ...,α_n that jointly entail β. Finding those propositions is a premiss-selection task. It is easy to generalize the proof example. It holds for any premiss search relative to any notion $\vdash_k$ of premiss-conclusion consequence.[44]

We see, then, that the premiss-conclusion pair engages the reasoner in two quite different ways, in consequence-spotting and in premises-spotting. Each is a task performed on a fixed element. In the first instance, the premisses are fixed and the question of the consequences to be drawn from them is open – that is, is the problem to be solved. In the second instance, the conclusion is fixed and the question

---

[44] Another example of premiss-searches relative to consequence relations of requisite strength are refutation arguments, in which the task of the premiss-searcher is to extract from his opponent concessions which entail or lend credible support to the contradictory of the opponent's own thesis. As Aristotle remarks (*Metaphysics,* K5, 1062ᵃ 2-3), this is not proof in the full sense but proof *ad hominem*. What Aristotle means that the premiss-conclusion connection noted by the refuting party reveals that the other party's retention of his thesis is inconsistent with his other concessions. It doesn't however, show that the thesis is false. Refutations are discussed in greater detail in chapter 11 below.

27

of what it takes to prove it is open. These are different situations, no doubt, but they are tied together in an essential way. In each case, the success of the outcome pivots on whether premisses, those presently both at hand and those yet to be found, do in fact bear to the conclusion the requisite consequence relation.

The obvious question is whether the essential link to the consequence relation is enough to qualify those tasks – the consequence-spotting task and the premiss-finding task – as falling responsibly into the ambit of logic. There is a good deal of opinion to the effect that premiss-finding is a psychological matter, that there is no such thing as the logic of discovery.[45] My response to this is twofold. Since the success of a premiss search is tied to whether the selected premisses stand in the intended consequence relation, and since consequence relations are a central business for logic, then logic has a stake in premisses searches. What is more, in as much as we have already psychologized our logic, it should not be foreclosed *à priori* that at least some features of the premisses-searches are fit subjects for logic.[46] Think again of the factors of consistency and inconsistency, relevance and irrelevance, and the plausible and the implausible.[47] So let us say that

Proposition 1.9
GENERATION AND RECOGNITION: *Subject to the indicated constraints, both premisses-generation and consequence-recognition are legitimate components of a duly psychologized logic of erroneous reasoning.*

---

[45] Reichenbach (1938); Popper (1934).

[46] Theorists of argument typically take on the broader task of pronouncing on the premiss-admissibility conditions for good arguments of all types, not excluding the dialectical and rhetorical varieties. While theirs is a more comprehensive task than I would propose for logicians of reasoning, there are some clear similarities between the two projects. The most thorough-going recent treatment of premisses-admissibility in broadly argumentative contexts is Freeman (2005).

[47] Consider also the question of syntactic constraints on inputs to the consequence relation, and, come to that, on the outputs too. Think here of the old saw that it is characteristic of deductive consequence that its inputs include general propositions and its outcomes singular or particular propositions; and the related idea that it is characteristic of inductive consequence that its inputs be singular or particular propositions and its outputs be general propositions. Think also – more seriously – of whether there might be syntactic peculiarities required for inputs to defeasible consequence. I'll come back to this in chapter 7.

28

The idea that relevance and plausibility are fit subjects for the attention of logicians has generated some notable work, especially in the case of relevance. In the modern mainstream, ensuing from the early papers of Anderson and Belnap is a large family of relevant logics, almost all devoted to relevantizing the consequence relation.[48] Two conceptions of relevance emerge in these writings. One is "containment" relevance, indicative of topic sharing and partially representable as the possession by two or more sentences of some common atomic element. The other is "full use" relevance, which is exhibited by proofs from hypothesis in which all the hypotheses have occurrences. Also important is the contribution of the linguists Dan Sperber and Deirdre Wilson, in which the basic idea is that information is relevant in a set of sentences called a "context" when it has "contextual effects" there; that is, when it combines with the context to produce deductive consequences of a certain sort, or when it contradicts something in that context, or when it reinforces or corroborates something there.[49]

The main contribution to date from the perspective of a naturalized logic is the "agenda relevance" approach I've developed with Dov Gabbay. Here the founding idea is that relevance is defined for quadruples $\langle X, I, B, A \rangle$ in which X is a practical agent, I is some information, B his background information, and A is his cognitive agenda. Thus information is relevant for an agent with respect to a given agenda when, relative to his background information the incoming information is processed by the agent in ways that produces a state of mind that advances or closes that agenda. A simple example: You want to know whether it snowed last night. You walk to the window and look down to the streets heaped with snow. Now you know. In its most basic sense, information is relevant when it is helpful. A further feature of agenda relevance approach is that relevance is a causal relation, rather than a semantic or probabilistic one. Information is relevant for a cognitive agent when in processing it his mind is caused to change in ways that are cognitively helpful to him. When you looked down into the street, you processed the incoming visual information in ways that caused

---

[48] The papers of Anderson and Belnap, and of some of their collaborators, are collected in Anderson and Belnap (1975, 1992). A recent survey is Restall (2006).
[49] Sperber and Wilson (1986/ 1995).

29

it to be the case that you don't any longer want to know whether it snowed last night.[50]

Next to relevance, plausibility is a new kid on the block. The most important work to date by a logician is Rescher's theory of plausible reasoning, which might profitably be considered in the light of criticisms by Gabbay and me.[51] Plausibility, is less developed than relevance. In part, this is because it has proved so far to be a tougher nut to crack than one might initially have supposed).[52]

A remarkable fact about the human reasoner is his impressive though imperfect facility with the evasion of irrelevance and implausibility. Beings like us are rather good at honouring Harman's Clutter Avoidance Maxim, which bids us not to clutter our minds with trivialities. Somehow, the human individual is a "filtration structure", screening out the junk that could violate Harman's rule. This leaves us with two good questions to ponder. One is, what is it that we are avoiding when we screen out the irrelevant and the implausible? The other is, how is it done? How does filtration work? Virtually all the attention to these matters by logicians has concentrated on the first of these questions, giving rise to the suggestion that the second is the proper business of psychology. Whether, and the extent to which, this is so depends on how a naturalized logic manages to negotiate the logic-psychology partnership.

1.10 *Semantics*

We come now to a complication. There is a standard way in which logicians have recognized the distinction between consequence-having and consequence-drawing, and the related distinction between their respective locales – logical space in the first instance and in-the-

---

[50] See Gabbay and Woods, *Agenda Relevance* (2003) and, for a critical discussion of relevance's causal status, Jacquette (2004). *Agenda Relevance* also includes detailed discussions of the contributions and Anderson and Belnap and Sperber and Wilson, among others. See also the special issue of the *Journal of Logic and Computation,* devoted to relevance and guest-edited by Ruth Kempson, Greg Restall and Woods. (Kempson *et al.*, 2011).

[51] Rescher (1976) and Gabbay and Woods, *The Reach of Abduction* (2005), chapter 7.

[52] Leading work in computer science is that of Friedman and Halpern (1995, 2001) on plausibility measures. I will have something to say of them later, in chapter 8.

head in the second. A theory of consequence-having is a semantic theory. It is a theory whose target properties are definable in wholly semantic terms. A theory of consequence-drawing is a pragmatic theory. By this I mean that at least some of its target concepts are not satisfactorily definable except by way of properties possessed by agents capable of symbolic-system processing. In the case of human beings, this includes language-use. Although a theory of consequence-having is disjoint from a theory of consequence-having, it is not excluded that a given consequence-having logic might subsume a theory of consequence-having as a sublogic. Indeed this is a possibility widely realized in the modern mainstream and (I say) in Aristotle as well. Support for these arrangements flow rather naturally from the standard understanding of what constitutes a logic's semantics. From Frege onwards, the dominant approach is a *referentialism*. In a referential semantics all target properties are defined over by way of relations between uninterpreted syntactic objects and abstractly set theoretic items, and functions of a similar character. When semantics is understood in this way, it is easy to see why consequence-drawing cannot be accommodated in a logic of semantic relations on pieces of syntax. In *any* treatment of consequence-drawing in which drawing reflects certain constraints on having, when those constraints require for their description reference to properties beyond those needed for having, drawing cannot be a semantic relation. In particular, theories that give a load-bearing role to the behaviour of linguistic agents exceed the reach of referential semantics.

Notwithstanding its modern dominance, referentialism is not an uncontested approach to semantics. In the logics that flow from Hilbert,[53] all target properties are definable in purely syntactic ways, and any "meanings" that a logic's language might come to have would be acquired by implicit definition by application of syntax-manipulation rules. Such logics are said to have a proof-theoretic, as opposed to referential semantics. The proof-theoretic treatment offers an interesting, and rivalrous, alternative to referentialism. A principal point of contention is whether there is an intelligible notion of truth for statements about set theoretical abstracta. Arguably, the most celebrated display of this rivalry is to be found in the clash between Wittgenstein and Gödel as to whether the Gödel sentence that gives the incompleteness results is paradoxical in the manner of the Liar sentence. It is unnecessary to explore the matter in any detail[54] before seeing that a proof-theoretic

---

[53] See, for example, Hilbert and Bernays (1939) and Carnap (1939).
[54] See here Berto (2009), chapter 12.

logic which admitted pragmatic properties could not, even so, be the same logic as a proof-theoretic logic that excluded them. So this would leave the having-drawing distinction sharply intact and unbreached.

What would it take to breach the having-drawing distinction? It would be by showing that the only (or anyhow the best) way of characterizing the target properties of a logic of *having* is by way of properties that include properties of language-use, especially properties that characterize the *argumental* behaviour of reasoning agents. The most notable modern expression of this idea is *game theoretic logic*, in which the quantifiers, connectives and other logical particles are defined by rules which regulate the performance of pairs of competing agents. Logics of this sort are adaptations of the mathematical theory of games originated by von Neumann and Morgenstern in1947. The first adaptation to logic was effected in 1961 by Henkin.[55] In the years that followed, the game theoretic programme gathered momentum, and is now a thriving stand-alone part of modern logic.[56]

Let us take it as given that arguing is a fact of human nature, and that various philosophers who have been interested in logic have also shown an interest in arguing. Perhaps there is something that we might call a logic of arguing. We might follow the tradition and call such a logic *dialectic*. Whatever else it is, a dialectical logic is a logic of consequence-drawing. But in the absence of contrary indications, dialectic can enjoy a prosperous theoretical life well-short of breaching the having-drawing distinction. A central part of his importance to logic was the emphasis that Aristotle would give this point. The logic of syllogisms, he insisted, was a logic of consequence having. Dialectic, he said, was (among other things) a theory of consequence-drawing. But the target properties of syllogistic logic would be wholly definable without reference to properties needed for the description of attack-and-defend encounters between human individuals.

If the game-theoretic approach is sound, Aristotle can't have been right about this. The right characterization of consequence-having could only be achieved in a dialectical logic (along with the other less bellicose versions of conflict management dialogue logics). It is not my purpose here to adjudicate the historical merits of this position[57], or to

---

[55] Von Neumann and Morgenstern (1947) and Henkin (1961). See also Hintikka (1968).

[56] See, for example, Pauly and Parikh (2006) and Rahman and Rückert (2001).

[57] There is some recent scholarship to the effect that the dichotomy-respecting position which I ascribe to Aristotle is not in fact Aristotle's view. On behalf of

assess the claim of game theoretic logic – anciently founded or newly arrived, as the case may be – to theoretical primacy in the treatment of consequence having. Suffice it to acknowledge that a game theoretic semantics cannot be a semantics in the referentialist/non-pragmatic sense. If the unbreakability of the having-drawing divide depended on the primacy of a pragmatic-free semantics, then the safety of the distinction and the safety of the primacy claim would stand in a tight reciprocity. Here too there is no time to take those matters further. So I will content myself with a decision to conditionalize the proposals of this book to the more traditional stance to semantics. Accordingly, whatever we end up saying about consequence drawing, consequence *having* is wholly definable without the need of pragmatic terms.

We lack a natural antonym for 'non-pragmatic". So let's agree that when, in the pages ahead, especially chapters 7 and 8, I ask how a certain kind of premiss-conclusion relation should be classified – is it a semantic relation or a pragmatic relation – what I mean by "semantic" is semantic in logic's traditional sense of pragmatics-free semantics.

1.11 *Inconsistency*

A main focus of this book is to tease out a logic of third way reasoning, that is, reasoning which, when good, is made so by circumstances other than deductive validity or inductive strength. This is far from saying that deductive considerations don't matter for human reasoning. There are lots of contexts in which validity *is* a requirement of right reasoning – of the reasoning, for example, that drives the proof of the Stone Representation Theorem. But even in third way contexts in which the drawing of a conclusion from premises or assumptions isn't deductively encumbered, it is necessary to know what to do should the premisses of this reasoning fall into inconsistency, that is, deductive inconsistency. When this happens, it matters greatly what follows deductively from what. My answer to that question will influence considerably the course of consistency restoration, and it will remind us of the desirability of premissory consistency in the first place.

Consistency is a premiss-selection filter. It is a device for what we might call the premises-management task. Given the interdeducibility of consequence and consistency, a logic of consequence-drawing is

---

this claim see Hintikka (1987) and (1997), Marion 2013, and Marion & Rückert, forthcoming. For responses, see Woods and Hansen (1997) and Woods (2013a).

directly tied to the logic of this filter. As we saw, in all going systems of mainstream logic, a quite general constraint on good reasoning is that it engender and maintain the reasoner's set of beliefs in a condition of deductive consistency. Dialethists apart, no one doubts the reasonability, or the applicability, of the consistency constraint for most comparatively small (or local) subsets of reasoners' beliefs. But it is out of the question for them all – that is to say, globally. Even if we restrict the constraint to elementary truth-functional consistency, the computational load is well beyond the reach of the human individual. It is another intractable task, embedded in a problem that is at least NP-hard.[58]

In classical (and, among others) intuitionist and modal logic there is a theorem called *ex falso quodlibet.* It asserts that a theory is negation-inconsistent if and only if it is absolutely inconsistent. A set of sentences is negation-inconsistent if and only if, for some sentence α, it and its negation hold true there. A theory is absolutely inconsistent if and only if every sentence of the language in question holds true. A logic that blocks *ex falso* is called a "paraconsistent" logic. A logic is "consistentist" if it honours *ex falso.* It is at this point that the distinction between consequence-having and consequence-drawing takes on a further importance. Even if consequence-having were consistentist, consequence-drawing could only be paraconsistentist. So let's make it official:

Proposition 1.11a
PARACONSISTENCY: *Consistency maintenance is not a wholly general requirement for inputs to the processes of reasoning. Individual reasoners operate perforce in paraconsistent contexts and are not intrinsically compromised by them.*

---

[58] "NP" denotes "nondeterministic polynomial time." Informally, NP is the set of all decision problems for which an affirmative answer admits of simple proofs of the fact that the answer is indeed affirmative. Such proofs are verifiable in polynomial time by a deterministic Turing machine. "NP-hard" denotes the class of problems at least as hard as the hardest problem in NP. The hardest problems in NP are called NP-complete problems, for which no polynomial time algorithms are known. A relatively untechnical treatment of these issues can be found in Cherniak (1986). For similar reservations, see Woods (1988a) and (1988b).

The question presently in view is of a kind with the constraint question for Bayesian priors. Informally put, the question can be understood as asking for conditions on premises-selection for Bayesian inference. For hard-nosed subjectivists the answer is that anything goes short of logical inconsistency. If this is right, inductive logic would be all about consequence-spotting and would have no role to play in premises-selection beyond unperformable consistency checks. This will strike some readers as over-austere, allowing for the Bayesian impeccability of inferences whose priors are, though deductively consistent, absurd or simply idiotic.[59] It will be readily agreed that a desideratum of our inferential practice is to reason from true or otherwise acceptable premises. No one seriously supposes that it falls to the logician to determine premises-eligibility concerning matters of fact – in microbiology, say, or historiography. But would it be going too far to expect some discouragement from logic regarding the selection of premisses "offensive to reason"?

I admit to little enthusiasm for such a proposal, chiefly on account of its vagueness. A possible contender, I suppose, is any sentence contradicting any analytic truth. In the last half-century or so, analyticity has taken something of a shellacking, so one must proceed with some care. For the moment I propose that on logic's behalf we accept the assignment subject to the requisite clarifications. I will return to this point.

I want now to touch on the connection between premiss-inconsistency and consequence-having. To do this without taking up more space than I can presently spare, I will draw attention to some historical cases in which theories have been discovered to contain inconsistencies. Let us begin with the observation that the most central objective of a theory of a given subject matter D is to achieve a knowledge of D, that is, to determine which propositions about D are true, as opposed to false, and which are false, as opposed to true. On the consistentist approach every sentence whatever is a consequence of an inconsistency. This, the *ex falso quodlibet* theorem, generates its own catastrophe thesis for theory construction. Suppose that our D-theory contains an inconsistency. Then, by *ex falso quodlibet*, every sentence of D is true and false. The theory is unable to determine with respect to any of its sentences that it is true *as opposed* to false or false *as opposed* to true. The central objective – indeed the very rationale – of the theory is

---

[59] An important contribution to the prior's literature is Suppes (2007). See also Binmore (2009), chapter 7.

dashed. It is impossible for it to achieve a knowledge of D – not a jot of it.

In 1902, Russell communicated to its author that Frege's masterwork, *Der Grundgesetze der Arithmetik*, contained an inconsistency. By catastrophist reasoning, there is absolutely nothing to learn about arithmetic or sets or anything else in the *Grundgeszetze*. Not one sentence of the *Grundgesetze* is true rather than false or false rather than true. Two and a half century's earlier, there was a similar situation in the arithmetic of small numbers. Calculus, the theory of infinitesmals, embodied a known inconsistency.[60] The inconsistency proved resistant to removal until the introduction in the 19th century of Wierestrasse's limits and, in the 20th, of Robinson's hyperreals. Again, the catastrophist line is that nothing about infinitesmals could be learned from Newton or Leibniz, that for the first 200 years of its existence there was no knowledge of anything – still less of infinitesmals, differentiation and integration – to be got from calculus.

There is no serious student of the history of mathematics who pays these dire sayings the slightest heed. Working mathematicians took it as given that there was lots to know about these things, notwithstanding their respective local inconsistencies. What this tells us is that in their workaday professional responses to local inconsistency, mathematicians implicitly reject the catastrophe thesis, and thereby also reject *ex falso*. It is not just that these mathematicians weren't *drawing* every consequence possessed by a local inconsistency, but rather that they weren't recognizing every sentence as a consequence it *has*. The moral of these historical cases is clear.

Proposition 1.11b
THE CONTRA CATASTROPHE THESIS: *If the mathematicians were reasoning properly in generating the knowledge those theories gave us, then it is an error of*

---

[60] Perhaps I am taking some historical liberties here. Leibniz regarded infinitesmals as mere fictions which stand in for very complicated exhaustion proofs and which can be shown to be wholly free of inconsistency. Newton on the other hand based his calculus on the notions of uniform motion and acceleration, a foundation that he held to carry no taint of paradox. Berkeley's position was, in effect, that both these inconsistency-free claims were mistaken, and that the calculus was indeed paradoxical. I haven't the means at hand to resolve this historical dispute. No matter, it suffices for my purposes to make do with the inconsistency of intuitive set theory.

consequence-recognition to suppose that every sentence is a consequence of a local inconsistency. That is, ex falso is false.

COROLLARY: *A paraconsistent logic is needed not only for consequence-drawing but also for consequence-having.*

Proposition 1.11b tells us, in effect, that inconsistency in reasoning contexts is not intrinsically intolerable. It goes a good part of the way in explaining why: Inconsistency in reasoning is not catastrophic. Reasoning can flourish even if tainted by inconsistency.

I am an error-theorist about error. If Proposition 1.11b and its Corollary are true, classical and consistentist accounts of deductive consequence-having and consequence-drawing are in error – all of them. If so, perhaps this is already interesting. But its real interest – its thick interest – resides in what, when it is finally worked out, the right account of deduction turns out to be. Paraconsistent logic is a growth industry, [61] and the jury is still out. The nature of deductive consequence is an open research-programme.

One of the trickiest unsolved problems is the Lewis-Langford[62] proof of *ex falso,* as follows:

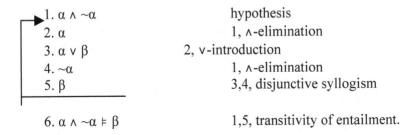

1. $\alpha \wedge \sim\alpha$		hypothesis
2. α		1, $\wedge$-elimination
3. $\alpha \vee \beta$		2, $\vee$-introduction
4. $\sim\alpha$		1, $\wedge$-elimination
5. β		3,4, disjunctive syllogism

6. $\alpha \wedge \sim\alpha \vDash \beta$ 1,5, transitivity of entailment.

If, as some would claim, *ex falso* is provably false, how can there be a proof of the opposite? Various attempts have been made by paraconsistentists to disarm the Lewis-Langford proof, none of them persuasive.[63] As long as it is kept in mind that the proof's connectives occur with their truth-functional meaning, it is impossible to find a

[61] Concerning which, see Priest *et al.* (1989), Woods, (2003), Da Costa and French (2003), Bertossi *et al.* (2004), Berto (2008), Priest (2007), Béziau *et al.* (2007) and Schotch *et al,* (2009).
[62] Lewis and Langford (1932).
[63] These are discussed in Woods (2003), pp. 58-79.

relevant and non-question begging way of blocking the consequence that a truth-functional contradiction entails any sentence whatever. And if we allow that every contradiction in the form $\ulcorner \alpha$ and not-$\alpha \urcorner$ entails a contradiction in the truth-functional form $\ulcorner \alpha \wedge \sim\alpha \urcorner$, it looks as if *ex falso* is robustly back in business. Wherewith a problem. Either my proof against *ex falso* is defective or Lewis' and Langford's proof for it is defective. Is there a principled basis on which to determine which is which?

My own view is that if we retain the fiction that a theory is a set of statements, whose function, among other things, is to partition its members into subsets of statements true, as opposed to false, and subsets of statements false (as opposed to true), there is no credible way of settling the matter. How much this would matter for reasoning from inconsistent inputs would depend greatly on how deeply we are prepared to commit to the distinction between consequence-having and consequence-drawing. Upon reflection, however, the view that the theories are proposition-sets might have to be re-thought. What if we said instead that theories are sets or complexes of *assertions*? Assertion and assertibility have not stirred much interest in the mainstream of logic, but Rescher has been thinking about it for a long time (1959/1961, 1968).

It bears on this also to mention that Jáskowski's founding investigations were directed to the logic of *discussion*. Jáskowski (1934/1967) called his systems "discussive" logics (and, in a variation, "discursive"). They are a branch of dialogue logic. Discursive systems are non-classical in certain predictable respects. In a conversation, one party might forward the claim that α, and another might assert a proposition incompatible with it. It might even be the case that there is a positive degree of reasoned support for each of α and $\ulcorner \sim\alpha \urcorner$. But there are two things Jáskowski won't allow. He won't allow that the combination of the reasons for α and the reasons for $\ulcorner \sim\alpha \urcorner$ constitutes any reason for $\ulcorner \alpha \wedge \sim\alpha \urcorner$. Neither will he allow that $\ulcorner \alpha \wedge \sim\alpha \urcorner$ has even *occurred* in this discussion (so the question of what are the reasons for it doesn't actually arise). Hence his n-party discussions are "non-adjunctive". The rule of adjunction is not a universally valid rule there.

Perhaps we now have the basis on which to mount a rejoinder to the contra-catastrophe thesis. Recall that if it holds, then any readiness to preserve *ex falso* truth for standard logic is doomed. The contra-catastrophe thesis presumes that a mathematical theory is a class of true sentences closed under logical or logico-mathematical operations. If $\mathcal{T}$ is an axiomatic theory, then what $\mathcal{T}$ tells us about its subject-matter is conveyed by those axioms and their closure under consequence (whatever the details). But if T's closure mechanisms fail to partition T's sentences in the requisite way, then whatever we might already know about T's subject matter or whatever we might come to know about it, this could not be a knowledge afforded by T. But again Rescher's insistence that his "semantic" rules are actually assertibility rules suggests a different conception of what a theory is. Contrary to what is assumed by the contra-catastrophe claim, a theory is not in the standard meaning of the term a semantic object, not a set of truth-valuable sentences closed under sentential relations. Rather a theory is a *pragmatic* object, that is, a set of assertions closed under its so-called "semantic" rules, which are actually rules of assertion. So we might think of the assertion-consequence relation as consequence in name only, as consequence in a manner of speaking. An assertibility logic can guard against catastrophe in two ways. It can forbid the assertoric aggregation of any assertion and its negation. It can also suppress *ex falso* for assertion.

What do we now make of the theories which provoked the case for *ex falso* – old calculus, old set theory and old quantum theory? How, we asked, could these inconsistent theories give us any instruction if deduction were closed under classical consequence? The answer that pends now is that this is all beside the point. Perhaps the old comprehension axiom does indeed entail everything whatever. But Rescherian theories aren't closed under standard logic. They are closed by the rules of assertion. Could such theories give us instruction about integration and differentiation about sets and about the interior of the atom? Yes, they could. I am not aware of any place in Rescher's writings in which he either identifies the contra-catastrophe defence of *ex falso* or marshals the present rebuttal of it. But if I am not mistaken, all the elements required for the rebuttal are prefigured in Rescher's nonstandard notion of a semantical rule.

Will it work? Not if you are a philosophically serious dialethist. Let α be the Russell sentence, "There exists a set R which is a member of itself if and only if it isn't." Philosophically serious dialethists are more than ready to assert this proposition, without the slightest inclination to see it as an assertoric blindspot. This reminds us of something important about cogent reasoning. One person's *reductio* is another person's demonstration of an utterly surprising truth.

Let us note that assertoric consequence is a nonmonotonic relation. We could have it that β is an assertoric consequence of α, but we would not accept that it is an assertoric consequence of $\{\alpha, \ulcorner \sim \alpha \urcorner\}$. I will say more of nonmonotonic consequence relations in chapters 7 and 8.

I close this chapter with a further word about the importance of paying attention. The informal logic community is one in which the contributions of modern formal logic are assayed unfavourably, or at least with a tightly selective circumspection. In some respects it is a justified resistance. Mainstream formal logics are logics of deductive validity. The extent to which the reasoning of interest to informalists is reasoning of the third way sort, the validity standards of mainstream logic fail to secure a purchase there. If validity is not typically the right standard for correct premiss-conclusion reasoning, deductive logics won't tell you much about the relation that underlies premisses and the conclusions rightly drawn from them. Premiss-consistency is another matter entirely. Third way logicians have a large stake in premissory inconsistency, and should have something principled to say about how it is managed by good reasoners. At the heart of it all is the fact that inconsistency is in more ways than one *deductively* significant. But the question of what that significance actually is is awash in controversy in logic's mainstream. The right story about inconsistency-management in third-way contexts demands telling, and its correct telling awaits a resolution in the coils of the foundations of deductive logic. There is not much attention paid to this troublesome turbulence by the informal logic community at large, and such notice as is given it on occasion is gestural. This is nothing but unfortunate. It is a matter that demands more attention.[64]

---

[64] I hasten to add that the informalists' resistance to mainstream logic has a two fold footing. One is that it is a logic of deduction. The other is that it is a mathematical logic of deduction. Though often jointly held, there are not equivalent suspicions. On the latter see Slater (2011).

CHAPTER TWO

ASSESSING

"Our inventions are wont to be pretty toys which distract our attention from serious things. They are but improved means to an unimproved end."   Henry David Thoreau

"[The error of modern logic is that] the people leading the masses hav[e] talents that lie, in broad terms, in numeracy rather than literacy."   Hartley Slater

"Sometimes you had to say *Stuff Logic* and go with the flow."
                                        Reginald Hill

2.1 *Scope*

The builder of a theory is faced with two quite general methodological tasks. He must say something about the theory's *scope*. He also must say something about its *conditions of adequacy*. There is a loose holism about such things. To a certain extent, we figure out a theory's proper scope and the conditions that make for its adequacy just in the course of building it. But it is also true that some measure of these determinations is available to us pretheoretically. The scope-question for a theory is a matter of both coverage and depth. A theory's coverage is the sum of the issues it raises – its collective subject matter. Depth reflects an expectation of how far the theory is to go in its treatment of the covered items. The depth of a theory varies between the exhaustive and definitive, and the provisional and promissory. A reasonable indication of this book's scope is furnished by its opening chapter. A thick logic of erroneous premiss-conclusion reasoning should do what it can to get to the bottom of the concept-list misalignment and cognitive virtue theses. It should seek for consequence relations, both in the spotting and the drawing, that range beyond the deductive and inductive paradigms. It should make up its mind about the having and drawing of consequences. It should have something to say about premises-management, especially as regards consistency, relevance and plausibility. It should try to solve the *ex falso* problem. It should make an effort to find a normative grounding for its evaluations, and in so doing

41

should take proper notice of the organic constitution, and the limitations of resource and design, of the individual practical reasoner. It should try to solve the concealment-detection problem for error. It should start developing a reflective and mature naturalism. It should find a plausible way to be psychologistic.

In most of these respects there is more to be done than can be done; that is, than can be done by me; than can be done perhaps by anyone within the span of a single book. This means that I am likelier to do better with coverage than depth. In much of what lies ahead, the emphasis will fall rather more on the to-be-done than the done, and diagnoses will outpace treatments. So the book, all in, is more a prolegomenon to a naturalized logic of error than the finished product. This is no bad thing. There is something to be said for charting the paths to be taken before rushing off in all directions and traversing them with unseemly haste.

2.2 *Adequacy*

The respect-for-data principle directs our attention to target phenomena – to the *data* a theory proposes to account for. But there is also the question of what further conditions need to be met of the account is to pass muster. It asks the builder of a theory of its target data to have a considered idea of what would make for a good account of them. It admonishes the theory-builder to arrive at some prior appreciation of a theory's adequacy conditions.

My main task in this chapter is to craft a response to this requirement. Reasoning is only a part of the human repertoire; it is but one of a large class of goings-on for which judgements of "well done" and "badly done" are not out of place. These are, let us say, our "normatively assessable" practices. Normatively assessable human performance is no stranger to theoretical attention. It is attention of a kind that generates its own questions about how to proceed and about the criteria by which such a theory could be judged. Here are some candidates to consider:

- *empirical adequacy*
- *mathematical adequacy*
- *normative adequacy*
- *conceptual adequacy*
- *descriptive adequacy.*

As we proceed, there will be ample occasion to take further notice of tensions that obtain between and among these conditions, and I will try my hand at bringing to the mix a degree of balance and mitigation. Most readers will be well acquainted with the standard rivalries hinted at a chapter ago – between the mathematical and the conceptual, between the empirical and the normative, and between the technically virtuosic and the empirically faithful. Later in chapter I will draw attention to a potentially corrosive tension between conditions of empirical and conceptual adequacy, triggered by the methodological decision to *naturalize*.

An empirically sensitive logic of reasoning is an investigation of the behaviour of practical agents, of beings who, in reasoning about the small things appropriate to their interests and capacities, pursue their ends in real time with the comparatively scant cognitive resources on which they are able to draw. Nearly everyone accepts that not every good theory has empirical entanglements. A theory of human reasoning is different; it is a theory of the behaviour of real people. There are empirical facts about such beings. In some cases, there are empirical laws about them too. It can only be expected that at least some of those facts and some of those laws would have to be heeded by any logic of reasoning with naturalistic ambitions. In mathematically virtuosic approaches, agents and contexts are sometimes taken into account, but are often much constrained by the austerities of formal modelling. Consider again the treatment of deductive reasoning by mathematical logic and of probabilistic reasoning by the mathematical theory of games of chance. There is a price to pay. It is theoretical overkill, the loss of empirical accuracy. It will come as no surprise that there are people who take a dark view of this. They think that the failure of a theory to be empirically adequate trumps its mathematical dazzle. More interesting, however, is the size and steadfastness of the contrary view, the view that in lots of cases the trumping relation should be reversed, that a theory's mathematical power and elegance should be allowed to override its empirical shortcomings.

There are well known suggestions for lightening the impact of theoretical overkill. One concerns the gap between what the theory demands and what actual experience delivers; and the suggestion is to blame the gap on *us*. Seen this way, the failure of real reasoners in their reasoning on the ground to conform their behaviour to the imperatives of this, that or the other mathematically virtuosic logic convicts them of substantial, perhaps even sweeping and systematic, illogicality. This is the *irrationality thesis*.

A less harsh version is one that offers misbehaving real-life agents the patronizing mitigation of approximation to the norms mandated by the logic they offend against. Accordingly, while there is plenty of evidence that real-life agents don't execute the rules of the logic, we are invited to suppose that they are able to deploy heuristics which, while a lesser thing logically speaking than the theory's norms, bring them to within striking distance of where they should have been in the first place. Although these methods aren't quite rational – after all, they are *shortcuts* – they are somewhat rational, and so likewise are the agents who avail themselves of their provisions. This is the *approximate rationality thesis.*

By far the dominant way of triggering the irrationality and approximate rationality thesis is a theoretician's invocation of *ideal models,* especially in those cases in which the model's provisions are forwarded with normative intent. In one way of proceeding, the model describes an ideal agent – a believer, a decider, a shopper, a union boss, a line officer, a university president, or whomever else – operating, if not perfectly, then at some theoretically endorsed level of rightness against which the adequacy of real performances on the ground is assessed.

In a classic *aperçu,* Nancy Cartwright (1983) observes an important distinction:

> There are the obvious idealizations of physics – infinite potentials, zero time correlations, perfectly rigid rods, and frictionless planes. But it would be a mistake to think entirely in terms of idealization, of properties which we conceive as limiting cases, to which we can approach closer and closer in reality. For some properties are not even approached in reality. (p. 153)

Although in this passage Cartwright is writing about physics, her distinction between approachable and unapproachable idealizations does useful work in the logic of reasoning. Something of this difference is reflected in the difference between the cardinalities of various normative shortfalls. If a theory contains the idealized norm of the closure of beliefs under consequence or of the knowledge of all logical truths, then notwithstanding the finite difference between Harry's and Sarah's performance scores, each will be infinitely short of what the norm requires. Sarah might well outpace Harry in these respects, but each of them fails the relevant ideals, impossibly and equally dismally.

Suppose, with Harman (1970), that Harry's belief set contains 300 beliefs and that Harry's task is to compute the probabilities of individual beliefs conditional on belief-conjunctions. It requires a finite though hefty number of calculations – namely 2^{300} – which is several orders greater than the number of elementary particles in the universe; and far too many to compute and far too many to store as innately secured primitive probabilities. Seen Cartwright's way, these are unapproachable norms, wholly inapproximable by anything that goes on in nature and, as she goes on to say, "pure fictions". If, on the other hand, the theory contains as idealized norms routines of less daunting complexity, the gap between norm and performance may be huge, but in some instances it will be finite. In Cartwright's terms, these are approachable idealizations, norms for which approximation relations are said to be definable in principle. Performances that fall short of unapproachable ideals lend themselves to the irrationality thesis. Performances that fall short of approachable ideals lend themselves to the approximate rationality thesis.

I am drawn to the view that the irrationality thesis is dismissible out of hand. Somewhat less easily disposed of is the slighter claim of approximate rationality. There are all sorts of things we are more or less at home with – whistling, the fox-trot, contract negotiations, linear logic, the lot. Why wouldn't we likewise be more or less adept at reasoning? Why wouldn't this lend some support to the approximate rationality thesis? Certainly everyone knows that when it comes to reasoning, human individuals perform more or less well, and that when it comes to rationality, some people in some circumstances exhibit more of it than others. But we should not for a moment think that the truth of the approximate rationality thesis implies the truth of just any old theory that implies it.

Consider now some of the standard objections to the entrenchment of the Bayesian protocols in an empirically sensitive, resource-bound logic of real-life error. A major complaint against it is mathematical. Running the model's algorithms is beyond the computational reach of beings like us, and – in various respects – actual human performance at its best fails even to approximate to Bayesian standards. The difficulty is not peculiar to Bayesian epistemology. It extends to virtually all model-based social science. Consider a case. In classical decision theory, a decision is the solution of a decision problem.[65] Such solutions are represented by right-to-left computational tree algorithms, called decision trees. Rules for the construction of decision-trees serve as the theory's decisional norms. It is widely, though not universally, believed that these are the rules that rationally bind the decision-maker. Decision-trees are outputs of procedures to maximize the agent's subjective expected utility (SEU). Decision-trees are "bushy" when they exhibit high degrees of computational complexity.[66] This is problematic. There is a level of complexity that strips the human individual's computational gears. Lots of decisions taken by individual reasoners violate the theory's bushy norms. But on the view presently in question, such defections are normatively tolerable. Here is William Cooper on this point:

> Of course, the organism's processing might not proceed in ways exactly analogous to [the production of right-to-left computational tree diagrams]. No one supposes that an organism will literally draw trees in its brain. It has only to execute some black-box approximations of that, with the processing giving rise to behaviour that looks *as if* a tree analysis had taken place. It isn't even clear that it must depend on the same general distinctions between choices, events, probabilities, consequences, and so on. The process need only

---

[65] Raiffa (1968). Here too there is a substantial and like-minded psychological literature, beginning with the review papers of Edwards (1954, 1961). The dominant position among the mainstream decision theorists is to be as Bayesian as possible. A notable defection is Dempster-Schafer theory (DST), which rejects the use of probability theory, favouring instead non-probabilistic *belief functions*. Even so, DST exhibits high levels of mathematical complexity, and in certain important respects it resembles the Bayesian approach. For example, informationally independent beliefs are aggregated by Dempster's rule of combination, which is a generalization of Bayes' theorem when events are independent. The classical works are Dempster (1967) and Shafer (1976). A leading critic is Pearl (1988). Another is Binmore (2009). Further efforts to lighten up, in the one case by fuzzifying probability measures and in the other by lessening the strictness of Bayesian standards are evident in the literature. See in the first instance Jeffrey (1984), van Fraassen (1990), and in the second van Fraassen (1984), Seidenfelt and Kadane (1989) and Mongin (1995). My view of the lightening up movement is twofold. It doesn't go far enough, and it claims a normative legitimacy it fails to establish. See the section to follow.
[66] Cooper (2001).

result in behaviour that is so interpretable to us as analysts accustomed to these concepts.[67]

Embedded in these reflections is a line of thinking which can be found running through model based social science. It is what we might call

Proposition 2.2a

THE HEURISTIC BRUSH OFF: *There are lots of cases in which the normative requirements for right reasoning are laid down algorithmically (or anyhow in a abstractly rule-governed way). Given that algorithmic (or rule-compliant) performance very often outpaces what actual reasoners are built for, there are normatively lesser procedures – usually involving a kind of shorthand – which give results that are approximately correct, or anyhow as correct as a human reasoner can attain. These are heuristic procedures.*

Proposition 2.2b

SEDUCTIVE SHORTCUTS: *An attraction of heuristics over algorithms is that they are natural and easy to work with. They are, so to speak, seductive shortcuts.*

COROLLARY: *Given that heuristics are by a theory's algorithmic standards a kind of error, and given that their*

---

[67] Cooper (2001), p. 58; emphasis in the original. For a decision theory that is sensitive to the real-life limitations of individual agents, see Weirich (2004) and Wakker (2010). Of course, it is quite widely recognized that real-life decision-makers discomport with the completeness, expected utility (SEU) and transitivity axioms, among others. Informally, completeness requires a rational agent to prefer one object to another or be indifferent to them both (Samuelson, 1938). Expected utility requires a rational decider to maximize his subjective utility (Kahneman and Tversky, 1979). The transitivity axiom requires that whenever one thing is preferred to another and it in turn to some third, a rational agent must always also prefer the first to the third (Loomes and Taylor, 1992). Descriptively motivated analysts have sought to reduce these inconsistencies of practice. For example, in Kahneman and Tversky's prospect theory the expected utility principle is modified by the introduction of two weighting functions, one for probability and the other for value. In quite simple situations, the modified principle does better on the score of empirical adequacy. But there is no evidence that in reaching their decisions in such cases, individual agents actually compute these functions. Complexity again is the problem. See again Wakker (2010).

seductiveness is widespread and their usage frequent, heuristics instantiate the traditional concept of fallacy.

I have opinions of this corollary. One is that it is absurd.

The sense of "practical" most widely invoked by philosophers is that which pertains to action *as opposed to* belief. It resembles the present –and also very popular –distinction between a heuristic procedure *as opposed to* a rule.[68] It is sometimes thought that these two instantiate a more comprehensive contrast between the useful and the true, as in a standard reading of instrumentalism *as opposed to* realism in science. In one of its forms, instrumentalism allows a scientific theory to be good – indeed very good; top of the line – when, although it doesn't get at the truth of things, it answers to other interests, such as predictive control. I have reservations about placing our three distinctions under this aegis. So interpreted, they are untenable dualisms; their purported exclusivities are not plausible. Perhaps it would be going too far to say that the contrasts they purport are false contrasts. But they are less load-bearing than is often supposed. They don't support the intended conceptual demarcations with definitive authority. Doings are sometimes believings; heuristics are often truth-tracking, and the usefulness of the useful is often a matter of the truths it uncovers. Concerning heuristics, there is something worthwhile to be learned. The idea of their intrinsic cognitive inferiority lacks a demonstrated foundation. In a well-known example from Gerd Gigerenzer and his colleagues, a heuristic method "is actually more accurate in classifying heart attack patients according to risk status than are some rather more complex statistical classification methods."[69]

Similarly, finite consequence drawing is less good than the full-bore regime mandated by the closure norm only if the betterness of the norm is established independently. Let us see about this.

---

[68] There is some variation of usage here. In computer science, "heuristics" originally denoted methods for finding mathematical proofs. Somewhat later a heuristic was a method for solving a problem in the absence of guaranteed success, hence a method that failed to be an algorithm. In time, a heuristic was an AI rule of thumb used by domain experts without the necessity of exhaustive searches. More generally, and more recently, heuristics are intelligent search strategies for computer problem-solving unattended by proofs of accuracy (Pearl, 1984, and Russell and Norvig, 1995, p. 94). In philosophical and psychological usage, heuristics are efficient rules of thumb for problem-solving – "fast and frugal" in the manner of (Gigerenzer, 1999).

[69] Gigerenzer *et al.,* 1999, pp. 4-5.

2.3 *Convergence*

It is not uncommonly proposed by the mainstream decision-theorists that the goodness of your decision, such as may be, lies not in following the rules of rational decision-making, which in large numbers of cases you can't manage, but in discharging routines which, while logically subpar, approximate in output to the output of the theory's mathematically modelled norms. The approximate rationality thesis invites us to suppose that our actual decision making heuristics are in turn approximations to what the rules require, that they are in some cases approximation to what a theory's algorithms produce. Thus there exists with some frequency a concurrence between what the mathematical model calls for and what the individual's heuristics enable him to do. This, it is said, is too much concurrence to be a coincidence. There must be a hypothesis that accounts for it. The preferred supposition is abductive. It turns on the conviction that the best account of our overall successes as decision makers is that the procedures we actually use are indeed approximations of these better procedures that we cannot implement. It is a platonic story, needless to say. A model's norms are its forms, and like forms of all stripes, it is they that get to wear the trousers. Heuristics trail along as a poorer relation. They are feeble instantiations, and even at their best the performances they facilitate are subpar.

Perhaps this is so. Perhaps this is the best way to handle the rough concurrence in outputs of a theory's algorithms and an agent's heuristics. But we should not accede to it in the absence of an independent specification of the postulated relation of approximation. Some apologists plead that all that the approximate rationality thesis really comes down to is that real-life human performance is not perfect.[70] The not-perfect thesis is supposed to leave it open that in actual practice we really do rather well by and large. How well is gauged by the degree to which we approximate to perfection. Although there are well defined approximation relations for physics, definitions of them are thin on the ground in agent-based logics.

Howard Raiffa is a well-known decision theorist and George Kennan was a well-known decision maker. As far as I know, Kennan, who was the chief architect of American Cold War policy[71], had no formal acquaintance with Raiffa's decision theory. If we put our minds to it, we could bring Raiffa's theory to bear on Kennan's decisions. We

---

[70] For example, Christensen (2004), chapter 6.
[71] "The sources of Soviet conduct", *Foreign Affairs,* July 1947.

might in so doing discover what these days is called a "disconnect" between some of Kennan's thinking and what Raiffa's theory calls for. In such cases, something will have gone wrong. Either Kennan made poorly reasoned decisions or the theory endorsed defective theorems. One thing is clear: Absent the patient examination of these disconnects, it would be foolish simply to defer to Raiffa in those cases in which the Kennan decisions strike us as sound. And it compounds the folly to ground this deference in the notion that since the theorems from which Kennan defects are derivable is something called an ideal theory of rational decision, they actually carry the day normatively.

It strains credulity to say that the rational superiority of the model's decisions derives from the mathematical virtuosity of classical decision-theory (but see just below). What is more, there is a *rival* of that version of the approximate rationality hypothesis that requires some attention. It proposes the opposite, that a better explanation of our decisional successes is that the heuristics that actually drive our decisions are the right ones – the rationally superior ones – and that the norms of classical decision-theory are logically subpar approximations of *them.* According to the first explanation, the heuristics that approximate to the model's norms are procedures that fall short of them. According to the second explanation, the norms that approximate to the on-the-ground heuristics are procedures that over-reach them. Seen the first way, the approximations don't go far enough. Seen the second way, they go too far. They put the lie to the adage that we can't get enough of a good thing.

The Raiffa-Kennan example should prompt us to wonder how a theorist of human practice knows whether his theory hits its normative targets? How does he know that his theory's theorems are normatively binding on actual practice? When we consult the mainstream theories, we can only be struck by the extent to which this question is not even posed, never mind answered. In some cases, contrary to what we said just now about "straining credulity", it appears to be assumed that the theory's normative legitimacy derives from, that is, *is constituted by,* the mathematical authority of its theorems, somewhat along the following lines:

THE MATHEMATICS ARGUMENT: *Is it not the case that mathematical truths are necessarily true? And is it not the case that what is necessarily true is trivially binding upon human conduct? (After all, we can't have any fact about human*

behaviour making false any law of mathematics, can we?) In
these cases, normative legitimacy "coat-tails" on the
mathematical necessity of the theory's theorems.

In other cases, we seem to have it that since the theory's norms are true in
the theory's model, they are analytically true.

THE ANALYTICITY ARGUMENT: *Is it not the case that what*
is analytically true is necessarily true? And is it not the case that
what is necessarily true is trivially binding on human conduct?
(After all, we can't have any bachelor making it the case that,
although a bachelor, he is concurrently a married man, can we?)
A related suggestion is that the theorems of a system of logic or a
theory of decision making, or more generally a theory of X-ing,
are the results of the meaning of the concept of good X-ing, as
specified in the model.

Lines of reasoning so deliciously defective should be attributed only with
care. I have no heart for lengthy search-and-destroy missions. Let me say
only that these arguments are howlers. It takes only a little technical
agility to negotiate into some or other mathematical model the negations
of the most obvious truths of normative good sense There is no trick to
getting the worst rubbish to come out true in an ideal model. Equally, if
someone tells us that it is part of what it means to be an able reasoner that
we must close our beliefs under consequence and know all the truths of
logic, I will simply tell him that there is not a contending theory of
meaning anywhere on this Earth that supports him in such a claim. Let us
leave it at that. If there are enthusiasts among logicians and decision
theorists and the like, for whom the normative legitimacy of theorems is
not merely assumed or not grounded in either of these mad *non sequiturs,*
perhaps they will be so kind to tell us in what this normative legitimacy
consists.

This, the problem of normative legitimacy, is arguably the central
problem of metaethics. It is an especially vexing problem for empiricism,
neatly gathered up in Hume's discouraging observation that values can't
be derived from facts. It would be churlish to chastise the ideal-modelling
crowd for its failure to solve the normative legitimacy problem. But
rebuke it we should when it takes no note of it, or, when having
recognized it, it offers carelessly light-weight solutions in the manner of

51

the mathematics and analyticity arguments. I have no interest in the outright denigration of the normative intuitions of this, that or the other ideal modeller. When from time to time I find myself disagreeing with those intuitions, the error may well be mine, not hers. But I will concede nothing to the silly idea that the very fact that her intuitions are ensconced in an ideal model gives them the requisite normative authority to underwrite either the irrationality or the approximate rationality theses.[72]

2.4 *Normativity and normalcy*

The favouritism extended to the precepts of mathematical decision theory is a particular case of a generally normative deference shown to a logic's rules.[73] Why would this be? The answer, I think, turns on the ambiguity of "rule". Not everything someone calls a rule is a rule on that account. But we might all agree that if something really is a rule, there is a sphere of conduct in which it has the authority to regulate behaviour. The rules of tic-tac-toe or the performance regulae of Beethoven's Ninth pack normative weight for as long as they are actually in force. Since nothing is a rule just on the sayso of an assertor of it, the inference "Since *R* is a normative rule, it is normatively binding" embodies a potentially disabling equivocation. The inference embodies a toadying deference and a massively dug-in affection, anchored by the presumption that the rules of *logic* have compelling evaluative force, that just as they stand they are axiologically decisive for empirical practice.

Some psychologists mark a distinction among three different models of cognitive performance. The "normative" model *N* specifies standards of optimal rational performance irrespective of the costs – including the computational costs – of compliance. The "prescriptive" model *P* attenuates these standards so as to make them computationally executable by individual agents. The "descriptive" model *D* gives a law-governed account of how individuals actually perform. It provides the

---

[72] Doubts about such approaches are advanced in a number of works, among them Cherniak (1986), Cummins (1998), Stich (1990), Gigerenzer and Selten (2001), Searle (2001), Woods (2003), chapter 8, Pollock (2006), Harman and Kulkarni (2007), and Morton (2013).

[73] This presumed normativity is also widely held by psychologists. Writing about natural sampling problems, Evans notes *en passant*, "[a]s with logic problems, these probability tasks are given [by experimenters] to people who generally lack training in the relevant normative systems of reasoning (*logic, probability theory*)" (Evans, 2007, p. 148; emphasis added)

results which a naturalized logic of errors of reasoning should expect to be influenced by. Of the three, D is the only model that takes seriously the evolutionary etiology of our reasoning practices. Following Stanovich (1999), it is possible to discern three different positions concerning how N, P and D are linked. The principle of linkage is nearness to the goal of right reasoning. On the "panglossian" approach, there is nothing to distinguish the three models in relation to this goal. At the opposite extreme is the "apologist" position in which N meets the goal, and both P and C fail it, and do so both seriously and next to equally. The *meliorist* approach takes up the middle position, according to which N meets the goal, P fails it, but not so badly as to preclude approximate realization, and D fails it outright.

The *NPD* trio captures some of our earlier distinctions. The irrationality thesis is apologetic and the approximate rationality thesis is melioristic. For reason already mentioned, I am disposed to reject N and the irrationality thesis. Similarly, computationally executability does not strike me as anything close to a convincing reason for accepting the P model. N purports to provide standards for optional rationality and P for computationally realizable optimal rationality (sometimes called "optimization subject to constraint" – OSC.)[74] D fashions its standards in ways that allow for the non-optimiality of right reasoning. D throws up questions of the first importance. One is whether reasoning goodness is compatible with non-optimality. The other is whether we might plausibly suppose that an individual *expresses* his rationality in the ways in which he manages to extract cognitive and behavioural advantage from conditions of suboptimality. I myself own to a strong inclination to answer both these questions affirmatively. Showing why in somewhat greater detail is something that lies ahead of us. For the present – and again in the spirit of cards on tables – I will until further notice make it a working assumption that

Proposition 2.4
CONVERGENCE OF THE NORMAL AND THE NORMATIVE (1): *As a first pass, and when there aren't particular reasons to the contrary, how we* do *reason from premises to conclusions is typically how we* should *reason. In other words, in matters of consequence-drawing there is a*

---

[74] See Stigler (1961).

trending convergence between the normative *and the* normal, *between* what is usually done *and* what is rightly done.[75]

The Raiffa-Kennan case has a bearing on NN-convergence. Consider again the example's imagined discrepancies between Kennan's actual decisions and Raiffa's decisional norms. Suppose again that you are required to select between the following two possibilities:

1. Give Kennan the benefit of the doubt.

2. Give Raiffa the benefit of the doubt.

Suppose further that you must make your choice without further examination of the discrepancies one by one. Those who favour (1) reflect the spirit of NN convergence. Those who favour (2) have a lower opinion of it. As far as I make out, Howard Raiffa knows no more about the Cold War than I do. My vote goes to Kennan. Why, in the absence of independent demonstration of the normative authority of Raiffa's stipulations, would we reject Kennan's apparently sound decisions for no reason other than the illness of their fit with them? Who made Howard Raiffa king of the castle?[76]

2.5 Accuracy and aptness

There are matters of normative significance which NN-convergence does not distinguish. In one sense, reasoning is normatively adequate if and only if it is *accurate*. In another sense, it is normatively

---

[75] Readers familiar with *The Reach of Abduction* may recognize in Proposition 2.4 what was there dubbed the Actually Happens Rule. (Gabbay and Woods, 2005)

[76] There is a story about Raiffa, too delightful not to be true, told by Paul Thagard in a lecture to MBR_2012 Italy, a conference on model based reasoning held in Sestri Levante in June 2012. Raiffa and Ernest Nagel were colleagues at Columbia. One day, Raiffa received an offer from Berkeley, and was in a quandary over whether to exchange the east coast for the west coast. "I don't know what to do", he nervously remarked to Nagel. "Why, that's an easy one", Nagel replied. "Why don't you construct a decision tree?" "Come on, Nagel", came the reply, "this is *serious*!"

adequate if and only if it is *apt*. "Apt" also means "suitable" or "appropriate". In some contexts it means "reasonable" or "smart".[77]

This is a distinction that runs somewhat orthogonally to those we have presently been discussing. Reasoning is accurate when it is rightly done and produces the right answer. Reasoning is apt when, in the circumstances of the case, it is reasonable or sensible – or even inspired – to reason in that way. Right reasoning can be inaccurate. It can be reasoning that gets the wrong answer. Heckle and Jekyll are both first class homicide investigators. Heckle examined the evidence and reasoned that Spike alone did the dirty deed. Jeckyll examined the same evidence and reasoned that Gertie alone did the dirty deed. In each case, the reasoning was excellent. In at least one of the two, the reasoning was inaccurate. How is this possible? It is possible because most excellent reasoning is not truth-preserving, and a good deal of excellent non-truth-preserving reasoning is reasoning from inputs that permit incompatible and equally well-reasoned outputs. Plausible reasoning is like this. It was plausible to suppose from the evidence that Spike alone did it. It was equally plausible on the same evidence that Gertie alone did it, hence not Spike after all.

We now have the means to add some clarification to the NN-convergence thesis.

Proposition 2.5a
THE CONVERGENCE OF THE NORMAL AND THE NORMATIVE (2) *When there aren't particular reasons to the contrary, it is typically the case that how we reason from premises to conclusions in real life circumstance is either* accurate *or* apt *or both.*

---

[77] My notions of accuracy and aptness intersect in only one place with a trio of nearly the same name advanced by Ernest Sosa and known as the AAA model. (Sosa, 2007) Sosa's AAA are accuracy, adroitness and aptness. For Sosa – and me too – a piece of reasoning is accurate when it gets at the truth of things (which is pretty much its meaning in everyday English). The meaning of "apt" in English is, again, *suitable* or *approximate.* But Sosa gives it a technical sense in which a piece of reasoning is apt when it is competently realized. I have nothing against competently realized reasoning. But Sosa's aptness is not mine. Sosa's adroitness is competence, whereas its more common meaning is skilfulness. I also have nothing against skillful reasoning, but I question its suitability as a general condition of reasoning rightly.

COROLLARY: *In the matter of drawing consequences reasoning rightly is more like breathing rightly, and less like interpreting quantum phenomena rightly.*

What draws me to the assumption of NN-convergence is the paucity of credible alternatives. We might say that in light of the failure of idealizers to establish the normative *bona fides* of their models, the consequence hypothesis recommends itself *faute de mieux*. Besides, it is what naturalism recommends. Propositions 2.5a is a default. It embodies a presumption. It expresses the idea that when it comes to drawing conclusions from premises, human reasoners are better at it than not in more cases than not, relative to the cases that matter. This furnishes some methodological guidance for our enquiry, particularly as regards the burden of proof. If the investigator brings to the theory of reasoning *NP* presumptions, actual reasoning is guilty until its innocence is shown. A *D* orientation, strongly embodied in the normativity of actual practice principle, reverses this onus. Again, actual reasoning is presumed innocent until proved guilty.

It is well to emphasize that the assumption of NN convergence wears its defeasibility on its sleeve. It is a generic claim. It says that taken collectively our reasoning practices are with a notable frequency accurate or apt. It does not say that they are perfect; it does not say that they are beyond practicable improvement; it does not say that they all pass muster; it does not say that the ones that do pass muster do so in all contexts; and, *basso profundo*, it does not say that by and large our reasoning strategies are error-free.

Proposition 2.5a formulates a quite *particular* claim about premiss-conclusion reasoning. It is not offered as a thesis about human cognitive practice in general. It leaves open the possibility that, as regards the various procedures with which we realize our cognitive ends, some have better track records than others. A case in to consider is belief. Comparatively speaking, most of our mistakes arise from misinformation – from false belief. There is a long story about this. A great deal of what we know we have been *told*. Epistemologists disagree about whether told-knowledge is reducible to non-told knowledge, that is to say, to independently and directly confirmable knowledge. This is a question I'll take up in chapter 9. For the present it is enough to point out that as a matter of fact – a practical matter of the availability of resources – such reductions would in the general case be beyond the reach of the individual agent, even if in some sense possible for him "in principle".

Proposition 2.5b

BETTER AT REASONING: *With regard to most of what he believes, the individual agent lies exposed to the prospect of misinformation on a scale that significantly outpaces rates of wrong reasoning.*

Why should we accept any suggestion that, by and large, or typically, our actual reasoning is all right? Doesn't the empirical record count against it strikingly? Isn't it a fact that individual reasoners are copious mistake makers? Isn't it a fact, that individual reasoners are notorious traffickers in *fallacious* reasoning? Doesn't this gravely jeopardize NN-convergence?

2.6 Reflective equilibria

The thesis of NN-convergence should not be confused with the reflective equilibrium thesis.[78] The reflective equilibrium thesis locates the normative legitimacy of our reasoning in the concurrent satisfaction of two conditions. One is that, when right, our actual reasoning conforms to our considered intuitions about what the rules of right reasoning are. The other is that what we take as the rules right reasoning reflect our considered intuitions about the rightness of our reasoning practices. We should note at once that there is nothing in the reflective equilibrium conception of normative legitimacy that lends the slightest credence to normative idealizations of the mainstream logics of reasoning and decision. For the practices which valid normative rules are in equilibrium with is *actual* practice. The logical omniscience and Bayesian proficiencies of good reasoners will, if true, have to be demonstrated in ways that have nothing to do with the reflective equilibrium thesis.[79] The question we are asking here is about the relation, if any, between it and our own thesis of NN-convergence.

The reflective equilibrium thesis provides that how we actually reason is normatively legitimate provided that our reasoning behaviour and our reasoning rules are in reflective equilibrium. NN convergence reverses the dependency. It requires that our reasoning practices be

---

[78] The name is Rawls's (1971). The idea is Goodman's (1954).

[79] Often the reflective equilibrium claimed by normative ideologues is that claimed to exist between norms advanced by their theories and their own normative intuitions. I hardly know what to say about this, beyond remarking its breathtaking tendentiousness.

equilibrial, but with a crucial difference. According to the thesis of reflective equilibrium, the reason that actual practice is normatively adequate is that it is equilibrial. According to NN convergence, the reason that our conclusion-drawing practices are equilibrial is that they are normatively adequate.

The reflective equilibrium thesis works better for some practices rather than others. English words are a case in point. Something is a word of English to the extent that it conforms to the provisions of the lexicon. Something is a provision of the lexicon to the extent that it conforms to lexical practice. A particular attraction of the reflective equilibrium approach is that the concepts that it regulates can be highly dynamic. In the years since *Beowulf* there has been frequent, brisk and collectively massive lexical change in English. English Today is very unlike English Then, but today's English is the same language as then. The reflective equilibrium approach is tailor-made for the normativities sanctioned by conventions. Conventions are solutions of co-ordination problems. They are solutions in which it is possible that shifting practice retires old norms and replaces them with new and incompatible one. There was a time when "There's a deer" would rightly report the presence of any furry-coated, four-footed denizen of the forest deep. This ceased to be so when in its settled use "deer" came to be reserved for deer.

Conventional activities are made distinctive by the fact that their normative soundness at a time is determined by settled practice at that time. Settled practice admits of description by way of reflective equilibria, but it is conventionality, not reflective equilibrium that wears the normative trousers here. The normative clout of reflective equilibria is entirely derivative. Reasoning would be subject to the normative clout of reflective equilibria if reasoning were a conventional practice. But this is the last thing that reasoning is. Reflective equilibrium doesn't work for reasoning. A speaker of English today is no more fluent a speaker of English than his 12th century compatriot, never mind that what they speak aggressively violates one another's respective rules. If the reflective equilibrium approach held for reasoning, we would have it in principle that even if today's modes of reasoning substantially violates those governing the reasoning of their 12th century counterparts, and vice versa, then today's reasoner is no better a reasoner than the reasoner of yore. But no one thinks that this could be true, even in principle. Correct reasoning, unlike correct speech, trends to the inertial even under massive changes in what we take for true. Reasoning is not conventional.

The present conclusion keys to Proposition 2.5, the better-at-reasoning thesis, in an interesting way. Proposition 2.5 says that we are more vulnerable to errors of misinformation than to errors of reasoning. If our present claim is true, it offers a degree of encouragement about errors of misinformation. It provides that, in matters conventional, conflicts between present and past practices are not error-producing. It provides equally that if patterns of reasoning-now conflicted with patterns of reasoning-then, at least one of the patterns would be defective. Such conflicts are not neutralized by conventionality. This leaves us with an empirical question. No doubt there are contextually sensitive niches in which today's reasoning surpasses yesterday's – proofs by mathematical induction, for example. But the main question is whether the empirical record reflects *en large* that the reasoning practices of yore were different overall from those that prevail today. If so, what would the historical evidence reveal? Would it reveal that, where we employ *modus ponens*, our forbears made do with affirming the consequent; or where we use *modus tollens*, they were happy with denying the antecedent? Or that, in the court of King Agamemnon, there was nothing identifiable as drawing consequences at all?[80]

2.7 *Charity*

Consider now what N.L. Wilson (1959) calls the Principle of Charity. Adapted by Quine (1960) and Davidson (1984) and applied in the present instance to reasoning, the principle bids us not to interpret people's behaviour, whether our predecessors or distantly apart contemporaries, in ways that make them noticeably less adept at reasoning about ordinary things than we. There is a sizeable literature about Charity which it is not to my purpose to sift through here.[81] I shall make do instead with the observation that if that principle is right about reasoning, it need not be right (or as right) about *information*. And if this were so, it would be wholly explicable that Proposition 2.5 should characterize us all as having a better record at drawing consequences than attaining true beliefs.

Some readers might be minded to resist the NN-convergence on the following grounds. As we saw, classical logicians of all stripes

---

[80] How about our 40,000[th] century BC cousins? Whether less or as good reasoners by and large, that would still not be a matter of convention. If the earliest practitioners of *technē* were a trifle on the slow side, evolution would put things right in its own sweet time. But evolution is not conventional either.

[81] But see, for example, Paglieri and Woods (2011) for a compact survey.

endorse the *ex falso quodlibet* rule, according to which deducing any sentence whatever from a contradiction accords with the requirements of deductive validity. Paraconsistent logicians of all stripes reject the *ex falso* rule. Notwithstanding the efforts of late in chapter 1, no one to date has figured out a way of making everyone happy about *ex falso*. In recent years, a clamorous pluralism has grown up around this and other disputed claims in the philosophy of logic. In one not uncommon approach, the phrase "a deductive consequence of" is now revealed to be ambiguous. Every sentence is a classical consequence of a contradiction, but not every sentence is paraconsistent consequence of it. *Ex falso* is valid in classical logic and invalid in paraconsistent logic. And since it is logic that determines what "valid" is to mean, and since this determination is bound by no known prior fact of the matter concerning the validity of *ex falso*, then the meaning of "valid" is conventional after all. But if validity's meaning is conventional, then any reasoning for whose rightness validity is a necessary condition will likewise be subject to the conventions that determine it as "valid" or "invalid". Hence the rules of deductive reasoning are conventional, and as such conform well to the reflective equilibrium thesis. Classical validity is what attains equilibrium in communities of users of "validity" according to the classical conventions, and paraconsistent validity is what attains equilibrium in communities of users of "validity" according to the paraconsistent conventions.

It won't work. It may well be true that "validity" is ambiguous in the ways suggested, hence that different sorts of reasoning will differentially pass or fail the validity standard, depending on which sense of "valid" is in the offing. But the question for the theorist of erroneous reasoning is not which sense of "valid" a piece of reasoning does or does not comport with, but rather whether validity in that sense is a condition on the reasoning's *rightness*. And that is not a matter to be settled by conventions which set the meaning of the word "valid". Equally, the word "right" is ambiguous, with different uses mandated by requisitely different conventions of use. Whether a piece of reasoning is right in one or more senses of the word "right", or in none, depends on the applicable sense of the word "right". But whether it is right in the sense of "right" by which the reasoning is rightly judged right is not set by conventions regulating the usage of "right" rather than "jolly" or "north of Cleveland". The conventions determine what to *call* it when it has the requisite characteristics. They do not determine whether the reasoning has the characteristics in virtue of which, thanks to the conventions for its

60

usage, it is appropriate to call it "right". Reasoning is not conventional. Accordingly, "... reflective equilibrium ... does not specify how to balance principles or judgments when they conflict. So it cannot settle serious disputes."[82]

The irrationality and approximate rationality theses have given mainline logicians less of a bad name than they deserve. Years ago, Richard Nixon's first vice-president, notwithstanding his unsatisfactory service in that office, gave voice to an alliterative rebuke unbettered by any American vice-president since. Scorning his and the administration's critics, he called them "nattering nabobs of negativism". Practical logicians drawn to NN-convergence could say the same of the mainstream nags – in logic and psychology alike – who make whole careers of telling us how hopeless we are at things.[83]

If he wanted to, Howard Raiffa could lay plausible claim to the Honourary Marshallship of the negativity parade. Speaking of Savage's own notion of expected utility, Raiffa opined:

> If most people behaved in a manner roughly consistent with Savage's theory then the theory would gain stature as a descriptive theory but would lose a good deal of its normative importance. *We do not have to teach people what comes naturally.*
> (Raiffa, 1961, pp. 690-691. Emphasis added).

The mind boggles.

2.8 *Conceptual adequacy*

Theories of human behaviour whose provisions are forwarded as normatively binding on actual practice owe to the practices it judges unfavourably the duty of antecedent establishment of their own normative legitimacy. In the absence of such a demonstration, behavioural

---

[82] Adler (2008), p. 21.

[83] Not to overlook the pungent words of the first Lord Halifax, who observed that the "vanity of teaching often tempteth a man to forget that he is a blockhead." See also the *Economist* newspaper of July 26, 2008, in which the columnist Bagehot attributes to behavioral economists the desire to "nudge" people who are "fallible: lazy, stupid, greedy and weak, loss-averse, stubborn, and prone to inertia and conformism ... poor decision-makers, often incapable of their own happiness."

disconformity underdetermines the question of wrongness, exposing two possibilities, not one. The behavioural disconformity may be wrong in relation to some norm whose force has not been established by the theory. Or it is the theory that is discredited by those behavioural deviations. That is to say, the theory may be exposed as both empirically and normatively inadequate.

By analogy with theories that are empirically sensitive, a "conceptually sensitive" theory is a theory one of whose functions is the elucidation or analysis of one or more target concepts. Basically this is the business of making meanings clear. A conceptually sensitive logic of semantic consequence is one that undertakes to produce a clear understanding of what entailment actually is, together with specifications of its key properties. A conceptually indifferent logic of consequence is one in which entailment is an artefact of theory, underdetermined by or independent of prior beliefs about entailment's true nature if it has one. Most mainstream logics exhibit at least a degree of conceptual sensitivity. There are at least some target concepts that they try to get right (or approximately right). Just as empirically sensitive theories are bound to a condition of empirical adequacy or bound to it to some contextually indicated degree, the same is true of conceptually sensitive theories. To some nontrivial extent, they are bound to a condition of conceptual adequacy.

The two requirements of empirical and conceptual adequacy intermingle in an interesting way. What is interesting about it is that they often tumble into conflict. Good theories contain *surprises*. They wouldn't be much good if they didn't. Of what use is a theory that merely reconfirms everything we knew about the theory's subject matter going in? Surprises override antecedent beliefs or antecedent dispositions thereto. Surprises subdue intuitions. Theories owe a good deal of their success to their capacity to tell us things we didn't know. Sometimes these things are underdetermined by the concept K under which they fall. Sometimes their effect is stronger. They require us to *rethink* the very idea of K-hood. For a very long time, physicists tried to conceive of elementary particles as small bodies, as exceedingly tiny Aristotelian substances. Matrix mechanics changed all that. Bodies are the wrong way to think of elementary particles. Elementary particles are nothing like bodies. They are "sets of quadruples of numbers according to an arbitrarily adopted system of co-ordinates."[84]

---

[84] Quine (1981), p. 17.

The last thing I want to propose is that we jettison abstractly modelled theories of human practice. No; models are essential to scientific success. My reservation applies to the conceptually distorting presumptions of the normative legitimacy of the orthodox ones. Nothing I have said so far counts against formal modelling as such. A formal model of a human practice is an abstract structure which serves as an idealized description of it, or more typically as an abstract structure of certain key aspects of it. "Idealized" has a twofold meaning which we should not be over-casual about. In one sense an idealization is an improvement of something. In another, it is a counterfactual way of thinking about it. Taken in the first way, ideal models are sitting ducks for the heresy of the normative superiority of models. There is little such encouragement when taken in the second way, which leads me to suppose that the second way of thinking about ideal models in the right way. This leaves the question of where, if not in its normative superiority, the value of modelling lies.

In a naturalized logic explication takes a particular form. Its analyses of concepts K are shaped in part by what we understand 'K' to mean. But they are also informed by what the natural sciences are able to tell us about how K-hood actually manifests itself on the ground. Accordingly, the analytical models that an empirically sensitive logician is interested in fashioning are models which imbibe the naturalized character of the logic. In the hands of practical logicians, an analysis of a concept is a naturalistically sensitive treatment of it. Beyond its strictly semantic component, models of this sort reshape the concept in question partly on the strength of what is known scientifically about its extension. Naturalized conceptualizations are part meaning and part fact.

Can we give ourselves some sort of antecedent guidance about how to impose formal constraints on conceptually analyzed inputs? Not perfect guidance. Not algorithmic guidance. But something more forthcoming than the casuistic indeterminencies of case by case cherry picking? In this I find myself drawn to Cohen's Rule, which I so name in appreciation of L.J. Cohen's contributions to matters of mutual concern.[85] As Cohen saw it, any theory of human practice – of decision making, of probabilistic reasoning, of contract negotiations, and so on – is an account that cannot attain the status of a scientific theory without the introduction of idealizations that facilitate the exposure of lawlike connections. In Cohen's hands, these idealizations are necessary for the

---

[85] Cohen (1986).

theory's *descriptive* adequacy. If we decided to pick the suggestion up, it would give the empirical sensitivity requirement:

Proposition 2.8a
COHEN'S RULE: *In the absence of particular indications to the contrary, formalization of an empirically sensitive theory should not introduce idealizations that greatly exceed those required for the theory's descriptive adequacy.*

Naturalization was a radical step in epistemology, and is no less so in logic. The naturalization of logic has sizable consequences for the intuitions approach to conceptual analysis – recall our observation earlier of the potentially corrosive tension between the conditions of descriptive and conceptual adequacy.

Proposition 2.8b
DOWNGRADING INTUITIONS: *At a minimum, the decision to naturalize the logic of reasoning is a decision to take into account well established lawlike results of the cognitive sciences. But established science has no fondness, and little respect, for the presumptions of one's (or "our") pre-theoretic intuitions about the matters it investigates.*[86]

On some understandings, a normatively assessable theory of human performance is empirically adequate to the extent that the actual behaviour of individual agents conforms to its provisions. But it is also sometimes supposed that a theory's empirical adequacy is not a matter of its conformity to actual behaviour, but rather is a matter of the theory's laws. The principal point of difference between these two senses of empirical adequacy is the factor of disconforming behaviour. Empirical adequacy in the first sense doesn't tolerate behavioural disconformity. But in its second sense it does; it positively invites it.

It is necessary to keep these two senses apart. Let us reserve "empirical adequacy" for the first sense and "descriptive adequacy" for the second. A decision to naturalize a philosophical theory of normatively assessable human performance is a decision to hold it to the requirement of descriptive adequacy. This is precisely the point, and the upshot which

---

[86] Readers of the AI literature will be familiar with a recent attack (Spohn, 2002) on Pollock's defeasible logic (Pollock, 1995), alleging that its normative presumptions rest on nothing more that Pollock's own *ad hoc* intuitions.

obliges a practical logic of cognitive systems to give due weight to the laws, such as may be, of empirical psychology.

It is a point of sufficient importance to bear repeating that descriptive adequacy is trouble for conceptual adequacy. Descriptively adequate science pays equivocal heed to pretheoretical common sense beliefs about its subject matter. Conceptual adequacy is all about the effort to have the theory preserve what everyone thinks he knows about its subject matter going in. Of course, no one doubts the practical necessity of having pre-theoretic beliefs. But when pretheoretic beliefs are common sense beliefs, or what everyone takes himself as knowing, there is little to prevent a descriptively adequate theory ransacking those beliefs in the fullness of time. Accordingly,

> COROLLARY: *Any theory of normatively assessable human performance which is subject to the requirement of descriptive adequacy is a theory for which conceptual adequacy is at best a requirement of lesser priority.*

> COROLLARY: *The greater the degree of disconformity between a descriptively adequate scientific theory and its pretheoretical common sense beliefs, the less the weight of the requirement of conceptual adequacy.*

The corollaries to Proposition 2.5b sound a useful warning. They carry the admonition *caveat emptor*. It advises that before "buying into" a naturalized logic of reasoning, the would-be theorist should be reconciled to the possibility of a significant foreclosure of his (and "our") pretheoretic intuitions. This is precisely the moral of the recent manifesto of the research programme in "experimental mainstream philosophy". Experimental philosophy stands to philosophy as empirically sensitive logic stands to mainstream logic. Concerning the intuitions on which conceptual analysts lay such store, the authors of the manifesto write as follows:

> ... [T]he first major goal of experimental philosophy ... [is] to determine what leads us to have the intuitions we do .... The ultimate hope is that we can use this information to help determine whether the psychological sources of [those] beliefs *undercut* the warrant for [them]. (Knobe and Nichols, 2008, p. 7. Emphasis added)

I have twofold opinions about the developments to date in experimental philosophy. On the one hand, I applaud the doubts it casts upon the imperiousness of pretheoretic intuitions in philosophy. On the other, I disagree in various respects with how it designs its experiments and how it interprets their results. Experimental philosophy is not our focus here. So my reservations needn't occupy us further.

2.9 *Virtuous distortion*

It is necessary to say a little something more about descriptive adequacy. On the face of it, it is an anomalous virtue, that is, no virtue at all. A descriptively adequate population biology is one whose well confirmed lawlike connections pivot on the idealization that populations are infinitely large. Of course, there are no such populations. Every biological group there is or ever could be is finite. Infinitely large populations are an unapproachable idealization. The largest biologically possible on earth approaches the ideal precisely as little as the earth's smallest biologically possible population. We have it then that a condition on the theory's descriptive adequacy is that it say what is utterly false about populations, made so by the utter disapproximation of nature's actual groupings. Similarly, there are whole classes of descriptive theories which propose causal explanations of the phenomena within their purview. On many such views, a causal explanation of a phenomena E is one that correctly identifies factors C_1, ..., C_n in whose presence E is invariably brought about, that is, factors that cause it. It is not all uncommon in theories of this sort that the demonstrated connection between the C_i and E requires that "surplus" factors U_1, ..., U_n be omitted from consideration. Indeed,

> This is the art of modeling: to know what aspects of reality one can *sacrifice* ... (Kokko, 2007, p. 5. Emphasis added.)

These factors are suppressed not because they don't occur, and not because they don't co-occur with the C_i, but because the sought for causal connection between the C_i and E doesn't obtain in their presence. Accordingly, in nature the C_i aren't the cause of E after all. They are the cause of E only in the theory's model. In the one case, a descriptively adequate scientific theory embraces the idealization that populations are infinitely large. In the second case, descriptively adequate explanatory theories simplify the scene they wish to explain by abstracting away from

66

what is actually the case what is "crucial to retain" (Kokko, 2007, p. 5). In the first case, the idealization says what is false. In the second case, the abstraction suppresses what is true. This is the anomaly, and with it comes a question: How can an empirical theory be descriptively adequate if it makes ineliminable use of falsehoods and implacably sacrifices relevant truths on the ground? And how, correspondingly, can it be right to subject our evolving logic of error to Cohen's Rule? Idealizations and abstractive simplifications are empirical distortions. How can an empirically distorting theory be descriptively adequate?

There is a substantial body of opinion among philosophers of science that idealizations and abstractions facilitate our *understanding* of his target phenomena, notwithstanding that they incorporate deliberate distortions of the empirical record. We might also say that one way of getting things right in science is getting things wrong, though not the same things at the same time. In population biology we get populations wrong, but in so doing we get natural selection right. In causally explanative theories, we get causes wrong, but in so doing we achieve an understanding of what the actual causal structure of our target phenomena is *like*; and given that knowing what something is like is a way of understanding it, getting the causes wrong is a way of getting an understanding of them right.[87]

The distortions of descriptively adequate model based science pose a fundamental problem for the epistemology of science.

Proposition 2.9
HOW VIRTUOUS DISTORTION IS POSSIBLE: *How, if successful science can't do without subscription to the false and suppression of the true, is a strict instrumentalism avoidable for such theories? Instrumentalism is strict when it detaches from science's success the duty to produce a knowledge of the objects of its assertions. In a typical case, some of its statements will honour the obligation, but many of the crucial ones will not.*

Suppose now that strict instrumentalism is true and that we decided to persist with Cohen's Rule. Cohen's Rule requires that we hitch the wagon of logic to the horse of empirical psychology, especially its laws. Cohen's Rule says that if we are to have models in our theory of

---

[87] For a more detailed discussion of cognitive value of the empirical distortions of scientific modelling, see again Woods and Rosales (2010a), Strevens (2008) and Batterman (2009).

reasoning, they should not exceed the models of our most descriptively adequate psychological theories. Since, for all their virtues, those theories are false, so too is our theory of reasoning. If this doesn't expose psychologism in logic as an utter nonstarter, what would or could? Why don't we just drop all this psychological silliness?

No. The question rests on a confusion. No one thinks for a minute that nature herself conforms to the theoretical assertions of population biology. No one thinks that one of the reasons population biology is good science is that the world's populations are infinitely large. If population biology is as good as we think it is, it will not owe its success to nature's conformity to the transfinite largeness of populations. Equally, if for the reasons presently in view it were thought that a good theory of reasoning must absorb at least some of the falsehoods of a descriptively adequate psychology, it would not follow from the goodness of the theory of reasoning that the false propositions of psychological theory describe what actually happens on the ground. It would not require the real-life reasoning individual to behave in ways analogous to the way of actual populations would behave if *they* verified the theoretical assertions of population biology.

In a way, this is all a bit of a wonder, epistemologically speaking. It means that it is all right to hold our theories of reasoning to standards a good deal less imposing than those that govern right reasoning *itself*. I don't mean to make light of this question. The question of how science gets to be cognitively virtuous in the face of the distortions necessitated by its models is an open question in the epistemology of science. I won't attempt to close it here. This is partly a matter of limitations of space. It is mainly a matter of not knowing what to say.[88]

The idealizations of normative models are a case apart. Except where normativity is descriptivized in the manner of NN-convergence, the empirical distortions wrought by a theory's empirically distorting norms lack observational payoff, they discomport with known facts. True, they comport with the theorist's intuitions, and perhaps with the intuitions of others as well. But in the world of philosophy naturalized, empirically challenged intuitions are debased currency and they dissipate in the embrace of something like Gresham's Law.

A principal motivation of the present chapter has been the mismanagement thesis, according to which mainstream logic has failed to secure an adequate theoretical grip on the general notion of erroneous

---

[88] Beyond what is already mentioned in Woods and Rosales (2010a), as well as in Woods and Rosales (2010b). See especially Woods (2013b).

reasoning. It is a failure in two related parts. The mismanagement of the first part was the decision to make invalidity and inductive weakness sufficient conditions of erroneous reasoning. The mismanagement of the second part was the decision to hold the theory of error – and the theory of rationality – to normative paradigms, based on a confidence owing more to mathematical virtuosity than respect for the evident successes overall thrown up by the empirical record.

2.10 *Third-way standards*

Neither validity nor inductive strength bears an intrinsic tie to right reasoning. Neither is inherently good-making. Neither as such is apt. Yet there is no reason to give up on validity and inductive strength or to misappreciate the theoretical light that logicians have already managed to throw on them. Sometimes, as we saw, they are precisely the standards against which reasoning as actually practised must be judged. If you want to construct a proof of the Stone Representation Theorem, your argument had better be valid. That's the apt way to go. If you and your team are running a drug-safety test for the Ministry of Health, you had better, as best you can, heed the standard of inductive strength. That's the apt way to go. But in other cases – most of them – these are simply the wrong standards. Trying to hit them would not be the apt way to go. They take too long to meet or are too complex to apply, or they don't apply at all, even in principle. This generates a major challenge for a logic of reasoning.

Proposition 2.10a
FINDING THIRD-WAY STANDARDS. *A logic of premises-conclusion reasoning must honour contextual variabilities in the application of the validity and inductive strength standards, and it must identify the further standards with which these two contextually alternate.*

There is no more important and widely open problem for the logic of error. It requires us to supply an answer to the question: "What is right about reasoning that is neither truth-preserving nor probabilistically clinching?" It might puzzle us that the question is still open. *Why* should this be so wide-open and difficult? Aren't there lots of candidates to consider? Aren't there already large literatures in which good reasoning pivots on nonstandard consequence relations – nonmonotonic consequence, defeasible consequence, default consequence, plausible consequence, presumptive consequence, abductive consequence, and what not? Isn't this an *embarras de richesses*? Why not simply link our question to the appropriate consequence relation and have done with it? Isn't this what the consequence rule tells us to do? I will return to this question in chapters 7 and 8. But let me now say what I will say at greater length there.

Most treatments of these consequence relations are more mathematically virtuosic than conceptually elucidating or empirically supported. A case in point concerns the foundational work in AI, aimed at formalizing common sense reasoning in relation to everyday problems (McCarthy, 1959). The basic formalisms of nonmonotonic reasoning are hardly credible as formalizations of common sense reasoning. More than one of nonmonotonic logic's founders attempted to capture its structure in classical and modal systems of deductive consequence. McCarthy himself always held that anything expressible is expressible in first order logic. He "considered this a kind of Turing thesis for logic ...."[89] So I repeat:

Proposition 2.10b
HARD AND OPEN: *The task of finding the conditions thanks to which most of right reasoning is right is at least as wide open as it is important and difficult.*

And I conjecture that

Proposition 2.10c
The difficulty in determining why it is so difficult to specify these conditions flows in no small measure from our failure to investigate the cognitive makeup of the human reasoner.

2.11 *Orthodoxies*

Orthodoxies are managed in much the same way that fashions are. They attract the fidelity of their respective "smart sets", of people in the swim of things. In English racing circles, the smart set calls the shots for hats. In logic circles, the smart set calls the shots for deduction. In the theory of ampliative reasoning, the smart set calls the shots for induction.

---

[89] Bochman (2007), p. 558.

No one should confuse the importance of hats with the importance of reasoning. Not every monomaniacal arbiter of how things go at Ascot is likely to think that a good hat is more important than a good piece of reasoning. Still, we should not underestimate the importance of hats in the hats' smart set.

In matters of fashion, standards flow from the trend setter's taste or, in more than a few cases, the trend setter's force of will. The magnet that draws a smart set around an arbiter's pronouncement is activated not just by a like taste but by the arbiter's success in selling it. In fashion nothing succeeds like success. It is something of the same way with intellectual orthodoxies. They attract their devotees by the strength of their track records. Here too nothing succeeds like success. Celebrity is its own biggest draw. Even so, there is a big difference between fashion and theoretical orthodoxy. Fashion is driven by progressive rhythms. In the world of fashion, change is good as such. Orthodoxy is driven by a conservative rhythms. In the world of orthodoxy, change should be resisted except for cause; but cause is precisely what the orthodox have little inclination to recognize. In extreme but far from uncommon cases, heterodox efforts to show cause against the orthodox are met with abuse and dismissiveness.

Lying behind the conservatism of orthodoxy is the Can Do Principle.[90] It bids the investigator to solve his problems with instruments that have stood the test of time and with which he is already familiar. It asks him to proceed on the basis of what he already knows how to do. It counsels against the reinvention of the wheel. Can Do is closely allied to a common idea about successful theories. A major mark of a good theory is the success it has in accounting for its target data. But a further factor is its success as the colonizer of data not originally in its sights and of theories purporting to account for them. So seen, the best theories are expansionist. They are what psychologists call "robust". They are takeover artists with good records in Mergers & Acquisitions. The expansionism of theories driven by Can Do plays a key role in science's enthusiasm for reductionism, and for unity of knowledge programmes more generally.

Can Do is a good thing. It underwrites a powerful conservative virtue. But like many virtues, too much of it can be a bad thing. So Can Do has a degenerate case. I call it Make Do. Make Do is Can Do stretched to what it is no longer good for. At its worst, Make Do is insulting and dismissive. It is an extreme case of *faute de mieux* theory

---

[90] Woods (2003), pp. 277 and 325.

making.[91] It makes do with the bad in the absence of the good. It answers to the silly idea that a bad theory is better than none at all. At bottom, Make Do is business as usual for no better reason than that it is a business the investigator knows how to operate. Orthodoxy is a conservative good. It is an orientation that minimizes start up costs. It is vouched for by Can Do and it is sustained by the expansionist rhythms of powerful science. But, for all their attractiveness, orthodoxies are sooner or later at risk for Make Do.[92]

When it comes to investigating the ins and outs of human performance, certain orthodoxies stand out. We couldn't miss them if we tried. In mathematical logic itself, the dominant orthodoxy is the first order predicate calculus. In theories of decision making, it is the optimization subject to constraints and the maximization of expected utilities paradigms. In theories of ampliative reasoning it is the probability calculus and the mathematical theory of statistical sampling.

We would be mad not to hold these orthodoxies in high regard. They have impressive and distinguished track records, especially on the score of mathematical virtuosity. Membership in the smart sets that crowd round these successes is not free on board. It has to be laboured after and paid for by performances of requisite distinction. Given the centrality of Can Do, it is a default-requirement that the investigator of human performance bring to his endeavours the heavy promise of the orthodox. But just as Can Do is itself a defeasible principle, so too is favouritism towards the orthodox. There has to be good reason to see in a given application of Can Do the degeneracy of Make Do. There has to be good reason to give up on the deduction paradigm or the maximization of expected utilities paradigm. But this is not to say that we should stick with them in all settings, no matter what. A visiting interplanetary anthropologist might ask himself, "Given that their theoretical target is the reasoning of human agents, why do human logicians invest so heavily in studies of the deductive consequence relation?" The answer is that for

---

[91] As witness Wakker (2010), p. 351: "[When the theory doesn't fit the data], [w]e then nevertheless continue to use our model if no more realistic and tractable model is available."

[92] A recent case in point is the wild ride that the standard semantics for modal contexts has had. The possible worlds framework has cropped up all over the place as the favoured instrument for handling an astonishing range of philosophical problems. For sober reflections on this fad, see Ballarin (2004). For an investigation of some Make Do uses of possible worlds, see Woods (2010d).

more than two millennia the theory of deductive consequence has been by far where logicians have made their deepest and most systematic progress. It is the right answer, well enough, but with an unmistakable whiff of Make Do.

2.12 *Data-bending*

The respect for data principle is more easily stated than followed. When a theory engages a kind K of normatively assessable behaviour, yet does so with an intended empirical sensitivity, the principle imposes two obligations on the theorist. It requires him to pay due heed to the particularities of how behaviour of that type plays out on the ground, not excluding the necessity to identify accurately the behaviour that actually instantiates the theory's subject matter. The principle also obliges the theorist to attend to, and adequately preserve, what is pretheoretically known of K-behaviour (We could call this the folk theory of K behaviour) When the theory is also normatively intended, there is the linked obligation to take note of what is pretheoretically known of the conditions that make for the rightness or wrongness of behaviour of that kind. We could think of these beliefs as the theorist's, or more generally "our", normative intuitions about K-behaviour. As already noted, the very idea of the legitimacy or bindingness of "our" normative intuitions has spawned a large and rivalrous and, by my lights, an inadequately realized philosophical literature. This rivalrousness is in turn a central interest of normative metaphilosophy. Less widely recognized are like frailties that attach to "our" pretheoretical intuitions about which things are and are not behaviours of the K-sort and of how those things are hooked up together in the ebb and flow of their actual involvements with one another on the ground.

It has been widely appreciated at least since Bacon's time that a physical theory's data – both its target data and those that are called into play in a confirmatory role – are frequently, if not typically, unavailable for theoretical engagement in the absence of some pretheoretical massaging. Massaging data is a way of prepping them for theoretical readiness. Massaging often takes the form of a simplifying redescription of facts on the ground and, therewith, some element of reconceptualization of what it is to be a K-thing. In a classic paper of 1962, Patrick Suppes makes the Baconian point that it is not untypical of model based physical theories to effect engagement not with the intuitive data of the theory's subject matter, but with *models* of those data.[93]

It is difficult to distinguish in a suitably principled way the boundary between profitable data-massaging and just making things up. It is the same difficulty as finding the boundary between clarifying or an old concept and making a new one up of that same name. A particular peril is the natural imperialism of a theorist's theoretical intuitions, including his normative ones. If he finds that data on the ground are not responsive to his theoretical insights, there are two bad options routinely open to him. He can subject the data to tendentious redescription. Or he can condemn them for their normative inadequacy. I have already had my say about the tendency of normative intuitionists to misplay their hand. I want now to say a bit more about data massaging on the descriptive side.

Our present approach to logic is preoccupied with the comparative paucity of cognitive assets on which resource bound agents can actually draw. The concept of rationality that I've assumed for such agents is that of the effective management of this scantness. Seen this way, logic is a disciplined description of how, with regard to the requisite aspects of their cognitive agendas, beings who are humanly rational employ their limited cognitive resources. As we saw, important examples of such resources are information, time, and computational capacity. To these I now add a fourth: *modes and methods of inquiry.* Following Gigerenzer, what I am going to suggest in this section is that for certain ranges of cases, how a theorist conceives of an individual's cognitive agency is often a function of the mismassaging of what is already discernibly true of it.

One particularly impressive example is the attempt to model discovery with machine learning programs, such as BACON, which undertake to extract scientific laws inductively from data.[94] But on another approach, discovery is a function of methods that produce and *process* data, rather than the data themselves. (For those who care about such things, these approaches to discovery are loosely congruent with the split between inductivists and non-inductivists concerning the theory ladenness of observational terms.)

One example of the modes and methods route to discovery has to do with the role of statistical inference and theory testing that has dominated the social sciences since the 1960s, applied to the study of cognitive processes. This development was a sequel to the

[93] Patrick Suppes (1962).
[94] Langley, Simon, Bradshaw, Zytkow (1987).

"institutionalization of inferential statistics in American experimental psychology between 1940 and 1955."[95] Under the influence of such methods, theories of cognition

> [w]ere cleansed of terms such as restructuring and insight, and the new mind has come to be portrayed as drawing random samples from nervous fibres, computing probabilities, calculating analyses of variance, setting decision criteria, and performing utility analyses. (Gigerenzer, 1996, p. 339)

As the inferential statistics became canonical for investigations in the social sciences, cognitive processes of various kinds were assumed to embody a kind of "intuitive statistics", and cognitive agents were viewed as executing its rules in the course of their cognitive engagement. Gigerenzer cites two examples.

Example one. Beings like us have the capacity to distinguish between the presence of a signal and the presence of mere noise. According to Tanner and Swets,[96] this is done in the same way that an investigator employs Neyman-Pearson statistical methods to adjudicate between two hypotheses. In Neyman-Pearson statistics, two hypotheses H and H′ are construed as sampling distributions. A decision rule is defined. This is conceived of as a likelihood ratio. In our present example, the agent computes two sampling distributions for "noise" (H) and "signal plus noise" (H′). He (or it) then fixes a decision criterion which takes into account the cost of the two possible decision errors. The brain takes sensory input which is transduced in such a way that enables it to calculate its likelihood ratio, which is then judged against the criterion, giving "signal" or "signal plus noise" as answer, depending on whether the ratio is smaller or larger than the criterion.[97] Neyman-Pearson techniques eventually evolved into a more general theory of cognition, including memory recognition,[98] eyewitness testimony,[99] and discrimination between random and non random patterns.[100]

---

[95] Gigerenzer (1996). See also Toulmin and Leary (1985).

[96] Tanner and Swets (1954).

[97] Gigerenzer (1996), p. 340.

[98] Murdock (1982) and Wicklegreen and Norman (1966).

[99] Birbaum (1983).

[100] Lopes (1982).

Example two. When an agent attributes a cause to an effect, it is proposed by H.H. Kelley[101] that the cognitive agent (or his brain) brings this off by producing an analysis of variance and testing null hypotheses. In Kelley's "attribution theory", the experimenter infers a causal connection between two variables from computing an analysis of variance in the manner of Fischer's analysis of variation (ANOVA) and then running an F-test. Similarly, the practical or individual agent makes causal attributions.[102] The significance of these examples lies not in the claim that the cognitive agent reacts to the cited data in the manner of intuitive statisticians, but rather in the claim that the agent as intuitive statistician reconceptualized the very data theories of cognition sought to take note of.

As Gigerenzer sees it, the construction of cognitive theories based on the assumption of the cognizer as intuitive statistician "radically changed the kind of phenomena reported, the kind of explanation looked for, and even the kind of data that were generated."[103] What is more, researchers who adopted the methods of inferential statistics were unaware of this change, since these methods had become canonical in psychology. If Gigerenzer is right, then a change in psychological methodology from that of inferential statistics to the methods of resource bound rationality should *likewise* change the kind of phenomena reported, the kind of explanation sought, and the kind of data generated.

Gigerenzer's reflections dovetail with a pair of matters of relevance to our own investigations. First is the Baconian idea that a theory's observational data are theory-laden or "massaged". This being so (if it is so), observational data have conceptualizations that bias what will count as candidates for the underlying theoretical targets. What is important about Gigerenzer's analysis is its exposure of the point that a theorist's conception of an appropriate investigative methodology can itself conceptualize the data in particular ways. So if we think that proper investigation requires that we see a practical agent as an intuitive statistician, then there is some likelihood that we will conceptualize the agent's behaviour accordingly. We will massage the data to fit the theory. This carries clear consequences for the respect for data principle.

A second point of connection is the Can Do Principle and its degenerate case, Make Do. Why, we asked, do researchers sometimes persist in using methods that seem inapplicable to the targets of their

---

[101] Kelley (1967).
[102] Fischer (1955), pp. 69-78.
[103] Gigerenzer (1996), p. 340.

inquiry? Our suggestion was that in lots of cases the researcher knows how to do the work that he actually ends up doing yet does not know how to do the work that he has set out to do. (So he pretends that there is a connection, or he presumes *à priori* access to the world of Ought.) The particular importance of the bounded resource analysis is that it gives a further explanation of why investigators persist with methods inappropriate to their theoretical objectives. The answer is that these methods have become canonical for these fields of inquiry. So, if we are obliged by the field's conception of what constitutes acceptable scientific practice to see cognitive agents as intuitive statisticians, then we can't help seeing the behaviour of such agents in that kind of way. And how we *see* the data will naturally influence our choice of hypotheses that best account for them. Clear consequences again for the respect for data principle.

What lessons might logicians of reasoning learn from this discussion? They should learn the following:

Proposition 2.12
DATA-BENDING: *Method influences the conceptualization of data. How the data are conceptualized influences, in turn, what counts as theoretically adequate responses to them.*

A good example of data bending is neoclassical economics, where utilities are idealized as infinitely divisible. Of course, the last thing that is humanly possible is that, for any given amount of pleasure, there are amounts of it smaller than it. Infinitely divisible pleasures are not a response to a theorist's normative intuitions about how the perfectly-made utility is, but rather it is because they enable the theory to engage the mathematical firepower of the calculus. Utilities are reconceptualized to enable the theory to mathematize its methods.

I said earlier that deviations from the orthodox are frequently met with abuse and dismissiveness. Deviations from the Bayesian orthodoxy have recently attracted this kind of disapproving assessment:

> That correct reasoning accords with Bayesian principles is now so widely held in philosophy, psychology, computer science and elsewhere, that the contrary is beginning to seem obtuse, or at least quaint. (Glymour and Danks, 2007, p. 464)

Just so. A dash of abuse and a daub of dismissiveness! As we have it now, Bayesianism isn't philosophy. It is theology.

2.13 *Making things up*

For most of its history since 1879, the year in which Frege's *Begriffsschrift* appeared, logicians have had a widely shared understanding of their duty to their principal targets – consequence, premiss-selection proof and the others. Consequence would be subject to two theoretical constraints. On the one hand, the concept of consequence would have to be mathematically well-defined. Its unpacking would have to be clear, precise, systematic, rigorous and general. Its definition would also have to facilitate recognizability, if not outright guarantee it. A treatment of consequence that met these conditions would be technically adequate. It would exhibit the right kind of mathematical virtuosity.

Logicians of the period also assigned themselves a second duty. Logic's mathematically adequate provisions for consequence would also have to serve as a conceptual elucidation of it, one which gave an improved understanding of it. When a logic met this further condition, its treatment of consequence would be a conceptually faithful one.

Applying formal methods to a concept constitutes a formalization of it. A concept's meaning in preformalized linguistic practice is its intuitive meaning or, equivalently, it reflects the intuitive concept. What is achieved by the formalization of a concept? Historical practice suggests four different answers to this question.

> *Analysis* The formalization of concept K explicitizes the meaning it has in pre-formalized linguistic practice; that is, it articulates K's intuitive meaning.
> *Explication* The formalization of a concept K preserves its intuitive meaning but does so in ways that gives to K a more aggressive clarity than it had pre-theoretically.
> *Rational reconstruction* The formalization of a concept K involves an additional attribution to K of features not present in pre-formalized linguistic usage, but in a way that retains enough of the intuitive concept to make it intelligible to say that the rational construction at hand is a formalization of *it*.
> *Stipulation* The formalization constitutes a nominal definition of a concept lacking a prior presence in preformalized linguistic practice, of concept, that is, that lacks an intuitive predecessor.

In the first instance, analysans and analysandum are taken as synonymous. In the next two cases, the equivalences are more a matter of partial synonymy. In case four, there is no trace of it.

The distinction between analysis and stipulation is roughly Kant's contrast between analysis and synthesis. Analysis, says Kant, is the business of making concepts clear, and synthesis the business of making clear concepts, that is, the business of making them *up*. Analysis is the purview of philosophy and synthesis the province of mathematics. Explication and rational reconstruction are hybrids, with explication more analysis-like and rational reconstruction tilting rather more towards stipulation.[104]

The war cry of formal philosophy in this period was, although not in these exact words, "Mathematical virtuosity is the preferred route to conceptual adequacy." A paradigm of this sentiment was the doctrine of contextual eliminability underlying Russell's theory of descriptions, itself – according to Ramsey – a paradigm of philosophy. The motto animated the practice of philosophy in the tradition of Russell, Carnap and Quine. It is an intriguing idea – akin to having your cake and eating it too. Most deductive logicians in the mainstream from Frege and Russell and well into the twentieth century, saw a well-made formal logic as engaging both sides of Kant's distinction. The same distinction is also present, though not in these terms, in Russell's *Principles of Mathematics* (1903).[105] Here, too, the name of the distinction is indiscernible in formal logic's conception of its own adequacy. Logic, tacitly or otherwise, would achieve its objective of making the concept of consequence clear by making up clear concepts such as the material "conditional". Accordingly, a well-made formal logic would be a combination of both technical and conceptual sophistication. It would attempt to come to grips with the fact that a concept gets to be philosophically interesting not only because of its intellectual importance but also because of its resistance to clear description. Lurking in this is a puzzle. Consider a purported clarification KC, of a philosophically important but unclear concept C. The problem is to determine whether KC is an analysis, an explication, or a rational construction of C, or the stipulation of a new concept under an old name. In Kantian terms, the problem is to determine whether conceptual elucidation is analysis or synthesis. Analysis, we may say, leaves everything in place. Everything else is, to one degree or another, is

---

[104] Kant (1764/1974), and (1800/1974).
[105] See here Griffin (1991), pp. 272-273.

79

tampering. It is adding what wasn't there before or removing what was there before.

If a philosopher at large were to reflect on logic in ways that take him beyond what he may think of it as a teacher of past or present elementary courses in it, he would likely be rather strong on analysis and somewhat grumpy about making things up. It is not, he would think, the proper job of the logic of consequence to stipulate one's theoretical outcomes, to achieve the gospel of entailment by inventing its provisions. Thus is exhibited a rather strong disposition towards conceptual conservatism. Ordinary language philosophers of the Oxford movement were perhaps the most conceptually conservative force in the broad and varied tradition of philosophical analysis in the century before this. Philosophers at large tend to also want their accounts of target concepts to be stipulation-free. Even so, in the period from Frege and Russell to well into that century, mainstream logicians have been *moderately synthetic* about their own subject matter. This synthetical side is reflected in a willingness to take liberties. The moderate side is reflected in a determination to preserve in its tamperings a large and discernible presence of the pre-theoretical concepts in question – to keep the tamperings within the limits of explication or, failing that, rational reconstruction.

It should be noted with some emphasis that logic in this century and the one before it is well-stocked with (ever enlarging) constituencies of theorists whose syntheticism is not moderate, but extreme. The presence of this radical syntheticism – the doctrine that logic is what you make it, that and nothing more – is discernible in what is now a quite standard reaction to the subject's pluralism. It is a nihilistic relativism according to which there is no fact of the matter about consequence except as provided for in some or other logical system, hence no fact about their respective target concepts that makes any one system a more accurate logic than any other, or its consequence relation more similar than others to how consequence really is. There is a shorter way of saying the same thing. Among practicing logicians there is a large and growing consensus that what truly matters in modern logic is mathematical virtuosity, not conceptual fidelity; that the goal of conceptual fidelity is sentimental *naïveté*; that, more centrally important, the job of logic is consequence*, not consequence.

It would be wrong to think of logic's relativistic semantic nihilism as an intrinsically corrosive abandonment of intellectual seriousness in logic.[106] An early adumbration of this nihilism is Carnap's Principle of Tolerance (1934/1937), according to which logic is both stipulationist and unattended by antecedent facts of the matter. For Carnap, logic is indeed what you make it, and the goodness or badness of its theorems owes nothing to their fidelity or lack of it to pre-existing logical facts.[107] But Carnap was far from permissive about logic. A logic's merits would be a function of what it was wanted for in the first place. For example, if your goal is the reduction of arithmetic to logic, there is reason to like the logic of *Principia Mathematica*. Accordingly, it is not quite accurate to deny the existence of facts of the matter about logic. There are instrumental facts and there are the facts achieved by stipulation. It is an instrumental fact that *PM* was good for logicism. It is a stipulated fact that the axioms of infinity and reducibility are true there.

This is a further variation of nihilism in logic, one that sets it apart from nihilism of the Carnapian sort. For lack of a name, I'll simply call it *experimental nihilism,* typified in nonlogical circles by circumstances attending Riemann's construction of non-Euclidean geometry, in which, among other things, the parallel postulate fails. No one, least of all its inventor, took Riemannian geometry as a description of the actual space of the physical world. Its virtues were technical, notably its proof of selective consistency. It was a significant piece of mathematical virtuosity but hardly up to the burdens of conceptual fidelity. Riemann's was not a semantic analysis of the concept space. But with the arrival of general relativity theory, it emerges that Riemann's space is the empirically well confirmed description of big, or global, space.

There is no reason to think that Riemann himself anticipated this Einsteinean application of his abstract theory. Even so, non-Euclidean geometry was ready to hand – on the shelf, so to speak – and was available for Einstein's later appropriation. But this couldn't have happened in the first place had it not been for the impressive technical

---

[106] Of course, there is in principle what we might call the option of mindless nihilism, according to which since there are no facts of the matter about what consequence actually is, all logics are false (for there is nothing that could make them true). There are lots of semantic nihilist logicians, but none of this mindless breed.

[107] Quine says that theories are conceptualizations "of our own making." (1981, p. 2) Eddington says that they "are put-up jobs".

virtuosity of Riemann's achievement. People would not have paid attention to it otherwise.[108]

Imagine a situation like Riemann's, but with a variation. A logician is building a logic and, like Riemann, his interests are mathematically virtuosic. He wants his logic to be sound and complete, or he wants its axioms to be categorical, and so on. In the course of realizing these objectives he has no expectation of conceptual adequacy, and no interest in it. But like Riemann, he publishes his logic on the understanding that in the fullness of time it may turn out to have an application. In that case, we could say that our logician's mathematical virtuosity was attended by an *experimental* intent. "Here is another logic", we can imagine him saying. "Who knows what its uses might turn out to have? Where's the harm in having it around, if needed?" Note, however, that when the paper is submitted to a journal the editors will accept it or reject it for its virtuosity or lack of it as a piece of mathematical contrivance. Applications, if such there be, are entirely the contingent byproduct of a favourable judgment of technical finesse.

How widespread is this experimentalism in logic? The short answer is: "Just read at random any published piece in any of the mainstream logic journals in the past forty years."[109]

I have nothing against making things up in logic. All I ask for is accuracy in labelling. When someone makes something up in population genetics, he knows he'll have to redeem himself at the empirical checkout counter. If someone makes something up in Bayesian epistemology, he

---

[108] I don't want to leave the impression that Riemann was simply playing about until one day, hey presto! the general theory of relativity popped into view. Riemann's geometry was a formidable piece of mathematics. It generalized Gauss's work on surfaces to n dimensional manifolds. When $n = 3$, Riemann's geometry is the result. Riemann didn't think the 3D case described physical space or – since scaling up is a dubious assumption in physics – that it ever would prove to do so. Still, he did think it a physically possible hypothesis, hence a description of how the world might conceivably be.

[109] A sociological observation: The *Journal of Symbolic Logic* used to make room in its reviews section for philosophical contributions to the subject. In time, this mandate was transferred to its sister journal, the *Bulletin of Symbolic Logic,* which was also soon dominated by mathematical work. A further adjustment, intended to accommodate such lingering philosophical interests as may be, was the *Review of Symbolic Logic,* which too, since its recent inception has also been dominated by mathematical pieces. Meanwhile, even the *Journal of Philosophical Logic*, which was another of the *Review*'s predecessors was rife with a mathematical emphasis. In a way, perhaps this is as it should be. It reflects the fact that in the past generation or so, logic largely ceased as a humanities discipline and has passed its remit to mathematics and computer science.

may try to redeem himself at the normative checkout counter. If someone makes something up in the absence of empirical and normative redemption, he invites the question of where the good of it lies. There is no general answer to this question. It only remains to wait and see. This is not fatalism. It is optimism.

Proposition 2.13
HOPEFUL CONTRIVANCE *It is a hopefulness that bets on the possibility that a theory may come along one day in fulfillment of two critical conditions. One is that the theory lucks out at the empirical or normative checkout. The other is that thing it makes up turns out to have been indispensable to those redemptions.*

COROLLARY: *In the absence of those conditions made up theories are, at their best, technically virtuostic play. They are the fruit of* homo ludens.

CHAPTER THREE

KNOWING (1)

"All men by nature desire to know."   Aristotle

"While man's desires and aspirations stir, He cannot choose but to err."   Goethe

"I have always said that a belief was knowledge if it was (i) true, (ii) certain, (iii) obtained by a reliable process."
                    Frank Ramsey

"I used for myself to collect my [logical] ideas under the designation *fallibilism*; and indeed the first step toward *finding out* is to acknowledge that you do not satisfactorily know already, so that no blight can so surely arrest all intellectual growth, as the blight of cocksureness."   Charles S. Peirce

3.1 *Epistemic sensitivity*

I keep saying that a naturalized logic of reasoning would be ill-advised to leave out of account what reasoning agents are actually like. What, then, are they like? That is to say, what are *we* like? Over time, various answers have been announced. We are rational animals; language-users; knowers of our own mortality; appreciators of our evolutionary fate, the only creatures who laugh; essentially bipedal if we are cyclists; essentially rational if we are mathematicians; anxious negotiators of our radical freedom; inference engines; maximizers of subjective expected utility; optimizers subject to constraints; intuitive statisticians; and heaven knows what else. Whatever the attractions of these bromides, a positive theory of error should not overlook the quite striking extent to which the human animal makes his way in life by *knowing things*. Human beings are cognitive beings, suffused with the drive to know and dependent for survival and prosperity on the conditions that satisfy it. Seen this way, reasoning is good when it engenders "good cognitive state transitions"[110], when it opens up "epistemic folkways".[111]

Individuals want to know what to think and what to do. They want to know what things are like. They want to know what others know. The only logician writing about error who comes close to a reflective appreciation of the drive to know is Rescher.

> The reality of it is that *Homo sapiens* has evolved within nature to fill the ecological niche of an intelligent being. The demand for understanding, for a cognitive accommodation to one's environment is one of the fundamental requirements of the human condition. Humans are *Homo quaerens.* We have questions and desire (nay, *need*) answers. The need for information, for cognitive orientation in our environment, is as pressing a human need as that for food (Rescher, 2007, p. 44).

To "be ignorant of what goes on about us is almost physically painful for us – no doubt because it is so dangerous from an evolutionary point of view".[112] So "… cognitive disorientation is actually stressful and distressing".[113]

Because reasoning is implicated in much of our cognitive practice, conferring benefits on it when it is good, and afflicting it with error when it is bad, a logic of reasoning should attend to the constitution of the human knower. It must try to say what it is like to know things, and what it is like to be a knower of them.[114]

---

[110] Goldman (1986), p. 82.

[111] Goldman (1992).

[112] Rescher (2007), p. 45.

[113] Ibid., p. 46.

[114] While it is true that mainstream agent-admitting logics tend to leave unexplored the tie between the theory of reasoning and the conditions under which cognition occurs, a number of informal logics are in varying degrees an exception. Blair (2012) imposes strong epistemic demands on argument and critical thinking. Freeman (2005) makes a thorough epistemological investigation of the factor of premiss-acceptability in the logic of argument. Pinto (2001) also gives epistemic considerations a load-bearing role in his approach to argument. Biro and Siegel (1997) develop an account of the fallacies as epistemic mistakes. See also Siegel (1988), Battersby (1989), Weinstein (1994) and (2013), as well as the 2005 and 2006 volumes of *Informal Logic,* which, under the guest-editorship of Christoph Lumer, dedicate two issues to "the epistemological approach to argumentation". As already mentioned,

It may strike some readers as perverse that, having shown some fondness for NN-convergence, I would now impose on reasoners the necessity to constrain themselves epistemologically. This is problematic, since most of the going epistemologies arrogate to themselves a normative authority that concedes nothing to the presumptive soundness of actual practice. What is wanted therefore is an epistemology of a kind that gives the NN-convergence thesis a decent shot at staying in business, of reconciling the normative with the natural. In that spirit, consider how the shift to naturalism might be effected.

3.2 *Cognitive ecologies*

Assume that for several years now a sizeable team of cognitive anthropologists from a distant and unknown planet has been hard at work in our midst, discreetly, undetectedly and fruitfully. Their earthly mission is the examination of human cognition. These cosmonauts from afar have come well-equipped for this task. Themselves organic beings, they have been able acclimate to the particularities of planet Earth. Themselves cognitive agents, they have a familiarity with how cognition works under the ecological constraints of nature. Then, too, they are accomplished field linguists and intelligent problem-solvers. I adopt the fiction of the visiting scientist from Bas van Fraassen's notion of an "epistemic marriage", which envisages an epistemic partnership between dolphins, extraterrestrials and ourselves. (van Fraassen 2005)

So far, our visitors have limited their enquiry to the cognitive functioning of individuals rather than institutions. A primary source of data for their enquiring is the knowledge-exhibiting behaviour of their subjects, supplemented by consultation of the disclosures of the earthbound cognitive sciences. However, in this initial stage of their enquiries, the visitors have decided to leave the offerings of earthbound *epistemology* for later consideration. In this first phase, they will read no

---

Finocchiaro (2005) places an emphasis on empirical logic and the epistemic contexts in which historically important arguments in science and politics have taken place. Adler (2002) is a monograph on epistemology, but it draws out several points of importance for the understanding of fallacies. Harman (1970) bears a subtitle which reads in part "A discussion of the relevance of the theory of knowledge to the theory of induction ...". Inductive logicians as a class are more open to epistemological considerations than deductive logicians are. But none of them, in my view, deductive or inductive, mathematical or informal, is sufficiently engaged by the psychological constitution of the reasoning agent.

philosophy. Theirs will be a scientific investigation of the cognitive achievements and limitations of the human individual – an exercise in descriptive epistemology. At a later point, their own epistemology and the epistemologies advanced by their human subjects could be compared and, where differences exist, adjustments considered. Part of their mission is to learn the languages of their hosts. This will involve some linguistic interaction with them. But once the language is learned, they will not interview their hosts semantically. They will not ask them to say what the word "knowledge" means. They will not ask them to give a conceptual analysis of knowledge. They will read the human sciences in this same spirit.

The visitors were quick to see that the organisms of earthbound nature are efficient processors of information, whether the amoeba or the prover of Fermat's Last Theorem. It was clear to them that a good part of what makes for the distinctiveness of the human individual is that he is an information processor with a drive to know, that it is in his nature to make his way in life by knowing things, and by knowing how to fix them when he falls into error. They saw the human agent as a cognitive system. Knowledge they took to be a product of nature, arising from interconnections of design, wherewithal, and place in the causal nexus. Human individuals are ecological entities. They are participants in webs of interrelations between habitats and inhabitants. They are also

Proposition 3.2a
COGNITIVE ECONOMIES: *An economy is an ecology for the generation and distribution of wealth. A cognitive economy is an ecology for the generation and distribution of knowledge.*

The team made special note not only of the capacities afforded, and the limitations imposed, by the human individual's cognitive make-up and the nature and availability of his cognitive resources, but also of the concomitant necessity for beings like that to advance their interests within those constraints. They were also impressed by the extent to which the processes of cognition are brain-supported undertakings and are transacted "down below", that is, out of sight of the mind's eye; and they saw at once the economic advantages of such arrangements, notably the computational fire-power of parallel processing.

The visitors saw the attachment to knowledge as an indispensable part of the human animal's evolutionary success. They recognized that knowledge is not a discretionary matter for beings like that. Knowledge is not a take-it or leave-it proposition. The human organism is built for knowledge, whose systematic failure to attain would be an evolutionary disaster unless compensated for in quite particular and tricky – some would say miraculous – ways. So the tie to knowledge is tight. Tight, but not direct. The human agent also is a representational being. His knowledge is achieved by the fulfillment of representability conditions. The team acknowledged that where there is representation there is also at least the prospect of misrepresentation, and that there are lots of contexts in which the distinction is not a self-announcing one. This gives rise to the possibility that what is taken for knowledge is not the real thing. That is, it raises the spectre of scepticism. I mean here big-box, large-scale scepticism – scepticism about the external world, about other minds, about the self, about the past, and so on.

Scepticism has a modest and innocent starting point. Its fulcrum is the plain fact of error and of humankind's susceptibility to it. This fallibility triggers a question. Granted that there is error in human life, isn't it possible that there is more of it than there appears to be? Isn't it possible that there is massively more of it, and more sweepingly more of it, than there appears to be?

A good deal of big-box scepticism is achieved by giving to the possibility of widespread error undue inferential weight. In this connection, the visitors made particular note of two features of how human *science* is done (minus its philosophical parts). One is that the possibility of error is readily conceded. The working scientist is by inclination and professional tutelage a fallibilist about science. The other is that this standing possibility of error is not given this same sort of inferential weight. This met with the visitors' approval. Big-box scepticism is not much of an option for human science. You don't find many physicists saying that a knowledge of space is not possible. Why, then, would big-box scepticism be an option for the *visitors'* science? Why would these scientists yield to it in arriving at their own working assumptions about human knowledge? Since what's sauce for the goose is sauce for the gander, they were quite prepared to default to the position that

Proposition 3.2b
THE COGNITIVE ABUNDANCE THESIS: *Human beings have knowledge, lots of it.*

Of course, scepticism has sources other than the attachment of evidential weight to the possibility of widespread error. Sometimes scepticism is occasioned by finding ranges of cases in which the facts of the situation combine to contradict the definition of knowledge. For example, if the definition imposes on one's knowledge of α the requirement that one have an indubitable certainty that α, it is easy to see how big-box scepticism would arise. It would be triggered by the fact that in most situations a human being's true belief that α is not attended by the sentiment of indubititability. Similarly, if the definition made the knowledge that α is incompatible with the possibility of error, the big-box scepticism arises quite naturally. It would arise because of the fact that most human beliefs are subject to the possibility of error.[115]

The visitors would not be much impressed by the idea that given the certain definition of "know" and the actual facts of human life, the extension of that term could be sweepingly less populated than customarily supposed. They would see in such tensions a competition between knowledge and a definition of knowledge. Anti-sceptics will incline to the view that any definition of knowledge that threatens the existence of knowledge to so sweeping an extent is an imperfect definition of knowledge. Anti-sceptics are more certain of the existence, abundance and sweep of knowledge than they are of a philosopher's definition. Sceptics are contrary minded. The definition is right. Knowledge is the way the definition says it is. If the way that knowledge actually is were to expose our ordinary epistemic beliefs to the necessity of wholesale abandonment, that wouldn't be down to the definition. It would be down to how knowledge actually is.

Disagreements of this kind instantiate what we ourselves might call Philosophy's Most Difficult Problem, lightly touched on – but not so-named – in chapter 2. It is the problem of finding an adequately general and disciplined way of distinguishing between two classes of good arguments, in each of which it is given that the argument's conclusion follows in some contextually appropriate sense from its premises and yet the conclusion is obviously false or absurd. In the first class, the arguments in question are *reductios* of something in or presupposed by their premiss-sets. In the other group, arguments are sound

---

[115] Some of the visitors saw in the inclination to give evidential weight to the possibility of error as a modal mistake. The possibility of error thesis with respect to a belief that α is not the thesis that it is possible that X knows α and that α is false. It is the thesis that it is possible that both X believes that α, and yet α is false.

demonstrations of something surprising or astonishing (and perhaps very disappointing).

Arguments for big-box scepticism draw clashing opinions about how they should be classified, with anti-sceptics plumping for the *reductio* option and sceptics favouring the surprising truth option. There is to date no systematic way of regulating this distinction which would offer assured guidance about how Philosophy's Most difficult Problem might be solved, both in the general case and in the particular case occasioned by big-box scepticism. When the time finally came to consult the human epistemological record, the visitors did not hesitate to express their admiration for those philosophers who pay the scepticism project the honour of reflective and subtle attention, and whose reflections reveal their respect for the difficulties and uncertainties that beset its deep engagement.[116] Yet they declined to be drawn. They were not minded to yield to philosophical conceptualizations of knowledge which put the existence of knowledge in such dramatic and wide-ranging peril. They had no liking for definitions of the predicate "knows" that dramatically shrivelled its extension.

The visitors' reluctance to be drawn is instructive. Anyhow, I myself take some instruction from it. I want to see what provision for errors of reasoning a naturalized logic of practical agency would make under the epistemological assumptions this book puts into play. Accordingly, I take the cognitive abundance thesis to be a *given*, under the protection of the respect for data principle.

If big-box scepticism is untenable, the scale and frequency of human error is noticeably less pronounced than it otherwise would be. Even so, it was evident to the visitors that in the informational ecologies in which human beings operate,

Proposition 3.2c
THE ERROR ABUNDANCE THESIS: *Human beings make errors, lots of them.*

To mitigate the apparent tensions between this and the cognitive abundance thesis, they ventured a third and fourth assumption:

---

[116] When our visitors began inspecting earthbound epistemology, it wasn't long before they noticed the influence of scepticism. They were soon of the view (rightly) that no philosopher of the present day is better on scepticism and on the metaphilosophical difficulties that attend its investigation than Barry Stroud. See, for example, Stroud (2012).

Proposition 3.2d
THE ENOUGH ALREADY THESIS. *Human beings are right enough about enough of the right things enough of the time to survive and prosper (and occasionally build great civilizations).*

Proposition 3.2e
THE KNOWING-WELL-BEING PROPORTIONALITY THESIS: *To a significant degree the scale of human flourishing matches the scale of cognitive well-doing.*

In support of Propositions 3.2d and e, the visitors were drawn to a further pair of considerations:

Proposition 3.2f
THE ADAPTIVE ADVANTAGE THESIS: *The human drive to know has a certain adaptive advantage. It would be evolutionarily puzzling if the advantages manifest in our survival and prosperity were not an outcome of it. If the human animal knew hardly anything, in what would the adaptive significance of the drive to know consist?*

Proposition 3.2g
THE VALUE PRESERVATION THESIS: *Theories of knowledge should try to preserve the presumption that knowledge is valuable.*

At this stage, the visiting team has read no human philosophy. It had arrived at these hypotheses on its own, stimulated by observation of what it is to be a cognitive being in the ecological circumstances of human life, and supplemented by consideration of the relevant human sciences, cleansed of any supplementary philosophical premises. It is important to note that members of the team were not of the view that these theses made a successful case against big-box scepticism. Anti-scepticism was their opening position, the stance from which they would make their investigations. Given that stance, they viewed the hypotheses that followed as well-founded, though not to say proved. They had no proofs to offer about the nature of human knowing. They had scientific hypotheses to offer.

Thus our alien friends formed an understanding of the human individual as a natural being in an informational ecology, an evolutionarily successful, goal oriented time-and-action cognitive system, a resource-limited organism, a cognitive economizer, a fallible being with efficient feedback mechanisms, an inhabitant of the causal order, neurally wired for knowledge and good at getting it – though far from perfectly. Human knowledge they also saw as a natural phenomenon, a partnership between the active and the passive, between doing and being done to. Some knowledge – for example, perceptual knowledge – is more a matter of being done to than doing. Other cases – for example, mathematical knowledge involved considerably more on the doing side.

3.3 Cognidiversity

The visitors paid particular attention to the "enough of the right things enough of the time"-clause of the enough already thesis. How was this to be interpreted? In reflecting on this, they came to the view that in its first occurrence "enough" had to do with the scope of the human animal's cognitive interests and needs. And what wide and varied interests they are, encompassing appreciations, both detailed and general, of the myriad contingencies of the passing scene, about its past and future, about its organizing principles, about himself and others, and about the good for humankind. So it is not just that the human beings have lots of knowledge; they have lots of knowledge about *lots of different things*. This led them to a further proposal.

Proposition 3.3a
THE COGNIDIVERSITY THESIS: *Cognidiversity is a fit and necessary adjustment to nature's diversity.*

COROLLARY: *There is no room in nature for the one-trick cognitive pony, for example, for humans whose knowledge of the world is exhausted by their acquaintance with sense-perception.*

There arose from these deliberations a ready acceptance of some methodological guidelines. We have seen that the visitors weren't much interested in what humans thought the concept of knowledge was or how the word "knowledge" was to be defined. It didn't take them long to discover that human beings were typically a good deal more adept at performing a task than defining the concept under which it falls. It is easier to speak English fluently than to specify what fluent English is. Speaking is easy (absent externalities). Linguistics is hard. Walking is easy (absent externalities). The mechanics of walking is difficult. Similarly, for large ranges of cases, knowledge is easy; epistemology is always hard.

Later, when they started reading some earthbound philosophy, the visitors were puzzled to discover that over its long history, epistemologies tend to be responses to a particular pair of motivating questions:

- What is the correct definition of knowledge?
- Given the definition, is knowledge *possible*?

The visitors readily allowed the legitimacy of these questions, but they were perplexed that should be asked so early in the enquiry. These were not, they thought, the right questions to start with. Their own start-up questions were different:

- What accounts for human cognitive abundance?
- What are its limits?

Noting the persuasive influence of big-box scepticism, the visitors came to think that mainstream epistemology's motivating questions embedded a harmful confusion. Normally, "Is it possible for us to know things?" is a silly question, attracting an obviously correct affirmative answer. Since philosophers aren't supposed to ask trivial questions, we may take it that when they ask whether knowledge is possible for us, it is a question with a different meaning, a meaning that rescues its rightful answer from triviality. But there is a cost to this semantic reconfiguration. The seriousness of the question it invokes indicates an openness to a scepticism which no visitor would tolerate so early in the proceedings.

Neither were investigators particularly interested in how they themselves would analyze the concept of knowledge or what they themselves meant by the word. The word "knowledge" was part of their working vocabulary as cognitive scientists. This would inform what they would decide about the ways and means and limitations of human knowing, but it was not itself a formal part of their enquiry. They were not in the business of asking whether they were working with the right notion of knowledge. They assumed that they were, all along. This motivates a procedural decision:

Proposition 3.3b
THE TAKING IT EASY ON KNOWLEDGE THESIS: *Any theory of knowledge which, in default of a strong case to the contrary, gave knowledge a hard time –e.g. by allowing it to fail Cognitive Abundance, Enough Already and Cognidiversity – should be resisted.*

COROLLARY: *Purported definitions of knowledge that gave their definienda a hard time should be resisted.*

3.4 *Knowledge-exhibition*

A challenging aspect of the visitors' investigation was the specification of what would count as knowledge-exhibiting behaviour. Quite early on, they recognized the importance of distinguishing between knowledge-exhibition behaviour and knowledge-attribution behaviour, especially self-attribution. Sometimes the presence of knowledge is taken verbal notice of, the primary means of which is the speech act of assertion in a non-epistemic idiom. The dominant way of communicating one's knowledge that α is to assert that α, not to assert that one *knows* it. In a great many contexts – not all, but certainly most – an utterance of "I know that Central Station is at the north end of Spuistraat" performs a function more specialized than "Central Station is at the north end of Spuistraat". More often than not "I know" has a dialectical or rhetorical significance. "Is it really that way?", someone says; and you say, "I know that it is". (Reassurance) "How do you know!" (Challenge). Whatever the details, the correct response, even if true, to "How do you know?"cannot be (for example), "Well, it's true, I believe that it is, and my belief was a justified one". Replying that way is, as the lawyers say, unresponsive.

Much of it hinges on how behaviour is recognized as belief-exhibiting. It is far from easy to say even when we ourselves, not our alien friends, are the ones trying to find out. This is especially so when the behaviour in question is non-verbal and not otherwise of deliberately communicational import. But here, too, express avowals of belief are less helpful indicators of belief than we might think. When Harry believes that Paris is the capital of France, there are quite ordinary and everyday contexts in which his saying "I believe that Paris is the capital of France" is a distortion of what he believes. No one on the team had what he regarded as a satisfactory systematic answer to the question, "What constitutes knowledge-exhibition behaviour?"

All the same, the visitors' observations and assumptions about the cognitive condition of humans reflected a rough correlation between earthbound knowledge-exhibition behaviour and what the investigators themselves independently knew to be the case. They were not wild guesses. This involves a capacity for recognizing a subject's goings-on as knowledge-exhibition behaviour. For this to be possible, a certain kind of inference must be available to them, loosely in the form:

> The team's subjects are behaving in a way that correlates with the investigators' own knowledge that α. So it may be hypothesized that to some approximation the visitors' subjects also know that α.

Of course, there is nothing in these brief words that conveys the logical structure of this kind of reasoning. Yet they all took it as given that everyone is adept at spotting cases of genuine epistemic attainment, that there are cases galore in which it is evident to observation that with respect to some proposition α, someone X knows that α. They reassured themselves with the sound observation that not being able to say how you do something is no reason for thinking that your doing it is in serious doubt.

3.5 *A causal response model*

The visitors discerned in the human condition a rapprochement between the objects and subjects of knowledge. Knowledge, they observed, is partly up to nature and partly up to the knower. For this reciprocity to obtain, the knowing subject must be built for the appropriate kind of congress with nature, and nature must be organized in ways that enable the subject's engagement of it. In time, the visitors attributed our cognitive successes to conditions whose fulfillment satisfies a *causal response* (CR) description of knowledge. The motivating factor behind that description is the plain fact that a human individual's knowledge is dominantly a matter of the states into which he has been put by his belief-forming mechanisms under environmental stimuli of various kinds. The qualification "causal" is intended to capture this being-put-into feature of cognition, and "response" to reflect the factor of environmental inducement. Accordingly, the visitors proposed as a first approximation,

95

Proposition 3.5a

A CAUSAL RESPONSE DESCRIPTION OF KNOWING: *A subject knows that α provided that α is true, he believes that α, his belief was produced by belief-forming devices in good working order and functioning herein the way they are meant to, operating on good information and in the absence of environmental distraction or interference.*

The CR-characterization was not arrived at carelessly. There was much discussion of its proposed conditions, and a general recognition that they offered a first word, not a last, about what it is for a human individual to be in a state of knowledge.

It is a point of some interest that these investigators didn't regard their CR-characterization as providing a definition of knowledge, that is, as specifying the conditions necessary and sufficient for its occurrence or for the correct application of the word "know". One reason for this was a quite general caution about defining the objects of a naturalistic enquiry. In the case of knowledge, the characterization they wanted would be true to the general case. The cited conditions would be close to necessary but not perhaps universally sufficient. Mathematical equations to one side, the generalizations of natural science hardly ever take the form of universally quantified conditionals. The sought-for a generality less strict, more in the manner of "Ocelots are four-legged" and less in the manner of "A triangle is a closed three-sided figure". The visitors advanced the CR-characterization in like spirit. They thought that as a plain matter of fact it was typical of knowledge to be constituted by true and well-produced belief. Strictly universal definitions are risky, made so by the brittleness of universality. Strict definitions are felled by a single counterexample. A counterexample requires any such definition to be abandoned. This, too, is problematic.

In the case of naturalistic properties, the strict universality of definitions by way of necessary and sufficient conditions exposes them to high rates of failure. Such definitions are hard on their definienda, not only by withering their extensions, but also by the low probability of defining them properly. As overturned definition hardly makes for a good theory of knowledge. So there is a clear enough sense in which

Proposition 3.5b

THE RISK OF STRICTNESS: *Strict definitions of knowledge carry a high risk of epistemological ungenerosity.*

Abandoning a definition calls into question our facility at finding necessary and sufficient conditions in the first place. It makes us ask, "In what are our candidates for necessity and sufficiency grounded?"

Here is an interesting point that caught their attention – a counterexample, as some would think. For various values of α, there is an evolutionary advantage in human beings believing that α. Suppose that α is one of those self-reflective attributions – "I am quite good-looking" is one of them – which is false of a good many more than not, but every now and then is true. Like the rest of the population, Sarah believes that α. But α happens to be true of Sarah. By construction of the case, Sarah would have believed α even if it were false, since doing so was part of what she was designed for, and was the product of belief-forming devices operating as they should. Since nothing in the case suggests that her belief-forming equipment was defective or malfunctioning, doesn't Sarah's case qualify as knowledge?

After some initial disagreement, the visitors came to think that Sarah's case motivates a distinction between beliefs that facilitate cognitive ends and beliefs that serve other good ends, in recognition of the fact that for certain values of α there is an adaptive advantage in their being objects of general belief and no particular advantage in (or possibility of) their being true in the general case. It would not be much of a contribution to evolutionary well-being if virtually everyone actually were good-looking. But nearly everyone's thinking that she is more good looking than she is might have a more significant evolutionary payoff. The visitors reasoned that there are two classes of good ends for which the mechanisms of belief are designed, the production of true beliefs and the production of belief whose advantages obtained independently of their truth. When a belief that happens to be true is produced by the properly operating mechanisms of alethically indifferent belief, the visitors concluded that, though performing properly, the mechanism that produced Sarah's belief was the result of evolutionarily advantageous alethic penetration. The truth of Sarah's belief was a fluke.

There was some discussion as to whether conative or affective epistemic penetration is always a bad thing. The group decided that it depended on cases. They noted the human susceptibility of bias, of a sort that compromised the likelihood of a satisfactory cognitive outcome. But they also found that some biases – a favouritism for loved ones, for example – possess adaptive advantages that could outweigh the good that would otherwise have been achieved by knowledge.

3.6 *Virtue and fallibility*

Later on, when the visitors started reading local epistemology, the visitors came to see that the CR-model offers its hospitality to one of the main strains of virtue epistemology (VE), initiated, as they discovered, by Ernest Sosa and developed by a number of philosophers since.[117] VE provides that knowing that α is a matter of believing truly that α as the result of exercising a cognitive skill. A person possesses this skill – this intellectual ability – when there is in him a comparatively stable disposition to respond with a notable frequency to the relevant stimuli with true belief, although VE-ers would add "to a degree consonant with the truth of fallibilism". When activated, the disposition implicates its possessor in a causal chain from stimulus to belief. Although there is plenty of room in principle for the skill-conception in other models of knowledge, the emphasis it gives to a causally dispositional understanding of that skill makes it a natural for the CR-model, something confirmed by Alvin Goldman's acceptance of it in his 1992 book. In fact, however, the visitors thought that the better way of describing the connection between VE and the "causal reliabilism" they wanted to endorse is "orthogonal". Unless one is prepared to water down the notion of skill so as to encompass mere causal responsiveness of the requisite sort, it seemed to them empirically untrue that all that an individual knows is produced skilfully. Suppose that Harry is looking out the window into the garden. He sees a robin in a tree. He knows that it's there because his perceptual devices were in good order and operating as they should. In achieving this knowledge, the last thing he did was exercise a skill.

The error abundance thesis gives expression to a form of fallibilism. Fallibilism is another of philosophy's big-tent notions, meaning more things than just one. In adopting it the visitors didn't want to embrace everything that fallibilism is said to be. They were not, for example, advancing the views of Peirce, Reichenbach and Popper – to name just three – for whom fallibilism is an extreme thesis about scientific knowledge.[118] So seen, scientific theories are mainly false and, even when replaced by better ones, are improved by theories that are also mainly false.

---

[117] Sosa (1991 and 2007) and, for example, Zagzebski (1997, 2005) and Greco (1999, 2010).
[118] Peirce, CP,1. 113-120, Reichenbach (1952) and Popper (1963). See also Rescher (1984).

There are a great many philosophers for whom scientific knowledge is the best that it ever gets in the knowledge game. Quine, himself a fallibilist, is clear on the point. If a set of beliefs fails to dance to the tune of our best methods for science, most of what we take for knowledge isn't.[119] Sometimes called scientism, this favouritism towards science produces severe consequences when adjoined to a strict scientific fallibilism. Crudely put, scientism says that if it isn't scientific knowledge, it's not knowledge; and a strict fallibilism says that if it's sanctioned by science it is hardly ever knowledge. Jointly, this is scepticism on a scale rivalling Sextus. There is a useful lesson in this for those who want to give scepticism a less free hand. To stop the double-whammy of strict scientific fallibilism and scientism combined, resist all temptation to give them your joint assent. Given the value of knowledge, the visitors rejected scientism; and theirs was both a gentler fallibilism than the strict version, and a rather more procedural one. It endorsed the execution of cognitive procedures, within science and out, known in advance to be attended by error with a notable frequency.

3.7 *The command and control model*

As the visitors came to appreciate, something quite like the CR-model has attained some mainstream respectability in the last forty years or so,[120] but it is far from being the dominant account for earthbound epistemologists. Its chief rival is what they called the *command and control* model (CC), according to which knowledge is the end-state of epistemic undertakings whose strategies for execution and whose conditions of success lie within the command and control of the would-be knower, and so are matters of the agent's volition.[121] Possibly the purest version of the CC-model – certainly the most interesting – is one in which knowledge is a case-making achievement, in which knowing that α depends on the knower's constructing a successful argument for it, or at least having the argument ready to hand and within his timely reach. The ancient and enduring idea that knowledge is justified true belief (JTB) is

---

[119] It is *conversation* or, as Rutherford once said, *stamp-collecting*.

[120] Armstrong (1973), Goldman (1967), Dretske (1981), Millikan (1984), Goldman (1986), Nozick (1986), Goldman (1992), Plantinga (1993), Trout and Bishop (2005). It all began with Ramsey (1931).

[121] There are also responsibilist varieties of virtue epistemology. So the VE-non-VE divide cuts across the CR-CC divide. See here Code (1987) and Montmarquit (1993).

a natural conceptual home for the case-making version of the CC-model. It offers a safe harbour to the idea that being justified in the belief that α is simply a matter of having a successful defence of it. This is justification as advocacy, and is important. It gives to knowledge a *dialectical* character.

In its earliest formulation, the CC-model enshrines the Platonic paradigm of *epistēmē* as *endoxon* + *aletheia* + *logos*. *Epistēmē* is genuine as opposed to apparent knowledge or, as is sometimes said today, "hard" knowledge as opposed to "soft". *Endoxon* is sincere and considered belief or opinion. *Aletheia* is truth. *Logos* is an account or demonstration or supporting theory. Thus a proposition known to be true must in some sense be *shown* to be true, or within the agent's means to show.

An important feature of the CC-model is its attachment to the idea that knowledge is a state an agent puts oneself into by the free exercise of his own intellectual powers, rather than a state that he is put into by the causal functioning of his cognitive devices. On the CC-model, knowledge is dominantly down to the knower. On the CR-model, knowledge is dominantly down to the knower's devices. Each model recognizes in the other features that it itself should accommodate. The CC-model recognizes the CR-characteristics of perceptual knowledge. The CR-model recognizes the CC characteristics of theoretical knowledge. The CR-model generalizes the perceptual paradigm to all of knowledge. The CC-model goes the other way. It generalizes the theoretical paradigm.[122]

The visitors thought a central difference between the two models that justification is a general condition on CC-knowledge and that it is no such thing for CR-knowledge. In their discussions of the CC-model the visitors fell into the habit of calling it the "No show, no know" conception of knowledge. It is a catchy phrase and a useful mnemonic, but not, as they realized, uniformly true. Still, if the CC model holds, in some cases it is literally true and in many others something like it is. Justifying, they reasoned, is a kind of showing. To the extent that knowledge is fact conditional upon justification, they concluded that showing that α is indeed a condition on knowing it.

---

[122] The CC-model has broad, though not universal, support among psychologists. Among the dissenters are R. Hassin, J. Uleman and J. Bargh (2005) and Bargh and Ferguson (2000) who regard it as an uncritical assumption and a false one. They see it as embodying the "illusion" that cognitive processes are freely willed and independent of deterministic mechanisms. See also Wilson (2002).

The visitors were troubled by the austerities of the CC model. They saw its treatment of knowledge as motivated by a premature question, by a question whose prematurity puts its answer at risk of violating Propositions 3.3b and 3.5b, which counsels that a definition of "knowledge" not be too hard on knowledge, and a violation too of the cognitive abundance presumption, 3.2b. The trouble, as they saw it, was CC justification-condition. There were simply too many cases of human knowledge unattended by any discernible presence of justification.

3.8 *Command and control reasoning*

Especially in its JTB form, the CC-model carries serious consequences for the logic of reasoning. It is widely agreed – by the visitors too – that a critical function of reasoning is to facilitate the attainment of an agent's cognitive ends, that is, ends that are realized by the achievement of knowledge. If justification is necessary for knowledge, and if justification is either case-making or the ready wherewithal for it, then the reasoning that facilitates the attainment of knowledge would itself hardly fail to have a case-making character. In this we see the ancient idea that *reasoning is argument*, and that even when transacted solo it always carries the potential for dialectical engagement. When Harry reasons to himself that β follows from a premiss α that he believes to be true, he is in effect presenting an argument to himself in which the propositions α and $^-$If α then β^- serve as a justification for the belief that β. If the argument is sound, the justification is impeccable. If the argument is plausible (enough) then the justification, though weaker, may still clear the bar of what is required for knowledge.

If, on the other hand, the CR-model is right for knowledge, and yet a primary function of reasoning is the achievement of it, it is much less plausible to endow reasoning with an intrinsically case-making character. A more natural model for knowledge-facilitating reasoning on the CR-model of knowledge is belief-change, or the revision or updating of an agent's belief-sets upon receipt of new information.[123] For all the comparatively little that is presently known of the dynamics of belief-

---

[123] See Harman (1986). The belief-change literature in computer and cognitive science, as well as various branches of agent-centred logic, is now vast. See, for example, van Benthem (2011) and the references there.

change, the reasoning that underlies it is not inherently case-making, indeed is nothing close to it.[124]

If this is right, it also carries consequences for the distinction noted in chapter 1 between consequence-having and consequence-drawing. It suggests that to a marked degree the recognition and drawing of consequences β of what one believes α is not a matter of arguing to oneself that since β is a consequence of α and α is true, then β is true as well. Even when constrained by what in one's belief set actually does follow from what, drawing consequences is with a notable frequency achieved by belief-change processes operating as they should, never mind that they operate in the absence of their owner's conscious awareness or capacity to articulate. If the CR-model is right for knowledge, then given reasoning's implication in the attainment of it, it is natural to suppose that reasoning too is best served by a CR-conception of it.

This is a matter of considerable importance for a naturalized empirically and epistemically sensitive logic of error. It is certainly true that such a logic can offer an account of a consequence relation – one or more as may be needed – independently of any particular theory of the cognitive make-up of the human reasoner. On the other hand, consequence-recognition and consequence drawing require a treatment that links in some systematic way to how the human consequence recognizer and consequence drawer is built, what he is good for and good at, what his limitations are, and so on.

There are lots of knowledgeable people who are fully seized of the historical fact that logic has mainly and, often enough exhaustively, been about the consequence relation. This consideration alone is sufficient to motivate a decision to reserve the name of logic for enterprises of just this character, and to call by another name the treatment of consequence recognition, consequence drawing, and the like. Subject to a caveat, this is fine with me. The caveat is: Stop calling logic a theory of reasoning. But, whether under the name of logic or some other, a CR-approach to human consequence drawing will embody roughly (very roughly) the following idea:

Proposition 3.8
CONSEQUENCE DRAWING: *An agent X draws β as a consequence of {α₁, ..., αₙ} when (1) his circumstances are such*

---

[124] For a more optimistic view, see Mercier and Sperber (2011) and the accompanying commentaries.

that his belief that α₁ ∧, …, ∧ αₙ causally suffices for his belief that β and his belief-forming devices are in good working order and operating here as they should, and (2) if asked what led him to believe that β, any disposition to reply would be a disposition to cite the αᵢ.

3.9 *Justification*

The two models also motivate quite different accounts of error. Seen the CC-way, we are largely responsible for what we believe. Our beliefs are the fruits of our own intellectual labour, redounding to our credit when they turn out right, and reflecting badly on us when they go wrong. CC-thinkers are beings whose epistemic achievements are the exercise of their cognitive autonomy. Just as Sarah's decisions are her own call in the aftermath of her own reflections, the autonomy of belief is reflected in like arrangements. What Harry believes is his own call, arising from his own cognitive efforts freely deployed. This suggests that error stands to cognitive virtue as sin stands to holiness.

One shouldn't over-press the kinship of error and sin, even by CC-lights. A defence against a charge of sin is that you couldn't help it. Likewise, a defence against a charge of error is that it couldn't be helped. But there is a difference. Not being able to help it precludes your having committed a sin, but not your having committed an error. Part of what is distinctive about the CC-model is its rather general hostility to the defence of not being able to help it. It is a hostility that makes particular use of the distinction between performance errors and competence errors. It will forgive you for errors committed when tired or drugged or in the aftermath of a stroke, but it is not much enamoured with the idea of your not being able to help it when there is nothing wrong with you. When there is nothing wrong with you, your errors are your own. Your errors are down to you. How could it be otherwise if your beliefs, including your beliefs about what should be concluded from what, are your call, arising from your own cognitive efforts freely summoned up? If your errors are down to you, how does it come about that you commit them? How does it come about that you commit them when you are not tanked up or strung out, deathly sick or dog-tired? You commit them because you are one or more of the following: careless, negligent, wilful, stupid, ignorant, irrational, duped.

3.10 *Earlier reliabilisms*

The CR-model is a variation of justificationist theories of causal reliabilism, minus the justificationism. It conceives of its commitment to empirical sensitivity as an impediment to the justification-condition. Historically, there are two distinct motivations for naturalization, one scientistic and the other anti-normative. The scientistic motivation has a positive intent, and the normative motivation a negative one. In the first instance, the objective is to replace philosophical ways of dealing with knowledge with ways furnished by the natural sciences. In a less austere variation, it is the decision to modify philosophy's contributions and supplement them with the disclosures of science. In the second instance, naturalization's goal is to close the gap between "ought" and "is". An ethical naturalist is someone who thinks that moral properties are natural properties, and thus amenable to empirical investigation. In the case of epistemology, the versions espoused by Quine and Kornblith[125] are advanced with replacement intent. The versions espoused by Goldman[126] and a good many other of his reliabilist allies were a variation of ethical naturalism. Here the intent was to de-normativize the theory of knowledge by de-normativizing the justification condition. In Goldman's eyes, justification is a normative undertaking on the traditional conception of it. In turn, the justification-condition makes knowledge a normative enterprise, lounging on the wrong side of the ought-is divide. If knowledge is to admit of empirical engagement, its normative defences must somehow be breeched. Goldman elects to do this by retaining the justification-condition for knowledge and yet breeching *its* normative defences. A belief is now justified when produced by belief-forming devices operating as they should. Justification is now causalized. Since knowledge's normativity was an inheritance of the normativity of justification, it is possible to de-normativize knowledge by trying to give it a wholly naturalistic construal. But it was never Goldman's position that philosophy had no place in naturalistic investigations of knowledge. Goldman's wasn't replacement naturalization, but rather what is called the "cooperative" variety.[127]

Problems arise. One is that if it is possible to de-normativize knowledge by de-normativizing justification, it ought to be possible to

---

[125] Quine (1969), Stich (1990)

[126] Goldman (1979), (1986).

[127] Goldman (1992), Stich and Nisbett (1980), Harman (1986), and even at times Kornblith (1994).

achieve the same end more directly, by de-normativizing knowledge itself. So justificationist reliabilism is needlessly complex, and the J-condition is redundant. The other worry is that the J-condition is an empirically tolerable condition on knowledge only by a stipulative reconceptualization of justification, by pleading ambiguities in the notion of justification which are far from apparent in actual usage. With the exception of Goldman's first crack at in in "A causal theory of knowing" in 1967, virtually all the going versions of reliabilism in one way or another retain a J-condition, which is the very thing that their CR model wants to reject as a general constraint on knowledge.

The visitors were not unsympathetic to the plight of the CC theorist. For most of Western philosophy's long history, a philosopher's interests were not considered achievable by the methods and presuppositions of science. How could it have been otherwise? Even now, cognitive science is still pretty much in its infancy. (For example, comparatively little is known at present about the conditions under which a person's cognitive equipment is working as it should. For another, it is only in our time that the psychology of perception has attained the status of a mature science.) The team concluded that the two millennia-long habit of philosophizing about human knowledge encompassed, *faute de mieux,* the analysis of concepts rather than the investigation of causes. And why not? The analysis of epistemic concepts felt a good deal more accessible to human effort than the investigation of epistemic causes.

On a literal reading of it, the J-condition is a deal-breaker for CC knowledge. Since justification hardly ever happens, knowing hardly ever happens either. This carries some sweeping costs. It costs us the cognitive abundance thesis. It costs us the cognidiversity thesis. It costs us the enough already thesis. It violates the condition that a theory of knowledge not be too hard on knowledge. The visitors saw these consequences as partitioning the CC-community, rendering it into three main trend lines made up of:

- Big-box sceptics.
- Normative self-insulators.
- Short-sellers of the J-condition.

Big-box scepticism: As already noted, in the first group it was accepted as demonstrably correct that for large classes of cases – the external world, other minds, the past, the future –we hardly ever attain a state of knowledge or that it is seriously in doubt that we do. No sufficient pause was given the possibility that if a theory of knowledge provided for generalized scepticism, that fact alone would discredit the theory. In some cases attempts were made to find replacement objects for the objects that scepticism denies us knowledge of. Leading the list of these prestidigitations is subjective idealism. (Thus, although chairs and people really exist, they are nothing like what we thought them to be.)

Normative self-insulation: The second group did not reject scepticism, but neither would it take our generally sorry justificatory performance on the ground as falsifying the CC-account. They insulated the theory by assigning it a normative authority. So, while we may be pretty bad at knowing, that is not the fault of the CC-definition. The fault is either our own or it is nature's. Of course, the visitors didn't much like group two's sceptical sympathies, but they were also troubled by the group's normative presumptuousness, as indeed we ourselves were in the chapter before this one. In what, the visitors want to know, does the normative authority of the analytical epistemologist consist, and whence his privileged access?

Short-selling: The third group tried for a footing midway between the two others. This they did by retaining the J-requirement but at the same time lightening up on its fulfillment conditions. This was done by trying to make of it something that's not in it to be. It was a form of philosophical bait-and-switch. The group three literature is very large, and it thrums with a nuance that rivals the tenderest subtleties of the High Middle Ages. In an effort to make their task a manageable one, the visiting team homed in on the main ways of making a radically unmeeetable condition a more easily met one. This was done by progressive equivocation, by attributing to "justification" some accommodating ambiguities.

3.11 *Externalism-externalism*

The visiting team allowed that there are indeed some ambiguities in the notion of justification. In one sense, justifications of α are furnished by considerations that would lead a reasonable person to believe that α. These we might call "reasonability" justifications. Justifications in a second sense would advance considerations that establish or provide grounds for the truth of α. These the visitors called "case-making" justifications.[128] Sometimes when a person advances considerations that

---

[128] See here Alston (1991): "To turn to justification, the first point is that I will be working with the concept of a subject S's *being justified in believing that p*,

106

he thinks attest to the truth of α, they don't. But that doesn't preclude those considerations making it reasonable for him or anyone similarly situated to believe that α. Thus a failed case-making justification can be a successful reasonability-justification. Each of these in turn is subject to an internalist-externalist ambiguity, landing the J-condition in a further muddle. Someone is an internalist about reasonability justification when he holds that considerations that make it reasonable to believe a proposition are those advanced by the believer in fulfillment of that function. An externalist will allow that the believer may have no idea of the considerations that make it reasonable for him to believe it, but that such considerations nevertheless exist. Similarly for case-making justifications. Internalist case-making requires only that cases be made on site (as it were). Externalist case-making requires that such cases exist. Whereupon we are met with more ambiguity still. It lies in the difference between *advancing* considerations with case-making or reasonability intent, and *there being* considerations of case-making or reasonability import. It all turns on an ambiguity of "justification" as between having one and there being one and, with it, on two readings of the J-condition. The externalist reading provides that the justification condition on knowledge is met by a justification in either of these senses. The internalist reading provides that the J-condition is met not by a justification in the possession of the would-be knower, but by the mere existence of it.

There is a further and related ambiguity, embodied in the difference between considerations that justify a person's believing that α and considerations that justify α itself, that is, are objectively evidential for or otherwise confirmative of it. It is not hard to think of cases in which considerations that justify Harry's believing that α fail utterly to justify what he believes.

This is getting to be a lot of distinctions to keep in play. From what our visitors were able to make of the western philosophical record, the J-condition originated with internalist case-making intent – or, we might say, internalist-internalist intent. So construed, the condition is

---

rather than with the concept of S's *justifying* a belief …. It is amazing how often these concepts are conflated in the literature. The crucial difference between them is that while to justify a belief is to marshall considerations in its support, in order for me to be justified in believing that *p* it is not necessary that I have done anything by way of an argument for *p* or for my epistemic situation vis-à-vis *p* ". (p. 71, emphases in the original)

107

distinguished by the comparative paucity of cases that fulfill it. Subsequent repairs were seen by the visitors as progressively venturesome changings of the subject, in a chain of amendments descending from internalism-internalism to externalism-internalism, and from there to *externalism-externalism*. Externalism-externalism with respect a proposition α is the doctrine that the J-condition is fulfilled if considerations exist which would induce a reasonable person to believe that α, and which lend to α objectively probative or confirmatory support. The visitors thought that this was latitudinaranism to no good end. It made the concept of justification unrecognizable.

They considered an example of the semantic erosion of "case-making". Having claimed that α, Harry goes on to make a spiritedly good case for it. The visitors conceded that this would be a clear case of both having a justification and there being one. Suppose now that Harry knows that α but has no case to make for α; he just remembers it. Suppose also that if challenged, and given a little time to gather himself, he would be able to construct the case for α. Then we might say that having a case to make is being able to make one more or less on demand. Suppose, on the other hand, that Harry couldn't make the case on demand, but could make it after a bit of research (say looking it up online). Then we might say that Harry satisfies the condition by being able to indicate how a case could be found. Or suppose that the nature of what Harry knows makes it impossible to recover or rebuild the case that he once had for α. (Perhaps α is the proposition that Harry cried quietly and out of sight when, at his third birthday, a hoped for present failed to materialize.) Perhaps the case exists in Harry's dormant memory. Then we might say that the condition is met by there being a case for α irrecoverably in Harry's memory. Or: Harry has reason to suppose that someone has a case for α or that a case for α exists somewhere, or that *God* has a case for it. Then if Harry does indeed know that α, Harry meets the showing condition if someone other than he has a case for α or a case for it exists or if God could make the case or – let us add (to no avail) – a case exists *in principle*. The visitors saw in this sweep from comparative at-handness to distant apartness, a semantic decay extending from case-making internalism to externalism.

The same holds for for reasonability-decay. Harry knows that α because he saw it. But Harry might have known that α because he remembers it, but without, then and there, attending to that memory. He might have known that α because someone told it, without having recalled the source of the telling, indeed without even taking note of its existence. Harry's knowledge that α might have been attended by considerations which, had he been aware of them, would have made it reasonable for him to believe it, never mind that those considerations did not in fact induce him to believe it. Here, too, the closer we get to an externalist reasonability-condition, the less that justificationism flags a recognizable concept of justification.

Perhaps it will be thought that externalism-externalism closes the gap between JTB-epistemologies and those of CR-intent. After all, if externalism-externalism allows for the sheer existence of considerations which, had Harry been aware of them, would have made it reasonable for him to believe α, to count as a justification of it, why wouldn't the sheer fact of the good working order and proper functioning of his belief-forming devices not also count as a justification, albeit of the externalist-externalist variety? Why wouldn't good order, proper functioning reliabilism carry an externalist-externalist significance? Indeed, isn't this precisely what justificationist reliabilists of virtually all stripes actually think? Wouldn't externalist-externalist reliabilism be the last resort for justificationism?

3.12 *In-name-only justifications*

This is how the visitors came to read it. Justificationist reliabilism is externalist-externalist justificationism *in extremis*. But they noted, even so, a gap that they wanted to make something of. It is a gap occasioned by the difference between knowing or reasonably believing that one's cognitive devices are in apple-pie order and knowing, or having some serious grasp of, the conditions in virtue of which this is so. It is a big gap. Comparatively speaking, not much is known of the conditions that make for the proper functioning of the mechanisms of knowledge. Finding out is the business of the empirical sciences of cognition. Even so, no one doubts that those factors exist, that there are yet-to-be discovered facts of the matter about how cognitive ship-shapeness is constituted. Perhaps it is true that if such factors were known to Harry, they would induce him to believe that α. Even if he already believes that α, presumably it remains the case that, if made known to him, there are considerations that would have induced the belief of α even had he not already had it. Knowing the mechanics of vision may or may not induce in Harry a reasonable belief that the bird he sees in the tree has a red breast. But it is laughable to suppose that it is a load-bearing fact that Harry doesn't know by looking that the bird's breast is red. He knows this because the mechanisms of vision are doing what they are meant for.

There are facts, some known and many not, about how these mechanisms operate. It may be that if those facts were made known to Harry, they wouldn't have induced that same belief. But it is going too far to make it a condition on his knowledge of the colour of the bird's breast that Harry know or have reason to believe that that eventuality didn't obtain on that occasion. Accordingly, the visitors concluded that

Proposition 3.12
IN-NAME-ONLY JUSTIFICATIONS: *Externist-externalist justificationist reliabilism is justificationism in name only. Its justifications are* façon de parler *justifications.*

The visitors readily conceded that theorists of knowledge were at liberty to contrive any new concept that proved congenial to their purposes. An assessment of the contrivance would depend on an assessment of the purposes. If the purpose of a theory is to furnish a descriptively adequate treatment of justification, it is necessary to note that justification is a pre-existing concept. If there is something it's likely to be a justification, it is something which a descriptively adequate account must respect. Failure to do so risks an account in which the concept of justification lacks a recognizable presence. It will be a theory of justification in name only. It will be at most a theory of justification*, where justification* is a concept of the theorist's own making. Recall here Quine's quip that one person's explication is another person's stipulation.

Clearly, the visitors were troubled by these equivocations. They thought that there were lots and lots of values of α, knowledge of which was a plausible thing to suppose, and yet the showability of which was not a plausible thing to suppose, save for these more attenuated and watered down interpretations of it. They thought of this as an "alienation problem". The further down we go in the chain of semantically decayed justifications, the more alienated is Harry from the case deemed to meet his obligation to make for α. Alienation, they thought, evacuates a condition of content. It makes it a phantom condition.

Perhaps there are considerations that would offer the concept of justification* secure employment. Perhaps, as mentioned in chapter 2, the day is coming when there will be reason to allow a Bayesian account to put present-day notions of knowledge into retirement. Perhaps what will be required is a new concept, knowledge*. In that case, it might turn out that this is a theory in which the concept of justification* is recognizably present. Perhaps this is a possibility that makes it reasonable to maintain an investment in theories of knowledge* and justification*. Who knows what the future may bring? But a theory of knowledge* need not and frequently cannot be a rival of a theory of knowledge. It is a theory with a different subject matter. As we have them now, justificationist epistemologies are theories of knowledge, not knowledge*. It is as such that they must be judged.

There was a dissenting rump among the visitors. Surely, they said, the cognitively competent individual must have *some* reason to suppose that the cognitive practices of him and his ilk are reasonable or reliable or at least reasonably reliable? Isn't there, in the corporate enterprise of human cognitive effort, ample evidence that we're pretty good at knowing? Isn't our reliance on being told things a good example of this? The answer, certainly, is yes. For example, if pressed about the epistemic comfort we derive from taking things on being told them, this would be the drift of our response, and we wouldn't be wrong in giving it. But what is this evidence *of*? It is evidence of the existence of the *presumption* of reliability. It is evidence of our disposition to persist with standard epistemic practice until there is particular reason pause or to cease. What it is *not* evidence for is that when our cognitive agendas are in process of execution, the proposition that we are doing things the right way occurs as a premiss of an inference whose conclusion is the thing we end up knowing.

3.13 *Evidentialism*

In reaching this conclusion, the visitors emphasized the importance of not losing sight of the ambiguity of "How do you know?" They readily allowed a similar ambiguity in "showing". It is the ambiguity of "show where you got it" and "demonstrate that it's true". Both sorts of showing have their detachment problems, needless to say, but compared to demonstration a proposition's truth, reporting our sources for it is a walk in the park. Accordingly, the visitors hypothesized that the more plausible the showing condition is, the likelier it bears the source-interpretation.

Something of this same ambiguity attaches to another epistemologically overworked concept – the concept of evidence. Sometimes "What's your evidence?" is fully answerable by "I heard it on the BBC". Still, in the long history of earthbound epistemology, the evidence condition is hardly ever advanced with the source-interpretation in mind. The dominant interpretation is contained in the epistemologists' notion of "evidentialism", to the following effect:

Proposition 3.13
EVIDENTIALISM. *A cognitive agent is justified at a time in believing that α just in case believing that α fits the total evidence he has at that time.*

The first thing the visitors noticed about this statement of evidentialism is that it make fitting with one's total evidence a condition on justification. Since they themselves had long since given up on justification as a general condition of knowledge, they weren't much inclined to give to this view – true or not – much mind. Even so, they were intrigued by an especially strong version of the doctrine offered by Jonathan Adler. It is, says Adler, a conceptual truth that people cannot "in full awareness" believe that α and also that they don't have adequate evidence for α.[129] The visitors were puzzled by this notion of conceptual truth. That is, they were puzzled by the "conceptual" part of it. But on thinking it over, they decided to focus on the "truth" part. They decided that, even supposing the justification-condition did hold for knowledge, this statement of evidentialism is either empirically false or vacuously true. It is false because there are large classes of cases in which something is known in the absence of the knower having any evidence for it. Harry knows that $e = mc^2$. He was taught it at school all those years ago. There is evidence that $e = mc^2$, but Harry has no clue as to what it is. True, Harry remembers that $e = mc^2$. But that is not evidence of its truth that is anything like the evidence available to relativity theory. On the other hand, evidentialism might be true but only vacuously so, by virtue of this same alienation that degrades the J-condition on knowledge, namely, that only alienated evidence, i.e. evidence*, stands any chance as a general condition on knowledge.

Some members of the group thought they discerned another ambiguity in evidence. In one sense, to be sure, one's evidence for α is the case that one is able to make for its truth. But in another sense, one's evidence for α is the information which, together with what else one also knows, causes one to believe it. It is an important difference, granting evidential status in the second sense to "I remember it" and withholding it in the first sense. And here too, they thought, and earlier point recurred:

---

[129] Adler (2002), pp. 31-32.

Evidentialism is a good deal more plausible in the second sense than the first.

The visiting scientists are a figment of my imagination, free of any expectation of fidelity to what is actually possible in a world such as ours. Its function is partly stylistic and partly evocative. Stylistically, the figure of the visiting explorers switches us from the declarative mode to the narrative. It gives to our exposition the feel of a story and, in so doing, it alters the focus from me to them.

The further purpose of the story is to stimulate a conception of how we ourselves might exercise our naturalism in our quest to arrive at an understanding of knowledge as freely as possible from the orthodox presumptions of epistemology. It is evoked here in the hopes offering an accurate picture not of the visitors but of us. Of course, this is not to say that such enquiries are philosophically unfettered in all respects. The visiting scientists have a certain feel for the conditions under which their theories are successful, for the logic, so to speak, of naturalistic enquiry. Moreover, given that we are not they, and that we *are* philosophers, it would be odd if their disclosures didn't attract our philosophical attention. A case in point is the susceptibility of their CR-model to Gettier problems.

CHAPTER FOUR

KNOWING (2)

"Our errors can be credited to two principal sources: The one source – the will's dissoluteness – which troubles and perverts the judgement is internal; the other source is external and lies in the objects we judge and in their being able to deceive our minds by false appearances".

Antoine Arnauld

Life, as far as it has a definite form, is but a mass of habits."

William James

4.1 *Gettier problems*

In 1963, Edmund Gettier made it harder than usual to stick with the JTB-model.[130] In what is arguably epistemology's most remarked-upon very short paper since the time of its appearance, Gettier constructed a class of counterexamples to it. Suppose that

> Harry has a friend, Sarah, who has driven a Buick for many years. Harry therefore thinks that Sarah drives an American car. He is not aware, however, that her Buick has recently been stolen, and he is also not aware that Sarah has replaced it with a Cadillac, which is a different kind of American car.

Then Sarah does drive an American car, Harry believes that she does and has good reason to think so. But if having good reasons for believing something is a justification for believing it, then Harry has a justified true belief that Sarah drives an American car, hence knows it. Although the present example is not Gettier's own, it nicely catches his intention.[131] Gettier and legions of other epistemologists take it as given that Harry does not in fact know that Sarah drives an American car, never mind that she does and that Harry with good reason believes that she does. So,

---

[130] There is a vast literature about Gettier problems, painstakingly reviewed in Shope (1983).

[131] The example is adapted from Weinberg *et al.* (2008), p. 29.

being a specification of necessary and sufficient conditions, the JTB model is overturned.

Gettier problems are a reason to doubt the sufficiency of the JTB-model, but it would appear that the CR model is subject to similar difficulties. As we have it at present, Harry knows that Sarah drives an American car if Sarah does indeed drive one, Harry believes that she does, and his belief arises in fulfillment of the other CR-conditions; in other words, is well-produced. So Harry actually does know that Sarah drives an American car. Anyone who thinks the Gettier examples are problematic for JTB-knowledge will likely have to the same opinion of CR knowledge. The question is, would they be right in so thinking? I myself think that the damage done to the JTB-model is remediable, if at all, only by imposition of further conditions on the definition of knowledge. I am a good deal less sure that the CR-model is, as it stands, vulnerable to Gettier difficulties. If so, the remediation question doesn't arise. How, then, would the CR-theorist defend against Gettier's challenge?

Various remedies suggest themselves. One is to interpret the condition that bans bad information as the way to handle the Gettier peculiarities. Harry's belief that Sarah's car is American is occasioned by out of date information. Sarah used to drive a Buick, but no longer does. Harry has no independent reason to believe that the car that Sarah now drives is not a Buick but a Cadillac. Had Harry's information been up to date, the belief set that contained the belief that she drives a Buick would have been revised to accommodate the new information that she now drives a Cadillac. The Buick belief would have been erased. If, in the presence of that new information, Harry's belief that Sarah drives an American car were occasioned by his old belief set and not the revised one, his belief-forming devices would not have been operating as they should. The old information would now be bad notwithstanding that, there and then, Harry is unaware of the fact. It is bad because it is now false. Harry's belief that Sarah drives an American car is still true, but the information on which that belief is grounded is not still true. It is now false. There are two things that properly operating belief-forming devices won't typically permit. It won't permit the simultaneous acceptance of new information and retention of old information when the new contradicts the old. It won't do it when the new information state is a response to a change in how the world is. Neither will it permit vanished information to ground or trigger belief. In a well-made doxastic system, the belief that Sarah drives an American car cannot be grounded in or triggered by erased information. Still, the central fact remains that in Harry's case the out of date information that triggered his well-functioning belief-forming mechanism was bad, made so by how the world had changed. So might we not say that whatever the fate of the JTB-model, Gettier examples cause no damage to the CR model?

Gettier problems teach a valuable lesson about defeasibility. In its broadest sense, defeasible reasoning is reasoning in the absence of guarantees. For anyone who is our kind of fallibilist, just about all the reasoning there is is defeasible, and the very name "defeasible reasoning" totters on the brink of pleonasm. Here again the distinction between consequence having and consequence drawing is a telling one. To the extent that good consequence drawing depends on true (or plausible or likely or probative) premisses, not even deductively valid reasoning will offer the kind of guarantees to qualify it as indefeasibly good. It may be that some premisses are immune from overthrow, and that some deductive patterns are so compelling as to make of their validity a sure-thing. But, equally, it would be folly to overlook the sheer volume of theoretical contentiousness among serious-minded logicians about these very issues. Ours is the Age of High Pluralism in logic, and is not a time to be playing fast and loose with presumptuous guarantees for logical truth and logical consequence.

The situation that Harry was in is one which, in the circumstances that then prevailed, he could not know that he is in. He could not have known that his information was out of date. Massively many of the information states we are presently in are subject to that same possibility. In Harry's case, new information falsifies the premiss of Harry's inference. It makes his reasoning unsound. In other cases, update information falsifies the conclusion of the inference and – supposing the reasoning to have been correct – falsifies the premisses as well, *modus tollendo tollens*. Still other cases exhibit a further difference. They are situations in which new information consistent with an inference's premisses and conclusion alike ruptures the premiss-conclusion link. In the first case, we have defeasibility of premisses and in the second the defeasibility of conclusions. The third gives the defeasibility of premiss-conclusion *links*.

The defeasibility of links creates a problem for CR's concept of good information. As we have it now, when Harry's belief-forming devices are operating on good information, the information is both accurate and current. But as the defeasibility of links gets us to see, it is possible that when Harry has a true belief produced by belief-producing devices operating on good information, the belief might be both impeccably produced and yet subject to information consistent with the present information and with the belief that Harry's devices produced, which nevertheless would stop that belief's production were it made known to Harry. The trouble is that the information on which Harry's belief-producing devices had been operating was incomplete in this particular way. So information that is accurate, current and incomplete is not only a defeater of premiss-conclusion links. It is an inhibitor of knowledge production on the CR model. So let us amend the information condition accordingly.

Proposition 4.1

GOOD INFORMATION: *Information is good for belief just when it is accurate, current and complete.*

It might be thought that this amendment alters this frequency of knowledge in a game-changing way. Given that nearly all information on which true well-produced belief is grounded is incomplete, wouldn't it only stand to reason that a good whack of it will be incomplete in this knowledge-inhibiting sense? If so, there would be two immediate consequences for the knowledge abundance and error abundance theses. The frequency of cases in which what we take ourselves as knowing is not knowledge after all would amply support Error Abundance. But wouldn't it at the same time show Cognitive Abundance in a bad light? Of course, we could always tough it out, and take refuge in Enough Already. It is true that our takings to know foster our survival and prosperity. But, as indicated in chapter 3, not everyone will be convinced that this shores up Cognitive Abundance on the desired scale.

It is a reasonable question. It is easy to see why one might think that the cognitive abundance thesis is imperilled by the hearty frequency of our epistemic mistakes. But the corrective offered by Enough Already isn't a measure of last resort. It is not epistemology's Hail Mary pass.[132] Couldn't we here re-invoke the earlier point that our theoretical provisions proved to be epistemologically ungenerous, epistemology would lose the assumption that knowledge is valuable, and thence a good part of the rationale for thinking that knowledge is worth all the effort to

---

[132] In North American football, a Hail Mary pass is a forward pass, often in the last play of the game, into the opposition's end zone in hopes that some or other teammate will manage to catch it. When caught, Hail Mary passes are often referred to as miraculous, though not with serious theological intent.

produce a decent theory of it. Varying what, in the *Republic,* Plato asks of mud, what would be the philosophical good of having a decent theory of knowledge? Besides, why would we want to endorse a conception of knowledge that proves hostile to the spirit of our epistemological naturalism? The opening position of an empirically sensitive theory of knowledge is that a load-bearing place not be reserved for the big box scepticisms. Error Abundance tells us, rightly, not to make light of our errors. Cognitive abundance tells us, rightly, not to make too much of Error Abundance.

But perhaps there is something more sure-footed to say. Complete information does not mean all information or even all information that bears on the events or facts, reported by the belief that α. It means all information causally necessary for the formation of that belief yet consistent with its truth at the time in question. It is not a condition on well-produced belief that it be served by complete information. To impose this would be tantamount to requiring that well-produced belief that it be served by *total* information. To impose it would be tantamount to requiring that well-produced belief be infallible. It would preclude the possibility of our arriving at well-produced belief on the basis of misinformation. It would convict our belief-forming devices of malfunction on a scale that itself defies belief. The requirement that the information be complete is a condition not on the well-formedness of belief, but rather on inference's knowledge-acquiring capabilities. Take Spike again. Given what we now know, it is reasonable to believe that Spike did it. That Spike did it would be a reasonable thing to infer. Indeed it would be what anyone having just that information would infer. At this stage, our belief-forming devices are operating as they should, hence, so, too, our inference-drawing mechanisms. Our premisses are true and our reasoning is just fine. But the reasoning isn't knowledge-generating. Its conclusion is contradicted by information we don't yet have which doesn't contradict the information we have now.

That well-produced belief and well-reasoned inference lack knowledge-generating transmitting guarantees is a foundational fact of our cognitive lives. How could it be otherwise? It is nothing short of what fallibilism requires.

4.2 *Is fallibilism paradoxical?*

There are two generic forms of fallibilism each of which comes in varying strengths. They are possibilist fallibilism and actualist fallibilism. Let K = {α_1. ..., α_n} be all and only the propositions I take myself as knowing at a time t. Then

Proposition 4.2a
WEAK POSSIBILIST FALLIBILISM: *There is at least one α_i whose falsity is not precluded by its membership in K.*

STRONG POSSIBILIST FALLIBILISM: *There are many α_i whose falsity is not precluded by their membership in K.*

WEAK ACTUALIST FALLIBILISM: *There is at least one α_i that is actually false.*

STRONG ACTUALIST FALLIBILISM: *Many α_i are actually false.*

HEDGED STRONG ACTUALIST FALLIBILISM: *The truth of strong actualist fallibilism is not precluded by what is currently known of the make-up and resource-exigencies of human knowledge.*

All versions save the third are modal theses. We should give their modality some attention, but not before considering fallibilsim's vulnerability, or want of it, to the Preface Paradox.

Everyone recognizes that a good deal of what he now knows to be true may cease being true at a later time. This is not fallibilism. It is acknowledgement of the fact today's date will not be tomorrow's. My kind of fallibilist asserts as a practical certainty that a good deal of what he thinks is true isn't and a good deal of what he thinks he knows he doesn't. This being so, a fallibilist should be prepared to concede, on the balance of probability, that he knows to a practical certainty that a good many of the propositions he now thinks he knows are now false. This begins to sound like the Preface Paradox.

The Preface Paradox originates with David Makinson's *Analysis* paper of 1965, and has taken many forms since then.[133] Here is one of them. You have written a scientific monograph, and have done your level best to get everything right. We might think of the book – the document – as one long conjunction, $\ulcorner\alpha_1, \wedge ...,\wedge \alpha_n\urcorner$, in which the α_i are all and only

---

[133] Makinson (1965).

the book's assertions.[134] Consider now a sentence β from the book's preface. β says that the author believes each of the α_i and yet that at least one of them is false. Some versions of the paradox are written so as to make β itself fall within its intended range of application. This produces, among other things, self-referential trouble. β is saying that it itself is possibly false. This is not the intended thrust of the paradox. So I'll not make β apply to itself.

In this form, the Preface Paradox purports to show that authors who are modest in this way hold inconsistent beliefs. Some logicians, Makinson included, understand the inconsistency as calling into question the conjunction rule for belief contexts:

(1) $(B\alpha_1 \wedge, ..., \wedge B\alpha_n) \rightarrow B(\alpha_1, ..., \alpha_n)$.

I myself am partial to this suggestion, but as a matter of psychology, not modal logic. Even if I believed that every statement of *Errors of Reasoning* is true, I couldn't believe any statement which is the conjunction of them all. It is too long for any being like me to get his mind around.

It may also be doubted on empirical grounds whether the conjunction rule holds even for low finite numbers of beliefs. Say that Harry believes that Paris is the capital of France. Suppose that later he came to believe that Sarah drives a Cadillac. If belief is a propositional attitude, there is no empirical encouragement for the inference that at that same time Harry believes that Sarah drives a Cadillac and *also* that Paris is the capital of France.

There may be something to be said for the idea that Harry's beliefs *commit* him to the truth of their conjunction. That is, it would be inconsistent with what Harry believes for him to deny the proposition that happens to be their conjunction closure. But for Harry to deny that $\ulcorner B(\alpha_1, ..., \alpha_n)\urcorner$ he must have it in mind, which is not anywhere close to being so in the general case.

Even waiving the problem of what it is possible for someone to get his mind around, failure of the conjunction rule for belief is not a complete remedy for preface problems. Makinson-prefaces occasion the appearance of further trouble; for example, the difficulty of quantifying into doxastic contexts. In particular, they present us with the following

---

[134] Including their logical consequences.

Barcan problem. What the modest scientist writes in his preface commits him to say something like:

(2) $B((\alpha_1 \wedge \dots \wedge \alpha_n) \wedge \exists \alpha_i (\sim\alpha_i))$.

The idea behind (2) is sometimes formulated as

(2') $\forall \alpha_i (\alpha_i \varepsilon D \rightarrow \beta \alpha_i) \wedge B (\exists \alpha_j \varepsilon D(\sim\alpha_j))$,

where D is the body of the book – the "document" in question. Here the RH belief-sentence has the belief operator in *de dicto* position. What we do not have, however, is the same sentence adjusted by removal of the belief operator from *de* dicto position to *de re*. Although (2') might be true,

(2'') $\forall \alpha_i (\alpha_i \varepsilon D \rightarrow \beta \alpha_i) \wedge \exists \alpha_j \varepsilon D(B \sim \alpha_j)$

is not. It could be true that I believe that there is a falsehood among the things I believe, but it needn't be the case that there is a given one of them of which I believe that it is false.

A Barcan modality is one which obeys the schematic condition:

(3) $\mathcal{M}Q\alpha \rightarrow Q\mathcal{M}\alpha$

in which $\mathcal{M}$ is a modal operator, Q a quantifier, and α the scope of (3)'s LH formula. When $\mathcal{M} = \Diamond$, the possibility operator, and $Q = \exists$, the existential quantifier, (3) becomes

(3') $\Diamond\exists\alpha \rightarrow \exists\Diamond\alpha$

or, concretely,

(3'') It might happen that someone runs a two and a half minute mile, then there exists someone for whom a two and a half minute mile is possible.

A good many readers of modal logic take (3″), hence (3′), to be a decisive *reductio* of Barcan's axiom (3).[135] It is one thing to believe that it could be that someone or other runs that fast. It is quite another thing to believe it of any particular person. Suppose now that belief is a modality. Then if Barcan's critics are right, (2′) could be true even though (2″) is false. So we could allow (2′) as a credible formulation of what the preface writers intended to say, without the embarrassing commitment to (2″) demanded by the Barcan formula as rewritten for belief. For, as rewritten, (2″) is certainly not what the preface writers intended to say.

Of course, Barcan's critics might be wrong, and the Barcan axiom might hold for possibility. But if belief is a modality with the formal semantic structure of possibility, belief would be a Barcan modality.[136] In that case, (2′) would entail (2″). What the preface writers intended to say would entail what they didn't intend to say. The moral of it all is that deciding whether (2′) is a credible formulation of what the preface writers are saying will require that we settle three disputed issues. Is the Barcan axiom true of possibility? Is belief a modality? If so, is it a Barcan modality? I regard each of these as still open questions. But my present inclination is that belief doesn't have the formal semantic structure of a modality, Barcan or otherwise.[137]

If Makinson-prefaces are not subject to Barcan difficulties, neither are the possibility versions of fallibilism noted at the beginning of this section. But it suffices for an *interesting* fallibilism that we pledge to the strong actualist form of it:

(4) Much of what I think I know is actually false.

---

[135] Apart from Barcan herself (Barcan Marcus, 1993), a notable exception is Simchen (2012).

[136] In alethic contexts $\ulcorner MQ\alpha \urcorner$ captures a modality's formal semantic structure, As before, let M be a modality, Q a quantifier, and α a sentence. Then a sentence $\ulcorner MQ\alpha \urcorner$ is true if and only if α is true in the quantification of a set theoretic structure. Informally, $\ulcorner$ Possibly α $\urcorner$ is true if and only if there is at least one world in which α is true. More formally, an interpretation is a sequence $\mathcal{W} = \langle W, A, v \rangle$ where W is a set of nonempty objects. A is a binary relation on W, and v a valuation function. Informally, any element w of W is a "world". Thus $\ulcorner$ Possibly α $\urcorner$ is true in a given set theoretic structure w if and only if there is at least one set theoretic structure w′ to which w bears the A-relation in which α is true.

[137] See Woods (2010d).

122

Of course, fallibilism gives trouble when asserted of a cited proposition.

(4') I think I know that α, yet α is false.

Fallibilism cannot be self-ascribed to cited propositions. (4') is, in Hintikka's words, a self-defeating thing to utter. But what (4') says of its self-defeating utterer could still be true of him. When someone other than he says it, it could be true. The same applies to Makinson-prefaces. If D expresses the author's conviction that he thinks that everything in it is true, he can't cite any proposition in D that he thinks is false. It is sometimes suggested that, although true, fallibilism is unspeakable – that is, that there is a pragmatic inconsistency in giving voice to it. If this is right, then expressions of fallibilism would have the character of Moorean blindspots, as with

(5) α, but I don't believe it.

A speaker of (5) *implicates* that he believes that α and *asserts* that he does not. Is an utterer of fallibilism in the same boat? He is not. Strong or weak, to assert fallibilism about your beliefs is not to give expression to them. When you assert (3') or (3") there is no proposition you advanced as something you say you believe, hence no proposition advanced as a proposition you believe of which you deny belief. So we may allow that fallibilist acknowledgement is perfectly "speakable". I conclude that if (4) is a credible formulation of an interesting fallibilism, we can cease worrying whether (2') is free of Barcan problems. For (2') is not the formulation of fallibilism that (4) is.

4.3 *Belief*

The CR-model reflects a certain picture of the human cognitive system. A human individual is a product of evolution. Included in his wherewithal are adaptive responses, and inheritances passed on by the mechanisms of early learning. He is, among other things, a causally responsive information-processing device instantiating a distinction between energy-to-energy transductions and energy-to-information transformations, and possessed of the means of uploading various sorts of information as belief. Belief is an essential response to the drive to know. A further part of the story is one that charts the link to the other conditions of knowledge, notably truth. But belief is also fundamental.

The human individual has a complex nature. He is an organism placed by evolution in a violently busy niche of the nature of things. He is also a social being, negotiating a plethora of interpersonal arrangements and playing a role in the collective manifestations of the human presence. The human organism is a child of nature socialized in a certain way. No one seriously doubts that his capacity for belief is rooted in his sensitivity to the provocations of the natural and social arrangements in which he is caught up. Most epistemologies claim for belief an indispensable role in knowledge. Much attention is paid in these philosophies to how beliefs are justified. But there are prior questions. How are beliefs *made*? What are beliefs *like*? What is it to *be* a believer?

All this talk of belief will not go down well in the halls of eliminative materialism. Eliminative materialism is more a lick and a promise than a settled thing. It is promissory-note materialism, spurred on by the application of the Can Do principle to strides made in the brain sciences. It is much too early to throw around accusations of Make Do, but for the same reason it is also too early to give up on belief and desire. Folk psychological explanations have been with us since days of yore, as Paul Churchland keeps on telling us. This bestirs critics to say that if after all this time we haven't been able to convert our folk psychological assumptions into a rip-roaring theory, so much the worse for those assumptions. But it is possible that practices of such longevity will have some (reciprocated) evolutionary influence on our cognitive machinery. In this I side with Peter Godfrey-Smith.[138] Our interpretations of one another's behaviour and other social relations are replete with folk psychological assumptions. Accordingly, our cognitive machinery has been exposed to natural selection in an environment shaped by these practices. As a consequence of this evolutionary influence, children re-wire the architecture of their social interactions in folk psychological ways. This happens before the age of four. It is a re-wiring which, from that point on, makes folk psychology true of them. So I'm not prepared just yet to capitulate to the smart set's nervousness about belief, or, in the poet's words, to make belief "in a most ruthless and piratical manner, to walk the plank." I am not minded to think that "the things of the mind [are] strictly for the birds."[139]

Part of what is valuable about Godfrey-Smith's remark is its openness to a further, not necessarily incompatible, possibility. This is a

---

[138] Godfrey-Smith (2005).

[139] Quine at B.F. Skinner's retirement party in 1974, recorded in Follesdal and D. Quine, (2008), p. 291.

view that emphasizes the sociocultural basis of our folk psychological proclivities. There is a particular example of this alternative approach which we might mention now. We will have occasion to return to in chapter 9, when we turn to the phenomena of told-knowledge. This is the narrative practice theory in the manner of Hutto (2008), according to which "children only come by the requisite framework for such understanding and master its practical application by being exposed to and engaging in a distinctive kind of narrative practice". (p. *x*)

To the decision to give to belief a load-bearing role in a naturalized logic of reasoning and knowledge there accrue all the problems that the notion of belief is known to attract, and some perhaps less apparent. One of the apparent one flows from the phenomenon of reasoning down below.

4.4 *Reasoning down below*

It is easy enough to speak of belief as something up and about and ready at hand to mind's inspection. If we thought of beliefs as nuggets of precious stones or clumps of sought after minerals, there would be a quite general tendency to think of our cognitive states and processes as in lodged in us in ways that permit surface-mining in the bright and safe light of day. This is an idea more congenial to the CC-model than its CR-vis-à-vis. In point of fact, it is not a view that either should have much time for at all.

Naturalized logic seeks a working relationship with the cognitive sciences. I have mentioned the importance to our project of an appreciation of what the individual cognitive agent is actually like. A good many logicians think that logic is a theory of language, albeit a highly specialized one. (Even Quine thinks it is "linguistics on purpose")[140] Empirically sensitive logic extends its reach more broadly. It encompasses the comparatively little that is presently known about sublinguistic and automatic reasoning, that is, "reasoning down below"[141] or what, in the case of perception, Humboldt called "unconscious inference", concerning which "Freud's views ... were premonitory".[142] Theories of down below reasoning are exemplified by

---

[140] Quine (1990). See also p. 67: "Logic chases truth up the tree of grammar".

[141] The term "reasoning down below" is coined in Gabbay and Woods (2003), chapter 3, and discussed further in, e.g., Bruza *et al.* (2007).

[142] Quoted from Edelman (2006), pp. 155-156.

connectionist/abductive logic of perception;[143] by representation-without-rules (RWR) approaches to cognitive modelling[144]; by integrations of connectionist and RWR orientations,[145]by offline non-representational systems developed for robotic implementation[146]; by theories of causal spread [147]; by aggregate systems of continuous reciprocal causation[148]; by evolutionary decision theory[149]; by neural-symbolic abductive networks[150]; by neuro-fuzzy argumentation networks; by semantic space theories of human information processing[151]; by theories of unconscious reasoning[152]; by nativist logics[153]; and by dual-process logics.[154] Since a cooperative by naturalized logic fashions its account of reasoning with due regard to the cognitive constitution of human reasoners, I don't want to invest over-heavily in models that are hostile to reasoning which is, in varying degrees and intermixtures, is unconscious, sublinguistic, inattentive, involuntary, automatic, effortless, non-semantic, computationally luxuriant, parallel, and deep.[155]

---

[143] Churchland (1989, 1995).

[144] Horgan and Tienson (1988).

[145] Guarini (2001).

[146] Wheeler (2001), Husbands and Meyer (1998).

[147] Wheeler and Clark (1999).

[148] Wimsatt (1986), Clark (1997).

[149] Cooper (2001).

[150] d'Avila Garcez et al. (2002), d'Avila Garcez and Lamb (2004), Gabbay and Woods (2005), chapter 6, d'Avila Garcez et al. (2007).

[151] Bruza et al. (2004), Gabbay and Woods (2005), chapter 9, Bruza and Cole (2005) and Bruza and Woods, to appear.

[152] Gigerenzer (2007).

[153] Hanna (2006).

[154] Evans (2007).

[155] Evans puts the distinction between up above ("system 1") and down below ("system 2") this way. *Reasoning up above*: linked with language and reflective consciousness; slow and sequential; linked to working memory and general intelligence; capable of abstract and hypothetical thinking; volitional or controlled – responsive to instructions and stated intentions; and high effort. *Reasoning down below*: unconscious, automatic; rapid, computationally powerful, massively parallel; associative; pragmatic (contextualizing problems in the light of prior knowledge and belief); does not require the resources of central working memory; functioning not related to individual differences in general intelligence; and low effort. (Evans, 2007, pp. 14-15) Nor are these characteristics strictly pairwise equivalent, as Shiffrin (1997), pp. 50-62 points out.

The modern camera is a bit of a marvel. For virtually all purposes, it delivers what is asked for when in its automatic or point-and-shoot-mode. On occasion, however, the picture will require its taker to play a role in adjusting the camera to the shot's special requirements. When this happens, the camera is put into manual mode. One of the advantages of the automatic mode is a high level of performance with little effort. One of the disadvantages of the manual mode is that in most situations it won't perform at a degree of betterness that would justify the bother (and time) required by its use. The automatic mode trades off efficiency for flexibility. The manual mode reverses the trade-off. For the most part, the camera should remain in automatic mode. But for certain classes of cases, nothing but the manual will do.[156]

In my appropriation of it here, the camera analogy plays out as follows. Reasoning down below is like a camera in automatic mode. Reasoning up above is more oriented to the manual mode. But reasoning above up – all of it – is rooted in an ecology of automatic processes down below.

4.5 Dark knowledge

A problem for the CC-model is accounting for our knowledge down below, or knowledge in automatic mode, which ebbs and flows and nourishes the whole project of knowledge unawares. So it is knowledge in automatic mode that is also dark. In so saying, I intend nothing menacing or troubled. Knowledge is dark when it is produced in automatic mode and out of sight of the mind's eye. The idea of dark knowledge stirs a certain resistance from critics. Various opinions are voiced. One is an aggressive scepticism: What we take for automatic mode knowledge isn't knowledge. Another possibility is to alter the CC-conditions on knowledge. Thus, in a simplified form, when you have the automatic mode knowledge that α, you have the dark belief that α and a

---

[156] I owe the camera analogy to Joshua Greene, whom I thank for helpful correspondence, and whose use of it serves a purpose somewhat different from mine. Greene's focus is moral judgement and its neural correlates. Judgements of intrinsic wrongness are the result of automatic processes lodged in the ventromedial prefrontal cortex. Judgements that require the calculation of consequences take longer to make and are more situation-specific, and are cited in the dorsolateral prefrontal cortex. The former judgements are often seen by moral philosophers as deontological in character, and the latter as utilitarian. See Greene (2011).

dark justification of it. There are problems with both solutions. If what we take for knowledge automatic mode or knowledge down below isn't knowledge, it is hard to see how a like rejection is avoidable for what philosophers of science call background knowledge and linguists call tacit knowledge. On the other hand, if dark knowledge is to count as knowledge and the CC-model is to be retained, we will have to make sense of automatic mode justification, of justification unavailable to awareness. (We can see externalism-externalism in justificationist epistemologies as a step in this direction). One promising way of making these construals is by likening automatic mode belief to a pre-existing causal disposition to form the corresponding transparent belief under appropriate stimulation or, even more basically, a pre-existing causal disposition to act or respond in appropriate ways. The mechanisms of automatic mode belief bear some resemblance to the mechanisms of dormant memory. A memory is dormant when there is a causal disposition to bring it to surface under appropriate stimulus. Roughly speaking, the mechanisms that bring dormant memories to the surface are those with a notable frequency that respond to the agent's cognitive interests. By and large, people don't pull memories out of dormancy just for the fun of it. They do so under conditions of relevance to agendas at hand. *How* they do so is a good and largely unanswered question.

No one should doubt the deep implication of knowledge down below in the project of conscious knowledge. Much new conscious knowledge is the product of the causal interplay – the ecological reciprocities – of the automatic and the manual. The nature and extent of this alliance are matters of fundamental importance for epistemology. Recognition of this causal integration has to be struggled for in CC-terms. But it is business as usual for the CR-model. This is a further reason to like it.

Proposition 4.5
UNIFICATION: *Once we have a better idea of how the relevant processes work, the CR-model offers a comparatively natural means of unifying the epistemologies of up and down, of the automatic and the manual, of the dark and the light.*

.

There is an important difference between automatic cameras and the mechanisms of cognition down below. The people at Nikon know what's required for generally good pictorial results even they have comparatively little understanding of the mechanics of a cognitive agent's goings-on down below. This turns out to matter for the rivalry, such as may be, between our two models of knowledge. A principal difference between the two is that the CC-model is open to the idea that some at least of the processes and conditions required for knowledge are subject to the agent's critical inspection. They are the monitorable aspects of knowing. Knowledge arises when the items on the agent's checklist "check out". It is harder to pitch this idea to the CR-model, never mind the phenomenal support our knowledge-having experiences sometimes appear to lend it. The CR-model says that it suffices for your knowing α, that you believe it and that your belief is well-produced and true. Even if CR-knowing is actually attended by the monitoring of what the agent takes to be a required checklist, and even if as far as the agent can tell the items on the list check out, there is nothing in the CR-model which in any direct way makes of this monitoring and checking an epistemic necessity. Indeed it is a hallmark of reasoning down below is that it is monitorless and uncheckable then and there by the agent whose reasoning it is. In some respects this makes error easy to understand; that is, it makes it easy to understand why beings whose cognitive devices serve them well overall should actually make errors. They make errors because they can't detect them, and they can't detect them because they are inaccessible to awareness. In other respects, errors of reasoning pose particular challenges for the theorist. If they are inaccessible to awareness, how is their presence felt? This was one of our questions in chapter 1, the concealment-detection question. If, in the agent's overt behaviour, we have the problem of how to negotiate diagnoses of error over the large gap between reasoning down below and behaviour up above, we are also met with the problem of determining how such errors are corrected. Almost certainly this mode of correction is often something that also occurs down below. This complicates the theorist's access.

4.6 *Integrating the models*

On the face of it, the causal response and command and control models are serious rivals. Certainly they are so when considered as general accounts of knowledge. However, various forms of reconciliation of this rivalry come readily to mind. One involves crimping their respective ranges of application, positing the CR-model for knowledge down below and what is often called "immediate" knowledge, such as the knowledge we have, or appear to have, of our own sensory experiences, and reserving the CC-model for "mediate" knowledge, in which knowledge above achieved by inference from down below "premisses" is considered paradigmatic. Another means of reconciliation is to give to our two models different subject matters. On this approach, CC is the right model for knowledge and CR is the right model for belief. Even if the CC-model holds as the canonical model of knowledge – especially in its justified true belief variation – it provides that knowledge has a constituent component for which the CR-model, not its CC rival, holds sway. These two modes of reconciliation are not disjoint. One could be a CR theorist about the knowledge of one's immediate experience, a CC - theorist about mediate knowledge, and a CR-theorist about the belief component of CC-knowledge.

Both models incorporate an investigative aspect of knowledge. They respect the fact that in large classes of cases knowledge has to be *quested for*, that a good deal of what we know is underlain by inquisitorial effort. One of the respects in which the two models differ has to do with how this inquisitiveness is to be understood theoretically. In the CC-orientation, investigation throws up candidates for selection, and selection is guided by the imperatives of case-making. Seen this way, it is not unnatural, as we have noted, to construe states of knowledge – certainly of mediate knowledge – as states in which a suitably inquisitive, reflective and critical agent puts himself by his own efforts, freely adopted and freely executed. But, again, the CR-model places a lesser emphasis on this volitional "free intellection" component of CC-knowledge, and concentrates more on the causal effects of enquiry. Although it is something of a caricature, it captures the basic idea of this difference to say that whereas on the CC-model the outcome of enquiry are propositions the knower selects, on the CR-model the endpoint of enquiry are propositions selected for him, that in the one model the knower acts whereas in the other he is acted upon.

In marking this difference, it would be appropriate to pick up on an earlier suggestion and give it formal expression. It is that nothing in CR-approach precludes the following kind of interplay between the two models:

Proposition 4.6a
CAUSALIZING THE CC-MODEL: *In determining whether α, it is possible for an agent both to attempt to follow and to experience himself as following the routines sanctioned by the CC model, and that doing so is sometimes a necessary condition of his actually doing well by the lights of the CR-model.*

COROLLARY: *There may be cases in which having a justification for an agent's belief α is causally necessary for the firing of the devices by which his belief is produced.*

COROLLARY: *In such cases, satisfaction of the J-condition is causally necessary for knowledge, but definitionally inessential to it.*

There is something to like about these observations. They defeat a certain misconception about causal theories of belief and reasoning. It helps us to see that being put into belief states is not in general something one experiences as merely passive. Beliefs do not always befall as the measles befall. Believing is not a matter of staying home in bathrobe and slippers, waiting to see what pops into one's head.

These past many pages, having noted that we are knowledge-driven beings, we have been wondering, in effect, how good are we at it? Let's sum up the main points of the discussion so far.

Proposition 4.6b
THE APPARENT GOODNESS THESIS: *It would appear that we do pretty well as knowers. It would appear that we haven't been built for knowledge to no good epistemic end. It would appear that we have some knowledge about most of the things we're interested in having knowledge of. Indeed it would appear that the cognitive abundance thesis is true.*

It is necessary to emphasize that Proposition 4.6b reports the appearance of things. Speaking for myself, I am content, until deprived of it, with the assumption that the appearances *are* fact. The decision to resist big-box scepticism invites it. But fact or not, they are interesting appearances which beg to be explained. At a certain level of generality, there are two ways of accounting for them.

Proposition 4.6c
THE TRUE APPEARANCE HYPOTHESIS: *A reason that this is the way things appear is that rather massively this is the way things actually are.*

Proposition 4.6d
THE FALSE APPEARANCE THESIS: *A reason that this is the*

way things appear is the sheer scale of our mistakenness. A mistake is an error attended by a contrary appearance. Accordingly, the more errors there are, the more there is the mere appearance of knowing.

There is a tendency to regard the true appearance and the false appearance theses as rivals. They shouldn't be. The reason why is that they can be true together. We do have lots of knowledge and we do commit lots of errors. Both facts together explain the range and intensity of the apparency of goodness. This is important. It gets us to see that the false appearance thesis is not, just so, ungenerous to knowledge.

The tie between good living and cognitive success provides that, up to a point, the better we are at knowing things, the better off we are. Doing well is a matter of both survival and prosperity. Survival requires the replication of certain traits in those of us yet to come. Prosperity – beyond the contingencies of food, shelter and security – is the satisfaction of conditions that answer to our interests. If this makes for a kind of instrumentalism about the cognitive virtues, it is not an implication to be scorned. Most of what is true would do us no good to know. Most of what we could in principle get to know wouldn't be worth the bother. It is true that some people are fonts of "useless information". Good for them! The truths thus grasped and the knowledge thus acquired satisfy conditions that answer to their interests. But there are limits on the good of curiosity. At a certain juncture, knowing lots and lots of things is pathological, and in some situations extremely bad for you.

4.7 Argument

A theory that accommodates dark knowledge and dark reasoning carries important consequences for the "argument-inference debate", briefly touched on in chapter 1. There is considerable support for the idea that there is a revealingly close kinship between forming the belief that β from the beliefs that $\alpha_1, \ldots, a_n$ and constructing an argument in which β is the conclusion and the α_i the premisses. The purported significance of the kinship is expressible as follows: There are certain good-making properties that reasoning has only if – in some versions, if and only if – the corresponding argument is good in that same way. There is some debate about what these good-making properties are, but one point of agreement is that both reasoning and arguing embed a consequence relation. For example, in deductive contexts this would be the idea that conclusion-drawing inferences and their corresponding arguments would be good only if the shared consequence relation were truth-preserving. The point would generalize to the counterpart property of any other consequence relation there chanced to be. This leads supporters of the view that inference is argument to the suggestion that whereas the similarities between inference and argument might not warrant their outright identification, it is similarity enough to justify the modelling of inference in an appropriately constructed logic of argument.[157] Thus on the affirmative side, the argument-inference question attracts one of two answers. Either inference is literally a form of argument,[158] or it isn't but it is enlightening to model it as argument. The negative side of the argument-inference question must marshall a convincing rebuttal of both these alternatives.

Counting in favour of the modelling option is the phenomenon of belief-change. Inference, *au fond*, is belief-change. Thanks to one of the ambiguities of "belief" talk, we have two different notions of change. One is causal. The other is propositional. There is reason to believe that both relations are abundantly instantiated in the cognitive lives of human agents, the former a good deal more so than the latter. Even so, this hasn't deterred logicians from modelling the former as the latter, with significant implications for the logic of belief. The model provides that whenever, in response to new information, someone's belief-set changes in some or other way, there is a proposition representing the new information, a proposition-set representing the old belief set, and a proposition-set representing the changed belief-set. There is also a relation on these proposition-sets such that, at a certain level of abstraction, the proposition-set representing the new belief set is a semantic consequence of the fusion of the new information and the proposition set representing the initial belief set. Whereupon we have it, at this same level of abstraction, that causal modification is modelled as semantic consequence. The CC-approach offers a welcoming conceptual home to this sort of modelling. The CR-approach is markedly less hospitable.

There are additional considerations that cast doubt on the inference-argument modelling decision. One is the palpable lack of empirical support. There are just too many conditions under which argument occurs that are unfulfilled when inference occurs. We have seen

---

[157] This view is widely endorsed in the argumentation community. See especially, van Eemeren and Grootendorst (2004). See also Walton (1990).
[158] See again Mercier and Sperber (2011).

that when a theory is formulated with normative intent, it supplies a motivation for downplaying empirical discomformity. I have already said this is a normative authority whose invocation routinely lacks an adequate defence, and that it threatens to convert a doubtful theory of reasoning into a stipulated theory of reasoning*. It changes the subject, taking the good of theoretical clarification to the extremes of make-believe. It is certainly possible to model reasoning as argument, and to make it the case that reasoning is good in the model if and only if the argument that models it is good in that same way. It is also possible to make it true in the model that the model's criteria for argument-goodness are binding on the goodness of the reasonings that arguments model. Then it will also be true that the inferences the model models will be good in the model only when the argument that models it is good in the model. But to cede to these facts the normative authority to direct traffic on the ground is rather breathlessly post modern for the likes of me.

There is a yet more central reason to reject even the modelling version of the inference-argument thesis. Our reflections of late indicate the centrality to reasoning up above of reasoning and knowledge down below. When reasoning and knowing occur down below, they are inaccessible to articulation. They are dark. They are present in an episode of reasoning in causally significant ways, but the reasoning agent is not able to give them voice then and now and, for a good many ranges of cases, ever. But if these episodes of reasoning were modelled as arguments, the arguments could not capture in their premises or presuppositions all the reasoning's constituents. The modelling arguments could not represent a good many of the features of background information causally implicated in the reasoning.

This is an omission the CR approach to reasoning seeks to avert. Reasoning is dominantly a causal response to information, some of which is consciously available to the reasoner, and some of which is not. The human reasoner has two basic kinds of access to this information. He can receive it consciously or he can causally respond to it unawares. There is no theory of argument known to me that treats argument in counterpart ways; and rightly so in my view. It is true that since ancient times logicians have recognized enthymemes, typically characterized as good arguments with missing premises, whose recovery would affect the nature of the tie between premises and conclusion. It would convert the absence of a consequence relation to the presence of one. Accordingly, these missing premises are deemed to occur implicitly in the arguments in which they are expressly absent, and the argument is deemed to be implicitly valid (or some such thing). Over the centuries, much ink has been spilt in an effort to characterize implicit validity and identify the absent premises on which implicity pivots.[159] It would not be far wrong to say that such efforts have yet to meet the standards that a seriously intended logic of argument would require. This is not to say that the logic of enthymematic argument is doomed to fail. But its prospects to date are far from good. Part of the difficulty is that as traditionally understood the missing-premiss component of an enthymematic argument is comparatively small, typically a matter of just one missing premiss. But the implicitness imposed by reasoning's down below components is typically much more extensive. Dark "premisses" vastly outnumber the missing premises typical of enthymemes. We are free, if we like, to model the reasoning down below as enthymematic argument, but such models will significantly under-represent the contributions of down below to reasoning on the ground.

Perhaps the pivotal point, most simply put, is that arguments, in both the wide and narrow senses distinguished in chapter 1, require a degree of premiss-articulation that reasoning on the ground stands no chance of achieving. There is a lesson in all this:

Proposition 4.7
BAD MODELLING: *Do not model premiss-conclusion reasoning in the logic of argument.*

It is advice easily heeded by the CR-model. The CC-offers only doubtful accommodation at best. I don't want to leave the impression that all is now well with the causalization of inference. Nearly everyone accepts that the notion of inference that attract the attention of logicians is a species of belief-change relationships. Inference is the drawing of *conclusions*, and conclusions are drawn from *premisses*. Doesn't the fact that inference is the drawing of conclusions from premises give us some reason to cut some slack to the inference-as-argument model? I think not. Even if we allow that a case might be made for the idea that arguing or case-making is an intrinsically CC-enterprise, reasoning even of the conclusion drawing sort is not. When, then, does reasoning as belief-change become reasoning as conclusion drawing? The answer is when the reasoning is from beliefs the conclusion-drawer takes for premisses. When is that? My answer is contained in clause (2) of Proposition 3.8.

---

[159] See Paglieri and Woods (2011) and citations therein.

Someone takes his beliefs $\alpha_1 \ldots \alpha_n$ as *premisses* for his belief that β when and to the extent that in answer to the question, "What makes you think that?" or "Where did you get that?", the conclusion-drawer would have some readiness to *cite* the beliefs he takes to have played this role.

There is one last point to consider. The reasoning we have been concentrating on here is reasoning of the premiss-conclusion sort. Part of what I have been saying all along is that there is plenty of empirical warrant for saying that reasoning of this sort is a quite early achievement of the human animal with virtually no tutelage, and that by and large it is something that the human animal is comparatively good at and at ease with. Reasoning well in this manner may not be as natural as breathing, but it isn't far off it. The human animal is also a born scrapper. We and our fellows love to sound off. We, like the nattering nabobs of negativism, take pleasure in telling our neighbours how wrong they are about things. When we give voice to these nay sayings there are lots of cases in which we try to draw conclusions from premisses. Perhaps these are the cases which best support the decision to model reasoning as argument. In my opinion they come nowhere close to lending it sufficient support. A large part of the reason why turns on a striking difference between premiss-availability in the two cases. When Sarah is arguing with Harry about whether α is the case, she will fail if her argument depends on premisses that Harry won't give her (not because he's mean, but because he doesn't agree with them). But if Sarah is simply trying to figure out whether α is the case, the premisses her reasoning depends on are virtually always premisses she's prepared to give herself. This alone makes arguing harder to pull off than reasoning, and because this is so, we are less good at arguing well than we are at reasoning well. That is a significant asymmetry. The wedge it drives between reasoning and argument is more than gappy enough to call into question the theorist's decision to model the one as the other.

4.8 *Traditions*

These reflections on argument remind me not to lose sight of the concept-list misalignment thesis. It is a thesis about traditions – the traditional *list* of fallacies and the traditional *concept* of fallacy. I have spent this chapter and the last in recommending against traditional epistemology, particularly the traditions in which justification is a wholly general condition on knowledge. I have said that it is the wrong epistemology for a naturalized logic of mistakes of reasoning. The fallacies traditions are somewhat differently positioned. A good part of what is required of the logic of error is that we make up our minds about the traditional list and the traditional concept. I am inclined to retain the concept and jettison the list. If I am right about this, then we could see that the way in which the tradition conceptualizes fallacy is ungenerous to its intended extension. The tradition I want to keep in play is the tradition that is ungenerous to the tradition I want to jettison. So it behooves me to be clear about what these are. I want, however, to guard against a misconception. When I say that my preferred course is to jettison the list, I intend nothing more than its abandonment as a candidate for membership in the concept's extension. There is nothing in this exclusion that demands a more sweeping condemnation. It remains entirely open to find the eighteen to be, as I myself do, extremely interesting features of reasoning and argument, and perfectly fit objects of theoretical attention. It also leaves it open that much of what is currently said about them in the argumentation community is both right and enlightening. This could be true even if it were neither enlightening nor right to say that they are fallacies.

By "traditional" I mean what has come down to us over the years in variations of some initial starting point. The starting point for the concept of fallacy is Aristotle. In 1970, the concept whose theoretical neglect Hamblin lamented was the traditional one. And since it had a dominant presence in 1970, it qualifies as the modern concept as well. A theory of the traditional concept must pay heed to its historical lineage, but it need not be the case – indeed hardly ever is the case – that when the lineage is a long one, a good theory of a traditional concept is unchangingly the same as the theory of its originating idea. So it should not be surprising that traditional fallacy theory is not Aristotle's. It started with Aristotle, and was passed down to us in the form with which Hamblin collided in 1970, and has lingered to the present day.

In *On Sophistical Refutations,* a fallacy is an argument which appears to be a syllogism but is not a syllogism in fact. Aristotle's list of these fallacies gives a "gang of thirteen". They are, as we saw: ambiguity, amphiboly, combination of words, division of words, wrong accent, the form of expression used, accident, *secundum quid, ignoratio elenchi* (misconception of refutation), begging the question, consequent, non-cause as cause and many questions. I say again that Aristotle's list is not the modern list, and that Aristotle's conception is not the traditional concept. But in each case there is enough of Aristotle preserved in both the traditional list and the traditional concept to justify our thinking of them as having descended from these origins. Aristotle's concept is preserved in the traditional idea of a fallacy as a piece of reasoning that appears to be good but isn't good in fact. Aristotle's list is preserved outright or in some recognizably similar way by the traditional list as regards the following: ambiguity (cf. the modern equivocation), combination of words (cf. the modern composition), division of words (cf. the modern division), *secundum quid* (cf. the modern hasty generalization),[160] *ignoratio elenchi* (cf. the modern straw man), begging the question (cf. the modern begging the question), and many questions (cf. the modern many questions).

The logician's notion of fallacy is a particular case of what is arguably the dominant usage in the population at large. By the conventions of everyday conversation, "fallacy" denotes a common misconception about something, that is, a false belief entrenched in what passes for common knowledge. ("Everyone knows that handling a toad causes warts.") This gives us two features to take note of. One is that fallacies are errors and the other is that fallacies are widespread, trending towards the universal. Fallacies in the logical tradition are – with possible exceptions already noted – fallacies of reasoning, not fallacies of false belief. But they clearly imbibe the two characteristic features of the core common sense concept.

The concept-list misalignment thesis is offered as part of an answer to Hamblin's Question as to why logicians have yet to produce a deep theoretical consensus about fallacy. I said before that there are numbers of ways in which the misalignment thesis could go wrong. I might have got the traditional list right and the traditional concept right, but be wrong about the non-instantiation claim. I might have got the traditional concept right, but be wrong about the traditional list. I might be right about the list but wrong about the concept. I am not much minded to think that I've got the list wrong. I said in chapter 1 that "the gang of eighteen" is a flexibly convenient designation, unaffected by the variations one finds between Hamblin's own list, and those to be found in

---

[160] In *On Sophistical Refutations, secundum quid* is the fallacy of "omitting a qualification", as in the inference of "Mr. Mandela is white" from the premiss "Mr. Mandela is white-haired" (*Soph. Ref.* 1955). So construed, it has no counterpart among the gang of eighteen, except possibly for composition in some cases. However, in the *Rhetoric, secundum quid* is something quite different, resembling closely the modern fallacy of hasty generalization (*Rhet.* 24, 1401^{b}, 30-31).

the primers of logic that triggered his concern in the first place, as well as in later textbooks, many of which were written at least partly in response to Hamblin's criticisms. Variations there clearly are, but there is a large commonality preserved in the majority of these listings, and the gang of eighteen is a credible representation of that core.

The traditional conception is a different matter and a more challenging one. As it has evolved historically, the logician's standard view of fallacies is that they are a special class of errors of reasoning, errors that are frequently characterized as "logical".[161] As already remarked, they are errors which are common enough to qualify for a kind of universality. They are also attractive; they are errors towards which people in general are drawn. They are also bad habits. They are difficult to break and so, in a sense, are incorrigible. A handy acronym, EAUI, picks up the initial letters of the adjectives "error", "attractive", "universal" and "incorrigible". "EAUI" doesn't come trippingly off the tongue letter by letter, but there is a perfectly pronounceable elision: "Yowee".[162]

The EAUI conception is present in, or intimated by, well-known characterizations of fallacies. Scriven (1987) observes that fallacies "are the attractive nuisances of argumentation, the ideal types of improper inference. They require labels because they are thought to be common enough or important enough to make the costs of labels worthwhile." (p. 333) Govier (1995) writes to the same effect: "By definition, a fallacy is a mistake in reasoning, a mistake which occurs with some frequency in real arguments and which is characteristically deceptive." (p. 172) Such, says Hitchcock, is "the standard conception of fallacy in the western tradition." (Hitchcock, 2006a, p. 1)

---

[161] See, for example, Finocchiaro (1987), p. 268: "[F]allaciousness is by definition the basic logical flaw of reasoning"; and Johnson (1987b), p. 246: It is one of the "governing presuppositions of FT [Fallacy Theory]" … that … "people do make logical mistakes in reasoning and argument"; and Finocchiaro (1987), p. 265: "[A] fallacy is a particular type of logic error."

[162] In recent writings, e.g., Woods (2008b), I expanded the acronym by prefixing the letter "B", suggestive of the assumption that fallacy-making is a *bad* thing. Badness is certainly part of the traditional understanding, but "BEAUI" is not as phonetically attractive as "EAUI". I will stick with "EAUI" and will assume that it tacitly embeds the badness condition.

We might take it as obvious that a satisfactory theory of fallacies will be one that gives principled accounts of the four defining features of them, – five if we include badness. Even so, notwithstanding the ebb and flow of its 2400-year presence on the research agenda of logic, we have as Hamblin notes nothing like a settled account. This should strike us as curious. For are not the defining traits of fallacies intuitively clear and rather commonplace concepts? Can't their connotations readily be specified? And since the connotation of "fallacy" is just the composition of these five, why shouldn't it also be possible to give it articulate expression?

The traditional conception of the fallacies is not the only conception of them to have achieved contemporary recognition and support. In chapter 15 I will examine two objections against the EAUI characterization: first, that what I take as the traditional concept of fallacy is not in fact the traditional concept; and, second, that regardless whether the traditional concept is or is not what I take it to be, the EAUI notion is not a fit target for a theoretically robust account of fallacy. But for the present I want to attend to a slightly different objection of my own: whether the traditional conception or not, the EAUI conception is not a sufficiently *clear* notion of fallacy to bear the weight of the concept-list misalignment thesis. If the EAUI conception is right, it takes quite a lot for a piece of reasoning to be fallacious. It must be an error that is attractive, universal, incorrigible and bad. Fallacy's five defining features gives to a piece of reasoning five defences against the charge of fallaciousness. Indeed it gives a piece of erroneous reasoning four methods of defence. No doubt this will give some readers pause. How can a piece of bad reasoning not be fallacious? My answer to this is that not being a fallacy is not enough to make a piece of reasoning good. Fallacies are errors of reasoning with a particular *signature*.

Very well, then. Suppose this is right. Suppose that the EAUI conception imposes on fallacies these five conditions. In so saying, attributions of fallacy are encumbered with a stiff burden of proof. The attributor must show that a putatively fallacious piece of reasoning has all five things wrong with it. The heart of the misalignment thesis is that, in one way or another, the eighteen don't meet these conditions. How is this claim to be evaluated? I say that it can't be evaluated until we get ourselves a lot clearer than we now are about how the EAUI conditions are to be understood.

140

On the traditional conception, there is a straight line from fallaciousness to badness. Fallacies are errors and errors are bad. So fallacies are bad. Our present reflections are a discouragement of this view. Here is why.

Consider our earlier distinction between inaccuracy and inaptness. Then there are two ways in which a piece of reasoning could go awry. In the inaccuracy camp, the consequence the reasoner draws might not be a consequence that its premisses have there might be implausibilies or irrelevancies in one's premiss selection; there might be relevant considerations which premiss selection overlooks; and there might be consequences which the premisses have and are relevant to the cognitive ends presently in view which, through inattention or carelessness, the reasoner fails to spot. In the inaptness camp, there are correctly reasoned consequences which are irrelevant to those same cognitive ends,[163] or which take too long to arrive at for their timely realization. The reasoning might have been shaped by inductive sampling procedures that maximize the avoidance of error, but which again take too long for the timely effectuation of the reasoner's goals. So it bears repeating that

Proposition 4.9
MAKING TOO MUCH OF ACCURACY: *Accuracy maximization is not an apt general objective.*

What our reflections show is that to get the badness condition right, it is necessary to permit the distinction between consequence-having and consequence-drawing[164] to influence our applications of it. There is no doubt that Aristotle honoured the distinction, though not in my words for it. It is also respected in a good many contemporary discussions of the various fallacies. But many of these same theorists do not honour in their discussions of the *definition* of fallacy. This has given rise to no end of confusion about whether begging the question, say, is or is not an error of logic. So the moral of the story is that the badness

---

[163] This is nearly enough, Aristotle's *sophisticus elenchus* and the modern red herring, or irrelevant conclusion, fallacy,

[164] Or in deference to our doubts about nonmonotonic consequence between conclusion-supporting and conclusion-drawing.

condition won't be properly understood unless the distinction between having and drawing is taken due notice of.

4.10 *The universality condition*

No one doubts that the universality condition is rather expansively formulated. Certainly it cannot bear anything like the logician's meaning of the word "universal". Something softer is intended. What the requirement suggests is a breed of generality that is not well-captured by the universal quantifier. It is generality of a kind that we associate with generic utterance. In this it resembles "Ocelots are four-legged", a true generalization that can remain true even in the face of true negative instances. Generic generalizations express what is characteristically the case. They do not express – and should not be judged as expressing – what is invariably the case. Part of what is meant by the universality of fallacious reasoning is conveyable generically.[165]

Proposition 4.10a
UNIVERSALITY AS GENERICITY: *It is* characteristic *of beings like us to commit such errors. They are errors that are committed, not always, but with a* notable frequency.

Part of what the universality condition wants to emphasize is that fallacy-committing is a substantially cross-typical affair. Although the actual record will reveal some fluctuations in rates of commission, they are not linked significantly to sex, adult-age, nationality, race, and the like. A theory of fallacious reasoning is required to attach this presumption to the generic character of the universality claim. One does this by taking due care in identifying the reference class of the generic assertion. This is done as follows: It is characteristic of (adult) human reasoners – not elderly reasoners, not Finnish reasoners, not Tory reasoners – to commit these errors of reasoning with a notable frequency. This leaves two tasks still unperformed. One, as we said, is to explain their "notable frequency". The other is to probe the question of why it should be characteristic of us to commit them. Whatever the answer may be, it can be expected that it will turn out to have some bearing on how to understand the attractiveness condition.

---

[165] Recall our doubts in chapter 3 about the brittleness of definitions by way of necessary and sufficient conditions specified with universalist intent.

What is a notable frequency of error-making? If we attended to the empirical record of the reasoning behaviour of individual agents, it would become quickly apparent that, with the gang of eighteen as our guide, beings like us hardly ever commit *them*. That is to say, their frequency in relation to conclusion drawing in general is exceedingly low. Similarly, for the premiss selection tasks falling within logic's ambit – inconsistency-management, irrelevance-screening, and the like – their frequency is also low. What we mustn't overlook, however, is that it is possible for errors that aren't very often committed to be committed nevertheless with a *notable* frequency.

Consider any situation in which conclusion-drawing or premises-management is a contextually appropriate thing for an agent to do. Appropriateness is influenced by his cognitive target at the time. It is also influenced by the set of premises or assumptions that present themselves for conclusional plumbing. Let us say that taken together these factors constitute an *occasion* for reasoning. Now take any of the eighteen – the *ad hominem* say – assuming for now that it really is fallacious. The vast majority of conclusion drawing situations in which you find yourself is not an occasion for reasoning in any way that instantiates the *ad hominem*. I think that we can say without much further ado that the notability of the frequency of a practice that hardly ever happens is an occasion-dependent trait. Another example: You don't commit the *ad verecundiam* fallacy except on occasions on which someone – some source – has told you something. You don't commit the *ad populum* fallacy except on those occasions on which we note that a given view is widely held. Or consider again our old friend: You don't commit the fallacy of hasty generalization in contexts that aren't occasions for it. Generalizing hastily requires that you regard the priors from which the conclusion in question is drawn as population *samples*. Accordingly, the universality of a reasoning practice is a matter of the frequency of its commission in relation to the frequency of its occasions.

An occasion for drawing a certain kind of consequence in a given way is occasion for "getting it right" or "getting it wrong". In many instances, as with a hasty generalization from an unrepresentative sample, the right move is not to generalize from it at all. What makes generalizing from it the wrong move is that, in those circumstances, the generalization can only be hasty. Or consider again the *ad verecundiam*. Suppose someone tells you what you already know with a serene and unshakable confidence. (Perhaps he tells you the date of your birthday). This would not be occasion for you to infer the date of your birthday from your informant's sayso, unless perhaps you had forgotten it. The question of fallacious reliance would not here arise. By and large, what is occasion for the commission of a given fallacy is *misoccasion* for commission of the others. Concerning our present examples, occasions for generalizing from samples are not occasions for deducing a component sentence of a conditional. So it bears tellingly on the fact that fallacies are hardly ever committed that most reasoning occasions are misoccasions for them. Accordingly,

Proposition 4.10b
OCCASIONED FREQUENCY. *What counts for the universality of a reasoning practice is relative frequency of occasions to commit it, or, as we shall say, its* occasioned frequency, *the frequency of its commission relative to occasions to commit it.*[166]

Nothing that is known empirically of the individual's cognitive make-up disturbs the assertion that the occasioned frequency of the eighteen is low. So we have not yet managed to acquire a serviceable understanding of the presumed universality of fallacious reasoning. Perhaps the difficulty might be overcome in the following way. Reasoning of the conclusion drawing sort is always occasioned. We might say that it is a principal function of an occasion for reasoning that it implicitly shapes a reasoner's space of reasoning options. Given that reasoning is a response to an occasion to reason, each case of reasoning behaviour involves the (typically dark) selection of an option from the space of reasoning strategies prompted by the agent's occasion. Since each such space contains the possibility of erroneous reasoning, we may speak in the abstract of an agent's rate of success over all choices made on all occasions presented. In populations that survive and prosper and occasionally fill up the Tate and the Prado, the frequency of success to error is satisfactory overall. So the frequency of error is correspondingly and *comparatively* low, say n. Then a frequency of error is "notable" to the degree that it exceeds n. Every space of reasoning options contains a

---

[166] Of course, I am taking liberties with the example of affirming the consequent. Comparatively speaking, not only is the real life human reasoner not often in an occasion to reason deductively, but when he is, and when he makes mistakes, affirming the consequent is hardly a dominant presence. Working mathematicians are routinely in the occasion of deductive misstep. It is said that on a good day a mathematician will make a dozen or more deductive errors. But they won't *ever* be the error of affirming the consequent!

wrong option. Wrong options are exercised with a frequency of n over all occasions. And, according to the traditional approach, they are exercised with a frequency greater than n when the option space contains any of the eighteen.

This does indeed make the universality condition something of an overstatement. It can be modified as follows:

Proposition 4.10c
UNIVERSALITY REFINED: *It is characteristic of individual reasoners to err in the manner of the eighteen more frequently when they have occasion to than to err in other ways when they have occasion to. The lion's share of universality is contributed by the first clause. The propensity to err in these ways is found in virtually everyone. But the frequency of error in the manner of the eighteen is still rather on the low side even in the circumstances that occasion it. So a fallacy is an error in reasoning that virtually everyone is disposed to commit with a frequency that, while comparatively low, is nontrivially greater than the occasioned frequency of their reasoning errors in general.*

Proposition 4.10c offers some useful advice for fallacy identification. Let r be a kind of reasoning concerning which the question of its fallaciousness has arisen. Then a question to ask is "Is r an error when there is occasion to commit it?" Do *not* ask: "Can you think of cases – even lots of cases – in which r would be an error *were* it to be committed?

4.11 *The attractiveness condition*

Aristotle's core insight was that a fallacy is a bad argument that looks good. More generally, a fallacy is an argument bad in a certain way that looks good in that same way. It may well be that in the general run of cases a piece of reasoning's looking good is implicated in the attractions it holds for us. It is certainly not a general condition. There are situations – addictions comes to mind – in which people are drawn to practices whose badness is manifest. Perhaps there are also pathological cases of reasoning in which a reasoner reasons against his will, fully seized of the badness of what he is doing or what is happening to him. In earlier chapters we noted the presence in a good many of the going theories of human performance – rational decision theory, theories of heuristics and bias, Bayesian epistemology, classical belief revision theories – of a defective-rationality assumption. Unless this is just a carelessly overstated way of ascribing errors of reasoning to these alleged misperformances, the accusation is a heavy one, carrying clear implications of pathological import. Judged by the norms of ideal models, ordinary everyday reasoners are with a notable frequency off their heads. There is value in repeating that the EAUI conception seems not to embed this assumption of pathological defect. If so, decision theory and the rest are operating with a tougher notion of fallacy and, correspondingly, with a gloomier appreciation of human misperformance, than one finds in the traditional idea. EAUI's attractiveness condition neither attributes nor presupposes pathological attachment to the fallacies. Accordingly, whatever EAUI's attractiveness turns out to be, it is not *that*.

Whether or not looking good admits of robust generalization, the link between the good-looking and the attractive is far from analytically guaranteed. Something is attractive only if it has the appearance of being good *for* something. There is a perfectly good proof of Lindenbaum's Lemma. Suppose you knew this. Suppose you could see that the Lemma is true. Suppose now that you wanted to prove the completeness of first order logic, and you see that Lindenbaum's Lemma is going to help you get there. I would say that in this second instance the proof of Lemma was attractive to you and that, in the first, it need not have been. In a rough and ready way, something is attractive to you if it appears to answer – or advance – your interests in some way. There are lots of cases in which being something that looks right in a certain way is not a condition on its seeming to answer to your interests in a certain, although not necessarily the same, way. That is, whereas a thing's rightness need not be an interest-linked property, its attractiveness is.

I think that this is right as far as it goes, namely not very. A little earlier I said that the fact that fallacies are such as to make it typical of us to commit them with a notable occasioned frequency is something that matters for their attractiveness. Typicality statements have the character of generic claims. It is typical of ocelots to be four-legged. It is typical of us to commit certain types of error when the conditions are right for it. The leggedness of ocelots is a fact about ocelots. The proneness to error is a fact about us. This takes us straight back to the design question. The four-leggedness of ocelots is a matter of how ocelots are built. The propensity to err is a matter of how we are built. Suppose that this were actually so. Then the attractiveness condition could be explained in a non-trivial, albeit partial, way by saying that the propensity to make errors is something for which beings like us have been built. This is an arresting suggestion. One might have thought that reasoning fallaciously stands to reasoning as Ozzie the three-legged ocelot stands to ocelots, *anomalies* in each case. One might have thought that errors of reasoning occur when the reasoner's reasoning apparatus is having an off day. But if it is typical of us to commit errors of these sorts, if committing them is what we've been built for, there is another arresting idea that we cannot long postpone. It is that, when the conditions are right and the errors are of the right sort, EAUI's attractiveness condition postulates, or at least suggests, adaptive significance. Accordingly

Proposition 4.11
ATTRACTIVENESS RECONSIDERED: *A not inconsiderable explanation of our attraction to the eighteen is that their occasioned performance is not typically the wrong way to reason. For example, it is not typically the wrong way to reason if it is part of our biological constitution to learn from experience by trial and error.*

COROLLARY: *This lends a degree of encouragement to proposition 1.2b, the cognitive virtue thesis. It asserts that to a nontrivial extent, the items on the traditional list are cognitively virtuous ways of reasoning.*

4.12 The incorrigibility condition

The attractiveness condition imputes to fallacious arguments and inferences the appearance of goodness. It is assumed moreover that the property of apparent goodness is not erased by successful diagnoses of it. It is possible that formal tutelage, or disciplined faithful practice, might occasion *some* dissipation of such appearances, but it is assumed even so that this would leave the presence of it undisturbed, and would affect only its intensity. There is an important literature that lends some support to these assumptions. Some of the best of this work, by Deanna Kuhn and Tim van Gelder, indicates that formal instruction in critical thinking tends not to improve students' levels of performance in this regard.[167] Some

---

[167] Kuhn (1991) and van Gelder (2005). For his part, van Gelder shows a degree of optimism. He conjectures that a software package known as "computer-supported argument mapping" might be refined in ways that work improvements

commentators infer from this the pedagogical inadequacy of courses on critical thinking (some of which are indeed truly awful). But it is also possible that something like the incorrigibility condition is here in play.

If it turns out that detection does not erase the appearance of rightness, that diagnosis does not cancel attractiveness, then we can conjecture that incorrigibility is the natural outcome of that state of affairs. If the attractiveness of a fallacy is what induces us to commit it in the first place, then why wouldn't its continuing attractiveness draw the same response? This gives us a further question to ask about attractiveness. What is it about the appearance of rightness that explains its causal allure even after detection?

The concealment-detection problem is a pivotal matter for any theory of erroneous reasoning. It is a matter of the dynamic interplay in contexts of error of apparency and inapparency. For large classes of cases, the inapparency of error is a condition of its commission, just as its apparency is a condition of its detection and correction.[168] There are examples galore in which the difference between commission and correction is easy to spot, as when new information erases the old. Other cases are harder to parse. Fallacies lead the list. An error's detection is typically sufficient for its non-recurrence. Once you see the error of your ways, your ways are mended accordingly. But if the error is a fallacy, and if the traditional conception of them holds sway, it is simply not true that detection stifles recurrence. Fallacies are incorrigible. Rates of recidivism are high.

Like it or not, the fallacies' record of re-offence calls back into play the possibility of reasoning pathologies. A pathology is a sickness. A mental pathology is a sickness of the mind. If in their incorrigibility the fallacies were evidence of pathological significance, then, contrary to our remarks just above, fallacy-makers would be off their heads. Since fallacy-makers are all of us, the whole species would stand convicted. I have a certain fondness for my species, but I am no chauvinist. Let the chips fall where they may, say I. There are several respects in which human misperformance is easy to understand. We can remember only so much at a time, handle only so much complexity at a time, draw only so

---

into ordinary critical thinking by transforming the constitution of the knowing agent, by changing his mind (literally). See here The Harrow Technology Report, http://www.The Harrow Group.com

[168] There are exceptions. Sometimes you know the error of your ways as they happen. Think of the champion skater who's having an off-night or the student who's blowing a final exam.

many consequences at a time, check the consistency of only so many premisses at a time, jump only so high at a time – the list goes on. Misperformance at the edges of literal capacity augurs ill for rehabilitation on any scale that matters. Stretched limits today are stretched limits tomorrow.

There are well-known theological traditions in which moral error attracts high rates of recidivism. There are some sins which we seem not to be able to keep from committing. The holy writings have it that a just man sins twelve times a day. But recurring sinfulness is usually chalked up to weakness of the will, and ongoing malfeasance is not untypically attended by guilty awareness. There is in this latter point a striking difference between sin and fallacy. It is that fallacies-recidivism is innocent and re-sinning is not.

When you see your first ocelot you know that ocelots are four-legged. Yours is a one-shot generalization, and you are right. Ocelots are four-legged. Even so, the tradition condemns the reasoning for haste. Suppose that, on having it pointed out to us, you accept the tradition's verdict. Suppose you resolve never to re-commit the ocelot error. This raises two interesting questions. How are you going to know that ocelots are four-legged – indeed, how does anyone get to know it? And assuming your unswerving fidelity to this pledge, what are you going to do when you spot your first lemur? I have no idea of how the first question might be answered. The second is easier. The answer is that you will see that lemurs too are four-legged. So what is the point of your resolve to suppress the same move with respect to ocelots? Accordingly, why don't we say

Proposition 4.12
INCORRIGIBILITY RECONSIDERED: *A not inconsiderable best explanation of our post diagnostic recurrence to the eighteen is that their occasioned re-commission is not typically the wrong way to reason.*

Three chapters ago, when I first floated the concept-list misalignment thesis, I said that I would try to keep a misalignment confirmation tally in which, one by one, the items on the fallacies list would be shown to be at least problematic instantiations of the traditional concept of fallacy. Since then other matters have deflected me from this course. Our present reflections are a context in which to get the implementation process started. In the little that remains of this chapter,

I'll content myself with a general observation about the eighteen. In the chapters to follow the misalignment tally will begin in earnest.

Here is the general comment. In the way we've developed it here, the traditional conception of fallacy creates a large burden of proof for sponsors of the eighteen. To convict a traditionally listed item of fallacy as traditionally conceived the ascriber must show that it instantiates reasoning of type which, when there is occasion to indulge it, is performed by the population at large with a frequency that exceeds that population's overall error-commission rate, and is subject to post diagnostic recurrence in that same population with something resembling that same notable frequency. It is a stiff burden, concerning which I shall content myself for the present with a simple admonition: It would be a serious mistake to underestimate it.

4.13 *The efficiencies of ambiguity*

Language plays a massive and indispensable role in the cognitive economies of those who have one. Languages are the principal means of knowledge-distribution. It is an interesting fact about human language that for it linguistically expressible meanings there is no relation that maps them one-to-one to the words that mean them. On the face of it, this is a telling shortfall, driving an unacceptably aggressive wedge between what can be thought and what can be said. It is not as bad as all that. Much of that shortfall is to be recovered by the simple device of ambiguity. An ambiguous word can express more than two meanings. A single word can express even more meanings than that. There are more ambiguous words in English than not. There is some good ancient advice about this: "Intelligenti pauca". Few words suffice for him who understands.

The value of ambiguity is economic. It endows a speaker with a large expressive capacity supported by a much smaller lexicon. Saying what you mean in English makes a lesser demand on lexical memory than saying what you mean in an unambiguous language (if there were such a thing). Efficiencies of like importance also extent to English's syntactic ambiguities. Ambiguity occurs in speech with a notable frequency. It likewise occurs in belief and belief-revision contexts and in the premiss-conclusion contexts that are our principal focus here. Whenever some statement is concluded from some premises, there is some positive likelihood that in the sentences expressing those premises and that conclusion an ambiguous term will occur or an ambiguity-making underlying structure. When in those premisses and that conclusion there is a further occurrence of a precedingly present ambiguous term, that would be some occasion for the second occurrence to have a sense different from the sense of its first occurrence. The traditional fallacy of equivocation is an informal way of committing the formal fallacy of four terms, typified by the inference

1. A bachelor is a man who has never married.
2. A bachelor is a holder of a university's first degree.
3. Mary is a bachelor.
4. So Mary is a man who has never married.

Of course, the example is more blatant than is usual in actual practice. Normally, premiss (2) wouldn't be stated but assumed. The fallacy of equivocation is a fallacy in the traditional sense only if it happens with a notable occasioned frequency that inferences containing two or more occurrences of an ambiguous term are inferences that instantiate the fallacy of four terms. There isn't the slightest reason empirically to suppose that this is true. So let us put it on the record, and add it to the misalignment tally, that

Proposition 4.13
NOT A FALLACY: *The traditional fallacy of equivocation doesn't instantiate to the traditional concept of fallacy.*[169]

The four-terms version of equivocation is a particular example of a more general pattern of ambiguity mismanagement. Its more general form is the illicit exploitation of the ambiguity of a term in a given context of reasoning or argument. It occurs when the (apparent) correctness of the reasoning or argument depends on the presence of a term shaving different senses from occurrence to occurrence in that same context. When this happens, let us say that the context is one on which its ambiguity *spreads*. In speaking of the traditional handling of the equivocation fallacy – the four-terms version – I said that it didn't meet the burden of proof required for indictment – it didn't meet the tradition's defining conditions on fallacy. I have also said that I am far from thinking that the traditional fallacy has a null extension, notwithstanding that the traditional eighteen are not to be found there. This would be a good point

---

[169] The same objection applies to inferences containing syntactically ambiguous sentences. I leave the details as an exercise.

to see whether we might get something useful from the joint consideration of these two claims.

Occasions of equivocation are contexts of reasoning or argument in which ambiguity spreads. Equivocation is the illicit exploitation of that fact. Equivocation is fallacy if it happens with a notable frequency in contexts of that sort. It is my position that there is nothing in the empirical record that lends credible support to the notable frequency claim. It is true that English is rife with ambiguity as a matter of course. But I am saying that, by and large, beings like us are competent managers of this rifeness.

By and large is one thing. Sometimes is another. I am open to the possibility that there are contexts in which such ambiguities are mismanaged with a notable frequency. What would bring such contexts about? What would be natural habitats for them?

It is clear that the likelihood of ambiguity-spread rises with the size of the context in question. The larger the context the greater the presence of ambiguity, and the more taxing the business of ambiguity management. It would also appear that a further contributory factor is the extent to which the contexts in question are case-making contexts. Perhaps this is most true of contexts where counterexamples are advanced against received opinions or support is marshalled for implausible conclusions. We might note in passing that effective ambiguity management embodies two agendas. One is the negative agenda of avoiding the wrongful exploitation of ambiguity. Another is the positive agenda in which the attribution of ambiguity is put to good tactical use. A classic example of the latter is compatibilism in the free will-determinism debate and a more recent one is the pluralistic response to conflicts in the philosophy of logic, for example, the conflicts about *ex falso quodlibet*. Come to think of it, philosophy is the quite general the business of rebutting received opinions and supporting implausible (or shocking) ones. Like it or not, in adversarial contexts such as philosophy (and law and labour-management negotiations) there is an abiding inducement to game the system. This is so, never mind the steadfastness of a practictioner's determination to resist it. Part of what makes for this is the size of the ambiguity-spreading context and the attendant difficulty of keeping the shifting ambiguities clearly in mind. A related factor is the force of the drive to prevail over adversaries.

A good part of epistemology's history charts the course of the justificationist wars. It is a huge context and has attracted a great deal of ambiguity. The history of the internalist-externalist competitions bristle with ambiguities mismanaged, as our earlier discussion attests. A case in point is the Gettier counterexample. In the *Analysis* paper, there are three examples of the JTB-view against which his counterexample is intended, one from Plato, a second from Chisholm and the third from Ayer. Gettier ascribes to Plato the view that the J-condition on S's knowledge is "S is justified in believing that P".[170] He ascribes to Chisholm "S has adequate evidence for P",[171] and to Ayer "S has the right to be sure that P".[172] Gettier thinks that his counterexample strikes equally against these three justificationisms. For this to be true, it must be the case that by "justified" Plato, Chisholm and Ayer mean the same thing. It is far from clear that they do. Plato's third condition demands for knowledge a *logos*. The word "logos" is heavily ambiguous in Greek, but here it means something like a theoretical demonstration. Chisholm's third condition demands evidence, but doesn't say whether evidence in his sense has the force of a supporting theory. Ayer's third condition requires that the would-be knower of α have a right to be certain that α. It is possible that Ayer himself thinks that in the absence of a theoretical demonstration the right to be certain is extinguished. If so, it is regrettable that he didn't say so. But there is no doubt at all that "has the right to be certain" has a significantly larger extension in English than "admits of a theoretical demonstration". The same is true of "evidence". While Gettier's example may be telling against Ayer, it is clearly not telling against Plato and it is doubtful that it succeeds against Chisholm. So we may say that, as applied to Plato, the argument mismanages its ambiguity-spread.

The question is whether it is a fallacy, that is, whether philosophical disputation is a natural occasion for it to occur and whether it occurs there with a notable frequency? Some people would be inclined to think that the answer is Yes each time. If they were right, grounds would exist for saying that equivocation stands a decent chance of being a fallacy in the philosophical population. I myself don't think so. It may be true that philosophers and people of adversarial profession more generally mismanage ambiguity-spread with a frequency that outpaces the population at large. But it can be doubted that even there the ratio of commission to non-commission exceeds the norm in a suitably notable way. Besides, the frequency of occasions to mismanage ambiguity is a function of the frequency of ambiguity-spreading contexts in the thought and discourse of the population at large. Those contexts – those occasions

---

[170] *Theaetetus,* 201
[171] Chisholm (1957), p. 16.
[172] Ayer (1956), p. 34.

– occur with a notable frequency in the population at large. But their mismanagement does not occur with a notable frequency in the population at large. So, again, equivocation doesn't qualify as a fallacy in the traditional sense.

CHAPTER FIVE

BELIEVING

"Fere libenter homines id quod volunt credunt."   Julius Caesar

"The action of thought is exerted by the initiation of doubt and ceases when belief is attained; so that the production of belief is the sole function of thought."   Charles S. Peirce

"The greatest obstacle to discovery is not ignorance – it is the illusion of knowledge."   Daniel J. Boorstin

5.1 *Belief-change*

Logicians have long recognized a distinction between categorical, conditional and hypothetical reasoning. Roughly speaking, categorical reasoning exhibits the form ⌐Since α, β⌐. Conditional reasoning exhibits the form ⌐If α then β⌐. Hypothetical reasoning exhibits the form ⌐Since α, it is reasonable to suppose (conjecture, hypothesize) that β⌐. Categorical and hypothetical reasoning is a matter of drawing consequences. Conditional reasoning is a matter of spotting consequences, not drawing them. Categorical reasoning maps belief to belief. Conditional reasoning engenders implicational belief. Hypothetical reasoning maps belief to supposition (conjecture, hypothesis). Since the belief is a constituent of reasoning in all these forms, it is only to be expected that it will have a role to play in the differentiation of right and wrong reasoning. A logic of reasoning should have something to say about this. Belief sits the saddle of reasoning both lightly and darkly, with the dark outpacing the light. Dark belief is belief down below. Light belief is belief up above. Roughly speaking, a belief is light when a condition on having it is one's awareness of having it. It is sometimes supposed that the distinction between light and dark belief coincides with the distinction between manual mode belief and automatic belief. This is a mistake. The manual-automatic duality is orthogonal to the light-dark pair. Most belief, light or dark, is on automatic mode.

In chapter 2 we discussed the Can Do Principle. This is the principle that instructs us to try to solve our theoretical problems in frameworks that are up and running and successful, using methods that are tried and true. I said that, provided it does not over-reach itself, Can Do gives methodological guidance of the first importance. For the better part of twenty-five years, belief-change theories have been a prominent part of the research programmes of AI[173] and formal epistemology (FE).[174] If reasoning involves belief-change, why wouldn't a theory of reasoning be an adaptation of a theory of belief change? Wouldn't a theory of belief-change be the natural place to look? Isn't this what Can Do would suggest? Similarly, belief has been the focus of attention of modal logicians for fifty years. Would a belief-logic also be a natural place for a logic of reasoning to seek instruction? Wouldn't Can Do also direct us there?

This is not the course that I am going to take. As will shortly become apparent, the concept of belief required for the account of error is not well catered for by these accounts. The purpose of the present section is to make a glancing review of the main theories of belief change and belief logics, pointing out reservations as we go.

A central challenge for AI is to provide an account of the competency of an artificial agent in amending its beliefs when its situation changes or when its knowledge of an unchanging situation increases. In this approach, a belief is modelled as a sentence that is believed true or false or is subjectively uncertain. It is postulated that an agent's beliefs are always consistent, and that the main part of the belief-change mandate is the preservation of consistency in the face of new information. Classical belief change theories have a mainstream cast to them, made so by the fact that the principal instruments for the management of belief-change are consistency and consequence, defined much in the same way as they are in classical deductive logic in variations thereof.

In addition to the classical AI/FE theories, there are psychological accounts of reasoning which have a bearing on belief-change. For the purposes of this section, I'll confine my remarks to the three dominant examples of the AI/FE orientation. They are the *syntax-based* approach, the *model-based* approach and the *hybrid* approach – and the two main orientations developed by psychologists. These are the *proof-theoretic* approach and the *mental models* approach.

---

[173] Important early contributions are the truth-maintenance system of Doyle (1979) and the data base-prioritization approach of Fagin, *et al.* (1983).
[174] The early philosophical contributions of note include Harper (1976), Levi (1977, 1980). For recent work, see Hendricks (2006), Symons (2011), and Furhmann (2011).

In the syntax-based AI approach, an agent's beliefs are made up of sentences, and his background beliefs likewise are sets of sentences. A belief-change is one in which an agent's current beliefs together with some background information deductively imply some new sentences. In this model it is not required that an agent actively draw any of these consequences, although in most versions immediate or obvious consequences would likely be drawn. On the other hand, in model-based accounts beliefs are equated with models, which are semantic interpretations – functions – making some designated group of beliefs true. Syntax-based theories deal with explicit beliefs. Roughly speaking, a consequence of an explicit belief is drawable only when it itself is an explicit belief. In model-based theories, beliefs include implicit beliefs. This motivates the "rationality postulate" that an agent will believe every deductive consequence of anything he believes. No one disputes the radicalness of the closure postulate. Human beings at their best come nowhere close to implementing it – they fall infinitely short of it – yet is widely held in the AI and FE communities that not only does the closure idealization simplify the account of belief-change, but it would be next to impossible to produce a decent formal theory without it. Syntax-based theories also reject the principle that if among the candidates for admittance to an agent's beliefs there are some logically equivalent sentences, then it is a matter of indifference as to which of them is chosen, hence that mere syntactic difference is irrelevant to belief-change. Model-based theories accept this principle.

The hybrid approach – also called the "theory based" approach[175] – blends various features of the syntax-based and model-based orientations. The best known hybrid theory, and in may ways still the core account, is the AGM model, so-named after the seminal paper of Carlos Alchourrón, Peter Gardenförs and David Makinson (1985). As with syntax based theories, in the AGM-model a belief state is a theory.[176] But, as with model-based theories, AGM-belief is closed under consequence. Hybrid theories also accept the principle of the irrelevance of merely syntactic differences to belief-change.

As we saw, a doxastic logic is a modal logic of belief-sentences. A belief-sentence is a sentence prefixed by the belief-operator B, which

---

[175] Elio and Pelletier (2008).
[176] In certain FE approaches, theories in this sense are "belief-sets" or "knowledge-sets" (and sometimes "corpora").

formulates "It is believed that …".[177] That B is a sentence operator is beyond doubt. When prefixed to a declarative sentence a further declarative sentence ensues. The dominant view among belief-logicians is that B is also a modal operator.[178] Nearly always, a logic of belief is an extension of some or other logic of the alethic modalities for necessity and possibility. Virtually all[179] the mainstream semantic treatments of the alethic modalities are, in one way or another, possible worlds theories, although at certain levels of abstraction, the name "possible world" names nothing essential – or even discernible – in the logic's formal interpretative structure.[180] The conversion of an alethic logic – S4 say – to a belief logic involves two crucial additions. One is that "believes" is made to be a logical particle. The other is that there are *people* in the ensuing logic. The people of doxastic logic – the entities who do the believing – resemble in key respects the agents of AI in belief-change contexts. They close their beliefs under consequence; they believe all the truths of logic; and they keep their belief-sets consistent.

Classical belief change theories of all types are highly idealized models of belief-change competence. The same is true of mainstream approaches to doxastic logic. Empirically psychological theories of belief are accounts of belief-change performance. Performance theories attend to the conditions under which human belief-change actually take place. Two of the most prominent performance oriented theories of reasoning have been developed by psychologists. One is the mental model approach of Johnson-Laird and his colleagues,[181] and the other is the proof-theoretic approach of Rips and others.[182] In Rips' approach a state of belief is a partial proof, and the transition operators that move states to states are a subset of the classical natural deduction inference rules. The operators are proof-building rules. Belief-change is then likened to proof-development or proof-completion.

On the mental models approach a state of belief includes an interpretation of a given set of sentences, Σ, each with a truth value

---

[177] "B_a" formulates "[agent] a believes that …".

[178] Doubts about whether beliefs have the semantic structure of modalities were briefly sounded in the preceding chapter.

[179] Jubien (2009) is a non-worlds approach to possibility.

[180] See here Blackburn and van Benthem (2007), in Blackburn, van Benthem and Wolter (2007). For reservations about the possible worlds approach to belief, see Woods (2010d).

[181] Johnson-Laird, Byrne and Schaeken (1992), Johnson-Laird and Byrne (1991).

[182] Rips (1983), Rips (1994), Braine and O'Brian (1991).

picked up by virtue of how the world is at that time. Operators map models to other models in fulfillment of the requirement to maintain consistency. Let α be any sentence not in Σ, but whose truth is a condition on the consistency of Σ's model. In each case, both proof-theoretic and mental model, the mechanisms of belief-change are search procedures for the identification of such sentences.

AI (or competence) theories and the psychological (or performance) accounts exhibit some interesting similarities and differences. AI's model-based and psychology's mental model approaches share the notion of model, but differ in other ways. AI model theories accept the irrelevance of the syntax principle, but mental model theories do not.

The aim of this book is to develop an empirically sensitive logic within which to advance a thick theory of mistakes of reasoning. As I've already had plenty of occasion to say, I take it that the commitment to empirical sensitivity mandates a resistance to accounts of human behaviour that are empirically hopeless, over-idealized and normatively dubious, and which place a premium on technical virtuosity over conceptual accuracy and behavioural fidelity. (One thinks here of Nietzsche's mockery of Spinoza's axiomatics as the "hocus-pocus of mathematical form.") It is true that a commitment to empirical sensitivity requires circumspect attention to empirical theories which, not untypically, make empirically distorting idealizations precisely for the purpose of facilitating the framing of lawlike connections in the service of high levels of observational payoff. We needn't and shouldn't be Luddites about these things. It is not open to us to reject the postulates of an empirical theory simply because of empirical distortion. But, to re-echo Quine, the returns had better be good. That is, the theory's descriptive laws must facilitate predictive control and the generation of confirmable observation sentences. Still, there is no theory of belief-change in AI and no logic of belief that pays serious mind to what beliefs actually are, and in particular to what it is like to *have* them. In the account we will unfold in this chapter, what believing is *like* plays a fundamental role. Let's begin with light beliefs, that is beliefs whose presence is accessible to the mind's eye.

5.2 *Cognitive saturation*

We are three kinds of beings at once. We are causal beings. We are dialectical beings. We are phenomenal beings. There is reason to think that our successes overall at knowing and reasoning would not have been possible save for the engagement of all sides of our trinitarian make-up. If this is right, it is natural to think that error might have something to do with the maladjustment of our tripartite selves. Error-correction, in turn, would be a matter of readjustment. An error theorist must try to solve the concealment-detection problem. He should say something about the conditions under which error-making is possible; he should attend to the conditions under which error-detection is possible; and he should strive for an explanation of why our facility at error-detection doesn't enable us to avoid making those errors in the first place. How, then, are the elements of our tripartite natures involved in this?

Here is a commonplace fact about the individual cognitive agent. Suppose that Harry's task is to know whether α. We can represent Harry's situation at this point by the 2-tuple $\langle H,A^{\alpha}\rangle$, where H is Harry, A is his agenda and the superscript α is the proposition the agenda is directed at. Suppose, whatever the details of it, that it came to pass that Harry is now in an "end-state" with respect to $\langle H,A^{\alpha}\rangle$. End states encompass two cases: Harry now thinks (in a certain way) that it is the case that α; or Harry now thinks (in a certain way) that it is not the case that α. The first is a positive end state; the other a negative end-state. (By and large, context will enable us to determine which is which.) Where Harry is not sure what to think, we say that $\langle H,A^{\alpha}\rangle$ lacks an end-state at that time. When Harry is fed up and doesn't any longer care whether α is the case, Harry's agenda has been "aborted". But when in an end-state, Harry thinks in a certain way that α or he thinks in a certain way that $\sim\alpha$. What is meant by the qualifier "in a certain way"? It means that the end-state achieved by thinking in that way is a "cognitively saturated" end state. If, with respect to his agenda A^{α}, Harry is in a state of cognitive saturation, he is in a state in which the cognitive impulse that drove his agenda has subsided. It is the attainment of a state that stills his cognitive yearning. The stilling of the drive to know is like the stilling of any other drive. Pathological cases aside, what a hungry person can't get enough of the repleted person can no longer abide. Of course, saturated end-points are typically for now, not forever. You might at present be in a saturated end state that binds you to "The Economic Development Officer is honest and forthright", only to learn next week that he is in the pay of developers.

There is a certain sense of "believe" according to which to believe that α is to be in a saturated end-state with respect to it. Other uses of "believe" have other connotations, at least one of which hovers at the far remove from saturating belief. In this second usage, to believe that α is to be in a non-saturating state with respect to it. If Harry's belief that α is non-saturating, it leaves A^α open. Accordingly, we could simplify our description by speaking of s-beliefs and non-s-beliefs. An s-belief is an "agenda closing" belief. A non-s-belief is an "agenda non-closing" belief. In what follows, I will adopt this way of marking that distinction.

There is another distinction which matters for error. It cuts across the distinction between saturating and non-saturating beliefs. It is the distinction between belief ascription from the first person perspective and belief ascription from the third-person perspective. Suppose that Harry has the s-belief that α is the case. When this is so, Lou might with perfect coherence assert (and it might be true) that although Harry believes that α, unfortunately α is false. It has been known for a long time that what is all right for Lou to say about Harry is not always all right for Harry about himself. Saying this is, for Harry, the way of Moorean blindspots: I believe that α, but α is false.

No one thinks that Moore's paradox is a contradiction. But it is widely acknowledged, and correctly, that the state of mind Harry is in were the Moorean sentence true of him would be *anomalous*. For the most part, beings like us aren't in such states – such is the connotation of "anomalous". It wouldn't be far wrong to say that for most of us and for virtually all values of α, being in a Moorean state is causally beyond our state-shaping capabilities.

Consider now that large class of cases in which,

(1) the belief that α is an s-belief

and

(2) the self-ascription of it won't take the clause "but α is false."

Let us say that beliefs that answer to these two conditions are saturating beliefs of first-person attribution, or "first-person s-beliefs" for short.

We should not think that the first-personhood of a first-person s-belief confers an epistemic advantage on it. First person s-beliefs, no less than the third person variety, are sometimes wrong. Indeed they are wrong with a notable frequency. First-person s-beliefs matter for error in a particular way. They make error unrecognizable. That is to say, they make α's error unrecognizable in the first-person s-case in the here and now.

Suppose now that the state that Harry is currently in is one of knowing that α. On the CR-model, for this to be so α would have to be true, Harry would have to believe it, and Harry's belief would have to have been well-produced. When Harry meets these conditions there is no general requirement that he knows that he has met them. Would this dislodge the KK-hypothesis?

Let's suppose that it did not. Then whenever Harry knows that α, the following conditions would also be met.

    1. He knows that α
    2. He believes that he knows that α
    3. The belief that he knows that α is a well-produced one.

Most people would agree that a degree of belief in α that satisfies the belief-condition on Harry's knowledge of α might not satisfy the belief-condition on Harry's knowledge of his knowledge that α. This is certainly right if his b-state closes his knowledge-seeking agenda with respect to α, but does not close his knowledge-seeking agenda with respect to whether he knows that α. In which case we seem to have lost the KK.

Consider now a re-description of the case. It is achieved by replacement of the second occurrence of the Kα-clause in ⌐Kα Kα⌐ with the CR-conditions on knowledge. In the schema just above (2) is replaced by

    2'. He believes that:
      i. α is true
      ii. he believes that α
      iii. his belief that α has been well-produced.

While it may be true that most people would reject line (2) in the first schema, it is likely that they would be a good deal happier with line (2') in the second. In that case, shouldn't we reconsider our rejection of the KK-hypothesis?

There are two options to consider. One is to accept the equivalence of (2) and (2') and disallow the substitution of (2') for (2) on the grounds that the intersubstitutivity of equivalents fails in modal contexts. The other is to reject the semantic equivalence of (2) and (2'), and with it the replacement of an occurrence of "knows" with the conditions that regulate its instantiations. Of the two, the latter is a better option. No minimally aware KK-theorist can have missed the point that in endorsing ⌐Kα → KαKα⌐ he has committed himself to the legitimacy of substituting ⌐Kα⌐ for α in the conditional's antecedent. So he will be disposed to think that option one begs the question against him. On the other hand, option two's denial of the equivalence of (2) and (2') is already our recommended course as regards the CR conditions on knowledge. They are not meant as giving the meaning of "know". They are generically intended conditions on when knowledge happens. So I am ready to say that although (2) is false and (2') is true, their purported equivalence fails, and with it too the KK-hypothesis.

A k-state is a special case of saturated end state. Harry is in a k-state with respect to α when he experiences himself as knowing that α. K-states are s-belief-states. They close agendas, and in self-ascriptive contexts they won't take the "but α is false" qualifier. We might also speak of b-states. As I will define them here, b-states are just like k-states, except for a difference in phenomenal texture. B-states are first-person s-states experienced not as a state of knowing that α, but rather as a state of *saturating satisfaction* which occasions closure of one's α-agenda.

Sometimes you will be in a k-state with respect to some α and α will not be true, or your k-state-forming devices will have been acting up. Not only can you be in a k-state about α without knowing that α, you can, when in it, be aware of this possibility. But if you *are* in that k-state, such awareness is experientially inert, then and there. It does not qualify, and does not compromise, either the repose that attends k-state possession or the cessation of the impulse to know.

In tackling the concealment-detection problem, a guiding principle, as we saw, is that cognitive agents enter the logic of reasoning as they actually are in reality, warts and all. In the next section, I want to consider one of the warts. It arises from another epistemological commonplace, and it bears an important link to the first-person/third-person asymmetry.

5.3 *The concealment-detection problem*

Proposition 5.3a
THE PSYCHOLOGICAL PRIORITY OF BELIEF OVER KNOWLEDGE:
There is a particular sense of the verb "to settle for" in which, for large ranges of cases, we quest *for knowledge, but* settle *for belief. We settle for belief when from the first-person perspective the*

Again, what this proposes is that being in a state of knowledge is not a condition of end-state saturation

Settling for belief instead of knowledge has a provenance that exceeds the present type of case. The *principium divisionis* is whether the settling part is known to the settler. There are lots of examples in this second category. Concerning your quest to know whether α, you might be short of time, you might have better things to do, you mightn't be able to think of what else to do. So you are ready to concede that you don't in fact know whether α, but you're pretty sure that α even so. You settle for this, and know it. These are not the cases envisaged here.

The thrust of Proposition 5.3a is nicely caught by Peirce:

> We may fancy that … we seek, not merely an opinion, but a true opinion. But put this fancy to the test, and it proves groundless; for as soon as a firm belief is reached we are entirely satisfied, whether the belief be true or false (Peirce, 1931-1958, CP 5.375).[183]

Thus the intent of Proposition 5.3a is that we settle for belief in that sense of "believing" which signifies a saturated end-state, that is, when one's belief is a k-state or a b-state.

Saturated end-states are essential for the work of this chapter. Here is one of the reasons why:

Proposition 5.3b
PHENOMENAL INAPPARENCY: *The invocation of s-states is meant to lay emphasis on the phenomenal unreality of the distinction between actually being in a state of knowing and being in a state one experiences as a state of knowing, and between being in a state of agenda closing satisfaction with respect to an α and bring in a state of knowing it.*

The phenomenal inapparency of false belief is a motivating datum for classical epistemology and metaphysics. Philosophers have invested

---

[183] Cf. 5.397: "And what, then, is belief? It is the demi-cadence which closes a musical phrase in the symphony of our intellectual life. We have seen that it has just three properties: First, it is something that we are aware of; second *it appeases the irritation of doubt*; and, third, it involves the establishment in our nature of a rule of action …." (Emphasis added)

heavily in strategies by which the individual cognizer might mitigate the entrapment of the first person, of what Ralph Barton Perry called "the egocentric predicament". By reflecting on what one thinks one knows, by subjecting it to the burdens of critical justification, by giving it dialectictical exposure, we sometimes see the error of our ways.

5.4 *Dialectical erasure*

We lightly touched on dialectical matters in chapter 2. It is time to give them further attention. This is tricky. There are classes of cases in which taking or responding to the third-person position with respect to your *own* k-state by considering whether the requisite justification exists brings about *dialectical erasure*. To see how dialectical erasure comes to be, suppose that Harry is in a k-state with respect to α, that he experiences himself as knowing that α. Suppose that the state that Harry is in fulfills the CR-conditions on knowledge. Enter Sarah. "I wonder how Harry knows this?", she wonders. So she asks Harry how he knows it. In asking this of Harry it is important to see that Sarah is *not* asking any question answerable as follows: "Well, α is true, I believe it, and my belief is well-produced". No, Sarah's question is a *challenge*. It calls upon Harry to *make a case* for the truth of α. Note that Sarah is not giving Harry notice of a challenge. She is not inviting Harry to go to the Library and look it up. Sarah wants Harry's defence here and now. Perhaps, given some time, Harry would be able to assemble a case for α – although as we saw in chapter 3 there are a great many values of α for which not even this is the case. Let's take a simple case. Let α be the proposition that Caesar crossed the Rubicon in 49 BC. Harry's reply to Sarah is: "Don't ask me. I just know it" (and he does) or "Everyone knows that" (and they do).

But Sarah rebukes Harry. "What a pitiful answer!" Of course, it *is* a pitiful *answer*, and Harry himself could hardly have thought otherwise. Suppose, as sometimes happens, that Harry's inability to satisfy Sarah's challenge throws him off-centre. Suppose, as a result, that Harry ceases to be in a s-state with respect to the date of Caesar's crossing. "Hmm", he says, "I wonder whether it actually *was* 49 BC?". Then perhaps Harry no longer fulfills the belief condition on knowledge. If so, he has undergone the dialectical erasure of the knowledge he once had. What he once knew he no longer does. It is a strange conversion. What would explain it? It is explained as follows: Sarah's challenge carries the CC-presumption that knowledge is not possible in the absence of a successful justification. In accepting her challenge, Harry takes that presumption on board. It is a fateful come-uppance. It costs Harry his knowledge of Caesar's crossing.

This reminds us of a lesson from chapter 2. It is that in our quest to understand what knowledge is and how it is arrived at, the ways in which we and the rest of us actually use the word "know" should be treated with a healthy caution. Suppose you're in a state of knowledge with respect to α. How would you report this fact? You could report it by uttering α with the assertive intonation contour. But you might misreport it if you uttered; "I know that α". "I know that α" sounds wrong here. It sounds defensive. Suppose that Sarah asks Harry whether β, and Harry happens to know it. Harry could tell Sarah what she wants to know with an assertive utterance of β or better, a simple "Yes". But here, too, even though Harry knows that α and even though his response to Sarah is predicated on that knowledge, saying "I know that α" is the wrong thing to say. In either case, Sarah might wonder, "What's bothering Harry? Why is he so defensive?" Accordingly,

Proposition 5.4a

DIALECTICAL RESISTANCE: *In large numbers of cases, it may be harmful to one's epistemic health to try to answer questions in the form "How do you know?"*[184]

It would be wrong to leave the impression that it is never appropriate (or safe) to respond to an epistemic challenge. There is a rough rule of thumb about this.

Proposition 5.4b

THE RESPONSE RULE: *The greater the responder's confidence in his accessibility to the requested justification the less the likelihood that his response is unsafe – that is, subject to dialectical erasure.*

---

[184] It is well to emphasize that not all uses of "How do you know?" are calls for an epistemic justification. We said so in chapter 3. In other uses, they are calls for an interlocutor's epistemic *source,* and are answerable along the following lines: "I heard it on the BBC", "I saw it with my own eyes", "Professor Zed proved it in Tuesday's class", "Barb told me so", and so on. Epistemologists sometimes think that identifying the source of one's knowledge is one way of justifying the claim to have it. Sometimes it is, but not always.

Important as it is, we shouldn't make more of dialectical erasure than is in it. Suppose Harry was in a k-state that fulfilled the CR condition on knowledge. Harry was challenged and dialectical erasure ensued. It would be unfortunate that Harry has lost the knowledge he once had. But we shouldn't think of such losses as irreparable. Harry's loss was then and there. In the fullness of time he might find the justification that his challenger called for and concurrently re-acquire the missing knowledge. Even if he tried to find it and failed, he might come to think that the call for it was misplaced; and his k-state might revive. Or, quite plausibly, after the initial spell of dialectical shock, the old k-state might re-emerge, all on its own, so to speak.

It would also be wrong to leave the impression that a person's k-states and b-states constitute or necessitate a state of close-mindedness. There is no inconsistency, logical or pragmatic, attaching to the s-states of a duly self-conscious fallibilist. Harry's agenda A^α might now be closed, but this does not preclude his concurrent awareness that he might be wrong. It suffices that he carries the concurrent disposition to reconsider his present position. One way in which to do this is by having a readiness to assume the third-person perspective.

5.5 Bubbles

Our focus here is error; how it comes about, how it is detected, and how it is corrected. Harry experiences himself as knowing that α yet doesn't know α. How, so situated, is error detection possible for Harry? We might say the following: Harry's readiness to take the third person point of view embodies the disposition to consider his position critically. By giving due dialectical weight to the critical views of others who in relation to him are in the place of the third person, he sometimes acquires the wherewithal to diagnose a mistake, to see that what he experienced himself as knowing he didn't. The whole dialectic of critical thinking is bound up with this view of corrigibility. Errors are detectable by assuming (or simulating) the third person position. Errors are avoidable by assuming the posture strategically – in advance of their commission, so to speak.

Even so, it shouldn't be overlooked that every cognitive perspective that is a third-person perspective with respect to you is the first-person perspective with respect to itself. This being so, cognitive agents operate within epistemic "bubbles", a notion we may find, albeit without the name, in an overlooked article of William Rozeboom.[185]

Proposition 5.5a

EPISTEMIC BUBBLES: *A cognitive agent is in an epistemic bubble with respect to proposition α if he is in a k-state with respect to α and the distinction between his knowing that α and his experiencing himself as knowing it is phenomenally inapparent to him in the there and now.*

COROLLARY: *The embubblement thesis also holds with respect to one's conclusion of β from α. That is, it holds with respect to one's k-states about the correctness of one's reasoning.*

COROLLARY: *The embubblement thesis, for both knowledge and right reasoning, also holds for b-states.*

With Proposition 5.5a at hand, we catch a glimpse of error-detection. If, having reflected on the matter or having listened to your critics, you now see that your former k-belief was mistaken, what you *currently* take yourself as knowing you equally might not. Current knowledge is subject to the same phenomenal inapparency of the distinction between knowing α and k-experiencing it. Equally, your critics and interlocutors, whose insights have led you to see your error, proceed on the only basis available to them. They proceed on what it is they take themselves as knowing, including what they think they know that you (now) know. But what they think they know they might not know. Discovery of the error you made *then* flows from the fact, as you *now* see it, that what you thought you then knew you only believed, that is, believed in the sense that now allows for the self-ascribed qualification "but α is false." In pressing this present view of the matter, you are taken over by what you think you now know. What you now think you know is incompatible with what you used to think you knew. Given what you now think you know, what you used to think you knew you now think you know to have been an error. Accordingly,

---

[185] Rozeboom (1967).

Proposition 5.5b
ERROR DETECTION: *For human agents, error detection when it is possible at all is an after-thought. It is rooted in their possession of phenomenal states the recognition of whose own erroneousness is not* concurrently *possible for them.*

The first-person/third-person asymmetry bites hard here. For the person who brings it off, error detection is a kind of coming to his senses. He comes to his senses in recognizing the incompatibility of what he now sees to be true with what he used to think was true. But the asymmetry is such that what is experienced in these ways may not be as those ways suggest. For one possibility is that, from the third-person point of view, it is no more than this: What one now believes seem to be incompatible with what one no longer believes. Since for significant classes of cases this seems to be so, it appears that we must reconcile ourselves to the following possibility:

Proposition 5.5c
THE NO ESCAPE THESIS: *Except for those limited ranges of possibilities in which knowledge is immediate, self-evident, or incorrigible, every k-state of a cognitive agent – even one that purports to ground the correction of a prior error or the present error of another person – is lodged irreducibly in an end-state bubble. Embubblement is not an escapable condition.*

Armies of epistemologists have poured their energies into attempts to evade the egocentric predicament. I myself have neither the time nor the inclination to join that fray here. What No Escape tells us is that fallibilism applies in two places. One is when we form the belief that α. The other is when we correct the belief that α with the now-formed belief that β. No Escape isn't scepticism. It is fallibilism. As it applies here, it tells something interesting about the concealment-detection problem and about error-correction as well. Sarah used to be in a k-state with respect to α. She now thinks better of it. She is now satisfied that α was an error, made so by its incompatibility with the β that is the object of her present k-state. What fallibilism says about this is that

Proposition 5.5d
NO PRIVILEGE: *While Sarah's present k-state may causally erase her prior k-state, there is nothing in this that constitutes*

her present state as epistemically privileged.

It is interesting to see how No Escape plays on our abundance theses and the enough-already principle that purports to reconcile them. As we have it now, the reconciliation achieved by the enough already principle provides for the concurrence of two abundances, one of knowledge and the other of error. But as we see, it is open to question whether enough already thesis strictly requires the cognitive abundance thesis to be true. It suffices that what we take ourselves as knowing should motivate the appropriate actions, the actions that bring about our well being. For do we not survive, prosper, and so, on by acting in ways that bring these things about, and do we not act in such ways when we are motivated by the requisite beliefs, by the requisite experiencings of ourselves as knowing? Why must those takings be the real thing? Wouldn't Stichian radicalism be the more prudent course to take?[186] Shouldn't we accept that "… it is hard to see why anyone but an epistemic chauvinist would be much concerned about rationality or truth."[187]

No. Here is why. It is the way of scepticism. It is an overreaction to the compatibility-in-principle between prosperity and ignorance. It is excessive. It is unnecessary. Apart from that, suppose that the advantages claimed for survival and prosperity could just as well be achieved by true belief. Consider now those masses of advantageous true beliefs that fall short of knowledge. What would account for this short-fall? On the CR-account it would be that those beliefs were formed by malfunctioning cognitive devices or under the influence of an anomalous, tricked-out environment. That is, the beliefs weren't well-produced. There is no reason to think that for the general run of such cases, our cognitive equipment is just broken, that virtually every advantageous belief was produced by cognitive devices that don't work. Nor is there reason to suppose that we live in a world of systematically hostile externalities, that is, in a world tricked out for generalized knowledge-failure.

Consider how this would bear on perceptual cognition. If none of our true perceptual beliefs amounted to knowledge, then one or both of two things would be true. Either our perceptual devices would be broken or they would be systematically overridden by epistemically hostile externalities. Whether these alternatives actually held would be an

---

[186] Or a Sartwellian radicalism. Sartwell (1991, 1992).

[187] Stich (1990), pp. 134-135.

empirical question. In the first instance, you would get your eyes checked. In the second, you would check for environmental irregularity or mischievous contrivance. I will take it without further ado that there is nothing in what is presently known empirically to suggest that these enquiries would yield those findings.

Our attachment to knowledge is instinctual. It rivals in primitive thrust the necessity to breathe. This gives to No Escape a sharp importance. It helps us see is that, although we have a drive to know, we are so constituted as to settle for s-states, that is, states of mind that still the impulse to know. We settle for belief in that sense of the word that honours this first-person/third-person duality. This is not because we are lazy or over-casual, but because we are subject to an involuntary epistemic limitation. We are embubbled cognizers. In our takings to know, we occupy the perspective from which belief does duty for knowledge. Judged from the third person perspective, this is a decidedly down-market form of satisficing. But the last thing it is, is satisficing *on purpose*. This gives us something to take formal notice of.

Proposition 5.5e
EPISTEMIC QUESTING: *For very large ranges of cases we could not* settle *for s-belief unless we* quested *for knowledge.*

As we now see, the word "satisfaction" does double-duty in a way that encompasses an important contrast. We should keep it in mind that:

Proposition 5.5f
THE AMBIGUITY OF SATISFACTION: *In its psychological sense, an agent's cognitive target is attained only when he is in some requisite k-state. In its epistemic sense, that same target is satisfied only when his k-state is a state of knowledge.*

Proposition 5.5g
BLURRING THE DISTINCTION PHENOMENALLY: *For these large ranges of cases, psychological satisfaction is experienced in the first person as epistemic satisfaction.*

At the bottom of all this is our occasional disposition to experience ourselves as CC-knowers. As we saw, in achieving our CR-successes we often see ourselves as trying to execute CC-strategies, and succeeding with them. That we see ourselves in this way is connected to the fact that what the word "know" *means* may be for us, fairly close to the CC conception of it.

Just who are these "us"? There are only two plausible candidates. Either they are philosophical theorists of knowledge. Or they are the man in the street. Then, in the first instance, what we mean by "know" is what we've been taught by our epistemological masters. As for the other, I myself have no very firm idea of what the man in the street means beyond what my dictionary tells me. What it tells me is what "know" means is "awareness of something by observation, inquiry or investigation" The philosophers' "know" carries more than a whiff of CC-bias, especially in its steadfast fidelity to the J-condition. The ordinary guy's "know" is CC-free.

5.6 Belief-change again

We have not yet got to the bottom of detection and correction. The detection and correction of error must be distinguished from belief-revision, of which it is a special case. It is here that the founding problem of AI achieves a certain resonance. This is the frame problem of McCarthy and Hayes, originators of the situation calculus.[188] The frame-problem, though not under that name, has a clear relevance to the out-dated information aspect of Gettier problems.

The frame-problem first arose in the theory of automated planning. The theory postulates an agent and attributes a goal to it. A goal is any statement of a kind that could be made true by the appropriate kind of action. Planning is a search for actions or sequences of actions that might achieve a given goal. In the STRIPS planner of Fikes and Nilsson (1971), planning has three elements. The first component is a given state of the world, called for convenience an "initial" stage. The second component is a set of operators taking world-states to world-states. The third element is the planner's goal. The planner's task is to find the operators that give the transitions from initial state to the state that fulfills the planner's goal. In the original theory, planners execute something quite similar to a logic of automated theorem proving. It was one of those many cases in which the Can Do principle directs the theorist to an already successful theory, in hopes that the new theory could be an adaptation of it, if not outright absorbed by it.

---

[188] McCarthy and Hayes (1969).

The theory of planning encumbers the agent with some troubling goal-facilitation tasks. In the process of mapping an initial state to a goal-realizing state, a planner must factor in changes in world states as the plan unfolds. He must also take due note of the things that do not change. For example, it is intuitively clear that formulating a good plan for changing the flat tire on the car down the road presupposes the car will still be there and that you will not be felled by a stroke. Your tire-changing plan must be sensitive to whether changes that would wreck the plan will actually occur. In the original theory, these predictions are formulated as "frame-axioms", or assumptions of inertia. In our present example, it is a frame-axiom that the car won't have been moved, and another frame axiom that the tire-changer won't have been knocked out of action by a stroke. Of course, even in situations as commonplace as tire changing, the number of changes that *could* occur is extraordinarily large. One of the problems created by this is that it is not in any sense practically feasible to identify all the possible changes whose complements could be written up as frame-axioms. Even if we could somehow get round this problem, the ensuing axioms would be very large in number and massively complex. This leads to a further difficulty. Suppose we inputted these axioms to an automated planner and instructed it to predict the outcomes of the plan deductively. In so doing, taking adequate note of what doesn't change would so deplete the resources of the system as to preclude its working out the other aspects of the plan. The burden imposed by the system to work through the frame axioms would exhaust its planning capacity. So we have it that:

Proposition 5.6a
THE FRAME-PROBLEM: *The frame-problem in AI is the problem of how to make a device's assumptions about nonchange* efficient.

COROLLARY: *Adjusted to practical reasoners, the frame-problem is the problem of how to make their assumptions of nonchange* possible.

Although it arose as a problem in AI engineering, it is clear that the frame-problem is also a challenge for epistemology. And since human beings are successful planners, it must follow that they have a solution to the frame-problem. Is it possible to identify this solution and to study its essential characteristics? I said that the original AI planner was an idealized executor of the routines of automated theorem-proving. It was a deductive system of a certain kind. It was not long before AI theorists and others were drawn to the view that frame-difficulties were occasioned by the deductive character of automated planning. Since humans are successful planners, it was concluded that they must employ non-deductive modes of reasoning. Starting in the early 1970s, people began proposing a more relaxed approach to the consequence relation. AI researchers began to take seriously the idea that, in the absence of indications to the contrary, it was reasonable to assume that things remain as they are unchanged.[189] Reasoning of this kind is called "temporal projection", and is an early form of defeasible and nonmonotonic reasoning. Here is an example.

Proposition 5.6b
PRINCIPLE OF TEMPORAL PROJECTION (RAW FORM):
Believing that α is true at t *is defeasible reason to believe that it is true at* t^* *($t < t^*$).*

Proposition 5.6b is very rough, needless to say. Even in this crude form, it carries implicit constraints on the values of α. For it is certainly not even defeasibly correct that tomorrow will be the same day as today, or that anyone living now will be alive in the year 2250. It is widely accepted that, as of now, providing necessary and sufficient conditions for the temporal projectibility of propositions is an unsolved problem.[190] This notwithstanding, it is easy to see the kind of conjectures that are intended for planning theory. It is widely accepted by anti-deductivists in the theory of planning that at least part of the reason why humans are successful planners is that their planning is defeasible rather than the deductive, that automated planning systems will not succeed until their deductive engines are replaced with defeasibilistic engines. I will revisit these conjectures in chapter 7. For the present let us reflect briefly on how closely linked, or otherwise, the frame-problem is to the problem of error-detection and correction.

One of the driving forces of belief revision is change through time, change that makes what was once true now false. It requires that the once true and now false be replaced by the now true. Error-correction is

---

[189] See, for example, Sandewall (1972), McDermott (1982) and McCarthy (1986).
[190] Equally so the problem of the projectibility of predicates. See Goodman (1983).

different. It requires that what was then false be replaced by what is true, both then and now. It is a nice question as to how this contrast presents itself to a cognitive system in the first person then and there. On the telling of preceding sections, it is simply a matter of now holding a belief believed to be incompatible with a belief previously held. But it is clear that this is insufficient to capture the distinction at hand. Believing now that α and believing that α is incompatible with a prior β leaves it undetermined as to whether the prior β was then true rather than then false. If true, then your present subscription to α and the concomitant dropping of β is a matter of providing a consistent update of your belief-set under conditions of worldly change. If β is false, then your present subscription to α and the concomitant dropping of β is a matter of your correcting the error that β was all along. So, again, one's present subscription to an α that one sees as incompatible with a prior β is not all there is to error-correction. Something more is required.

Does the principle of temporal projection provide us with this further guidance? It does not. It provides that α's truth at t is reason to assume its truth at t*, unless we know better. "Knowing better" is knowing or having convincing reason to believe that α at t* is not true. But the determination of α's falsity at t* leaves entirely undisturbed α's truth at t. This being so, the principle of temporal projection gives us systematically unsatisfactory guidance about error-detection. α's falsity now is never adequate reason to suppose that α was also false then. True, we could always postulate a "backwards"-looking temporal projection rule, which would tell us that Ozzie the ocelot's not being four-footed now is indeed reason to suppose that neither was Ozzie four-footed then. Like the original principle, the backwards-looking principle works plausibly for certain specific cases. The trouble with both principles is that they are not available to us in convincingly general formulations.

Still, fundamental to the distinction between belief-update and error-correction is our present contrast between change and nonchange. When you replace your old true belief that the booming bittern is on the lawn with your currently true belief that the now-silent bittern is in the tree, two changes have occurred, one in nature and one in you. The change in nature is the location of the bittern, there then and now not. The change in you is a change in belief, the displacement of what you thought true then by what you think true now. Fixing errors is different. Beliefs are changed, but nature isn't. If your belief was that the bittern is in the tree and you see now that the bittern wasn't in the tree after all, there is no change of location corresponding to your change of belief. Memory has a constructive role to play here. Whether something has changed in nature depends, for you, on how memory operates. In the case of update, you remember two things. One is where the bittern was and the other is what you then believed about it. This is supplemented by new information. If the information leaves the memory standing, it provides the occasion for replacing the old belief with a new one. But if the new information prompts the correction of an error, it is essential that the old memory be retired. New information can be memory preserving or memory erasing.[191]

5.7 *Popping the bubble?*

If error arises from thinking we know things we don't or from believing to be true things that aren't, and if the state we are in when we think we know or believe to be true are states which give the appearance of knowledge and truth while at the same time disguising their absence; if, in other words, being in error-disguising k-states or b-states is a large part of the story of how our errors come about, would we not be better advised to stop *being* in those states? Would it not be better never to experience oneself as knowing that α, or as wholly satisfied that α?

Perhaps the degreed character of belief will give us a way out of error-occasioning difficulties. If having s-states is a large part of the problem, why not, as we say, stop being in s-states? Why not substitute for s-states beliefs of weaker measure? Wasn't this the advice of Groucho Marx's doctor to the famous funny man? Groucho consulted his doctor about a sharp pain in his shoulder. "When I move my arm like this, it hurts like the very devil", complained Groucho. "My advice", said the doctor, "is don't move your arm like that. That will be $500.00, please." Wouldn't the same advice be helpful for reasoning agents?

Proposition 5.7a
GROUCHO'S DOCTOR'S PRESCRIPTION: *In matters of what you seek to know, stop being so sure of yourself.*

Proposition 5.7a recommends that we stop having beliefs whose strength occasions the phenomenal inapparency of the distinction between having the belief that α and knowing that α. Since beliefs are said to come in degrees, this is something that we might consider.

---

[191] For more on erasure see Gabbay *et al.* (2002), (2004), and D'Avila Garcez *et al.* (2007).

Case one: Consider the belief condition on knowledge, a condition held in common by the CC and CR-models. If the belief-condition with respect to your knowledge of α requires you to think that you know that α, that is, to be in a s-state with respect to α, then any degree of belief weaker than this fails to fulfill the belief-condition on knowledge. So the extent to which we comport with Groucho's doctor's prescription with respect to α, knowledge of α is out of the question. The course of case one is the course of a rather hefty scepticism about knowledge. It provides that lowering one's sights with respect to α wipes out the possibility of our knowing it.

Case two. The belief-condition for knowing that α does not require that you think that you know that α, that is, that you be in a s-state with respect to α. All that's required is some degree of pro-attitude towards α above a certain point. Let us stipulate *n* as that point. Then any belief satisfying the belief condition on knowledge will be of strength $n + m$, for some non-zero *m*. But, given the reasoning of case one, no belief of strength $n + m$ constitutes a s-state. Accordingly, case two provides that there is nothing of which the following are both true: First, that you know it. Second, that you think that you know it. Thus, the course of case two is the course of what we might call Knowledge-Lite, in which knowing something is never something you experience yourself as doing. Given the outlyingness of truth, there are lots of philosophers who in their theoretical moments are entirely reconciled to the prospects of Knowledge-Lite. Why would they not welcome it for case two?

I said earlier, and it bears repeating here, that nothing in the empirical record lends the slightest support to the suggestion that preventing ourselves from being in s-states is something that lies within our capacity to bring off an a scale that would qualify it as a general policy. Possibly we might train for it in some Pyrrhonian monastery in the wilds of a distant land, re-programming ourselves to think lite. Perhaps one day before long the same adjustments might be available to us surgically. There might be package specials – a tummy tuck and a k-state suppression procedure for the price of the tuck alone.

Still, the idea that suppressing our s-experiences is not incompatible with knowing the very things we might think not raises an obvious question. It is the question of the degree to which, if any, the project of Knowledge-Lite spares us the trouble of being in states of mind that disguise the presence of error. The answer is, "To no degree that matters." It may be that being in a state of belief of degree $n + m$ with respect to α does not create the felt experience of being in a state of knowledge about it. But it remains the case that if α is indeed an error, believing that α with a strength sufficient to fulfill the belief condition on knowledge will disguise that fact. We may even allow that where α is believed to degree $n + m$, $\ulcorner \neg \ \alpha \urcorner$ might concurrently be believed with degree $1 - (n + m)$. This might happen when one is of two minds about α, when one is able to see the pros and the cons. But, again, if believing that α to degree $n + m$ is sufficient belief for knowledge; and if error is present, it will be inapparent to the believer.

As we have it now, n is the arbitrary point above which belief is strong enough to fulfill the belief condition on knowledge. But how high is this? Some further guidance is available to us in the idea of *reason for action*. Intuitively, a belief of degree n or higher is a belief of sufficient strength to serve as a reason for acting on it. The "on it" qualification is important. Acting on a proposition – a belief, a hypothesis, an assumption – is both treating it as true or defeasibly true and re-organizing one's behaviour accordingly. So if you believe to a degree that satisfies the belief-condition on knowing that Barbara's birthday is on the 3rd, you will be moved now to buy a present or at least a card.

There is an obvious distinction between thinking that something gives one reason to act on it and its being the case that it gives a reason to act on it. If being in a belief state sufficiently strong to fulfill the belief condition on knowing that α is also sufficient for thinking there is reason to act on α, then thinking so will disguise the absence of that reason in those cases in which it is indeed absent. Again, weak belief does not solve the concealment of error problem.

Proposition 5.7b
THE FUTILITY OF THE ADVICE THAT GROUCHO GOT: *The expungement of a cognitive system's saturated end-states is not presently possible or forseeably likely. Even if it were, weak belief with respect to both α's truth and reasons for acting on it does not erase the phenomenal inapparency of error.*

5.8 *Assent*

Before bringing this section to a close, it remains to say a brief word about acceptance or its more expressly linguistic variant, assent. It is sometimes held that assenting to and dissenting from propositions is a way of putting them in and out of play without the burdens of having to believe or disbelieve them. A good many systems exist in which assent, dissent and other speech-act types play load-bearing roles. Logics of assertion[192] and the many systems of dialogue logic stand out in this regard. One thing to like about these approaches is that they afford no formal occasion to recognize the private states of agents. They are logics without s-states. They are doxastically neutred. They are logics whose concept of agency is stripped of its phenomenal aspect. Some of these systems are highly developed mathematically. Some of them simulate more or less well the behaviour of real life agents. They say their piece about reasoning without the encumbrance of their agents' dark interiors. Clearly, then, logical theories of the reasoning of doxastically neutred agents are possible; and assent is belief-like without the necessity of being the real thing. So why don't we drop belief? Why don't we de-mentalize our logic? Why don't we re-pledge to Frege's resistance to psychology in logic?

The short (and now old) answer is that we want a logic that is naturalized, one that takes official note of how beings like us are actually constituted. This gives rise to the warts-and-all principle. In generating an account of how beings like us make errors of reasoning, we admit ourselves to theory, warts and all. One of the warts is that we are not doxastically neutred beings.

A related consideration is that when it comes to how the human individual actually operates, his own assenting and dissenting behaviour is very often driven by the extent to which it accurately reflects what he believes and does not. Even in those cases in which a person assents to something he disbelieves (or fails to understand), the assent is *un acte gratuit* unless attended by the belief, or the disposition to believe, that in the circumstances at hand that was the right, or the appropriate, or the prudent thing to concede. When it comes to beings like us, there is no getting rid of belief. Accordingly,

Proposition 5.8a

S-STATE INELIMINABILITY: *Aside from* l'acte gratuit, *when a cognitive agent has a reason to assent to α (or dissent from it), even a merely prudential reason for so doing – even indeed an epistemically disreputable reason – there will be an agenda for the achievement of that end and an agenda closing s-state to the*

---

[192] See Rescher (1968), chapter 14. Jary (2010) is a philosophical book about assertion, but it is not a contribution to the logic of assertion.

effect that assenting to α (or dissenting from it) is in the circumstances at hand is the way to go.

COROLLARY: *In at least the general run of cases, even the most highly qualified expressions of epistemic uncertainty are causally anchored to underlying s-states. Thus "I really haven't the slightest idea about whether α" is underlain and causally supported by "I am in an s-state with respect to the agenda of knowing my own mind as regards my doxastic relations to α".*

What would it be like never to have beliefs? By this I mean, what would it be like never to be in s-states? "Well", some people will say, "it wouldn't be all that bad. We would simply go through life with a Peircean lightness of touch". But this is a way of proceeding which not even Peirce would have thought uniformly available, especially in matters of "vital importance". Not having s-experiences at all would in the most radical of ways be not having a clue. It would be a disorientation so total as to defy the victim's conceptualization of it. Lots of things we think we know we don't know, and more still we think we might not know. But knowing that there are lots of things that you don't know is *itself* a s-state, and so is suspecting that you might not know. "I know my own mind", as the saying has it, and it is not all wrong. It is the cornerstone of philosophical idealism and of much of the foundationalist impulse in classical epistemology. If we didn't know our own mind there would be no mind. Suppose, contrary to what Peirce allows, that it were always possible for us to avoid being in s-states with respect to vital affairs. Suppose again that after putting yourself in the expert hands of a venturesome neurosurgeon, you are able to suppress your s-experiences of the ordinary world in favour of non-s-experiences. So instead of experiencing yourself as knowing the woman across the room is your wife, you experience yourself as merely believing that she is. Suppose that you are keen to kiss your wife warmly. Will you now proceed to kiss the woman across the room? Will you do it warmly? Will you do it *ardently*?

Suppose now that your accommodating neurosurgeon went further than he intended. Suppose he completely extinguished your capacity for s-states. Then, again, you couldn't know your own mind. You couldn't even know that you're practically certain that the person across the room is your wife; you couldn't know that you're pretty sure of it. In losing the capacity to know your own mind, you also lose the wherewithal for responding to the drive to know. You are alienated from your nature as an epistemic yearner. You've become a sublimater of what St. Augustine calls "the *eros* of the mind". Even in these surgically diminished circumstances, wanting to know couldn't plausibly be replaced with wanting to have a belief.

It is necessary to flag a possible confusion. It is a confusion we first encountered in chapter 4, when we considered the suggestion that fallibilism commits a form of the Preface Paradox. Whatever your s-state may be, you are always free to *report* it as a non-s-state. It is useful to call to mind the dialectical character of our social natures. I said before that when the report of a s-state is challenged, the human individual often has a disposition to take a CC position towards his own assertion. For how is such a challenge to be understood if not as the demand for a case to be made for the assertion that evinced it? It is one thing to accept the appropriateness of a challenge and quite another thing to make the case that answers it. There are lots of cases in which, having conceded the former, we'd prefer to delay the latter. So we rephrase and down-grade: "Well, that's what I *think*, anyhow". This is not yet dialectical erasure, although it is a near thing. Dialectical erasure occurs when, in his inability to mount a convincing case for it here and now, an agent's prior k-state is extinguished. But acknowledging one's fallibility certainly need not be dialectically erasing. For someone in a s-state, the fallibilist concession "But, of course, I might be wrong" is not a *psychological* hedge, whatever it might be dialectically.

Re-reporting is sometimes the surrender to dialectical erasure. In other cases, it is just a manner of speaking. What matters here is that talk is cheap. Re-reporting a k-state as a mere belief-state need not be a way of making it not to have been a k-state after all. It is not even a way of transforming it into a mere belief-state. Our question is not one of how an agent may or not report his states but rather of how it might come about that beings like us stop having k-states altogether.

5.9 *Mental health*

On reflection, it may strike us as odd that epistemologists, especially those of naturalistic bent, are little interested in exploring the connection between cognitive functioning and mental health. It is common knowledge that mental illness can occasion cognitive deficits. For example, our personal identity claim receives substantial support from what we know about schizophrenia. The main feature of logical processing among schizophrenics is *context- blindness*.[193] Context-blindness exhibits itself in a number of different but connected ways. One is a difficulty in being guided by, or even recognizing, common knowledge. Yet another is a difficulty in synchronizing repetitive actions. Another involves the mismanagement of part-whole relations, and the confusion of essential and contingent properties. A fourth is a tendency to misgeneralize against type. Relevance is also a problem for schizophrenic patients.

> Normal individuals perceive mainly task-relevant information. A schizophrenic individual, however is hypothesized to little relevant, or too much irrelevant information.[194]

There is little doubt that schizophrenic patients experience "persistent and systematic cognitive disequilibrium" which in turn has serious implications for personal identity. Diminished identity is a recognized feature of the disorder:

> Schizophrenia often involves a profound experience of one's identity as diminished, which complicates adaptation to the demands of daily life (Lysaker and Hermans (2007), p. 129).

Schizophrenia is diagnosed on the basis of positive and negative symptoms. Although diagnostic criteria are fully laid out in the *Diagnostic and Statistical Manual of Mental Disorders*, it would be in order to say something about the positive symptoms of schizophrenia. They include: (1) *thought disruption* (disorganized and illogical thought); (2) *delusions* (the holding of false and bizarre beliefs) and (3) *hallucinations* (perception of things that don't exist, particularly hearing voices). In a CC-way of putting it, each of these symptoms represents a failure on the part of the schizophrenic individual to regulate his epistemic states, particularly belief. But the CR-idiom also comes trippingly to the tongue, as witness Langdon and Coltheart (1999):

> [I]f patients cannot reflect on beliefs as representations of reality, then the distinction between subjectivity and objectivity collapses, leading to maintenance of delusions (p. 44).

---

[193] Selesnick and Owen (2012).
[194] Hirt and Pithers (1991), p. 140.

Schizophrenic language has all the hallmarks of cognitive disequilibrium. The patient below has just been asked how he came to be living in a particular US city. His response displays the intrusion of irrelevant thoughts, as if he is incapable of damping down the 'constant changes in informational flow-through: He sees the tie and shell and is immediately compelled to introduce them into his reply:

> Then I left San Francisco and moved to…where did you get that tie? It looks like it's left over from the 1950s. I like the warm weather in San Diego. Is that a conch shell on your desk? Have you ever gone scuba diving? (Thomas, 1997, p. 41)

In clanging or glossomania – a feature of formal-thought disorder in schizophrenia – there is no convincing semblance of the patient being in the right sequences of classes of epistemic states with requisite efficiency (glossomania is also found in mania). His thought process pursues meaning and sound associations that are cognitively unproductive and inefficient. For example, a subject is asked the colour of an object. It is salmon pink. The patient responds:

> A fish swims. You call it a salmon. You cook it. You put it in a can. You open the can. You look at it in this colour.

> Looks like clay. Sounds like gray. Take you for a roll in the hay. Hay-day. Mayday. Help. I need help. (Cohen, 1978, p. 29)

Essential to successful cognition is the capacity to process relevant information and the related ability to ignore informational clutter. What these studies indicate is that impairments of these capacities could well be pathological.

5.10 *Shrinking-violet fallibilism*

If our present reflections are allowed to stand, there is a version of fallibilism that can't be made to work for individual agents. For lack of a name, I'll call it "shrinking-violet fallibilism". Shrinking-violet fallibilism comes in two parts. It shares its first part with all forms of fallibilism. It asserts, as usual, that our cognitive procedures are not only not error-proof, but are attended by nontrivial frequencies of mishap. Its second part separates it from the version that we have been promoting here. The second part looks with favour on dumbing down. It asserts that given the practical unavoidability of error, together with error's concealedness, it is better not to believe the disclosures of our cognitive procedures, but rather to accept them with a tentativeness appropriate to their vulnerability. As we saw, what this second condition provides is that cognitive agents stop having s-states, and substitute for these something like weak belief or mere acceptance. In a gentler and more realistic variation, what shrinking violet fallibilism requires is that cognitive agents stop *acting* on s-states, reserving their actions for those that are sanctioned by the requisite "acceptance"-states.

It is historically interesting that the version of fallibilism to which the likes of Peirce, Reichenbach and Popper have been drawn is the shrinking-violet variety. They were drawn to it in the course of reflecting on the methodological requirements for epistemically responsible science. In that very respect, the interest they evince in fallibilism proceeds from an interest in the behaviour of *institutional* agents, of the various disciplines and collective enterprises that science is made of. At a certain level of abstraction it is easy to see that science can manage very substantial levels of tentativeness, well beyond what an individual is built for. An individual marine biologist can go to bed each night crammed with s-states about the ecology of the Pacific Rim, just as an individual mathematician can go to bed each night with the uneroded belief that Goldbach's Conjecture is true. But they will be well-advised to constrain themselves utterly in exposing these convictions to the scrutiny of a journal editor. In this regard, they are able to accept the admonition of shrinking violet fallibilism. They are able to do so as publishing professionals. They are able to restrain their *publication behaviour* in the requisite ways. But they are not capable of such restraint across the board, not even in matters relating to the ecology of the waters that ebb and flow over the beaches of their city or the interdependencies that yoke together the universe of prime numbers.

On some tellings, Peirce is a strict shrinking-violet fallibilist about science. So are Reichenbach and Popper and, in some of his manifestations, Feyerabend. A certain support for such conservatism, for such risk-averseness, is that institutional entities don't have psychologies, except in the metaphorical sense of some imagined relation of supervenience. They aren't subject to the provisions of the CR-model of knowledge, except again metaphorically. And they aren't in s-states, except with the same proviso.

Even if he could stop himself from being in s-states, and, if not, even if he could sever the causal tie between such states and the actions they give rise to, and reserve all his actions for those sanctioned by acceptance-states or states of partial belief, the individual would place himself at a staggering disadvantage. He would lose *range* and *timeliness,* what with very large classes of actions now either forgone or left too late or both. In the grim and unforgiving econologics in which the individual's cognitive dynamic plays out, this is diffidence on a scale that would kill him. Shrinking-violet fallibilism is not on for individuals. Something less battened-down is required.

5.11 *Whiggish fallibilism*

I don't think that the concealment-detection problem is in any very direct fashion an easy matter. At least, this is what I think of the detection part. End-state embubblement is the problem. It suggests that in the matter of error detection we are no better off than with error-avoidance, and that with regard to *that* we are no better off than Error Abundance and Enough Already jointly allow. As applied to history, "whiggism" denotes the proposition that when governments behave themselves, the human good improves progressively with the passage of time. As applied to epistemology, it denotes the view that when the methods of enquiry are respected, our knowledge of the world improves progressively with the passage of time.

Epistemological whiggery suggests a way to handle the concealment-detection problem. The shrinking-violet fallibilism of a section ago helps set the stage. The prominent forms of shrinking violet fallibilism are also forms of whiggery. Whiggish fallibilists think that, properly conducted, enquiry is *progressive*. The scientific fallibilism of Peirce and others provides that ultimate answers are impossible or, if possible, at all, attainable only at some theoretical limit. But these philosophers are also progressives. They think that we get closer and closer to ultimate answers – or more carefully – that we get better at achieving a partial knowledge of things over time. If this is right, we have a loosely expressible principle that lends weight to the betterness-over-time of holding with the new and dispensing with the old when the new conflicts with the old.

Proposition 5.11
FAVOURING THE NEW: *If our cognitive devices are functioning as they should, they will over time favour the new*

185

over the contrary old.

The scientific fallibism in which this Whiggery is embedded is a philosophical thesis about scientific knowledge. The fallibilism I am plumping for has a wider scope, and it entails neither radical scientific fallibilism nor its negation. Proposition 5.11 is offered in like spirit. It is a claim about the cognitive enterprise in all its variations. This being so, we cannot ground Proposition 5.11 in the progressivism of scientific fallibilism. We need a wider basis for it. What might this basis be?

I seek a progressivism that lends some support to Proposition 5.11. Since the early parts of the book it has been ready to hand. No one disputes that favouritism towards the new is a fact of our cognitive lives, and NN-convergence proposes that this alone is defeasible reason to think that the practice is right. Enough Already reports a further fact. Our cognitive practices are right enough enough of the time about enough of the right things for us to survive and prosper. Progressivism towards error-correction is simply too embedded in human cognitive practice for it not to be subject to the provisions of Enough Already. On the other hand, Error Abundance records, if not a fact, then a universally held s-belief that we make lots and lots of errors. This couldn't square with Enough Already unless as a matter of course the human individual were graced with a disciplined and efficient "error management system". He is. A marked difference between individual and institutional agency is that individuals have quick and reliable feedback mechanisms and institutions by and large don't. It is one of those comparatively rare cases in which a cognitive resource is more available to the individual agent than to the institutional agent. An individual's feedback mechanisms are an integral part of how he manages error, making possible some striking economies. It is quicker and less harmful to make correctable errors than at all costs to avoid them altogether. We see in this the embodiment of an ancient and wise idea: It is better to learn from experience than in advance of it. Some people will be of a different mind. Better, they will say, to work very hard and sharpen our cognitive skills so as to make error-correction unnecessary. Perhaps this is so in some idealized model. But, causal forces being what they are, it hasn't a chance of being true on the ground.

When one forms an s-belief in the first place, one is operating within an end-state bubble. It is true that, when correcting that belief, one is in a later end-state bubble. Embubblement is a permanent condition. There is no escaping it. This is no cause for alarm. Embubblement helps explain how error is possible in the first place. Its inescapability is especially important for fallibilism. A fallibilism that endows beliefs with a certain susceptibility to error is not credible if it fails to extend the same propensity to new beliefs purporting to correct the errors of the old. Embubblement at both ends of the error-making/error-correcting process is a good way of stating a credible fallibilism. Again, fallibilism is an equal opportunity employer. But how, it might be asked, does this allow for progressivism? A basic part of the answer is that error correction at second remove doesn't restore original belief. That is, if β corrects α and, later, λ corrects β, it is not in the general case true – in fact it is hardly ever true – that α is thereby reinstated. Correction of correction doesn't work like Double Negation. Accordingly, there is a good deal of stability to error-correction. To a considerable extent, once an error, always an error. This is in substantial measure the good of progressivism. As time passes, the list of permanent errors increases with a striking stability.

5.12 *Detection Again*

Have we solved the detection problem? Let us see. Error-detection has two main components. Let "E" denote an error, that is, a proposition it would be an error to believe true.

1. *Incompatibility.* There must be one or more true propositions with which E is seen to be incompatible.
2. *Then-and-now.* Given such an incompatibility, we must honour the distinction between the lapsing of E's former truth and the discovery of E's falsity all along. While it would be a mistake to persist with a proposition which is no longer true, the crucial factor here is how to mark the difference between the now false and the false all along.

It is true that the progressivism of a section ago provides an answer to the embubblement problem. Why, when they conflict, is it good to favour the embubbled new over the embubbled old? Because, we said, in so doing we increasingly enlarge our grasp of beliefs that shouldn't be held. Doing so makes us growingly savvy about where the epistemic bodies are buried. But the progressivist argument does nothing to solve the now-and-then problem. Granted that E is now false, how are we to tell in a suitably general way whether it was also false then? If it wasn't false then, our recognition of its falsity now wouldn't count as detection of an error. At most, it would be detection of a potential for error – the error of holding the now false now true.

Waiving the problems that evidently assail the incompatibility problem, what are we to say of the now-and-then problem? If we bring to this task a pure form of the CC-sensibility, we are likely to look for criteria or rules, $R_1, \ldots, R_n$, such that the following expresses a true bi-entailment.

> (*)With respect to a proposition α, an agent X is able to distinguish between its now-falsity and then-truth and its now-falsity and then-falsity if and only if he is able to show that his procedures for so doing satisfy $R_1, \ldots, R_n$.

This imposes two tasks on X, neither of which is hardly ever remotely performable on the ground. He must specify the R_i. And he must establish that his procedures satisfy the R_i shades of our earlier discussion of the J-condition in chapter 3. Softer versions of (*) are ready to hand, whose softness is proportional to their ludicrousness. Giving softness at first remove is the requirement that X specify the R_i and that his procedures satisfy them, even though X might be unable to show that they do. Softer still is the requirement that X's procedures satisfy the R_i, even though X is unable to specify them or to show that his behaviour squares with them. In these last two cases, dark accomplishment is offered as a way out. True, X may be able to give express specification of the R_i and to show that his behaviour respects them, but this is precisely what he does do darkly, in the tacit precincts of down-below. If it is a fact that we make lots of errors, and if it is also a fact that we are impressively good at detecting them after the fact, then it couldn't be that we are not adept solvers of the then-and-now problem. These, then, are facts. However their resolution may be brought about, there is little question of its being a matter of implementing CC-protocols on any honest reading of their CCness.

The CR-model beckons from the wings. With respect to present matters it tells us that what makes it possible to solve the then-and-now problem and to set up the bona fides of error-progressivism is that (a) that we have error detection devices and (b) by and large our devices work as they should. Since there is nothing in the CR-model that requires us to specify, or be able to specify, the cognitive procedures that serve us well, this is not a Molière-answer to our question, in the manner of the dormitive virtue of sleep. Instead, the error detection "problem" is more realistically understood as a "task-identifier": Here is what we must be able to do to detect error. It is more a deconstruction of error detection than it is a recipe for how to pull it off. It answers the question, "What are the subtasks of the task of detecting error?" It does not answer the question: "How are these tasks performed?"

No one should think that "How is it done?" isn't an interesting question, made so in no small measure by our inability to answer it. Of course, the answer rests with the natural sciences of cognition. It is an answer for the lab to produce, not the sherry-soaked senior common room. One of the reasons to like the CR-model is that it is so at home with the idea of how little our adeptness at knowing things owes to a stateable awareness of how to do it. And, again, on any fair reading this is a hard pill for CC to swallow.

Have we now solved the then-and-now problem? In a way we have. We have shown that knowing how to solve it is not a condition of solving it, notwithstanding that it is a problem we all solve all the time. Beyond that, if the nuts and bolts of this wherewithal have yet to be ascertained by the cognitive scientists, it is a mistake to press the logician for them in the here and now, if ever.

CR is a way of redeeming our CC-shortfalls. The CC-model's performance standards are more honoured in the breach than not. All those CC doings that lie beyond our reach are either not doable at all or are doable by our cognitive devices working mainly in the dark and in causal nexi that are only now starting to be picked up on. What Harry has a terrible time doing in the CC-way, Harry's *devices* have a bang-up time doing in the CR-way.

CHAPTER SIX

ECONOMIZING

"There is nothing to be learned from the second kick of a mule."

American saying

"Unfortunately, the difference between 'optimizing' and 'satisficing' is often referred to as a difference in the quality of a certain choice. It is a triviality that an optimal result in an optimization can be an unsatisfactory result in a satisficing model. The best thing would therefore be to avoid a general use of both these words."

Jan Odnoff

"[I]nformation, rather than knowledge or belief, should be the most basic concept of philosophy."

Jaakko Hintikka

6.1 *Errors: in and of*

An ecology is a dynamic system of interdependencies between organisms and their environments. An economy is an ecology for the production and distribution of wealth. A cognitive economy is an ecology for the production and distribution of knowledge. Economies of both kinds are subject to some quite general constraints. They define circumstances in whose absence the desired product is not produced or not as abundantly as producers would have wished. They are circumstances in whose absence abundance is compromised, if not outright precluded, by inefficiencies in the means of the production, brought about by cost overruns and production delays. These same considerations apply with equal force to an economy's product distribution side.

All economies are prone to error as a matter of course. Error is a fixed fact of life in human affairs. Part of an efficient abundance-producing economy is its system of error-management. An efficient error-management system must achieve an affordable balance among different but important goals, of which the most important are avoidance, detection and correction. Error correction procedures are more than the replacement of false beliefs with true and of bad reasoning with good. They are also harm-mitigation strategies. They help producers recover

190

from the damage done by error. In some cases, recovery strategies also can be production *enhancers*. In cognitive economies this happens when knowers learn from their mistakes. An important error-management objective is the timely detection and repair of errors of greatest productive damage. Some errors hardly matter – indeed lots don't – but the ones that do call for timely and efficacious redress. The deployment of these strategies are subject to ecological constraints. They are unexecutable if they demand of producers more than is in them to produce. Here, as everywhere, you can't get blood out of a turnip.

Proposition 6.1a
THE APTNESS OF SOME ERRORS: *Errors are sometimes committed when the agent's cognitive devices are functioning as they should. It is sometimes reasonable to use procedures that lead to error. Blanket error avoidance is not, therefore, a general condition on cognitive success.*

COROLLARY: *There is cognitive good to be achieved by the engagement of cognitive procedures that let us down with notable frequency. Such letdowns are occasion to learn from experience. They are fruitful contexts for trial by error.*

Why should we accede to Proposition 6.1a? In no small measure, it is because of an important difference in cognitive type between individuals and institutions:

Proposition 6.1b
FEEDBACK: *By and large, individuals have speedy and reliable feedback mechanisms. In contrast, institutions fare less well. Indeed, in lots of cases, this is an understatement.*

Propositions 6.1a and b make an instructive point. They encourage us to acknowledge a distinction between errors made in the course of reasoning and errors made by the reasoning itself, or as we could say, errors *in* reasoning and errors *of* reasoning. It also calls attention to a limitation on error-avoidance. Building an error-free system often has advantages that an error-permitting system lacks. But it might also happen – indeed it often does happen – that overall the error-permitting system is superior. Such is the view of evolutionary biologists and any number of

191

economists. In evolutionary biology, it is necessary to take into account the overall fitness of an organism, which might have to make trade-offs between rival goals. One of these might be the goal of error avoidance. Relatedly, another might be the avoidance of that cost, owing to selection pressures, by building error-permitting systems. The trade-off is, of course, blind, driven by the mechanisms of differential reproductive success. Whatever the details, I see cognition as a flexible adaptive strategy in complex environments, subject to constraints imposed by phenotypic trade-offs in energy investment. This view underlies much work in cognitive ethology[195] and was first applied to an evolutionary approach to cognition by Godfrey-Smith in his Environmental Complexity Thesis.[196] For my purposes, what matters is that error avoidance is not necessarily the only option in a complex environment. Sober also writes to this same effect in the case of learning.[197] The adaptive wiring of simple model animals is well discussed by Braitenberg and Churchland.[198] There is a large parallel body of work on the psychology on implicit and explicit learning. Still important, in a literature that turns over rapidly, are Dienes and Perner (1999) and Wilson (2002).[199]

So, then, sometimes errors of reasoning are better to commit than avoid. This is so when it is cheaper and more productive to make and correct them than making them in the first place. A theory that honours the distinction between errors in reasoning and errors of reasoning is one that would try to divide reasoning errors into two possibly overlapping sets. The partition would give us an obvious question to reflect upon. Might it not be the case that, comparatively speaking, the cardinality of the first class (*in*) is notably larger than the cardinality of the second class (*of*). I have an opinion about this:

Proposition 6.1c
SMALL: *In relation to the human reasoner's total error-space, the number of errors of reasoning is comparatively small.*

If Proposition 6.1c can be made to stand, it will turn out that we are not notably deficient in our reasoning practices. At a minimum, incorrect

---

[195] Dukas, (1998), (2009).
[196] For elaboration of the ECT see Godfrey-Smith (2002), and Sterelny (2004).
[197] Sober (1994).
[198] Braitenberg (1984) and Churchland (1985).
[199] See also Vokey and Higham (2004).

reasoning would be far from being the primary source of our errors, including the errors we make when we are reasoning. In this respect, reasoning is more like perceiving than believing. It is widely acknowledged, and rightly, that our perceptual errors significantly lag behind our doxastic errors. The same, I keep saying, is true of errors of reasoning. Proposition 6.1c harkens back to Proposition 2.4b. Proposition 2.4b says that in relation to errors of belief errors of reasoning throw a smaller shadow. Proposition 6.1c says that in relation to errors in reasoning, errors of reasoning cast a likewise smaller shadow.

The distinction between errors in reasoning and errors of reasoning links to the contrast between inaccuracy and inaptness. When Harry makes a mistake in reasoning, he is in error about something. If the error is not an error of reasoning, a deficiency in the reasoning itself, then there need be nothing inapt in Harry's having resorted to it.

6.2 *Monstrous belief*

"I'm all right, Jack" is a colloquial British saying which, with a complacency that belies its confidence, conveys that all's well with me, never mind the rest of you. What whistling past the graveyard means for courage is false courage, and what "I'm all right, Jack" means for confidence is false confidence. On a colloquial reading, the enough already principle might also belong to this grouping, signifying an attribution of well-doing that owes more to bravado than to fact. Certainly any endorsement of Enough Already, made without recognition of the stupidity and evils that scatter through human life in all its social arrangements, would likewise score better for bravado than factuality. The intention of Enough Already is to call attention to our successes overall in the face of our day-in, day-out stupidities and our not entirely infrequent collapses of judgement, will and moral dignity. In the form we have it here, Enough Already is not meant as the boastful "I'm all right, Jack" vulgarism. It is intended to record the empirical observation that, while our errors sometimes lead to staggering unpleasantness and depravity, the species *Homo sapiens* is still vigorously in business.

In chapter 1, we gave some thought to those conditions of premiss adequacy on which it would be reasonable to expect some suitably reflective contribution from logic. We said that we might turn to logic for guidance about the consistency, relevance and plausibility of premises. What we didn't consider is the avoidance of premises which in Berkeley's words are "repugnant to reason". Although the idea of repugnancy to reason doesn't strike one as all that obscure, determining its instances is a more difficult task than might have been imagined. The denial of the law of identity would make the list rather comfortably, but not so easily the denial of the law of non-contradiction, thanks to the exertions of our dialethic friends. "$2 + 2 = 2$" might also make the cut, as well as metaphysical amusements such as "There is nothing" and ethical lunacies such as "Water-boarding infants is perfectly all right if you're really into it." There is, even so, something unsatisfying about the exercise. It all seems so trivial.

Suppose now that we expand our repugnancy-to-reason list by addition of beliefs actually held and acted on by individual agents. Of particular note are offences that cause great suffering or other unbearable costs. They are two depravities at once, one intellectual and the other moral. Let us call them what they are: They are *monstrous beliefs.*

It is sometimes remarked, with the requisite incomprehension, that the Third Reich displayed a masterly command of the logistics that drove the Final Solution. Extinguishing the Jews from the face of Europe was a daunting task, and it took masses of clever people to bring the programme as close to completion as it got. There was much excellent reasoning involved in this – excellent engineering, excellent chemistry, excellent transportation policies, excellent opinion making, excellent personnel management, and so on. They brought the technology of mass execution to a whole new level. If reasoning were simply a matter of accurate or apt consequence-drawing, we would be landed in the unpleasantness of having to say that, from the point of view of logic, there is little bad to say against the Final Solution, that the Nazis were first rate reasoners, notwithstanding that they were simply monstrous believers.

Defective reasoning is not just the drawing or failing to draw consequences that should not have been drawn or should have been drawn. Defective reasoning is also about the acceptance of premises that offend reason and outrage decency. In extremis, when mistakes of this kind are made, we say that their committers have lost their minds. So, in a way, they have. But it is an especially chilling way to lose one's mind, given the radicalness of the gap between being bad at premiss-selection and being sharp as a tack at conclusion-drawing.

The phenomena of monstrous belief aggregate with a notably frequency to create perils of local import, some of which stretch the very meaning of "local". The fall of Athens changed the then known world, but it was a local calamity even so. No logic of error can be indifferent to

194

the sheer scale in human affairs of the breakdown of reason. There are social policies, movements and ideologies galore which no one in his right mind would touch with a barge pole. Yet these repugnancies to reason proliferate and flourish. There is much in the ideological exuberances of the human race that exceed the theoretical reach of logic, no matter how aggressive its naturalization. But there are also immediate consequences for what we are doing here. One of them is that the assumption of NN-convergence requires an overhaul that honours the asymmetry between our adeptness at premiss-selection and our adeptness at conclusion-drawing.

Proposition 6.2a
REVISING THE NN-CONVERGENCE THESIS: *Except where there are indications to the contrary, the conclusion-drawings and the conclusion-drawing omissions that actually occur are in a broadly typical way either accurate or apt or both. Similarly, except where there are indications to the contrary, premiss-selections that are actually made are not by and large repugnant to reason or common decency.*

COROLLARY: *However, indications to the contrary in premiss-selection contexts typically exceed those that attend conclusion-drawing contexts, and do so rather extensively.*

Again, rotten belief outpaces rotten reasoning by a country mile.

6.3 Satisficing

When your camera is on automatic it is a satisficer. When it is on manual it is a maximizer. Doubts about the wisdom of investing our theories of human performance with maximization and optimality assumptions in the manner of the OSC and SEU orientations is necessary occasion for reflection on the alternative. A good many economists would attribute the successes which our cognitive practices bring us to a form of satisficing, a suggestion that goes hand in glove with the enough already principle. Indeed, virtually the whole heuristics and bias literature in psychology is, in one way or another, an exploration of satisficing.[200]

---

[200] Pioneering work on the satisficing character of human rationality is Simon (1957) and (1979). See also his (1983) and (1996).

Human individuals operate in cognitive economies. Economies are value-added enterprises advanced by prudent use of resources for the realization of reasonably selected ends. Its participants transact their cognitive agendas under press of scant resources. Their cognitive tasks are small – not small in importance, but small in regard to the heft of the resources needed for their attainment, and to enable their performers to operate as welterweights rather than super-heavyweights. No one in boxing circles thinks that the best welterweight is a less good boxer than the best super-heavyweight. Not infrequently it is the other way round. But there are certain assets that welterweights have less of – weight is one and, relatedly, punching power. Although not invariably so, it is typical of the human individual to make his way through life's cognitive thickets with deficits of information and time, and storage and computational capacity. To the extent that this is so, it is necessary to be a cognitive economizer. It is well to re-emphasize that the scantness of the resources available to individual cognitive agents is a *comparative* matter in relation to the cognitive wherewithal of institutional agents, such as the Ministry of Health, MI5 or Her Majesty's Government. Agents of both types are frequently pressed for information, and time, and frequently are up against capacity limits on memory and computation. To that extent, agents of both types are required to economize. Still, comparatively speaking, individuals have less information, time, and storage and computational capacity than institutions. Their need to economize is comparatively greater. Their need to be resource allocators is in that same regard more pressing.

Attractive as it is, the metaphor of satisficing should be approached with care. It has more meanings than serve our purposes here. In one sense, a satisficing decision is one that settles for present gains over the possibility of future better ones that lie within our grasp. In another sense, a satisficing decision is one that accepts present gains over gains which, while better, lie beyond our grasp. Seen the first way, satisficers forgo a possibly better outcome – achievable if it exists – as a hedge against a future that could actually turn out badly. Satisficing here is quitting while we're ahead. Conceived of the second way, satisficing is doing as well as we possibly can. It is the best humanly performable approximation of some ideally imagined conception of the best. Between quitting while we're ahead and the best that is in us to do is a gap big enough to make something of.

The good-enough version of satisficing charts a particular kind of connection to our interest in good practice. It reminds us that our cognitive efforts, while they carry the promise of success, are bound by the necessity to draw down the assets required for their deployment, and that the associated costs often have a way of exceeding benefits. That the cognitive agent is a satisficer in this second sense is far from a trivial thing to say about him. It captures an essential condition on his successes, especially by the lights of Enough Already. It calls attention to an important limitation:

Proposition 6.3a
SATISFICING: *Our cognitive successes demand the judicious husbanding of the resources required to achieve them.*

It is a truism that there are lots of things we cannot do. The fabled cow may have been able to jump over the moon, but not we. Even the best high-jumper in the history of the world wouldn't dream of organizing his training around an attempt to emulate Bossie's lunar achievement. It is not much different with cognition. Given the size of the human brain, there is a lot that exceeds its reach. Evolution has decreed that the human brain must fit a head capable of negotiating its mother's birth canal. So we must eschew theories that require us to have brains "the size of a blimp."[201]

The individual cognizer manages his affairs with an eye on costs and benefits. There is a considerable literature supporting mathematically aggressive cost benefit theories. Factored into such analyses are considerations of resource allocation, marginal expected reliability, marginal utilities, start up costs, opportunity costs, cost benefit curves, and so on. Cost benefit theories tend to fall within the *NP* sector of the *NPD* troika discussed in chapter 2. They emphasize optimality or computably executable optimality over empirical accuracy. They re-start the debate about how to deal with the significant discomportment of these theories with actual practice. They call into play the standard moves: the irrationality thesis and the approximate rationality thesis. Cost benefit theories often have insightful things to say about satisficing strategies, but rarely are the decision mechanisms for the selection of a satisficing strategy within the actual control of reasoners on the ground. What is missing in mathematical theories of costs and benefits are empirically correct descriptions of how human individuals manage the economics of their tasks as actually transacted. Accordingly,

---

[201] Stich (1990), p. 27.

Proposition 6.3b

COST-BENEFIT ANALYSES: *The standard approaches to mathematically virtuosic analyses of costs and benefits are ill-adapted to the practical concerns of real life reasoners and give a distorted picture of how they actually perform as cognitive satisficers.*[202]

6.4 Proportionality

I want now to bring to the fore an idea that has been hovering since the book's opening pages. It is the idea of proportionate cognitive targeting. A sensible thinker, like a sensible athlete, will be disposed to set targets which, with some frequency, he has some chance of hitting with resources at hand or available to him with appropriate effort. So cognitive agency is governed by an important pair of constraints:

Proposition 6.4a

TARGET-PROPORTIONALITY: *In the environments in which they find themselves individual agents have a disposition to make their cognitive targets, their cognitive agendas, proportional to the cognitive resources available for their attainment.*

Proposition 6.4a introduces a second proportionality thesis, a companion to the knowing-doing well thesis. The present one, though similar, is different. The first linked our well-being to our cognitive achievements. This one links our cognitive achievements to realistic target setting. Accordingly,

COROLLARY: *Our doing well is linked to realistic target setting.*

---

[202] Cost benefit analyses are a hot ticket in philosophy and the social sciences. The *loci classici* are von Neumann and Morgenstern (1947), Arrow (1963) and, again, Raiffa (1968). Aside from the applied mathematics of economic reasoning proper (see also Mishan, 1971), cost-benefit postures are struck in linguistic pragmatics (Prince and Smolensky, 1993), sociobiology (Wilson, 1975), law (Posner, 1973, Coase, 1960 and Calabresi, 1970), philosophy of economics (Hansson, 2006), philosophy of science. (Mayo and Hollander, 1991) and quantum computing (Shor, 1994). For a sceptical reaction, see Gigerenzer (2002).

We also have it from the target-proportionality thesis that something is an error only relatively speaking.

Proposition 6.4b
RELATIVITY OF ERROR: *Something is an error only in relation to the success-standards encompassed by the agent's cognitive target.*

It is a fundamental insight of a resource-based approach to human cognitive practice that for the most part an agent's resources do not enable him to meet the standards of either mathematical proof or scientific confirmation. We may say that, given the resources that he commands, the human reasoner is alienated from those standards by and large. This is critically important for both the interpretation and assessment of reasoning, recalling a pair of claims from chapter 1.

Proposition 6.4c
INTERPRETATION: *It is a default-condition on the interpretation of an individual's reasoning on the ground that it* not *be in the service of cognitive targets that call for either validity or inductive strength.*

Proposition 6.4d
ASSESSMENT: *It is a default-condition on the assessment of an individual's reasoning on the ground that its invalidity or inductive weakness are* not *grounds to judge it negatively.*

The cognitive abundance and error abundance theses now have some room to breathe. That most of right reasoning is invalid or inductively weak removes an orthodox impediment to Cognitive Abundance. Most of what we know we know in ways that embody invalidity or inductive weakness. It also allows Error Abundance to confirm something important about error. It supports the claim that for the most part bad reasoning is constituted by factors other than invalidity and inductive weakness. It also gives us occasion to amend fundamentally the RR-rule-violation conception of errors of reasoning. Since most good reasoning violates the rules of right deductive and inductive reasoning, we must now say that what the present conception intends is that

Proposition 6.4e

RR RULE-VIOLATION AMENDED: *Something is an error of reasoning only if it violates a rule of right reasoning that is* contextually in force.

Something else of interest drops out. With a scattering of possible exceptions, logic's traditional assessment of the traditional fallacies is that they are errors of reasoning occasioned by invalidity or inductive infirmity. Thus the fallacy of affirming the consequent is constituted by its invalidity, and the fallacy of hasty generalization is constituted by its inductive weakness. There are those that think that some few of the traditional fallacies aren't so occasioned. This would be because, like the fallacy of many questions, they are thought to be dialectical infelicities rather than errors of reasoning. Still, these possible exceptions aside, the traditional fallacies are put in an odd position by our recent reflections. By the assessment-default, the traditional fallacies don't deserve our low opinion of them. One should not judge them negatively for their invalidity or inductive weakness except where there is particular reason to do so.

We might think that the standards to which a reasoning task is plighted is implicit in the linguistic behaviour required for its performance, notably in the reasoner's use of so-called "conclusion-indicator" words: "since", "hence", "therefore", "it follows that" and so on. It is true that these indicate that some consequence-spotting or conclusion-drawing is going on, but they underdetermine whether it is deductive consequence, or inductive, or some other variety. Similarly, as readers of the Sherlock Holmes stories well know, it is entirely possible to characterize your task as "deduction" without its being deductive in the logician's sense or to describe what's going on as "induction" which isn't induction in the logician's sense. This is something that the respect for data principle obliges us to pay attention to. Not only is it possible to say "therefore" without meaning the deductive logician's therefore, and to say "deduction" without meaning the deductive logician's deduction, and so on, but in light of the target proportionality thesis –Proposition 6.4a – it is not typically accurate to attribute to actual individuals reasoning objectives that are bound by the deductive logician's standards of deduction or the inductive logician's standards of induction. By these lights, there is something amiss – certainly something untypical – about the commission of any fallacy whose fault is thought to lie in its deductive invalidity or inductive frailty. It is true, as we have said, that there are types of cognitive activity for which these are the right standards – proofs in topology and drug approval tests, and so on. But relative to the whole space of cognitive tasks open to a human individual, these types are not, so to speak, typical.

6.5 *Rules*

All along we have left it open that, in some manner or other, errors of reasoning are violations of rules of right reasoning. This, the deontological view of right reasoning, is kith-and-kin of the justificationist view of knowing. I have already expressed my reservations about justificationism. It is now time for rules to go.

In its customary employment, the term "rule" invites CC-construal. Sometimes that is the right way to take it. If Harry is producing a natural deduction proof of something, he must cite the rules that justify the proof's successive lines, and those lines must arise in ways that satisfy those rules. Similarly, if Sarah wants to find a cure for cancer, we may grant that doing so is in her interest, that is, that Sarah, like the rest of us, is favourably disposed to be in cognitive states that would conduce to an understanding of these things. But she is hardly likely to *get* a cure for cancer unless she puts her mind to it, unless she follows the right experimental procedures. In such cases, Harry and Sarah are in do-it-yourself mode; their cognitive devices are on manual.

Rules are not always like this. Sarah is a good driver with a perfect record. She is good at obeying the rules set out in the Highways Traffic Act. But Sarah has never seen those rules, and would have an awfully hard time of it to cite them if asked. When it comes to good driving Sarah is in point-and-shoot mode; except for complications, her cognitive devices are on automatic. If Sarah's good driving is a matter of compliance with the rules, the question of how she manages to engage them takes on a certain complexity.

There is something further to say about rules. In any honest sense of the word, there are many fewer of them than would be required if all good reasoning were rule-following and all bad reasoning were rule-breaking. Suppose that Harry has just now completed some reasoning. It is sometimes assumed that, on the rules approach, Harry's reasoning cannot have been good in the absence of a rule that tells him that it is necessary (or permissible) to reason in the way that just did. But this is carelessly wrong. The more reflective claim is that Harry's reasoning cannot have been good in the absence of a *correct* rule that makes it so.

What, we might ask, is a correct rule of reasoning? It is a rule made correct by the rightness of the reasoning it mandates. So it is not the rule that makes Harry's reasoning right. It is the reasoning that makes the rule correct. "Rule" is all right as a *façon de parler*. But it is not much of a load-bearer in the theory of right and wrong reasoning. Rules don't wear the trousers there.

This, if true, is a setback for the RR-violation conception of mistakes of reasoning. Of course, it is possible that when a logician characterizes a bit of incorrect reasoning as reasoning that breaks a rule, his use of "rule" is a loose convenience. It is possible that what he means is that the reasoning in question disconforms to a *condition* necessary for its rightness; that is, is necessary in the context at hand. In such cases, a rule is a condition in a manner of speaking. It is indeed a *façon de parler*. I see no harm in it, except that, so-used, "rule" does not mean what it means in ordinary speech. These aren't rules. They are rules*.

Proposition 6.5
DROPPING THE RR RULE-VIOLATION CONCEPTION: *Except as a* façon de parler, *the RR rule-violation conception seriously misconstrues errors of reasoning.*

6.6 Dumbing-down

The theses of proportionality, relativity, interpretation and assessment are the undercarriage of the account of error that I am proposing here. If these claims weren't able to withstand critical attack, I should have lost my bearings in a major way. So it behooves me to give some further thought to what I take to be one of the most serious-seeming objections to these claims.

The main moral of the target-proportionality thesis is that the individual agent is disposed to eschew tasks that get him into cognitive trouble, that he tries to be a "fair-weather" reasoner. It might be objected that we are giving too much sway to how the ordinary human reasoner transacts his cognitive tasks in actual practice. It may well be a fact that reasoners on the ground generally set the bar of their cognitive endeavours low enough to make validity and inductive strength the wrong standards. But why make a virtue of this lowness? Why not say instead that, in dumbing-down, the fair-weather reasoner ducks the honourable discipline of attainment standards of a decent and measured rigour, that dumbing-down is a kind of cheating? Why not say that the project of knowledge would be better served by smartening up rather than by dumbing-down? People who dumb-down solely for the sake of error avoidance are Cliffordians of radical stripe. They so value error avoidance over the attainment of truth that they suppress their disposition to get to know what it is in their interest *to* know.

My answer is that ordinary reasoners are not in this sense dumbers-down. What shows this to be so are the theses of cognitive abundance and error abundance in combination with Enough Already. We have already seen how this goes. Error Abundance tells us that even though our cognitive objectives are usually set at levels that don't call for validity and inductive strength, we still make lots of errors. So setting comparatively modest cognitive targets is not a particularly effective Cliffordian strategy. Cognitive Abundance tells a similar tale. Although our ordinary practices are peppered with error, this is no impediment to their substantial cognitive success. They are practices that give us knowledge, lots of it. Enough Already provides the clincher. It tells us that, over-all, the admixture of error and knowledge redounds to our net collective advantage, for it produces the wherewithal for survival, prosperity and even, on occasion, high cultural attainment. Overall, what we see is that individual cognizers pursue satisficing strategies that repay them with collective success.

This recalls to mind Groucho's doctor's advice: If you want to pop your epistemic bubbles, stop being so sure of yourself, that is, replace strong belief with beliefs of lower intensity. Against this I proposed that the toning down of intensity is not causally available to us as general strategy of belief-state regulation. The smartening up proposal pulls in the opposite direction. It recommends that we upgrade our cognitive practices, that we so tailor our reasonings as to enable them to fulfill stricter conditions of rightness. I have the same thing to say against the proposal. It is not causally available in the general case.

The dumbing-down objection embodies a further misconception. It is the suggestion that, irrespective of what real life reasoners are up to on the ground, truth preservation is a *worthier* cognitive target and, when attained, a *better* cognitive result than strict scientific confirmation, which in turn, is a worthier thing to aim for and a better thing to attain than targets that respond well to the call of, say, third way reasoning. By parity of reasoning, the attainment standards embedded in these targets are rank orderable, with validity topping inductive strength, and inductive strength topping third way effectiveness. Logicians are greatly tempted by such orderings, and rightly so if the betterness of a target and its standards for attainment are equated with strictness. But if betterness is reckoned up with reference to what human individuals are capable of and the comportment of their cognitive behaviour with the enough already thesis, then truth preservation and experimental/statistical confirmation typically aren't the best targets, and validity and inductive strength typically aren't the best standards. Similarly for the strategy of smartening up.

Proposition 6.6

CONTRA SMARTENING UP: *It is not in the general case preferable – indeed it is not smart and not even possible – to upgrade our cognitive targets in ways that favour truth-preservation or experimental/statistical confirmation as general cognitive strategies.*

So, then, as in the economic domain, so too in the cognitive, our satisficing proclivities betray no fondness for underachievement.

6.7 Scarcity

The individual agent often finds it necessary to transact his cognitive agendas in the absence of information it would be good to have. When this happens, information is scarce. Scarcity should not be confused with scantness. Scantness is a comparative measure. The scantness of the information available to Harry on any occasion typically contrasts with the abundance of information available to MI5 on any occasion. Harry has less of it, lots less, and MI5 has more of it, lots more. Scarcity is orthogonal to this distinction. Never mind its quantities, whether for Harry or for MI5, when a cognitive agent is operating with scarce information, the information he (or it) possesses lacks data it would be helpful (or necessary) for him (or it) to have. This, we might say, is missing information. Whether in large quantities or small, what matters about missing information is the absence of the helpfulness its presence would have provided. Of course, it cannot plausibly be said that what distinguishes Harry from MI5 is that Harry has missing-information problems and MI5 does not. Given the duties it is charged with and the information-gathering difficulties counterespionage routinely encounters, it should not astonish us to discover that there are times when relative to the tasks in play, MI5's missing-information challenges dwarf Harry's own. So I say again that the scarce information-sufficient information distinction cuts across the individual-institution distinction, and that the scant information-abundant information distinction closely parallels it.

The idea of dark knowledge, dark belief, dark inference – and the having of interests darkly – complicates the present picture of cognitive agency. Harry, I say, is an information-processor capable of belief. Consider Harry's sensorium, the informational nexus of the five senses combined. It contains 11 million bits of information at a time, of which 40 make their way into Harry's consciousness at that time. Let $\alpha_1, ..., \alpha_n$ be all the conscious beliefs had by Harry on account of the information then contained in his sensorium. It is natural to think that some of that information was somehow causally constitutive of Harry's having those beliefs; that is, that it makes a material contribution to them. Harry could not on that occasion have the conscious belief that there is a bittern in the tree except on the basis of a good deal of materially supporting information of which he has no awareness. Let us say that the conscious information that materially supports Harry's conscious belief on that occasion is "cognitively engaged" on that occasion. When on the basis of that cognitive engagement Harry sees that the bittern is in the tree, there is more information in Harry than is in his sensorium. Perhaps some of that non-sensory information was also causally implicated in Harry's observation of the bittern. Certainly there will have to have been certain informational exchanges going on for it to have been the case that Harry had a functioning sensorium at all. But it seems right to say that at a certain point, at a certain depth, there is information in Harry in whose absence Harry couldn't have seen the bittern in the tree, but which nevertheless is not cognitively engaged in that perception; that is, is not a material part of it. So a question for the logic and epistemology of information systems capable of belief is how to draw in a disciplined way the distinction between unconscious information causally necessary for the observation that α and unconscious information not only causally necessary for it but also functioning as a material part of it.

The phenomena of dark cognition raises a similar question for cognitively engaged information. Consider the information that's in Harry at the moment of his believing that the bittern is in the tree. Some of that information is cognitively engaged with respect to that. How much of it is dark belief? In other words, is it true – and if so, to what extent – that all the unconscious information that is materially part in Harry's observation takes the form of dark cognition? If all cognitively engaged information is itself dark cognition, then dark cognition is several orders more plentiful and a good deal more deeply situated than is usually supposed by tacit-knowledge hypothesizers. If, on the other hand, some of the cognitively engaged information implicated in Harry's observation does not itself rise to the level of tacit cognition, then it becomes a pressing matter to find the boundary between cognitive and non-cognitive engagement. If dark cognition doesn't go all the way down, at what point does it stop, and why?

We now see that with respect to a given cognitive target there are at least four ways for information to be missing. It might not occur in any of the agent's information-processing sectors. It might occur in an agent's information-processing sector but not in its cognitively engaged regions. It might be a part of the individual's cognitively engaged information but not itself darkly cognizable. It might be darkly or tacitly cognized but not lightly or expressly so. There is a rough contrast between the first way of being missing and the other three. The first, we might say, is "external" missing, and the challenge it creates is to get some new information into the system. The second is "internal" missing, and the challenge it creates is to bring deep information closer to the surface. For the first challenge the appropriate means of address is *discovery*. For the second the appropriate means of address is *reflection*. There are limits, of course. No amount of thinking will bring causally implicated but cognitively non-engaged information to the attention of the mind's eye. No amount of thinking will bring uncognizable cognitively engaged information to the surface. No amount of thinking can bring all one's dark cognition to the surface – there is too much of it and the surface is too small. Even so, though there are no general guarantees, sometimes thinking about things will indeed be rewarded by bringing to the fore selective fragments of dark information. Remembering is sometimes like this. You think you know where you left your laptop, but you can't quite put your finger on it. So you think. You try to remember. You retrace in your mind the events of the past hour. Then it comes to you. You left it at Harry's house. Of course. You knew it all along.

6.8 *Interests*

A satisfactory understanding of the individual cognitive agent requires that we make some effort to harmonize his cognitivity with his practicality. Harry is a cognitive being. He wants to know things. He wants to know what is the case. He wants to know what to do. He wants to know what others know. Harry is also a practical being. He executes his cognitive routines with due regard for the comparative scantness of resources available to him. Harry is an economical knower. Important to

Harry's cognitive economy is the factor of *interest*.[203] Roughly speaking, what Harry ends up knowing on a given occasion will be conditioned by what it is in his interest to know. It is possible in a general sort of way to indicate what is in Harry's interest to know quite independently of a familiarity with the particularities of Harry's situation here and now. One of the things that it is in Harry's interest to know is what is relevant to his purposes and inclinations. Thus, as we said earlier, Harry's interests activate, or themselves function as, an irrelevance filter. More broadly, Harry has an interest in knowing what would be bad for him not to know. Knowing things has adaptive significance for Harry – not just any old thing, but things that are good for him to know. Harry, like the rest of us, is pretty good at knowing things that are good for him to know.

The concept of interest hooks up in an interesting way with the concept of agenda, which is the central notion of the agenda theory of relevance. In Gabbay and Woods (2003) agendas are abstractions from plans,[204] intentions,[205] strategies,[206] functions,[207] programs,[208] scripts,[209] matrices[210] and inclinations.[211] Sometimes a person's agenda is wholly transparent to him. Otherwise they are implicit or, as we have said, dark.

Proposition 6.8
DARK AGENDAS: *Dark agendas are closed or advanced in some or all of the following ways: unconsciously, automatically, inattentively, involuntarily, non-linguistically, non-semantically, and deeply.*

There is reason to think that these factors are not equivalent or even extensionally coincident.[212] But when some are present, typically others are too; and where this is the case, and some cognition is going on, this is dark or implicit cognition; it is knowing "down below". It is only natural

---

[203] The role of interests in framing and advancing our cognitive agendas is ably discussed in Habermas (1971).
[204] McCarthy and Hayes (1969), Fikes and Nilsson (1971), and Bratman (1987).
[205] Montague (1974).
[206] Wylie (1967) and Jauch and Glueck (1982).
[207] Lycan (1991).
[208] Harel (1984) and Rosenschein (1981).
[209] Johnson and Reeder (1997).
[210] Gabbay and Woods (2003).
[211] Gabbay and Woods (2003).
[212] Shiffrin (1997), pp. 50-62.

to extend the metaphor of down below to agendas. Often an agent is fully aware of his agendas and disposed towards their advancement in the most psychologically transparent of ways. But sometimes it seems right to think of a person's agenda as implicit and of its advancement or closure as happening under conditions that escape his notice. Equally, since information is relevant for an agent when it acts upon him in ways that facilitate the closure of his agendas, it seems right to leave it open that the relevance of information for an agent is something of which he is sometimes, perhaps even typically, unaware.[213] Allowing for the darkness of agendas is one way of emphasizing the patiency of knowledge, its being-done-to aspects. Accordingly,

> COROLLARY: *For the most part, agenda framing and agenda closing is a point-and-shoot enterprise.*

As I conceive of them here, interests are an extension of this general notion of agenda. An agent's interest on an occasion is what it would be in his interest to know. More generally, an agent has an interest in being in a certain state or in acting in a certain way if it would be in his interest to be in that state or to be behaving thus. The states it would be in Harry's interest to be in and the actions it would be in his interest to perform are states that are good for Harry to be in, states that matter for his doing well, and actions that matter in this way too. We can now say Harry *has* those interests to the extent that he is disposed to be in conditions it would be in his interest to be in. Thus Harry has an interest in breathing. He has an interest in peristalsis. He has an interest in knowing that water can't be respirated. And in the appropriate circumstances he has an interest in knowing the Stone Representation Theorem.

Like agendas, interests are frequently dark. For it to be true that something is in Harry's interest to know it need not be the case that Harry is aware of it. Harry's being pretty good at knowing things that are good for him to know doesn't require that he be conscious of either the interests or the knowings. Harry's knowings sometimes obtain darkly. Equally, having an interest, like having knowledge, is sometimes as much or more a matter of what happens to you as it is a matter of what you bring about by your own efforts freely and deliberately transacted. Patiency also matters here. For a certain level of abstraction an interest is

---

[213] See again Gabbay and Woods (2003), chapter 8.

a causal disposition to be done to in ways that promote the actualization of what it would be good for one.

In various places I have represented cognitive agency as a device for setting and trying to meet cognitive targets. Where, we might ask, do targets stand in relation to agendas and interests? The answer is that they stand to them identically, and that, as used here, the role of the word "target" is lexical variation. In most places here, "target" and "agenda" are interchangeable. In others, a target is an interest in the broader sense of a disposition to be in conditions that it would be good for one to be in. Context will be our guide.

In the contexts of practical reasoning, this notion of interest resembles Pollock's concept of *liking,* of which he distinguishes two varieties.[214] Pollock's main idea is that it is the job of an agent's rationality to make things more to his liking. For this to be a performable job, agents have a way of appreciating how likeable a situation is. Pollock thinks that this is achieved by having the requisite feelings. Feelings of likeability are open to introspection. Likings of this kind Pollock calls "situation-likings". "Feature-likings" are different. They are "quick and inflexible modules" that provide efficient and timely information about which features are and are not causally relevant in bringing about likeable situations. Humans use these modules – heuristics – to calculate the comparative expected value of plans. Some of the features of Pollock's likings seem to me to be excessively CC-like: the introspectability of situation-liking feelings and the calculability of comparative expected value. But the main idea is right. The reasoning we engage in is driven (and focussed) by our responsiveness to what it is in our interest to know, what it would be good for us to know. It bears repeating that most of what is true we'll never know. Very many of these truths lie beyond our cognitive capacities. Of those that lie within our reach, most remain unknown because knowing them would not answer to our interests. Interests are not only an irrelevance filter; they are also an inhibiter of naked epistemic ambition, a discourager of promiscuous cognitive yearning.

6.9 *Information-stacks*

The idioms of up and down and surface and depth suggest a vertical and layered organization for information. This encourages the

---

[214] Pollock (1995).

idea of the individual cognitive agent as an "information-stack", which in turn is a stack stacked on top of some causal stacks. I have already mentioned the difficulty in marking the boundary at which energy-to-energy transductions become energy-to-information conversions. There is also the problem of finding the divide between the information one has at a given time and what one believes at that time. Then comes the question of marking the beginning and the end of the dark and the light. These are not easy questions to answer. A good part of the reason is the unruliness of the concept of information. This has prompted Hintikka to observe it is

> far from clear ... what (if anything) is meant by these different "informations" – or whether they are related to each other at all. These questions seem to mark a most urgent challenge to philosophical analysis. (Hintikka, 2007, p. 189)

Not a unitary idea, it is more nearly correct to think of information as a family of not always happily reconcilable notions. It has been well said that "information is notoriously a polymorphic phenomenon and a polysemantic concept ....".[215] There are four especially influential members of this family. I will call them the "epistemic" sense, the "probability" sense, the "complexity" sense and the "military" sense.[216] In its epistemic sense, information is something taken in by an agent. It is a representation of what is the case. Information is what informs us about how things actually are. It is what is conveyed by informative answers. Information in this sense lies in the investigative domain of logic and epistemology.[217] In its probability sense, information is what is channelled from a source to a receiver. The source emits signals with a certain frequency, and the information picked up by the receiver is conceived of as the expected reduction of probabilistic uncertainty. Agency is not a necessary factor in the transmission or reception of

---

[215] Floridi (2008), p. 117.

[216] Perhaps the best two single volume treatments of the current state and future prospects of the philosophy of information are Adriaans and van Benthem (2008) and Floridi (2003). Also important is Dretske (2008). For applications to psychology, see Attneave (1959).

[217] See Bar-Hillel (1964), Dretske (1981) and (2008), Nozick (1978) and Floridi (2008)

information in this sense. Its principal means of investigation are probability theory and physics.[218]

In its third sense, information has to do with codes. The informational value of a code-string is the algorithmic or Kolmogoroff complexity of the string, which is defined as "the shortest program that computes it on some fixed universal Turing machine."[219] Complexity information is studied by theoretical computer science, probability theory, statistics and physics.[220] A problem posed by the first two senses is that information in the first sense needn't be true, yet information in the second sense can't not be true.[221] On the other hand, there is a route from the complexity conception to the probability conception via the set of all prefix-free programs under provisions of a technical result known as Kraft's Inequality.[222]

The fourth notion of "information" derives from its use in intelligence and counter-intelligence work. According to the CIA's World Fact Book:

> Information is raw data from any source, data that might be fragmentary, contradictory, unreliable, ambiguous, deceptive, or wrong. Intelligence is information that has been collected, integrated, evaluated, analyzed and interpreted.[223]

This, the military sense of information, is widely used in computer science or informatics. It is prudent to keep in mind a marked tension between the epistemic and military conceptions.

In my remarks in preceding sections there is evidence of the presence of the epistemic and probabilistic senses of information. If we say, with Dretske, that the individual cognitive agent is an information-processing system capable of belief and that belief is information in digitalized form, information is (mainly) epistemic. When I speak of the quantities of information in Harry's sensorium, I am speaking mainly of

---

[218] The *loci classici* are Hartley (1928), Shannon (1948) and Shannon and Weaver (1949).

[219] Quoting Adriaans and van Benthem (2008), p. 12. See also Solomonoff (1997), Chaitin (1987) and Li and Vitányi (1997).

[220] See Grünwald and Vitányi (2008). Also logic: Chaitin (1982).

[221] Well discussed in Dretske (2008).

[222] Adriaans and van Benthem (2008), p. 12.

[223] https: //www.cia.gov/library/publications/the-world-factbook/docs/history. Quoted here from Adriaans and van Benthem (2008), p. 7

probabilistic information. Some of the central issues for probabilistic information, such as the reliability of information transmitting channels, are also of importance for epistemic information, as the chapter will try to make clear. There are circumstances in which the various notions of information admit of explanatorily useful integration, and contexts in which they pull to opposite effect. There is at present no known way of charting these harmonies and dissonances with any completeness.

6.10 *Light and dark*

We are awash in information. It is a wonder that we are not in a state of chronic overload. Overload is the exception, not the rule. By and large, individuals are able to process masses of information to their advantage, converting some of it to belief when it suits their circumstances. A good part of what makes this possible is our evolved facility at discrimination. Lying along side is the capacity to evade informational clutter, to filter out what's relevant from what's not. A further part of the mix is our talent for abduction, for forming coherent conjectures about the passing scene. Also involved is our adeptness at recognizing plausibility, our responsiveness to "the ring of truth". To a considerable extent these are processes that operate darkly; they leave little phenomenal trace. They are prime candidates for CR-engagement.

A deep advantage of dark reasoning are the economies it achieves. Even when applied to light reasoning – to reasoning above above – the CR-model preserves some of that advantage. The causal processes that fix belief and drive the rhythms of reasoning operate unconsciously for the most part. Consciousness is an expensive endowment. Consciousness has a narrow bandwidth. It processes information very slowly. The rate of processing from the five senses jointly, as we just saw, is in the neighbourhood of 11 million bits per second. For any of those seconds, something fewer that 40 bits make their way into consciousness. Consciousness is highly entropic, a thermodynamically costly state for a human organism to be in. At any given time there is an extraordinary quantity of information processed by the human system, which consciousness cannot gain access to. Equally, the bandwidth of language is far narrower than that of sensation. A great deal of what we know – most in fact – we aren't able to tell one another. Our conversational exchanges run at something like 16 bits per second.[224]

---

[224] Zimmerman (1989).

It is now evident that we must refine the claim that individual agents operate with comparatively scant quantities of information. In dark states, human systems are flooded with information. Consciousness serves as an aggressive suppressor of it, reserving radically small percentages that are available to consciousness. To the extent that some of an individual's thinking and decision making is subconscious, it is necessary, as we have said, to postulate devices that avert the distortion, indeed the collapse, occasioned by informational overload. Even at the conscious level, it is apparent that various constraints are at work to inhibit informational surfeit. The conscious human thinker cannot have, and could not handle if he did have, information that much exceeded the limits we have been discussing. Human beings make do with slight information because this is all the information that a conscious individual can have. As we might say: *Light reasoning – slight information.*

This gives us two rather different perspectives on the cognitive economy. At the dark levels of down below, informational tsunamis are controlled and filtered to productive ends by massively parallel and computationally powerful sorting and storing procedures. Here the economy is the efficient handling of huge inputs. Up above is different. Consciousness makes for informational niggardliness. The economies here are two. If consciousness could absorb a larger informational bounty, it would pay correspondingly high tariffs of time, of actions and decisions slowed down considerably. The other factor is satisficing. Given that the scantness of the information a conscious human system is able to hold at a time, care must be taken to get the best one can out of the little one has.

As we have it now there are two pairs of contrasts that converge on the distinction between individual and institutional cognitive agency. One is that individuals are conscious and institutions aren't. The other is that is typical of individuals to make do with less information than institutions do. Given that consciousness is an aggressive suppressor of information,

Proposition 6.10
LINKAGE: *Consciousness is a vital link between individuality and informational scantness.*

6.11 *Premiss-selection*

It falls within the province of logic to take note of, and where possible, provide accounts of various filtration-conditions on premiss selection. As we saw, these regulate the influence of relevance, nonredundancy, noncircularity, finite manageability and so on. It is hard work, landing its employees well short of consensus. Whatever the details of these filtration constraints it would appear that a properly functioning inferential mechanism will heed them with a notable effortlessness and frequency. At least, it will heed them if we allowed for the requisite linkage between the NN-convergence thesis and the idea of good working order. If so, we would have it that a properly functioning human inference engine is with a notable frequency one that hits the required spots (never mind that theorists struggle to give them precise articulation.) Here again one's *devices* are more "with it" than one's theoretical speculations.

Some readers will dislike the immodesty of the present suggestion, that – never mind the theorist's difficulties in producing analyses of them – the premiss-screening requirements of nonredundancy, relevance, finite manageability and the like are typically catered for automatically by our premiss-selection devices when operating as they should. There are two parts to the suggestion, each open to critical resistance. One is that the routines of premiss screening are dark. The other is that they owe their reliability to the good working order of our cognitive devices. My advice is to relax. Consider again the case of irrelevance filtration. Logicians and linguists – to say nothing of legal reasoning theorists – have been labouring without success for a half-century and more to produce a theoretically respectable consensus about relevance. Yet it is as clear as the nose on one's face that something the human individual is extremely good at is keeping irrelevancies at bay. What would explain so striking a discrepancy between theory and practice? One possibility is that relevance theorists are stupid or lazy, that they simply aren't paying attention to the conditions that actually prevail when irrelevance is evaded. Of course, that is uncharitable. It is also possible that when irrelevance is evaded the conditions which enable this to happen are of a different character than the relevance theorist might suppose. Suppose that what the theorist wants to know are the *criteria* of relevance that an agent needs to take note of if he is to design and execute the task of irrelevance filtration? The second possibility is that this is the wrong thing to look for; that irrelevance evasion is not a matter of applying the criteria of relevance to the design and execution of irrelevance-evasion procedures. Irrespective of how relevance is to be defined, irrelevance-evasion and the other requirements of clutter avoidance, are not down to us, but rather down to our devices. What, then, *are* the conditions that obtain when irrelevance-evasion is in progress? Empirical details are thin on the ground. But it is possible to say in a quite general sense that when irrelevance evasion is underway one's cognitive devices are operating as they should.

Human beings have a natural tendency to anthropomorphize things. Psychologists – and I too – find it attractive to say that in our perceiving moments, our devices "analyze" information and "propose" its best interpretation to conscious experience. We say that in memory, our mechanisms "flip through" stored-up inventories and "map" specific elements to our current interests. We say that in processing new information, we "scan" the resulting belief-set for consistency. We "frame" agendas, "set" targets, and "coordinate" targets with interests. We say that when we reach a decision we "construct" a decision tree; and on and on. Anthropomorphism is a natural friend of the active voice, of "I do" rather than "I am done to". Employment of the active voice considerably out-distances its literal applicability. There is much more that is said in the active voice than is true in it. When appended to the CR-model this gives rise to a paradoxical-seeming state of affairs. It is that our cognitive successes are a matter of our doing the right things without having much of a clue as to what those things are. Once we subdue the impulse to anthropomorphize, the appearance of paradox subsides, and rightly. Since, beyond the bland generalities we have already canvassed, the average human knower knows virtually nothing about what it takes for his cognitive devices to be working as they should, his epistemic successes will not be graced by much at all, awareness, if any, of that to which his success is owed. In quite wide ranges of cases, we are rather massively "strangers to ourselves".[225]

6.12 *Generalizing*

Whereas the dominant interest of mainstream logicians has been the consequence relation, the dominant interest of theorists of reasoning has been the drawing of consequences. One of the standard ways in which human reasoners draw consequences is by making generalizations from samples. It is easy to see that projecting from samples is one of the most labour-saving devices in the cognitive ecologies of being like us. So we should pause awhile over it. There is a large consensus a among

---

[225] Wilson (2002). See also Wegner (2002). In like vein, see Nietzsche (2002) and (1974).

mainstream logicians and psychologists alike that we aren't very good at drawing consequences of this sort, and that we are too prone for our own rational good to "jump to conclusions" or "yield to stereotypes". My own view is that this is itself a hasty generalization.

A focal theme of the default logics of inheritance reasoning is the *generic proposition*.[226] To see how, let us briefly inspect the Tweety Triangle and the Nixon Diamond. Default logics will occupy us further in the chapter to follow. For now, all we need are the examples. Tweety's situation is that although birds fly and Tweety is a bird, Tweety is a bird of the penguin kind; so Tweety doesn't fly. Nixon's situation is that he was both a Quaker and a Republican. But Quakers are pacifists and Republicans not. In each case, the looming inconsistency is abetted by postulating that the links of birdness to flying, penguins to not-flying, Quakerhood to pacificism and Republicanism to non-pacifism is one of *default-consequence*. This stirs up two problems. One, by the consequence rule, is to characterize the default-consequence relation. The other is to figure out how it can be true that the correct default from Tweety's birdhood is that Tweety flies, that the correct default from Tweety's penguinhood is that Tweety does not fly, and that Tweety's birdhood and penguinhood don't add up to a contradiction? It is widely assumed that in the Tweety example, "Birds fly" and "Penguins don't fly" are forwarded as *generic* claims. In the Nixon example, the same is true of "Quakers are pacificists" and "Republicans aren't pacifists".

I will say in due course why I think these are suspect assumptions, but there will be no harm in staying with them for now as an expository convenience. It will also prove instructive to examine this idea of genericity in the context of a kind of case discussed by Gigerenzer and typified by the following example.

> Consider a 3-year-old who uses the phrase "I gived" instead of "I gave". A child cannot know in advance which verbs are irregular; because irregular verbs are rare, the child's best bet is to assume the regular form until proved wrong. The error is "good" – that is, useful – because if the 3-year-old did not try out new forms and occasionally make errors, but instead played it safe and used

---

[226] Generics also come in non-general forms, as with the kind-sentence, "The BlackBerry was invented in Waterloo.", as well as generic sentences about individuals ("Sarah always goes to bed at 10:00"). An important discussion of the link between generics and default reasoning is Pelletier and Asher (1997).

only those words it had already heard, she would learn a language at a very slow rate.[227]

Economic considerations are at work here. It is much more expensive to learn the irregular verbs "up front" and in one fell swoop (indeed for a three-year-old it is impossible) than to learn them one by one. A rule that purported to cover all the exceptions would be punishingly complex and almost certainly wrong. Thanks to the efficiency of feedback, learning from one's mistakes is better economics than learning that is error-free; and in large ranges of cases is not an inferior breed of learning. A further consideration involves how individual agents handle generalizations. When the little girl said "I gived", she was instantiating a general rule – the stem+-"ed" rule for the past tense.

There are a good many logics of nonstandard stripe that tend to assume to the standard definitions of the logical particles. In the present instance, quantification is a case in point. Logicians have long since adapted their thinking to a particular view of generalizations. They see them as the universal quantifications of conditional propositions. Achieving true universal quantifications is risky business. In chapter 4, we remarked that part of what makes this so is their *brittleness*. Universal quantifications are felled by a single true negative instance, that is, by the slightest deviation from invariability. By contrast, generic propositions can tolerate false instantiations. Consider a case. You are tramping in the breathtaking wilds of Brazil and you espy your first ocelot, four legs and all. You say to your companion, "How interesting! I always imagined that ocelots were two-legged." As you now see, ocelots are four-legged, not two. Of course, some ocelots aren't four-legged, perhaps as a result of congenital mishap or injury. Suppose this other ocelot, Ozzie, turns out to be three-legged. Does this true negative instance falsify the claim that ocelots are four-legged? It does if your generalization is the universal quantification of the requisite conditional. It does not if your generalization is a generic proposition. Unlike universally quantified generalizations, which are brittle, generic generalizations are *elastic*.[228] There are true negative instances in the face of which they remain true. Elasticity is essential to economical generalization, to both inferences from and inferences to. Accordingly,

---

[227] Gigerenzer (2005), p. 196.
[228] Compare here David Armstrong's characterization of laws of nature as "oaken" rather than "steel". (Armstrong, 1997, pp. 230-231).

Proposition 6.12a

FUNDAMENTAL LAW OF BRITTLENESS: *Let α be a sentence in the form* $\ulcorner \forall v (Gv \supset Hv) \urcorner$ *and β an instantiation of it. Then β's falsity entails α's falsity.*

Proposition 6.12b

FUNDAMENTAL LAW OF ELASTICITY: *Let α be a generic claim in the form* $\ulcorner Gs\ H \urcorner$, *and let β be an instantiation of it. Then β's falsity does not entail α's falsity.*

Default logicians have a large stake in elastic generalizations. They play an essential role in side-stepping the fundamental law of brittleness. The semantics of elasticity is an open problem in the genericity research programme. Getting the *falsity* conditions right for "Ocelots are four-legged" lies at the heart of this enterprise.[229]

As a strictly practical matter, generic generalization offers a number of advantages. One is that it makes available to the reasoning agent an attractive way of reducing rates of error. When his instantiation from a universally quantified generalization is false, he has made *two* mistakes, one of which might be quite difficult to correct. He is mistaken about Ozzie's four-leggedness and he is wrong about "For all x, if x is an ocelot, then x is four-legged". So he is wrong in particular, and he is wrong in general. On the other hand, if his generalization is generic, he is wrong about Ozzie but not about ocelots. He is wrong in particular – which is a comparatively easy thing to correct – but he is right in general, leaving nothing that needs correcting. So there is a clear cognitive, as well as economic, advantage in not generalizing to universally quantified conditional propositions.

6.13 *Haste*

When it comes to generalization, nothing has held the attention of inductive logicians more than the haste with which it is so frequently achieved. Hasty generalization – or what social scientists call "thin-

---

[229] Carlson and Pelletier (1995) is required reading. Also important is Pelletier (2010), of which the editor's "Generics: A philosophical introduction" is a valuable primer.

slicing"[230] – has been on the not-to-do list of logicians for nearly all of that subject's modern history. I myself am not so sure.

What would it make sense for beings having our interests, talents and wherewithal to generalize *to*? Universally quantified generalizations serve us well in contexts in which literal invariability is required for generality. These are comparatively rare contexts. They encompass mathematics and the branches of science that harbour ambitions for mathematically strict lawlikeness. For most of the living world, however, nontrivial literal universality is scientifically unattainable.[231] Its lawlike regularities are better served by and more faithfully rendered as generic generalizations on the model of the four-footedness of ocelots.

A common setting in which individual agents generalize is one in which projections are made from samples. It is an established conviction among mainstream inductive logicians and philosophers of science that generalizations are worthless when they are projected from unrepresentative samples. They commit the dread fallacy of hasty generalization. It is also widely accepted that, beyond a certain point, the smaller the sample, the greater its unrepresentativeness. Even in polling contexts, where samples can be very small in relation to test-populations, they must be quite large in comparison with the samples that very often trigger generalizations in everyday life. The requirement that a generalization be deferred until a sample of appropriate size and randomness is assembled is a costly one. In practical terms, as we saw, it would mean that it is hardly ever the case that an individual agent gets to make a generalization when he needs to have it. Inductively robust generalizations are expensive for the likes of us, even when it is possible to pull them off. This is not significantly less true of generic generalizations than of universally quantified ones. Even so, holding the generic four-leggedness of ocelots to the strict standard of inductive strength is hardly less costly than doing the same for their universally quantified four-leggedness. When one reflects on the empirical record, it is clear that individual agents on the ground routinely omit to pay this cost in either case.

Generic claims are but one form of non-universal generalization. Similar advantages also accrue to other forms of it, both those that carry quantifiers expressly ("For the most part, Gs are Hs") and those that are

---

[230] Ambady and Rosenthal (1993), Carrière and Gottman (1999).

[231] For a damaging critique of the idea that scientific generalizations require or admit of formulation as universal quantifications, see the unjustly neglected Wilson (1979).

modified by non-quantificational sentential adverbs ("Normally Gs are Hs", "Usually Gs are Hs") Even so, there are significant differences here to which we will recur a bit later.[232] For now let us give formal recognition to a pair of central points:

Proposition 6.13a
ECONOMICAL INSTANTIATION: *There are cognitive economies involved in instantiating from non-universal generalizations.*

Proposition 6.13b
ECONOMICAL GENERALIZATION: *There are kindred economies involved in generalizing to non-universal generalizations.*

Thin-slicing turns essentially on our ability to recognize patterns. In its hasty generalization form, pattern recognition is tied to our ability to recognize natural kinds.[233] Among philosophers such as Quine there is a certain scepticism about natural kinds.[234] But there are convincing approximations of them well-attested to in the empirical literature, whether *frames*[235], *prototypes*[236], *scripts*[237] or *exemplars*[238]. There is also some significant work in linguistics in which the semantics of natural kind noun phrases is intimately bound up with factors of genericity. It behoves a naturalized logic not to give the idea of natural kinds too short a shrift.[239] Even so, the idea of natural kinds is also linked to the temporal predictability of propositions. This implicates hasty generalization in the frame problem. The question whether it is all right to generalize on the proposition "This ocelot is young and frisky" resembles the question whether it is all right to temporally project it. I mentioned earlier that the

---

[232] For plural quantification and like matters, see Barwise and Cooper (1981), Sher (1991), van Benthem (1999) Peters and Westerstål (2006), and Westerståhl (2008), and citations therein.
[233] Which, as we saw a chapter ago, is one of the things schizophrenics have difficulty with.
[234] Quine (1969).
[235] Minsky (1975).
[236] Smith and Medin (1981).
[237] Schank and Abelson (1977).
[238] Rosch (1978).
[239] Krifka (1995). See also Prasada (2010).

problem of temporal projection has no known solution in the logic of defeasibility. Analogously, the same applies to the thin-slicing/natural kinds problem.

It is a matter of fact that individual agents are hell-bent-for-leather thin-slicers. On the received view, this convicts them of substantial error, if not of massive irrationality. But if the NN-convergence thesis is given any standing here, we are wrong to confine our generalizations to those that arise in an inductively strong way from the triggering samples. Certainly, it is difficult to believe that when the little girl of Gigerenzer's example generalized to the stem + "ed" rule she got the rule wrong or was in any way irrational. Such cases are legion. They are cases in which our hasty generalizations turn out to be *right*. Not every hasty generalization is right. But some are. In fact, *lots are*. There is a significant correlation here:

Proposition 6.13c
A BENIGN CORRELATION: *The hasty generalizations that we are* right *to draw are by and large the ones that we* do *draw*.[240]

COROLLARY: *Hasty generalization is not a fallacy in the traditional sense.*

COROLLARY: *Hasty generalization is a cognitively virtuous way to generalize.*

Seen this way, one of the individual agent's strongest contributions to his success in the cognitive economy is his capacity for determining when to thin-slice, when to leap to conclusions. This

---

[240] The cognitive science literature recognizes a related distinction between exploration and exploitation. When in an explorative state, an agent runs some experiments, tries to understand, asks (at times) stupid questions, and behaves suboptimally when judged against a short-term horizon. But this can be considered as an investment in a better understanding of the situation which will then – now adopting a long-term perspective – pay off in the future. The Secretary Problem (Gilbert and Mosteller, 1966; Ferguson, 1989) is famously one such example. In the first phase, people acquire information which will be used to set a threshold, and only in the second phase will they make a decision based on comparisons with objects against this threshold. It is suboptimal if phase one is too short or phase two is too long. The transition from exploration to exploitation is, however, quite often mathematically tractable.

involves a capacity for seeing in samples that are sometimes as small as they get the requisite representativeness. One of the tasks of cognitive psychology and cognitive neurobiology is to uncover the mechanisms that provide for the individual reasoner so providentially – to expose the design of our cognitive mechanisms in the point-and-shoot mode. Even if it is conceded that for every mistake there is a perspective from which it can be seen as wrong, there are many cases in which the collateral benefits of actually committing them outweigh their wrongfulness. The little girl's error is a case in point. Her selection of "I gived" instead of "I gave" represents a default instantiation of the stem + "ed" rule. Considered generically, the rule is right. It is in this particular default that her mistake lies. But drawing it was not the wrong thing to do. It was the right thing to do. It is the only feasible way for a three-year-old to learn the irregular verbs.

Thin-slicing teaches a similiar lesson. The rule of which "I gived" is a default is something the little girl generalized to, on the basis – at her tender age – of a smattering at most of correct utterances in the form "I [stem]-ed." In generalizing to this rule, the child did the apt thing. In generalizing to it hastily, she revealed her adeptness in discerning generalization in small samples. Had she set out to find the universally quantified rules of lawlike linguistic correlations, she would never have learned her language. Accordingly,

Proposition 6.13d
THE ECONOMICS OF FEEDBACK: *Learning from feedback is often cheaper than (and just as good as) learning by universal generalization.*

The vulnerability of haste is compensated for by the corrective wherewithal of feedback. In so saying, we are reminded of our earlier observation that the fallacies embed a notable resistance to such correction. This now requires some qualification. Consider the large class of cases in the hasty generalizations of individual reasoners are benign but to which some exceptions have been found. With respect to those true negative instances – Ozzie, for example – the mechanisms of feedback operate smoothly and efficiently. We began by instantiating the four-legged generalization to Ozzie. When it turned out that Ozzie's four-leggedness is an error of fact, we fixed it. Ozzie isn't four-legged; Ozzie is three-legged. In this there isn't the slightest indication of laggardness in our error detection and correction capabilities. On the other hand, where our feedback mechanisms would be notably laggard – to the point of downing tools – is in the discouragement of the hasty generalization that ocelots are four-legged and of the default inference that any given ocelot is that way too. Clearly, it *would* take training, lots of it, to get people to stop doing these things. But this is as it should be. They shouldn't stop doing these things. Putting ourselves in manual mode here would be unbearably expensive.

6.14 *Causal disreputability?*

It is a fact of some interest that our two models of knowledge are framed in terms on which philosophers and some logicians have heaped their harshest scorn. The CC-model is set out in the language of justification and evidence, and yet the concept evidence, especially inductive evidence, has attracted the corrosive scepticism of philosophers as formidable as Hume and Goodman. Causality is the idiom of choice for the CR-model, and it too has been buffeted by Hume's scepticism. Not to be outdone by disgruntled empiricists, traditional fallacy theorists have also weighed in. We are, they say, bad at causal reasoning; bad enough indeed to merit the indictment of fallacy. It is instructive to compare the reservations of Hume and the fallacy crowd. Hume is an empiricist of the old school. *Nisi in intellectu nisi prius in sensu;* only the observably-grounded can be understood. His complaint against causality pivots on the element of necessitation. If necessary connections occur in nature they are not open to human observation. Accordingly, judgements in the form "C caused E" cannot be understood. So one of us knows any fact in the form "C caused E." For beings like us, causal talk is incoherent, all of it.[241]

The case pressed by fallacy theorists, while not as sweeping, is still a substantial bill of particulars, and a large puzzle if true. The indictment is that we are defective attributors of causation on a scale large enough to fuss about. The problem, they allege, is that there are two junctures at which our cause-ascertaining powers let us down with a notable frequency. One is the juncture at which we differentiate between positive correlations and causal connections. The other is the juncture at which we differentiate between temporal succession and causal connection. The first failure is the fallacy of false cause. The other is the infamous *post hoc, ergo, propter hoc* fallacy.

---

[241] Ignore here the lurking paradox. If Hume doesn't understand "necessary connection", he can hardly understand that causes must be necessary connections.

There remains a significant gap between Hume and the fallacy theorists. If Hume is right, no improvement of our causal judgement making is possible. No matter our efforts, we'll never be better at judging causes than we are now; namely, *dead wrong*. The fallacy theorists are somewhat more sanguine. Greater care, is their message: Take greater care and you'll make fewer causal mistakes. What would it take for this to be true? It would take two things. It would have to be the case when faced with a positive correlation-causal connection problem or a temporal succession problem, we'd arrive at the wrong solution, not just occasionally, not just often, but with a frequency and generality that is characteristic of a mistake bad enough to be a fallacy. It would also have to be the case that, on a scale sufficient to offset the sheer sweep of fallacious performance, there are learnable techniques for improving our facility with causal thinking. My view is that the realization of this pair of conditions hasn't the remotest prospect of fulfillment. I turn to this matter now, leaving Hume's nihilism to stew in its own juices.[242]

The human reasoner is a natural object. He is an organism interacting with an environment. He is a denizen of the causal order. It is hard to overstate the extent to which even a rudimentarily adequate life involves the negotiation of causal forces, the harnessing to good effect of nature's myriad powers, and the management of the slings and arrows of outrageous fortune. Causal reasoning is the human individual's stock in trade. Those who charge us with false cause and *post hoc* allege that it is typical of us to mistake, with a notable frequency, correlations for causes and temporal successions for causal dependencies. They say that it is typical of us to do these things when reasoning causally. No one will deny that noticed constant conjunctions and noticed constant temporal successions can be causal-question triggers. How could they not be, given that in suitably qualified formulations they are necessary conditions of causality? For the traditional accusations to stand, it would have to be

---

[242] Except to say that it is the wrong answer to a question asked at the wrong time. We said in chapter 4 that the first question for an epistemologist to ask is not whether knowledge is even possible, but what makes it happen so abundantly. Of course, there are contexts in which whether knowledge (of X) is possible is an askable question, but when X is "causes" the answer must do what Hume's answer fails to do, namely, solve an instance of Philosophy's Most Difficult Problem. It must show that any argument whose conclusion is that a knowledge of causes is not possible is justifiably a sound demonstration of a surprising truth and not a successful *reductio* of something in or presupposed by its premises-set.

typical of us all, when these causal trigger-questions arise, to give them the wrong answer at a rate that out-distances the frequency of our error-making in general.

The mere fact of our survival and prosperity in a causally hostile world is enough to show that we are good at causal recognition, and that we are good at it on a scale that keeps nature's destructiveness in surprisingly good check. Since causal adeptness implies, among other things, a good record at distinguishing correlations from causes and temporal successions from causal dependencies, it is hard to see how it could be typical of us to mismanage these discriminations with a frequency that outruns the general rate of error making.

We should not think that the determination of causes is a piece of cake. In lots of cases it is extremely difficult to get the causal story right. This is especially so in science, where the sought-for causes are most hidden and the agreed methods for their discernment the most demanding and hardest to implement. There is a large literature about the travails of causalists in the philosophy of science. Of course scientists are often wrong in their causal attributions. But when this happens, it is never the case that their reasoning is in the form:

1. There is a noticed constant conjunction between C-events and E-events.
2. Therefore, Es are caused by Cs.

Or in the form:

i. Noticed C-events have always to date followed upon E-events.
ii. Therefore, Es are caused by Cs.

Inferences in these forms are too stupid for words. It beggars belief that anyone draws them with anything approaching a notable frequency.

It is true that there are cases in which, when we get things wrong, we make causal judgements that over-stretch our causal recognition capacities. This happens rather routinely when correct answers require the rigours of expert or technical thinking and we are not experts. But even here ambiguities lurk. When Sarah reasons that cigarette smoking increases the likelihood of lung cancer then, however she assembles it it will not meet the standards imposed by randomized experimental design tests. This is because ethical consideration affecting the testing of human subjects preclude the use of these methods, as do technical difficulties affecting the testing of animals. Cigarette companies are notorious for insisting that it has not been scientifically established that cigarette smoking is a causally positive factor for lung cancer. If this means that the link has not been established by randomized experimental design tests, it is trivially correct. If all science's causal inferences – or our own – were held to this standard, the causal thinking that actually occurs in human life would be massively mistaken. There are a great many scientific contexts in which demands on causal thinking are less strict, flowing from the methods of prospective testing, retrospective testing and surveys. So care should be taken not to import into those proceedings unrealistic performance standards.

Proposition 6.14

MORE MISALIGNMENT CONFIRMATION: *False cause and* post hoc, ergo, propter hoc *aren't fallacies as traditionally conceived – that is, false cause and* post hoc, ergo, propter hoc *fail to instantiate the traditional concept of fallacy.*

CHAPTER SEVEN

THIRD-WAY REASONING

"[L]ogical formalizations convert what in many [situations] is a fast and computationally trivial check for [the] presence and absence of attitudes into a computationally difficult or impossible check for provability, unprovability, consistency or inconsistency. This inaptness seems especially galling in light of the initial problem-solving motivations for nonmonotonic assumptions ... which ... served to speed inference, not slow it."
Jon Doyle

"But I do claim that logic has a long-standing art of choosing abstraction levels that are sparse and yet revealing."
Johan van Benthem

"[P]erfect reasoning is a perilous plan for living. Perfection has no safety net. One slip and it shatters."
Reginald Hill

7.1 *Defeasibility*

We are trying to get to the bottom of third way reasoning. Of particular interest is its defeasibility.[243]

When premiss-conclusion reasoning is correctly performed there is a relation R on what is reasoned from to what is reasoned to, a relation which varies in type according to the type of reasoning at hand. There is a large family of premiss-conclusion reasonings of which the following two conditions hold.

1. The reasoning in question is correct, that is, a properly executed example of its type.

---

[243] In the original sense of the term, introduced into the philosophy of legal reasoning by H.L.A. Hart (1949), "defeasible" is itself something of a neologism, carrying the legal meaning of "open in principle to exceptions." Its nearest term of antecedent usage is "defeasance", also a legal word, denoting the action or process of rendering something null and void. "Defeasance" derives from the Old French verb *défaire/désfaire,* meaning "to undo".

2. The underlying R-relation is subject to rupture upon the addition of true premisses consistent with the old ones and also with the conclusion.

Any R which meets these conditions is a nonmonotonic relation, and any reasoning which implements it is defeasible. Needless to say, there are more ways than one for a piece of premiss-conclusion reasoning to go wrong. The premisses might be false, not true. The conclusion might also be false. The falsity of the conclusion might establish the falsity of premisses or it might defeat the presumed link between premisses and conclusion. In all these situations it would be correct to say that the reasoning in question has met with considerations that defeat it.

This tells us something of interest. There are plenty of run-of-the-mill cases of belief-revision occasioned by new information in which

Proposition 7.1a
THE BREADTH OF DEFEASIBILITY: *Defeasible reasoning needn't be nonmonotonic. Think, for example, of a syllogism in* Barbara *with false premisses.*

None of these run-of-the-mill examples captures the feature that distinguishes the cases conditioned by (1) and (2). In the run-of-the-mill cases, the common factor is what we might call the stress of input-output *falsity*. A premiss is false (so soundness is defeated); a conclusion is false (so either soundness or the premiss-conclusion link is defeated); or the premisses are true and the conclusion false (so the link is defeated). But the cases captured by conditions (1) and (2) are genuinely different. The Rs conditioned by (1) and (2) – the defeasible relations – are subject to rupture in a quite particular way. The contexts in which they rupture are *alethically benign*: The link breaks notwithstanding that the premisses stay true, and the conclusion too.

There is a further kind of rupturable relation, typified by the Ozzie case. Of course ocelots are four-legged. So it only stands to reason that Ozzie the ocelot is also four-legged. But alas, he isn't. His three-leggedness ruptures the link between the four-leggedness of ocelots and the four-leggedness of this, that and the other ocelot. As with the Spike case, there is always room to wonder whether, given the actual state of the world (including the bits not yet present to us), the relation between the old evidence and the conclusion that Ozzie is an ocelot was ever instantiated in the first place. In the Ozzie case, we expected that he would turn out to be four-legged, and we were mistaken. He isn't now and wasn't then. The prospects of a link present then and absent now by virtue of what we now know and didn't then depends on how that link is constituted, on how its make-up is to be characterized. But leave that for now. The peculiarity that currently occupies us is that the proposition that ruptures the Ozzie-link falsifies the conclusion and does not falsify the premises, not even the premiss of which it is a true negative instance. Peculiar as this decidedly is, it is more a peculiarity of generic generalization – of the generic use of "all", we might say[244] – and less a peculiarity of the its intolerance of its link to the Ozzie conclusion conjointly with its refusal to be disturbed by it. The peculiarity, in other words, is the peculiarity of *generically impervious* consequence-rupture.

So, then, we have two properties of note – alethically benign rupturability and generically impervious rupturability. It matters that they are disjoint properties. It means that logics that do more or less well for the one property might not enjoy a like success – or any – with the second.

Proposition 7.1b
TWO TYPES OF RUPTURE: *Alethically benign and generically impervious rupturability are disjoint properties of R-relations.*

COROLLARY: *Logicians should take care not to allow their theories to override their disjointness, except for weighty cause.*

I have been saying all along that in most cases of good reasoning the standards of deductive validity and inductive strength are neither met nor in play. The goodness of reasoning when it is neither valid nor inductively strong resides in its success with a standard peculiar to reasoning of that type. This we've been calling third way reasoning, but without the presumption that there is some one thing that it is.[245] A

---

[244] There is plenty of room for this in English: Sarah is always in bed before 10:00. Harry is never late for an appointment. That is to say, it is characteristic of Sarah to be in bed by 10:00. Were it to be otherwise, something would be amiss or there would have been some disturbance of routine. Were Harry late for an appointment something would have gone wrong. The city would have been paralyzed by a snow storm, or he'd have had a stroke; or he wouldn't have been told of the appointment.

[245] On Reiter's reckoning, the reasoning that is underlain by nonmonotonic consequence is actually *fifth*-way reasoning – standing apart from "probabilistic

central task for a logic of reasoning is the theoretical elucidation of this third type. A central task for a naturalized logic of reasoning is to make its elucidations with due regard for descriptive and empirical adequacy. As mentioned earlier, there seems to be no want of candidate logics for the analysis of third way reasoning – nonmonotonic logics, truth maintenance systems, defeasible inheritance logics, default logics, autoepistemic logics, circumscription logics, logic programming systems, preferential reasoning logics, abductive logics, theory-revision logics, belief-change logics and whatever else.[246] The Can Do Principle instructs us to begin our search for the logic of the third way with an examination of the possibilities already at hand. This will involve us in a fairly sustained examination of what these logics have to say about the consequence relation.

From the very beginning, logic has a foundational interest in consequence and in properties interdefinable with it. As originally conceived of, consequence is a relation that binds premisses to conclusions under conditions it was the business of logic to make clear. Whence the importance and the ubiquity of the consequence rule. Of course, since everyone knows that when premiss-conclusion reasoning is performed correctly, there is a premiss-conclusion relation R which obtains between what is reasoned from to what is reasoned to, how could logicians not know this too? And since consequence engages their attention as a premiss-conclusion relation, why wouldn't logicians suppose that defeasible premiss-conclusion relations – that is, the R-relations defined by (1) and (2) or typified by the Ozzie case are relations of *defeasible* consequence? Everyone agrees that the premiss-conclusion relation in which logicians took a founding interest, are relations that would warrant the attention of logicians. For they were, were they not, relations of *logical* consequence? This might lead us to wonder. Are logicians interested in premiss-conclusion relations when they happen to be relations of logical consequence? Or does a relation become a relation of logical consequence just by being of interest to logicians?

The dominant position is that premiss-conclusion relations are indeed consequence relations and that, being so, they are rightly of interest to logicians. If this is right, defeasible relations between premisses and conclusions are likewise consequence relations, are

---

reasoning", "fuzzy reasoning", "inductive reasoning" and "deductive reasoning". "The most recent addition to this list is nonmonotonic reasoning, the study of which seems unique to AI". (Reiter, 1987, p. 147)

[246] I reserve discussion of autoepistemic and abductive logics until chapter 11.

likewise relations that warrant the attention of logicians, are likewise – at least in some minimal sense – relations of logical consequence, and yet are subject to alethetically benign or generically impervious rupture.

By the thinking that lies behind the consequence rule, an important part of what has been taken as the adequacy standard of third-way reasoning is the existence of a consequence relation of requisite type between what is reasoned from to what is reasoned to. Accordingly, I will say that a *D-logic* is one of that cluster of nonmonotonic logics for the investigation of (one or more types) of defeasible inference. Similarly, with a diffidence born of Proposition 7.1b and its corollary, I will speak of the consequence relations postulated by *D*-logics as relations of *D-consequence*.

A founding motivation of these logics was the interest taken by AI theorists in the logic of planning. An early difficulty was the frame problem, the problem of how to deal with inertial reasoning.[247] We have already seen how the frame problem is implicated in distinguishing simple belief-revision from error-detection and correction. Belief-revision is required when new information contradicts an old belief which is no longer true. Error-correction is required when the new information shows that the old belief never was true, never mind that it might be true now. Relatedly, it is the problem that computers have in determining where in the system's knowledge-base to make updates or revisions when new information is made to flow through. The problem also takes an action-theoretic form. When an action is performed, some things change and some don't. The problem here is twofold. How do we distinguish changing from unchanging facts? And can we represent the preservation of those unchanging facts in a natural way?

A second motivation is the "qualification problem" or the problem of *ceteris paribus* qualifications.[248] Imagine that we wanted to

---

[247] The frame problem is one of a trio which are "the three main problems in formalizing action: .... Other than that, no worries." (Ginsberg, 1993) A second of these is the *knowledge-organization problem*. It is the problem of deciding on the various patterns in which knowledge should be stored, and has a clear bearing, for example, on how information gets organized around natural kinds. The third is the *relevance problem*. It is the problem of determining what information in a knowledge base would or might be of assistance in handling a task that has been presented to a computer. We should note that what philosophers mean by the frame problem is what computer scientists mean by the relevance problem. See Gabbay and Woods (2003), pp. 110, 152-153.

[248] McCarthy (1986).

robotize a certain fairly simple task, such as the stacking and unstacking of some boxes in a warehouse. The problem besetting the program is that stacking and unstacking are open-ended processes. Any proposed procedures for performing these tasks will be subject to arbitrarily many possibilities of override. The question to which this commonplace fact gives rise is, given that one cannot embed all these possibilities of override as working qualifications of the system's algorithms, how does one determine the procedures which the stacking-unstacking enterprise should follow?

A third motivator is "closed-world reasoning", briefly met with in chapter 4.[249] In closed world reasoning, it is assumed that any information that would weaken or cancel the premisses' connection to the conclusion is already contained in the premisses. Of course, as we have seen, except for classically deductive contexts it is possible that the assumption is almost always false. So the problem is to determine whether the assumption is ever actually available to the reasoning agent and, if so, what are the conditions under which this is so.

A fourth inducement arises from "default inheritance" difficulties associated with certain kinds of information-taxonomies.[250] These are exemplified by the now iconic Tweety Triangle and the Nixon Diamond. We saw that, although birds fly and Tweety is a bird, Tweety is a bird of the penguin kind; so Tweety doesn't fly. Nixon's situation is that he was both a Quaker and a Republican. But Quakers are pacifists and Republicans not. In each case, the looming inconsistency is averted by the fact that the links of birdness to flying of penguins to not flying, of Quakerhood to pacifism and of Republicanism to nonpacifism are those of nonmonotonic consequence. This activates two tasks. One, by the consequence rule, is to characterize the consequence relation. The other is to show how inconsistency is averted by conditionals with true antecedents and mutually inconsistent consequents. How can it be true that the correct inference from Tweety's birdhood is that Tweety flies, that the correct inference from Tweety's penguinhood is that Tweety does not fly, and yet that Tweety's birdhood and penguinhood don't add up to a contradiction?

---

[249] Reiter (1978). See also Etherington and Reiter (1983), McDermott (1982). The closed world assumption is also a motivation for the logic of circumspection originated by McCarthy (1980). See also McCarthy (1986). Minker (1997) is a good, more recent, survey.
[250] Touretsky (1986) and Horty, Thomason and Touretsky (1990).

Perhaps the best known formal approach to defeasible reasoning is the default logic pioneered by Ray Reiter. In Reiter's approach, a default logic is an extension of first order logic, supplemented by default rules. Informally, a default rule is something like "From x is a bird, infer the default that x flies."[251]

7.2 $\mathcal{D}$-consequence

Let k denote a family of types of reasoning, of which the following is true. Reasoning of type k is consequence-drawing. Drawn consequences are consequences of a nonmonotonic consequence relation. Although the types of reasonings in k may exhibit various differences with one another, they are all anchored in this basic nonmonotonic substructure. Accordingly, variations in the types of reasoning in k reflect variations in their respective consequence relations. A central task of a $\mathcal{D}$-logic is to map the peculiarities of a particular type of k-reasoning to particular properties of its embedded consequence relation.

Here again, without regard for redundancies, are the $\mathcal{D}$ community's paradigm examples of reasonings of the k-type.

Typicality reasoning[252]
    1. Birds typically fly
    2. Tweety is a bird
    3. So Tweety flies.

    5. Quakers are usually pacifists
    6. Nixon is a Quaker
    7. So Nixon is a pacifist.

Generic reasoning[253]
    i. Ocelots are four-legged

---

[251] In the AI literature, default logics – including Reiter's own – often embed a fixed point (or cognitive equilibrium) orientation, which is an anticipation of sorts of the cognitive texture of reasoning developed here in earlier chapters. McDermott and Doyle (1980) is a modal fixed point logic. Connections to default logics are worked out in Lifshitz (1994). Also important are model preference approaches developed by McCarthy's circumscription theory. Significant variations are Krautz (1986) and Shoham (1987).

[252] See, for example, Horty (2002).

[253] See, for example, Carlson and Pelletier (1995).

ii. Ozzie is an ocelot

iii. So Ozzie is four-legged.

Autoepistemic reasoning[254]

1. The departure board shows no flights from Vancouver to London after 6:00 p.m.

2. So there aren't any.

Circumscription reasoning[255]

1. All normal birds fly

2. Tweety is a bird

3. Tweety is not abnormal

4. So Tweety flies.

Nonmonotonic inheritance reasoning[256]

1. Mammals don't fly

2. Bats are an exception

3. Baby bats are an exception to *this*

4. Belinda is a baby bat

5. So, while bats fly, Belinda doesn't.

Default reasoning[257]

1. In default of indications to the contrary, birds fly

2. Jeremy is a bird

3. There are no indications to the contrary

4. So Jeremy flies.[258]

---

[254] Moore (1983, 1988). Variations include logic programming (Baral, 2003).

[255] McCarthy (1980).

[256] Antonelli (1997); see also Horty (1994).

[257] Reiter (1980).

[258] Other developments within AI include truth maintenance systems (Doyle, 1979) and its extension to abstract argumentation theory (Lin and Shoham, 1989, and Dung, 1995). The importance of abduction for AI is already evident in Minsky (1974) and given more general treatment by Reiter and de Kleer (1987). The connection of abduction to logic programming is explored in Brewka and Konolige (1993). The application of nonmonotonic factors to causal reasoning is widely investigated, beginning with Lifschitz (1989) and Geffner (1992). See also McCain and Turner (1997) and Bochman (2004).

If we think (as we should) that there are non-trivial respects in which these several types of reasoning differ from one another, then on the present assumption those differences must vary systematically with variations in the $\mathcal{D}$-consequence relations in which they are respectively embedded; and we may speak correspondently of different types of $\mathcal{D}$-consequence – typicality consequence, generic consequence, autoepistemic consequence, and so on. This would make it reasonable to suppose that

Proposition 7.2a
CONSEQUENCE TYPE DETERMINES DEFEASIBILITY TYPE: *The defeasibility of defeasence of a given type is a function of properties peculiar to the nonmonotonic consequence relation in which it is grounded.*

Proposition 7.2a imposes a particular form of the consequence rule.

Proposition 7.2b
THE DEFEASIBLE CONSEQUENCE RULE: *A logic of defeasible inference is required to specify the properties of nonmonotonic consequence which reflect the distinctive character of the reasoning in which it is embedded, that is, reasoning that is vulnerable to alethically benign and generically impervious breakdown.*

The principal difference between $\mathcal{D}$-consequence and classical consequence pivots on the monotony property, which classical consequence has and $\mathcal{D}$-consequence lacks. Let Γ and Δ be sets of sentences, α, β, γ, ... be individual sentences, and $\vdash$ a classical consequence relation. Then

Monotony: If $\Gamma \vdash \alpha$ and $\Gamma \subseteq \Delta$, then $\Delta \vdash \alpha$. (Informally, monotony provides that new information added to old doesn't pre-empt any consequence of the old information alone.)

Other classical properties are:

Reflexivity: If $\alpha \; \varepsilon \; \Gamma$ then $\Gamma \vdash \alpha$. (Informally, a sentence is a consequence of any set of sentences of which it itself is a member.)

Cut: If $\Gamma \vdash \alpha$, and $\Gamma \cup \{\alpha\} \vdash \beta$, then $\Gamma \vdash \beta$. (Informally, if α is a consequence of Γ, then β is a consequence of Γ together with α, provided only it is also a consequence of Γ alone).[259]

Compactness: Whenever $\Gamma \vdash \alpha$ there is a finite subset $\Gamma' \subseteq \Gamma \vdash \alpha$.

Another property of $\vdash$ is:

Superclassicality: If $\Gamma \vDash \alpha$ then $\Gamma \vdash \alpha$. (Informally, if Γ classically entails α, then α is a $\vdash$ consequence of Γ. Since $\vdash$ here just is $\vDash$, superclassicality holds trivially.)

It is clear that defeasible reasoning fails monotony. New information added to old can override consequences drawn from the old alone. Even so, there are special cases of monotony that most $\mathcal{D}$-logics are drawn to consider. One is

Cautious monotony: Putting $\vert\sim$ as $\mathcal{D}$-consequence, if $\Gamma \vert\sim \alpha$ and $\Gamma \vert\sim \beta$, then $\Gamma \cup \{\alpha\} \vert\sim \beta$. (Informally, if α is a $\mathcal{D}$-consequence of Γ and β is another $\mathcal{D}$-consequence of it, then it is also a $\mathcal{D}$-consequence of Γ and α together.)

Cautious monotony is the reverse of cut. Cut says that adding to a set of sentences any of its consequences doesn't boost that set's inferential capacity. Cautious monotony states that a set of sentences together with any of its consequences doesn't depress that set's inferential capacity. If we put cut and cautious monotony together, we have it that if α is a consequence of Γ and β is any statement then β is a consequence of Γ just in case it is a consequence of Γ and α. It is widely accepted that reflexivity, cut[260] and cautious monotony[261] are necessary features of

---

[259] Cut is not the Gentzen rule of the same name. It is cumulative transitivity, whose initials "C.T." are pronounceable as "cut", albeit with a foreshortened "u".
[260] Semantic inheritance networks lack cut.
[261] Cautious monotony doesn't hold in semantic inheritance networks. However, the system of Horty, Thomason and Touretsky (1990) mandates special versions of cautious monotony and cut. Default logics in the manner of Reiter (1980) and Etherington and Reiter (1983) also fail cautious monotony. Preferential logics

most respectable relations of $\mathcal{D}$-consequence.[262] To these it is also customary to add supercalssicality.

Another special case of monotony is

> *Rational monotony:* If $\Gamma \not\!\!\sim \alpha$. And yet $\Gamma \mid\!\sim \beta$, then $\Gamma \cup \{\alpha\} \mid\!\sim \beta$. (Informally, provided that its negation is not a consequence of it, any sentence may be added to a set of sentences without disturbing any consequence that that set already has.)

Rational monotony is subject to a well-known counterexample by Robert Stalnaker (its examination needn't detain us here) and is usually not included among the defining properties of $\mathcal{D}$-consequence.[263]

Although reflexivity, cut, cautious monotony and superclassicality are typically taken to form the core of $\mathcal{D}$-consequence, there are further properties that are sometimes invoked. These include

> *Cumulativity.* If α is a consequence of Γ it is also a consequence of $\{\alpha\} \cup \Gamma$.

> *Disjunction in the premisses:* Whenever $\Gamma \cup \{\alpha\} \mid\!\sim \gamma$ and $\Gamma \cup \{\beta\} \mid\!\sim \gamma$ then $\Gamma \cup \{\alpha \cup \beta\} \mid\!\sim \gamma$.

> *Full absorption.* Let $Cl(\Gamma)$ be the classical closure of Γ and let $K(\Gamma)$ be the set of Γ's $\mathcal{D}$-consequences. Then $Cl(K(\Gamma)) = K(\Gamma) = K(Cl\Gamma)$.[264]

> *Distribution.* Everything common to $K(\Gamma)$ and $K(\Delta)$ is in what is common to the set of defeasible consequences of what's common

---

(Shoham, 1987) fails cut and cautious monotony unless they have a limit assumption which rules out the possibility that the logic not have a most preferred model. ($\Gamma \vdash \alpha$ just in case α is true in all the most preferred models of Γ.) Systems with the limit assumptions also have full absorption and distribution. Such systems are also supraclassical: If β is a classical consequence of α, it is also included in α's defeasible consequences.

[262] See here Gabbay (1985).

[263] Stalnaker (1994).

[264] Full absorption provides that defeasible consequence handles sets Γ of implying formulas and the classical closure of Γ in the same way.

to the classical closure of Γ and the classical closure of Γ and the classical closure of Δ; i.e., $K(\Gamma) \cap K(\Delta) \subseteq K(Cl(\Gamma) \cup Cl(\Delta))$.

Conditionalization. If α is a defeasible consequence of a set $\Gamma \cup \{\beta\}$, then the material conditional $\ulcorner\beta \supset \alpha\urcorner$ is a defeasible consequence of Γ itself.

In addition to rational monotony, a number of disputed properties have also been considered for defeasible consequence. Two them are

* *Disjunctive rationality.* If $\Gamma \cup \{\alpha\} \nvdash_D \beta$ and $\Gamma \cup \{\lambda\} \nvdash_D$, then $\Gamma \cup \{(\alpha \vee \lambda)\} \nvdash_D \beta$.[265]

* *Consistency-preservation.* If Γ is classically consistent so is $K(\Gamma)$.[266]

Given what's demanded by the defeasible consequence rule, we now have at hand some vital information. The core properties of *D*-consequence are reflexivity, cut, cautious monotony and superclassicality. By Proposition 7.1a, these are the properties centrally implicated in the defeasibility of defeasence. If Proposition 7.1a is true, then we should expect it to be the case that the distinctive characteristics of the paradigm cases of defeasible reasoning derive mainly from the fact that their embedded consequence relations satisfy those four core conditions. This is a bold claim, and rather remarkable if true.

7.3 The influence of classical consequence

With the main exception of autoepistemic logics,[267] the *D*-logics presently under consideration all arise from classical logic by attaching to

---

[265] Some preferential logics lack disjunctive rationality, rational monotony and consistency-preservation.

[266] Defeasible consequence in Lehmann and Magidor (1992) lacks consistency-preservation, as does the 0-entailment relation in extreme probability systems in the manner of Pearl (1990). Circumscription logics in the style of McCarthy (1980, 1986) lack rational monotony but satisfy consistency-preservation.

[267] Autoepistemic logics are made distinctive by the addition to their classical connectives of the so-called *introspective* modal operator. Two other exceptions are the theory of defeasible inheritance sets and the abstract theory of argument defect. In these cases, the role of logical connectives is much diminished. A third

classical consequence various and shifting families of constraints. In these systems, connectives are also interpreted classically. So it would be as wrong to think of these logics as rivals of classical logic as it would be to think of S5 as a rival of the classical logic of propositions.

David Makinson (2005) shows nonmonotonic consequence to be a relation at two removes from classical consequence; or, as we might say, is a second cousin of it. In a first variation, classical consequence becomes paraclassical consequence, which in a second variation becomes nonmonotonic consequence. Motivating the bridge between classical and nonmonotonic logic is a reasoner's desire to obtain more from a set of premisses than could be obtained solely from classical consequence. Makinson develops this bridge in the three critical areas of *assumptions, valuations* and *rules*. In each case he defines a paraclassical consequence relation and then transforms it into a form of nonmonotonic consequence.

For our purposes, it will be enough to sketch the first and simplest case, in which inferences engage with background assumptions. Let Σ be any set of formulae, and α an individual formula. The language is that of classical logic, and let's call the set of all its formulae L. Let B be the set of background assumptions, and $B \subseteq L$ be fixed set of formulae. Then a *pivotal-assumption consequence* α is a consequence of $B \cup \Sigma$. Pivotal-assumption consequence is supraclassical (classical consequence is a proper subrelation of it), reflexive, cumulatively transitive and monotonic, which makes it a closure relation. Possession of these properties renders the relation paraclassical. Also satisfied are disjunction in the premisses and compactness.

How do we obtain nonmonotonicity from paraclassicality? We do it by varying the background assumptions B with the set of premisses Σ. This is achieved by imposing a consistency constraint. When the constraint is violated, the usable part of B is diminished. The decisive step towards nonmonotonicity is then to use the maximal subsets B' of B that are consistent with Σ and take as output what their separate outputs share in common. (Note that a maximally consistent subset cannot be a proper subset of any other subset consistent with Σ).

Makinson calls the ensuing relation *default-assumption consequence*. Default assumption consequence is the consequence

---

exception, on its face at least, are systems of logic programming with negation as failure. But Makinson claims, and I agree with him, that there are ways to "re-classify" these rules. (Makinson, 2005, chapter 4) relation that obtains when α is a consequence of B′ ∪ Σ for every B′ ⊆ B that is maximally consistent with α. Default assumption consequence relations are nonmonotonic. In the case at hand the result of varying B with Σ is that we gain premises but, because of the consistency condition, we lose background assumptions.

Default-assumption consequence relations are supraclassical, and satisfy cut or cumulative transitivity and disjunction in the premises. They also satisfy cautious monotony but fail compactness. So our question recurs, which of the features distinctive of the $\mathcal{D}$-community's paradigm cases of defeasible reasoning are adumbrated by these core properties?

7.4 *Defeasibility conflicts*

An important consideration for $\mathcal{D}$-logics is how conflicts between defeasible conclusions are to be handled. Here are two cases to consider. In the first, new information overrides a defeasibily drawn consequence. In the other, a defeasibly drawn consequence conflicts with another defeasibility drawn consequence from respective premises the former of which are true if the latter are also true. Virtually all $\mathcal{D}$-logics handle the first kind of conflict by deferring to new information, not because it is new but it is presumed to be "hard"; whereas the defeasible consequence it overrides is "soft" – that is plausible, but fallible, tentative, presumptive, open to correction, and so on. Tensions of the second kind are subject to two different kinds of treatment, of which the Nixon inference is a standard example. If Harry's information is that Nixon is both a Quaker and a Republican, then which nonmonotonic consequence should he draw? That Nixon is a pacifist? Or that he is not? If Harry is a "bold" reasoner and believes that he has no rational basis for selecting the one consequence over the other, he will nevertheless opt for one and reject the other. If he is a "cautious" reasoner, he will desist from drawing either consequence. Sometimes the cautious reasoner is characterized as "sceptical". This is a mistake. His caution isn't borne of not knowing which consequence should be preferred, but rather by a rather antique notion of negation, according to which "Nixon is a pacifist" and "Nixon isn't a pacifist" simply cancel one another out. Hence, there is nothing to *be* decided. At present, the defeasibility literature hasn't reached a consensus about which, if either, of these options to adopt (and when).

Like any of the others, $\mathcal{D}$-logics are judged against recognized standards of adequacy. One is conceptual-empirical adequacy, also called in the $\mathcal{D}$-literature material adequacy (in the sense of Tarski's material adequacy condition for theories of truth.) A second requirement is mathematical virtuosity, also called formal adequacy. A third consideration (which can also be seen as intersecting the prior two) concerns the complexity of the reasoning routines mandated by the $\mathcal{D}$-logic in question. Perhaps it will come as no surprise that $\mathcal{D}$-logicians recognize the tension between the requirements of material and formal adequacy. In a number of respects, the AI literature has shown a readiness to favour the material over the formal. A case in point is cautious monotony which, as Makinson points out, is sometimes forgone in favour of our defeasibility intuitions.[268] Trumping considerations also flow the other way. For example, it is difficult to believe that reflexivity holds for $\mathcal{D}$-consequence. We may allow that every sentence is its own deductive consequence. But how can it be that a sentence follows from itself nonmonotonically? How can it be true that, with α holding true *as a premiss*, α *as a conclusion* might concurrently have to be given up in the face of defeating information?[269]

I have already said that my general view of the $\mathcal{D}$-logics of AI is that while they make some considerable display of mathematical virtuosity, they are less impressive in matters of conceptual and empirical adequacy. Again, McCarthy's Law bids the logician of common sense reasoning to eschew the stipulation that the human individual closes his beliefs under consequence. It is a prohibition which AI theorists routinely ignore. John Pollock writes to the same effect:

> Unfortunately, their lack of philosophical training led AI researchers to produce accounts that were mathematically sophisticated, but epistemologically naïve. Their theories could not possibly be right as accounts of human cognition, because

---

[268] Makinson (1994). As for complexity, let us simply note that, even in comparison with classical first order logic, $\mathcal{D}$-logics are combinatorial nightmares. (This is Doyle's point in one of this chapter's epigraphs.)

[269] I don't want to make more of this point than is in it. A $\mathcal{D}$-logic is a formal theory of (a type of) defeasible reasoning. Formal theories distort their target data as a matter of course – see again chapter 2. It matters that all formalization is distortion. But it matters more whether the attendant distortion is compensated for by the theory's provisions overall.

they could not accommodate the varieties of defeasible reasoning humans actually employ (2008, p. 452).

Pollock adds that although "there is still a burgeoning industry in AI studying nonmotonic logic this shortcoming remains to the present day." The majority of the defeasibility systems are elaborations of quite simple formal languages – more often than not propositional languages. The reason for this is that, with the exception of Pollock's OSCAR (1995), automations of defeasible reasoners have not been successful for systems richer than propositional logic or some of its even less rich sublogics. In this regard OSCAR is a standout. OSCAR is an AI architecture for knowers – for cognitive agents – and can be thought of as a general-purpose defeasible reasoner. Even so, OSCAR cannot to date handle defeasible reasonings that vary in degrees of goodness or strength. Indeed

> There are currently no other proposals in the AI or philosophical literature for how to perform defeasible reasoning with varying degrees of justification. (Pollock, 2008, p. 459).[270]

7.5 *Misconceiving some interconnections*

$\mathcal{D}$-logics are replete with assumptions about third way thinking, one of which is that in its nonmonotonicity defeasible reasoning differs in a significant respect from deductive reasoning. If $\mathcal{D}$-consequence is classical consequence at substantial second remove how different can deductive and defeasible reasoning be? With this in mind, I want now to examine the logical connections, or lack of them, between and among the three $\mathcal{D}$-properties of nonmonotonicity, defeasibility and defaults, and the two classical properties of validity and soundness. We will soon see some of the sources of the conceptual confusions of which Pollock complains.

Nonmonotonicity and deductive validity. The universally accepted definition of a nonmonotive logic is one whose consequence relation is not monotonic. It is commonly held that monotonicity is a universal feature of deductive consequence. This is a mistake. Let S be a

---

[270] But see here Horty (2012), pp. 138-141. *Reasons as Defaults* reached me just as my own book finally left my desk. I regret that I was not able to have benefited from Horty's insights and for the philosophical sophistication with which he deals with this subject.

syllogistic logic and $\vdash_s$ the syllogistic consequence relation. One of the conditions on α, $\beta \vdash_s \gamma$ is that α, β entail γ. Another is there be no redundant premisses. So, although $\vdash_s$ is a relation of deductive consequence, it is nonmonotonic as well. Similarly, in Anderson and Belnap (1975) there are logics in which a proof of β from hypotheses α_1, ..., α_n is a relevantly valid proof only if each of the α_i has an occurrence there.

Let $\mathcal{L}$ be a linear logic[271] and $\vdash_L$ the relation of linear consequence. In $\mathcal{L}$ nonmonotonicity is imposed by definition: a

α, α, $\alpha \rightarrow \beta$ does not prove β, because there are here two copies of α. The second occurrence of α is not a participant. This spoils the proof. Here, too, we have a deductive consequence relation, but not a monotonic one.

Proposition 7.5a

MISCONCEIVING MONOTONICITY: *Monotonicity is a typical but not intrinsic property of deductive consequence relations.*

COROLLARY: *Making monotonicity a defining feature of it embodies a conceptual confusion about deductive reasoning.*

Anyone mindful of the importance of the difference between consequence-having and consequence-drawing will already be predisposed not to draw every consequence he is able to spot. Sometimes, to be sure, the addition of new information to a premiss set (or data bank or knowledge base) should change our minds about what now follows. In those cases, the failure to be a consequence is trivially a matter of whether it is a consequence anyone should (or could) draw. But a much commoner case against consequence drawing has nothing to do with consequence having. Perhaps the clearest example of this is one in which new information falsifies *without need of compensation* a premiss in an otherwise sound argument. By and large, we don't want (categorical) arguments from false premisses. A valid argument from true premisses is

---

[271] Linear logics in the modern sense originate with Girard (1987) in a semantics for System F of the polymorphic lambda calculus. Connections to computer science were first discernible in the Curry-Howard isomorphism (Howard, 1980).

sound. But, even in those systems in which validity is monotonic – namely, most of them – soundness is not. Other such properties (at least in their intuitive senses) are: plausibility, probability and possibility.

It is simple economics that in the general case a reasoner's first task is to test for unsoundness, leaving a second task – checking on the consequence relation – to be performed only after a positive finding with respect to the premisses (i.e., they are true, or plausible, or probable or some such thing). Why would this be so? Because premiss inadequacy is decisive against consequence drawing.

Proposition 7.5b

THE PRIORITY OF UNSOUNDNESS: *In the logic of reasoning, bad news about premisses trumps good news about consequence having.*

7.6 Generically impervious rupture

As we are telling it here, instantiations of generic generalizations are defaults. In the AI literature defaults are sometimes the major premisses in default inferences, that is to say, the generic propositions themselves. Here they are their conclusions. In some ways, the AI convention is unfortunate. It obscures the distinctive character of the relationship borne by the conclusions of default inferences to their premisses. In many writings, as we saw, a default takes the form of a *rule*: "Given that α, conclude that β unless there is information to the contrary."[272] Seen this way, the default is not the tail of the rule. The rule gets its grip in the generic fact that ocelots are four-legged. But the generic claim is not the default either . For suppose that it were. Then it would be covered by a default rule, by something in the form ⌐Given α, conclude that ocelots are four-legged except where there is particular reason to the contrary.⌐ There are problems with this. One is that if the rule held for "Ocelots are four-legged", it would hold for all propositions.

---

[272] See, for example, Horty (2001). In Reiter (1980) there is a consistency clause: "If x is a bird and if it is consistent that x can fly, x flies." For technical reasons, Reiter's rules aren't rules of inference. The reason is that any model theory for default logic must, in some way or another, already refer to defaults. See Reiter (1978), (1980) and Łukaszewicz (1984). Gabbay points out in conversation that defaults can also be represented as constraints on Henkin models.

There is little to be gained from terminological wrangles. The habit of calling generic premisses "defaults" is fairly well established. So is the habit of calling defaults "rules", of seeing them as generic instantiations, and also of seeing the link between a generic and its instantiation as default-consequence. Fortunately, the uses are linked. A generic claim is a default because it mandates a default rule. An instantiation of a generic claim is a default because it conforms to a rule which the generic claim mandates. The tail of the default rule is a default-consequence of its head. Better, then, that we follow the course of least resistance, and allow for the fourfold use, leaving it to context to sort out which is which in particular cases.

Still, there is something troubling about the Ozzie case. I mean by this that Ozzie creates a problem for premiss-conclusion rupture. It is true that anyone who now becomes aware of Ozzie's three-leggedness would no longer infer his four-leggedness even though it remains true that ocelots are four-legged and Ozzie is an ocelot. This makes the information that Ozzie is three-legged consequence-severing, and it invests the premiss-conclusion relation severed by it with epistemic significance. Depending on how we think the relation is structured, it might also allow the default relation to pop in and out of existence on the wings of varying information. It would vary the existence or non-existence of default-consequence with individual knowers.

On the other hand, if it is facts rather than information states that sever default relations, then the relation between the premisses and conclusion of the Ozzie inference has been severed for as long as Ozzie has been three-legged, never mind who may or may not have known it. If this is so, the default relation between "Ocelots are four-legged and Ozzie is an ocelot" and "Ozzie is four-legged" exists only if Ozzie is four-legged. But this is a stiff requirement, one that trivializes the distinctive character of default reasoning, which is that it is safe until found to be otherwise. Accordingly, the more intuitive course might be to concede that

Proposition 7.6
DEFAULT-RELATIONS AS EPISTEMIC: *Since the defeasibility of default-inference is its vulnerability to R-rupturing information, then, as we have it here, the relation of default-consequence is an epistemic not a semantic relation.*

COROLLARY: *The presence or absence of default-consequence in a piece of reasoning at a time varies from reasoner to reasoner and with what they respectively know then.*

As developed in chapter 1, a key factor in the disjointness of consequence-having and consequence-drawing is that drawing is an intrinsically pragmatic relation and having is intrinsically not. Epistemic relations put a strain on the purity of this dichotomy. In some of its CC-manifestations, epistemic relations are definable without appeal to agents and their symbolic-processing doings. In the JTB variation, belief is usually a proposition, a resident of logical space, and evidence is a neighbour, a proposition or set of them. Whereupon the relation of justification or justifiability is itself a logical-space relation, a relation from proposition (sets) to propositions. Whatever one might think of it just as it stands, this is a harder story to tell in the defeasibility environments of nonmonotonicity, in which well-instantiated evidence-for relations are ruptured by new information. But new information is definable only in relation to those who didn't have it and now do. New information is key to nonmonotonicity and symbolic systems-processing agents are key to new information. So the epistemic evidence-for relation is not a semantic relation.

7.7 *Informational semantics*

This would be a good place to recall a reservation that also arose in chapter 1 concerning the purported assertoric consequence relation of Rescher's assertion logic. We noted that there too assertoric consequence is a pragmatic relation, not a semantic one. We also noted its nonmonotonicity. This stirred the idea that it might be expressively maladroit to call this assertoric relation by the name of consequence. It suggested that assertoric consequence might be consequence in name only. The $\mathcal{D}$-relations currently in view are also nonmonotonic. So the same thought stirs. Might it not be that they are consequence relations in name only?

We might note here an objection which arises from informational semantics, whose principal purpose is to provide an explication of the consequence relation different from both its model theoretic and proof theoretic treatments. Mainstream model theory is usually implicated in the defence of classical logic, and proof theory in the defence of intuitionist logic. Informational semantics was purpose-built for such

246

things as relevant and other sorts of substructural logics.[273]

The basic idea of informational consequence can be set out as follows.

Proposition 7.7

INFORMATIONAL CONSEQUENCE: *A conclusion β is a consequence of premisses Γ if and only if the information content of β does not exceed the combined information content of the premisses in Γ.* (Copeland, 1979)

Of course, any relation of purported consequence that satisfies the conditions of Proposition 7.7 is a monotonic, not nonmonotonic, consequence relation, and as such is not germane to our interest in whether nonmonotonic relations qualify for consequencehood. On the other hand, the initial target of relevant and substructural logics was not the specification of an epistemic relation between premisses and conclusions. But if the relevant and the substructural tie between premisses and conclusions is indeed informational, as opposed to truth-conditional (model theoretic) or derivative (proof theoretic), how unfeasible could it be to epistemicize the informational character of the premiss-conclusion tie in the cases presently under review?

It all depends on how we interpret the notion of information content in the informational consequence rule. If information is taken in an objective sense, as something present in the environment independently of anyone's awareness of it, then the consequence relation may be all right, but it is not the epistemicized relation exhibited by the $\mathcal{D}$-relations under discussion here. If, on the other hand, information is taken subjectively, as information that is believed by some or other agent to be sound, then informational consequence falls into the epistemicized camp.

---

[273] Dunn (1976) combines Routley's first degree entailment with Barwise and Perry's situation semantics (1983). The resulting system was subsequently adjusted to defend against the objection that the Routley-Meyer relational semantics didn't add up to a genuine semantics. Restall (1995) brought to this defence a version of Barwise's channel theory Barwise and Seligman (1997), and Mares (1997) a theory of information flow arising from Israel and Perry. Other developments could also be described as informational; for example, Wansing's approach to substructural logic (1993a, b), the informational treatment of sequents described by Paoli (2002), and the Boolean informational approach described by John Corcoran (Saguillo, 2009).

Kit Fine has recently advanced the idea that difficulties such as those embodied in Frege's Paradox of Identity call for a Copernican Revolution in logic.[274] It calls for the abandonment of the idea that the domain of semantic facts is closed under classical consequence. If this call were heeded, the domain of semantics would become one of information rather than facts. The truths of semantics would become epistemic requirements for the understanding of language, and the domain of semantic facts would properly understood be closed not under consequence but under a weaker relation – an epistemicized relation of "manifest consequence." The main idea is that an inference is manifestly sound if it can be drawn "regardless of the identity of the objects in the hypothesis [premisses]". (p. 48)[275]

We see that Fine's manifest consequence and Weiss' weak consequence have a considerably different motivation from our purported relations of $\mathcal{D}$-consequence. Manifest consequence is a response to a problem affecting any logic that admits the identity relation. $\mathcal{D}$-consequence is a response to alethically benign and generically impervious rupturability. Fine's and Weiss' project is more radical than the $\mathcal{D}$-project. But there is a point on which Fine (especially) and I are of one mind. The R-relation which is an epistemic, not semantic, relation. It is not a semantic relation of any kind, no matter how weak.

7.8 *Re-thinking the consequence-rule?*

I have no knockdown case to advance against the very idea of $\mathcal{D}$-consequence. But logicians, like everyone else, learn from experience. Logicians have spent most of the last 2400 years tilling the soil of deductive consequence. Their endeavours, especially in the past century or so, have met with a dazzling success. Along the way, methods of investigation have become firmly established and have been rewarded with high payoff. Ideas have formed about what a logic of consequence could be, and with it the idea of what it takes to *be* a consequence. It is a fundamental fact about enquiry of any kind that, up to a point, what you

---

[274] Fine (2007).

[275] This is subject to counterexample. (Weiss, 2013). Weiss goes on to revise manifest consequence into a weaker relation than classical consequence "of *independent* interest, for it might be held to constitute a cogent standard of reasoning that proceeds under a deficient grasp on the identity of objects". (2013, abstract; emphasis added)

find is what you expect to find. What you expect to find reflects your expectation of what is there to be found. Over time, logicians of deductive consequence have learned what to look for. They take it as given that any relation of deductive consequence, whatever its specific peculiarities, will have what it takes to make what logicians look for the right sort of things to *be* looked for. So tools are perfected for the task of searching out the properties that supply answers to those questions. The relevant literature leaves us in no doubt as to the $\mathcal{D}$-community's expectations. Theirs, too, are expectations which encapsulate a certain conception of what it is to be a rupturable R-relation. It would be a relation with respect to which there are answerable questions appropriate to ask of it. Does it have reflexivity, cut, cautious monotony superclassicality, and so on?

      Suppose that we agreed that part of what it takes to be a consequence relation of any kind is that it admits of answerable questions of these kinds. If that were so, then we would have it trivially that the $\mathcal{D}$-logics are indeed the right sorts of instruments for the exposure of $\mathcal{D}$-consequence. The fact remains, even so, that the dominating *motivations* of the creators of those logics has been to get to the bottom of phenomena such as alethically benign and generically impervious rupturability. But there is not much advancement of that agenda in unearthing of properties that make of conclusionality relations relations of consequence, if that indeed is what they are. Even if core properties accounted in large measure for the vulnerability of $\mathcal{D}$-relations, it isn't to them that the paradigm cases owe their peculiarity, but rather it is to the special character of their premises. This is true by inspection. What, we ask, is distinctive about the core cases that absorb $\mathcal{D}$-logicians? The answer is clear: typicality *premisses*, generic *premisses*, autoepistemic *premisses*, default *premisses*, and so on.[276] This is a strange development. Given that motivation, $\mathcal{D}$-logics are imperfect instruments even though they are the right instruments for $\mathcal{D}$-consequence if such there be. That is a reason to think that our rupturable R-relations aren't after all relations of $\mathcal{D}$-consequence. If so, there is a useful moral in it for the naturalized logic of third-way reasoning.

---

[276] So, too, in the case of nonmonotonic abductive reasoning premisses (crudely) of the form "If α were true then α would account for what we want to know about" play the pivotal role. The same is true of theories that treat causal inference as a kind of abductive reasoning.

Proposition 7.8
NEW METHODS: *Logicians should consider developing new methods for the treatment of defeasible reasoning.*

COROLLARY: *Consequence is the wrong paradigm for R-conclusionality.*

Everyone agrees that when someone rightly concludes from α and β that γ, there is a relation between α, β, and γ in whose absence the reasoning could not have right in the way that it actually is. Call this a conclusionality relation. All third way logics recognize this relation, and there is virtually no one there who doesn't identify it as a kind of nondeductive consequence. The conclusionality between α, β and γ is *vulnerable*. It is vulnerable to rupture without damage to the truth of true α and β. One of the questions before us is whether a consequence relation can consistently be called vulnerable. Variations in what we know and changes in how the world goes effect the presence and absence of the $\mathcal{D}$-relation in question. They are, so to speak, "fugitive" relations.[277] Consider again the relation $\mid\sim_e$ of being empirical evidence for. $\mid\sim_e$ is a vulnerable relation, susceptible to rupturing evidential shifts. If we understand evidence as what is presently known of the relevant aspects of the world, then $\mid\sim_e$ is an epistemic relation. On the other hand, if we understand evidence as how the relevant aspects of the world are independently of what may be known of them, $\mid\sim_e$ is an ontic relation. Since what we know of the world is depends in part how the world is, even the epistemic reading of $\mid\sim_e$ has ontic significance.

The epistemic (or epistemico-ontic) character of $\mid\sim_e$ relations calls into question the wisdom of tying their specification to variations of variations of the epistemically neutral relation of classical consequence. Again, what is it about reflexivity, cut, cautious monotony and superclassicality that gives to $\mid\sim_e$ its epistemic character, that makes its presence or absence at a time a matter of what we know at that time? If these variations do in fact generate evidential consequence from classical consequence, this is transmutation of high order, on the scale of rendering of base metals into gold. It is the transformation of an invulnerable intrinsically non-epistemic relation into a vulnerable intrinsically

---

[277] In the sense of "fugacious" or "fleeting".

epistemic relation. [278] The point easily generalizes to any $\mathcal{D}$-consequence relation made vulnerable by its susceptibility to rupturing information.

What now of the tie between the core properties of the purported consequence relations of $\mathcal{D}$ and the peculiarities of the paradigm cases? The answer is "Not much". The premiss-peculiarities which absorb the attention of $\mathcal{D}$-logicians certainly merit the attention of anyone who has an interest in the logic of third way reasoning. I would not want to leave the impression that these premiss-peculiarities should be omitted from consideration. Far from it. The moral to draw from our present discussion is that a logician of the third way has two major programmes to run, not one. He must come to some principled and suitably settled position on the vulnerability of R-relations. He must also work up an improved understanding of premiss-peculiarities. This latter will be the business of what remains of this chapter. In chapter 8, I'll re-open the question of whether a credible and load-bearing role for $\mathcal{D}$-consequence relations in the discharge of these tasks might be found. It might strike the reader as perverse that, having taken pains to raise doubts about $\mathcal{D}$-consequence, I should now want to reconsider them. Perverse is just about the last thing this is. Having raised some doubts, I want in chapter 8 to reflect on how they might be answered. There is nothing wrong with reconsidered judgement, especially in philosophy. Besides, wouldn't it be significant if our attempt to renew the license of $\mathcal{D}$-consequence *failed*?

7.9 Adverbial non-quantifications

Premiss-peculiarity is our focus here. True generics have true negative instances. True non-universal plural quantifications also have true negative instances. It is right to think of the first as exceptions. But this is not the right way to think of the other. A true negative instance is a counterinstance of a generic; it is not an exception to a non-universal plural quantification (indeed it is expressly catered for by its quantifier). Exceptions to generics have an interesting feature. They are, to some nontrivial degree, *anomalies*. They are indications that somehow things are out of joint. Ozzie the three-legged ocelot is an ocelot-anomaly; he is

---

[278] The epistemicization of consequence is by no means the exclusive preserve of $\mathcal{D}$-logics. Hintikka (1962) gives an epistemic adaptation of S4 which, among other things, gives a pragmatic characterization of logical truth. The epistemic turn in AI, also a variation of modal logic, begins with Moore (1985). See also his (1993) and (1995).

one of the ways that ocelots aren't meant to be. Tweety the penguin may be a comparative rarity for birds, but he is not an anomalous bird.

The gap between "Ocelots are four-legged" and "Birds fly" is rather fuzzy. In between this pair we find the adverbally qualified non-quantifications – "Typically Fs G", "Characteristically Fs G", "Normally FsG". Lacking quantifiers, adverbially qualified generalizations resemble generic generalizations. But since true negative instances are catered for by their adverbial qualifications, they also resemble non-universal plural quantifications. This intermediacy is reflected in the character of their negative instances. Italians are typically ebullient, but Guido is not. There is more than a statistical significance to Guido's subduance. His father worries that Guido is not a "real Italian", that he might just as well have been an Englishman. But it would entirely wrong to think of Guido as an anomaly in the manner of Ozzie. There is something wrong with Ozzie, but saying this of Guido would be going too far. Even so, there is something more to Guido's case than comparative statistical rarity. Not anomalous, and not merely rare, Guido's situation goes somewhat against the flow.

One of the differences between a generic truth and a true non-universal plural quantification is that generics invest their corresponding conditionals with nomological force. Non-universal plural quantifications can be entirely free of nomological presumption. Adverbially qualified non-quantifications fall in between. They have a nomological character intermediate between these two extremes – hence, while not anomalous, their true negative instances go somewhat against the flow. Provided we are careful about it, we could say that they are in some sense *odd*.[279] The differences between anomalies, oddities and rarities are far from precise. There is little prospect of giving them mathematical expression to conceptual advantage. But they are important distinctions. They warn us against theories in which, fuzzy though they are, they aren't honoured.

7.10 *Resolving Tweety*

Suppose there were a species of three-legged ocelots. Call this species *quocelots*. Then we would have it as a generic truth that quocelots are three-legged. If so, we would *not* have it as a generic truth that ocelots are four-legged. Quocelots are ocelots that are three-legged; and since

---

[279] In Gabbay (2007), α nonmonotonically implies β if the English sentence "α but not-~β" sounds odd.

species are not anomalies of their genera, "Ocelots are four-legged" would be falsified by non-anomalous counterinstances. So "Ocelots are four-legged" would not be a generic truth.

This tells us something important. It tells us that, although "Penguins don't fly" expresses a generic truth, "Birds fly" does not. "Birds fly" expresses a truth, but the truth it expresses is not a generic. It is better to express "Birds fly" as a proposition to the effect that

By far, most species of birds fly.

Note that this is a *hybrid* of a non-universal plural quantification and a generic sentence,

By far, most species x of birds are such that xs fly,

in which the italicized clause is a generic open sentence whose variable is bound by the non-universal quantifier, "Most by far x".

How then should the tension be resolved between Tweety's flyingness *qua* bird and his nonflyingness *qua* penguin? Of course, the well-travelled answer is that we pick the inference with the less specific major premiss. This is an answer both satisfying and not.[280] It is satisfying in so far as it picks the winning alternative. It is not satisfying in so far as it leaves it unexplained as to why specific information should be allowed to trump more general information. Indeed, why should it?

Consider the two sentence-pairs:

1. Tweety is a member of a species of a genus of which most by far of its species fly.
2. Tweety doesn't fly.
   a. Tweety is a member of a species of non-flyers.
   b. Tweety flies.

In each case there is a tension between the first member of the pair and its second member. In the first, we could say that the truth of the first member makes the truth of second significantly *unlikely*. In the second case we must say something different, and stronger. For here the truth of

---

[280] On the technical side, default logics don't allow general defaults to defer to more specific ones. Attempts to adopt such a principle are notoriously complicated and computationally discouraging. See here Asher and Morreau (1991) and Horty (1994).

the first member of the pair makes the truth of the second member at least *anomalous*. Since in real life, unlikelihoods significantly outrun anomalies, we could say that unlikelihoods give less offence than anomalies to the natural order of things. Accordingly it might strike us as appropriate to consider the following resolution rule:

Proposition 7.10a
THE TWEETY RESOLUTION RULE: *Concerning Tweety's flying or not flying,* accept *premises (1) and (a) and* reject *the conclusion whose truth would give greater offence to the natural course of things.*[281]

If this were the right rule, people who cite specificity as a tie-breaker would be right to think that the relata of the specific-nonspecific relation play a role in Tweety resolution. But there would be a sense in which they would have got the *flow* of the relationship wrong. Generics trump non-generics. Sentences backed by generics trump competing sentences backed by non-generics. Tweety's not flying is backed by "Penguins don't fly", whereas Tweety's flying is backed by "By far most species of birds fly". "Penguins fly" is a generic, true even in the face of anomalies. "By far most species of birds fly" is a non-universal plural quantification that is falsified by the mere absence of the requisite species. There is a general-specific lesson to be learned from this. It is that in a competition between not-flying and flying, pick the option backed by the more generic premiss.

 The Tweety case is but one of a number of cases in which new information is added to old in ways that influence a conclusion backed by the old. Here is a further pair of them:

 The old information is that Zoë is a bird, and the conclusion indicated by it is that Zoë flies. The new information is that Zoë is an Antarctic bird, and the conclusion it indicates is also that Zoë flies. But there is a difference. The new information backs that same conclusion more weakly – we might say more defeasibly – than the old alone. So we have

---

[281] I speak loosely and, as it may seem to some, across the grain of my own scepticism about a central role for rules in the analysis of right and wrong reasoning. The tension is misjudged. For me, rules are a *façon de parler*. The real import of the Tweety "rule" is that it tells us which of two inferences a human being's well-made inference-engine would make.

Proposition 7.10b

THE ZOË RULE: *Adjust your judgement about whether Zoë flies to all the relevant available information, not just proper subsets of it.*

Consider now the situation in which the old information is that Fred is a bird, and the conclusion indicated by it is that Fred flies. The new information is that Fred is an aviary-caged bird, indicative of the same conclusion. Here, too, there is a difference, but a different difference. The new information backs the conclusion more strongly – less defeasibly – than the old alone. But the covering rule is the same

Proposition 7.10c

THE FRED RULE: *Adjust your judgement about whether Fred flies to all relevant available information, not just proper subsets of it.*

COROLLARY: *Be circumspect about closed-world presumptions.*

There are instructive examples. They get us to see that the better rule for Tweety is a variation of the rules for Zoë and Fred. They get us to see that Tweety, like Zoë and Fred, is but one of a class of cases governed by what lawyers call best evidence. It is true that in the examples at hand the new information implies the old, hence is more specific than it. But this seems not to be the decisive factor. There are lots of cases also governed by the principle of best evidence in which the new information is not more specific than the old.

7.11 Resolving Nixon

Nixon requires separate treatment. It lacks the specificalities present in the cases we've been just now examining. That Nixon is a Republican doesn't imply that he is a Quaker. That he is a Quaker doesn't imply that he is a Republican. But here, too, a variation of the revised Tweety (and Zoë and Fred) rule seems right by and large, that is, defeasibly right:

Proposition 7.11

THE NIXON RULE: *In default of contrary indications, try to* acquire *information that will break the pacifist-nonpacifist tie.*

COROLLARY: *Neither the bold option nor the cautious option is a sound general policy.*

The Nixon rule is a special kind of variation of the revised Tweety rule. The Nixon rule doesn't embed the revised Tweety rule. Rather it bids the would-be reasoner to acquire the information that would enable him to apply it.

What would the Nixon rule's contrary indications be? Consider a variation of the Nixon example. There is evidence that strongly indicates that Spike committed the crime. New evidence is found that weakens that inference. If you are a detective with CID, the Nixon rule sets the right course of action. You should continue with the investigation. But if you are a juror at Spike's trial, you must vote for acquittal. And if you are a casual and largely uninterested reader of the morning paper in a part of the country far-removed from the scene of that crime, you might better forgo judgement altogether.

There is no one-size-fits-all resolution for Nixon cases. The right way to proceed is a matter of context, and the contexts matter most are those that reflect the tie-breaker's personal circumstances. Thus Nixon resolutions respond, quite properly, to considerations *ad hominem.*

7.12 *How to falsify a generic*

We saw that if quocelots were a species of ocelots then "Ocelots are four-legged" could not be a generic truth. "By far most ocelots are four-legged" would be true, but "Ocelots are four-legged" would be false. Let us say of any two genericizations "Gs H" and "Js K" that the "head" of the first properly includes the "head" of the other if and only if all Gs are Js and some Js are not Gs. Similarly let us say that "Gs H" and "Js K" have incompatible "tails" if and only if K entails not-H. We can now say some thing official about falsifying the generic.[282]

---

[282] Consider the following cases from Carlson and Pelletier (1995), p. 44: "A bird lays eggs", "An *Anopheles* mosquito carries malaria" and "A turtle lives a long life". Less than half the birds lay eggs; no more than five percent of *Anopheles* mosquitoes carry malaria; and hardly any turtles have long lives, owing to the efficiencies of anti-turtle predation. Are these cases *anomalies*? I would hardly say so. The better option (by far) is to rewrite: Thus birds are an

256

Proposition 7.12a

FALSIFYING THE GENERIC: *Let α and α′ be generic generalizations. Then α′ falsifies α if α′ is true, α's head properly includes α″'s head and α and α′ have incompatible tails.*

There is a valuable lesson to be learned from this. Let α and α′ be two purported genericizations that meet the present conditions Then if α′ is true α is false *if it is a generic.* There lies in this an economical strategy for saving false purported genericizations: *Re-issue* them *as* adverbial non-quantifications – e.g., replacing "Ocelots are four-legged" with "Ocelots are typically four-legged". Alternatively, they could be seen as non-universal plural quantifications-generic hybrids – e.g., as "Virtually all species of ocelots are four-legged"

Proposition 7.12a successfully captures one reading of the intuition that deference be shown to the more specific of any two incompatible generalizations. But, as we see, the deference that is due it is not to its specificity but rather the fact that it is a *falsity-maker* of the more general claim. But 7.12a also leaves the question of what *else* would falsify a generic. Let us see.

There is something not quite right about the present condition. It seems to be open to a class of cases that pull its teeth. We say that "Ocelots are four-legged" is a true generic. If so, the claim is not defeated by its anomalies. It is not defeated by the congenitally compromised Ozzie. It is not defeated by any three-legged specimen made so by surgery. But consider now the class of those very anomalies. We have it that leggedly anomalous ocelots aren't four-legged. If this is a generalization at least as strong as a generic (indeed it makes a fair claim to be a universally quantified conditional), then it certainly falsifies "Ocelots are four legged". Anomalous ocelots are properly included in the ocelots, and not-four-leggedness is incompatible with the four-leggedness.

We may take it that no one seriously believes that anomalous ocelots overturn the generic four-leggedness of ocelots. If Proposition 7.12a is to hold, we will have to exclude this kind of case. Can we do this in a plausible way? It would appear that we can. We could determine that, given their incompatibility more specific generics overturn more expansive ones. What the present case suggests is that we cash an

---

egg-laying species; *Anopheles* mosquitoes are capable of carrying malaria; the turtle is genetically equipped for a long life.

intuition of paragraphs ago. So I venture to say that only a generic can override a generic. What is more, since anomalous ocelots are not a natural kind – certainly not a species – their non-four-leggedness is not a generic trait of them. Equally, since universally quantified conditionhood is not sufficient for genericity, even the true sentence "$\forall x$ (x is an anomalous ocelot $\supset$ x is not four-legged)"[283] need not overturn the generic truth that ocelots are four-legged.

This way of proceeding has some welcome advantages. One is that we have yet another shot at understanding when a false instantiation of a true generalization is a mistake of reasoning. The mistake in the reasoning that takes us from the true premises "Birds fly" and "Tweety is a bird" to the false conclusion "Tweety flies" is now apparent. Contrary to what we (and the AI literature) had been supposing, the general premiss is *not* true. It is falsified by a sentence one of whose default consequences is itself the negation of "Tweety flies". It is not in general sufficient for bad reasoning that it embody a false premiss. But when the way in which the premiss is false provides, just so, for the negation of its conclusion – call this "clean-sweep" falsification – might we not fairly judge the reasoning as bad? I think not. There was a time when Europeans were unaware of Australia's black swans. Jake is a swan paddling about in the pool in front of the Parliament in Canberra. Would a European have committed an error of reasoning had he predicted that Jake is white? Since when is provincialism an error of reasoning?

I will say again that the general question of premiss adequacy is much larger than any logician can or can be expected to pronounce upon. Most of the premisses of molecular biology, the theory of rents, and all the rest of the special sciences lie outside his ken and his job-description. All the same, there are premissory lapses for which reasoners at large would seem to be accountable. But I cannot find such a fault in the Jake case. There are lots of snappy reasoners who are biogeographical ignoramuses.

Another way of understanding the inference of Jake's whiteness is to re-interpret the premiss. Either the reasoner was not intending "Swans are white" as an unqualified generic, but rather as something in the form "Local swans are white"; or he was intending it non-generically, as something in the form "For the most part, swans are white". The first

---

[283] Cf. the circumscription default rule, wherein "ab_1", "ab_2, ... as anomaly or abnormality predicates: For all x, (x is an ocelot $\wedge$ $\sim ab_1(x)$ $\wedge$ $\sim ab_2(x)$ $\wedge$ ... $\wedge$ $\sim ab_n(x)$) $\supset x$ is four-legged." (McCarthy, 1980, 1986; Lifschitz, 1988).

possibility serves as a useful reminder of the appeal of Grice's Quantity Maxim: Don't say more (or less) than is necessary for full and effective communication. Don't over-commit; don't indulge in premissory over-kill. The value of the second interpretation is that it turns our attention to the importance of property-choice for generic utterance. It is well known that colour is not invariant under speciation, but limbedness, for example, is. The safest context in which to thin-slice is one in which the triggering property is invariant under speciation. Of course, it is a large, and still largely open, question as to how we go about making these determinations.

General propositions line up in a natural way depending on their degree of susceptibility to deviant instances. Another way of saying this is that general propositions are roughly ordered by their brittleness, ranging from most to least – universal quantifications, the various forms of non-universal plural quantification themselves ordered by quantificational range, generics statements and adverbial non-quantifications. The less brittle a generalization is, the less likely that a deviant instance will overturn it. This gives rise to a sensible rule.

Proposition 7.12b
MODEST GENERALIZATION: *When an agent's cognitive task requires making or instantiating a generalization, he should opt for a generalization of the least brittleness compatible with the task's satisfactory performance.*

In other words: *Do not generalize beyond necessity.*

7.13 *Ceteris paribus*

Sometimes a generic claim is taken as a universal quantification conditioned by a *ceteris paribus* clause. It is a suggestion of little merit. The *ceteris paribus* clause is an artificial universality-preserver. On a common reading, it is a place-holder for a principled and properly stand alone specification of a generalization's exceptions. Its whole semantic content can be summed up as "except for exceptions". So it is a gesture. The *ceteris paribus* device actually has two disadvantages, both of them serious. One is that it is a vacuous qualification. The other, paradoxically, is that it is too strong a qualification. If told that all birds fly except for those that don't, we are given no indication of the statement's exceptions. (Vacuousness) If told that all frogs cause warts except for those that

259

don't, we are told something that makes a mockery of the very idea of exceptions to the rule. And we are told it in a form that makes every false generalization true. (Triviality)

Devices similar in function to the *ceteris cerberis* clause have commended themselves to the favour of default logicians.[284] More's the pity, since they are heavily implicated in a bad confusion about generic defaults. Consider again "Ocelots are four-legged" and "Birds fly". These tend to receive a common classification in the AI literature. They are both considered impervious to certain exceptions. The four-leggedness of ocelots is proof against falsification by anomalies (e.g., as in a non-four-leggedness arising from congenital defect or injury) or irregularities (as in departures from the stem + "ed" rule). Whereupon we have it that, except for anomalies or irregularities, ocelots are four-legged and the past tense is the infinitive stem + "ed". The exceptions to which the flyingness of birds is considered impervious is birds that don't fly, the nonflyingness of which is no anomaly, either with respect to the fliers or the non-fliers. The trouble with the anomaly qualification is that, unlike the *ceteris paribus* qualification, it is both too strong and too weak. It is too strong in the sense that it obliterates the distinction between universally quantified and generic generalizations, leaving us nothing to choose between "Except for anomalies, for all *x*, if *x* is an ocelot then *x* is four-legged", and "Except for anomalies, ocelots are four-legged". It is too weak in the sense that it cannot save the flyingness of birds: "Except for anomalies, birds fly" is false. Penguins are birds. They are not anomalous birds. And they don't fly. This is not to say that "Birds fly" can't be saved exceptively (albeit incompletely). For "Except for penguins, ostriches, etc., etc., birds fly" is true. The trouble with this is the "etc." part. It is a gesture.

Attached to the question of what *ceteris paribus* clauses are wanted for (never mind their failure to deliver) is the whole – and almost as badly neglected – question of the role of background knowledge in the practical reasonings of individual agents. I will come back to this in chapter 11.

---

[284] It arises in McCarthy (1986) as the qualification problem. See also Lifschitz (1987).

7.14 *Denying the antecedent*

As frequently supposed, the character of defeasible reasoning can be outlined in three stages.[285]

Stage one: For Σ a set of sentences and α a sentence, $\Sigma \mid\sim \alpha$.

Stage two: For a new fact β, $\Sigma \cup \{\beta\} \mid\!\!\not\sim \alpha$.

What this shows is that since the consequence relation instantiated by the pair $\langle\Sigma, \alpha\rangle$ is ruptured by the information imparted by β. The conclusion α now "has to be retracted". Note that the retraction of α can be schematized as

Stage three:
$$\Sigma \mid\sim \alpha$$
$$\Sigma \cup \beta \mid\!\!\not\sim \alpha$$
$$\text{So } \sim\alpha,$$

which has the look of a denial of the antecedent, another of the traditional fallacies. It is not quite this, since it is not Σ or $\Sigma \cup \beta$ that are being denied, but rather that α is a *consequence* of $\Sigma \cup \beta$. There are two further differences. One is that the fallacy of denying the antecedent is defined for deductive consequence, not $\mathcal{D}$-consequence. Another is that the fallacy treats "not" as sentence-negation. But in its occurrence here, "not" expresses retraction not negation. This, "not" has a pragmatic function, not a semantic one.

Most good reasoning is at risk for consequence-rupturing information, and much good reasoning eventually runs foul of it. Included among such cases are those in which the information that eventuates in the rupturing of the consequence relation instantiated by $\langle\Sigma, \alpha\rangle$ also falsifies a sentence in Σ. Thanks to β, two things have gone wrong. Σ has been falsified; and α has to be retracted. Let's now go back to Harry and Sarah.

Sarah: β; so something in Σ is false.
Harry: Damn!

---

[285] See Reiter's remarks in the epigraph of chapter 8.

Sarah: β; so α has to be retracted.
Harry: Damn!

How can this not be a case of denying the antecedent? The antecedent *is* denied. And the consequent *is* given up. The answer is that although the antecedent is denied, this is not the basis on which the consequent is given up. It is given up because the information that chanced to falsify Σ broke the consequence relation between Σ and α. So the schematized retraction wouldn't commit the fallacy of denying the antecedent even if, contrary to fact, retraction were negation.

Retraction-moves of this sort are liberally instantiated in real life, sufficiently so to give the error abundance thesis a robust grounding. Such cases, we may say, exhibit a notable frequency of occurrence. They have the look of denying the antecedent, but they are no such thing. To deny the antecedent, three things are required. The antecedent must be derived by affirmation of its negation; the consequent must be denied by affirmation of *its* negation. And its negation must be drawn as a deductive consequence of the antecedent's negation.

The interpretation thesis requires that we not be careless in attributing these features even to contexts in which they may give the appearance of occurring. The assessment thesis requires that we not negatively judge a context before giving it its interpretational due. I venture to say that

Proposition 7.14
ANOTHER NON-FALLACY: *This effectively puts denying the antecedent out of the fallacy business. In the contexts in which it seems most likely to give occurred – defeasible retraction contexts – it doesn't. How, then, could it meet the requirement of universality, that is, of notable occasioned frequency? Denying the antecedent is not a fallacy as traditionally defined. It is not in the extension of the traditional concept of fallacy.*

COROLLARY: *It is additional encouragement of the concept-list misalignment thesis.*

CHAPTER EIGHT

FOLLOWING FROM

"A *logos* rests on certain propositions such that they involve necessarily the assertion of something other than has been stated, through what has been stated."
                                                                                          Aristotle

"The province of nonmonotonic reasoning is the derivation of plausible (but not infallible) conclusions from a knowledge base viewed abstractly as a set of formulae in a suitable logic. Any such conclusion ... may have to be retracted after new information has been added to the knowledge base."
                                                                                          Raymond Reiter

8.1 Adverbs

What would it take to show that a conclusionality relation is or isn't a consequence relation? Is there an antecedent fact of the matter? If there is, how would we go about gaining access to it? This is not the line of enquiry that I wish to pursue. A more manageable (some would say less fanciful) task is to ask questions carrying a reasonable expectation of non-question-begging answerability. A good deal of the discussion in the chapter before this one centred around a question of this sort, albeit not explicitly formulated. We can repair that omission now.

GOOD FOR WHAT? Third-way $\mathcal{D}$-logics are theories of premiss-conclusion reasoning underwritten by benignly and imperviously rupturable R-relations. They are logics that set for themselves certain tasks – notably a measured appreciation of the impact on such reasoning of the premiss-peculiarities we've been discussing here. It is natural to ask what contribution is made to the performance of these tasks by the analyses of $\mathcal{D}$-consequence? And, given the peculiarities of our own enterprise here, it is also necessary to ask whether the good that is done by those $\mathcal{D}$-logics is adaptable to the needs of a *naturalized* logic.

We can ask these same questions obversely.

WHAT HARM? What *harm* would be done a naturalized logic of third-way reasoning if it were denied the assumption that conclusionality relations are relations of logical consequence?

My present answer to the first question is that the cited properties of the $\mathcal{D}$-consequence relation aren't of much load-bearing value in elucidating of the premiss-peculiarities which were, in the first place, a large part of the $\mathcal{D}$-logics' motivation. The implied answer is that nothing of consequence (no pun intended) for third-way logics would attend the suppression of the equation of conclusionality with consequence. Let's now see what might be said against these answers.

It is customary for logicians to bring to their work three rock-ribbed assumptions. One is that consequence is the converse of following from. Another is that α is a consequence of $\sum$ just when the truth of $\sum$ is sufficient for the truth of α.[286] It is also widely, though not universally, understood that since there are different senses of sufficiency – logical sufficiency, physical sufficiency, causal sufficiency, and so on – the concept of consequence reflects a corresponding ambiguity. The third assumption is that whenever the truth of $\sum$ is sufficient for the truth of α there is a sense of "if … then" in which if $\sum$ is the case then α is also the case; that is, the conditional sentence ⌐If $\sum$ then α⌐ is true.

The notion of logically following from was introduced by Aristotle in the early volumes of the *Organon* as one of the defining conditions of syllogisms. At *Topics* 1, 100^a 25-27, Aristotle says that consequence is a necessitation relation.

> A syllogism is a *logos* in which certain things having been supposed, something different from these things supposed results of necessity because these things are so.[287]

This is not the full definition. There are arguments whose conclusions follow of necessity from their premisses, but they are not syllogisms unless further conditions are met. For example, conclusions must not repeat a premiss, and premisses must not be redundant. At *Prior*

---

[286] I take liberties. When I say that the truth of $\sum$ suffices for the truth of α, this is short for "the joint truth of the premisses in $\sum$".

[287] This is repeated at various places in the *Organon*, including *On Sophistical Refutations* 1, 165^a 1-3 and *Prior Analytics* A 24^b 19-22. Note the similarity to this chapter's first epigraph.

Analytics A 32, 47ᵃ 33-35, Aristotle draws an important distinction. He says that a valid argument is an *anagkion,* and a *syllogismos* is a special case of it. We find in this distinction two notions of following from, one which I'll call "straight necessitation" and the other "syllogistic necessitation". Corresponding is the distinction between straight consequence and syllogistic consequence. In Aristotle's earlier logic, straight consequence is a theoretical primitive.

These contrasts are instructive in a number of ways. Since Aristotle doesn't define it, it may be impossible to say with certainty whether straight consequence is monotonic. But it is clear that syllogistic consequence is not. If an argument is a syllogism, its conclusion is a syllogistic consequence of its premisses. If a new premiss is added, then by the requirement that syllogisms not have redundant premisses, the conclusion may still be a straight consequence of those premisses, but it cannot be a *syllogistic* consequence of them.

It is also noteworthy that Aristotle's nonmonotonic consequence relation arises from his (possibly monotonic) straight relation by the imposition of further conditions, much as $\mathcal{D}$-consequence relations, which are nonmonotonic, arise from classical consequence, which is monotonic, by the imposition of conditions on conditions on the classical relation. Thus neither the notion of nonmonotonicity nor its tie to monotonic consequence is new or even recent. Each arose with the very founding of systematic logic.

Also of interest is Aristotle's use of the adverbial constructions "of necessity" and "necessarily" as qualifiers of the connection between a set of premisses and its consequences, straight or syllogistic as the case may be. Thus γ follows from α, β if and only if it "results of necessity" from α and β, in which case, α and β "necessarily involve" γ. Aristotle makes a number of attempts in the *Prior Analytics* to extend syllogistic logic to expressly modal contexts, one of which is a fair anticipation of Lewis' S5.[288] But even in the earliest of his logical writings, Aristotle is alert to the utility of adverbial qualification in characterizing the consequence relations which hold his interest. We might say, then, that the *modal adverbs* are load-bearing in Aristotle's logic.

Perhaps it is only to be expected that the adverbs of necessity would be of help in characterizing relations of deductive consequence which, after all, is the relation Aristotle himself calls necessitation. It is interesting to consider whether adverbs of non-necessity might also play

---

[288] Corcoran (1974), pp. 202-203.

a load-bearing role in characterizing relations of non-deductive consequence – adverbs such as "probably", "plausibly", and – of course – "defeasibly". Let us look into this a bit further. For ease of reference, I'll call these the $\mathcal{D}$-adverbs.

A standard way of saying that you are reasoning defeasibly is to report the reasoning with the help of the adverbs of defeasibility. AI theorists make ample and undisciplined use of them in their writings. There are at least four ways in which these adverbs operate. They can occur as *sentence-operators,* as in "Defeasibily, Ozzie is four-legged". They can occur as *operator-operators,* as in "Ocelots are four-legged and Ozzie is an ocelot; so, defeasibly, Ozzie is four-legged." They can occur as *relation-operators,* as in "That ocelots are four-legged and Ozzie is an ocelot defeasibly implies that Ozzie is four-legged." They can occur as *conditionality-operators,* as in "If ocelots are four-legged and Ozzie is an ocelot then, defeasibly, Ozzie is four-legged." The logician of defeasible reasoning has an interest in knowing what makes it defeasible and, when good, what are the standards required for it. The logician's quest is semantic. As applied to reasoning, he is looking for truth conditions for the predicates "is a defeasible consequence of", "is defeasibly correct", and so on.

8.2 *Irregular adverbs*

The standard modalities, necessity and possibility, are expressible in well-behaved ways by adverbs and adverbial clauses – "necessarily", "it is necessarily the case that", and so on. It falls to these adverbial constructions to ascribe to semantic properties and relations further semantic properties. Thus, in the paradigm case, "necessarily" takes "true" to "necessarily true", and necessity is a way of being true, a modality of truth. When functioning in the manner of the alethic modalities, we might say that the adverbs of necessity are *regular*. That is, they ascribe to properties of a given type characteristics of that same type.

It is reasonable to ask how much of a paradigm we should make of the regularity model. To what extent, if any at all, does it behoove us to hold the $\mathcal{D}$-adverbs to the regularity requirement? To what extent, if any at all, should we tolerate adverbial irregularity in the logics of $\mathcal{D}$-consequence? The easier part of the answer can be given now

Proposition 8.2a

IRREGULARITY: *The $\mathcal{D}$-adverbs are irregular.*

Whether this matters, or how, is the harder thing to answer, as we shall see.

Adverbs are tricky in logic. There are lots of logicians who wouldn't give them the time of day. The terms that concern us here all have uses is which they are modifiers of speech acts, not, as some would say, carriers of propositional content. When Harry tells Sarah that he will probably go to the high school reunion, he is giving a qualified or weakened commitment to attend. What he says gives her a justified expectation of compliance but not unfettered grounds to rebuke him for not showing up. Logicians who take a hard line on the hedging adverbs and deny them a place in logic will usually grant their pragmatic character, and on that basis relegate them to linguistics.

Of course, this is not entirely accurate, as the alethic models clearly attest. When α is, of necessity, a consequence of $\sum$, that is, when α follows from it, of necessity, this is a conclusion that you might have occasion to draw. You might also have occasion to say so. But, in giving voice (or keystroke) to your conclusion of α from $\sum$, you needn't qualify your "so" by "of necessity". Your reasoning would be modally adequate had you merely said, $\ulcorner\sum$, so $\alpha\urcorner$. True, it is open to you to add the qualifier "necessarily". If you did, you would be lending your report some emphasis. Instead of weakening your commitment to α, you would be strengthening it. In so doing, "necessarily" would serve a pragmatic end. In that particular case, "necessarily" would be functioning irregularly. But it wouldn't follow from this that modal adverbs are semantically inert, that they have no place in a logic's modal theory. The adverbs of necessity fairly bristle with semantic import, and modal logics are a thriving industry. The modal adverbs have a long and settled history of regularity. But our target here are the adverbs of defeasibility, the $\mathcal{D}$-adverbs. We want to know how their irregularity denies them a load-bearing role in the analysis of $\mathcal{D}$-consequence. We shouldn't expel them from logic without giving "defeasibly" a fair hearing.

I want to proceed in the spirit of the Can Do. I'll begin, as noted, with the adverbs with which we have the most pretheoretical familiarity. First in line is "necessarily", and then "possibly" and "probably". Next comes "defeasibly", made tricky by its slight presence in pre-theoretical usage, and, finally, "presumably" which, notwithstanding a comfortable presence in everyday speech, will prove a harder thing to deal with than the others.

Schema one
α
β
So, necessarily, γ.

Schema two
α
β
So, necessarily (γ).

Schema three
{α, β} necessarily-

γ

        implies γ.

Schema four
If α, β then, necessarily,

Right away two points stand out. One is that the distinction between operator-operator (schema one) and sentence-operator (schema two) appears to be well entrenched and theoretically well understood for "necessarily". The other is that the two schemas are inequivalent. The conflation of the second with the first is the modal confusion known as Sleigh's Fallacy. One of the reasons for its commission is ambiguity. The vernacular sentence, "So necessarily γ" is ambiguous as between the conclusion of schema one and the conclusion of schema two, and sentences punctuated in the manner of the former occur in actual speech with a much smaller frequency than those punctuated in the manner of the latter. This is not strictly speaking a lexical underdetermination problem. It is a *punctuational* underdetermination problem. It has a not inconsiderable provenance in $\mathcal{D}$-logics.

In the standard modal systems, there are principles that tie together schemata one, three and four. The deduction metatheorem, which provides that $\alpha \vdash \beta$ if and only if $\vdash (\alpha \supset \beta)$, suggests the equivalence of schema one and schema three. Then, too, $\alpha \vdash \beta$ only if, for a requisite sense of "if ... then," if α then β. Accordingly, it seems reasonable to suppose that the application of "necessarily" to a consequence relation, as in schema three, would be recapitulated as a conditionality modifier in four.

In the mainstream modal logics "necessarily" occurs only as a sentence-operator (even in *de re* contexts it modifies open sentences). Our schemata show a contrary disposition. Except for schema two, which is invalid in all modal systems, "necessarily" occurs non-sententially. In one it modifies the conclusional operator "so"; in three it modifies the implication relation; and in four it modifies the conditional connective "if

... then". Why, then, doesn't this wreck the equivalency suggestion with respect to one, three and four? Part of the answer is that in three and four "necessarily" is subject to contextually eliminating paraphrase. With "⊢" for "implies" and "—3   "for" if ... then", schema three becomes the metalinguistic ⌐{α, β} ⊢ γ⌐ and four becomes the object-linguistic ⌐(α, β) —3   γ⌐. Schema one resists such paraphrase, but this is for want of conclusion-indicator expressions ("therefore", "hence", "so") in modal logic, not the failure of "necessarily" to qualify them. In other words, if "so" were part of the language of such systems, why wouldn't we make room for "so, necessarily,"? But it isn't, so we don't.

Our present question is whether a semantics of the adverbs of defeasibility would help elucidate the conditions for the truth and correctness of defeasible statement and inference. The same question arises for necessity. If, in the case of necessity, we think that a central question for any logic is the nature of its consequence relation, then it would be a fact of some relevance, if fact it be, that adverbial schemata one, three and four all converge on the notion of consequence – with schema two the outlier. If so, this would give rise to two interesting suggestions about $\mathcal{D}$-consequence.

Proposition 8.2b

CONCLUSIONS AS $\mathcal{D}$-CONSEQUENCES: *As with the monotonic cases, here too concluding β from α is a matter of drawing it as a consequence of α.*

Proposition 8.2c

THE CONDITIONALITY OF $\mathcal{D}$-CONSEQUENCE: *A proposition β is a consequence of α if and only if, there is a requisite sense of "if ... then" in which if α then β.*[289]

---

[289] The conditionality requirement arises in Chellas (1975) as a condition on nonmonotonic consequence, in which conditionality obeys left-nonmonotonicity (⌐If α then β⌐ doesn't imply ⌐If α ∧ γ then β⌐). Others were late in seeing the connection between their own work and like developments in conditional logic, but have more than made up for it since the early nineties. See, for example, Gärdenfors and Makinson (1994), Pearl (1994), Gabbay (1995) and Asher (1995).

COROLLARY: *If α is a 𝒟-consequence of kind* k, *and 8.2c holds true, then the requisite sense of "if … then" can be called its* k-*sense.*

8.3 *Possibility*

That is the way things appear to be for "necessarily", an adverb with a stably regular provenance. Is it the same with "possibly"? Here, too, we see an at least fourfold usage. Here, too, there are pragmatic functions to perform (chiefly weakening). But, here too, it would be quite wrong to deny them a semantic significance, that is, a significance characterizable without the need to invoke pragmatic features. In which case, there would be ranges of use in which "possibly" would be a regular adverb.

Schema (a)	*Schema (b)*
α	α
β	β
So, possibly, γ.	So, possibly (γ).

Schema (c)	*Schema (d)*
{α, β} possibly- implies γ.	If α, β then, possibly, γ.

Intuitions flow less smoothly here. Mainstream modal logics can represent (b) as the deduction ⟨α, β, ◊ γ⟩, but the others are hopeless. Not only are the other occurrences of "possibly" non-sentential, there are for schemata (c) and (d) no contextually eliminating paraphrases such as those that do the job for necessity. (See just below).

Standard modal logics aside, what of the equivalence of schemata (a) and (b)? In my idiolect they are not equivalent. They ring differently in my ear. Schema (a) represents a qualified argument for the truth of its conclusion, but schema (b) represents an argument as strong as you like for the possibility of its conclusion. In the first instance we have a weak argument for a strong conclusion and, in the second, a strong argument for a weak conclusion. But it is certainly no condition on an argument for a proposition's possibility that there be any argument, no matter how qualified, supporting that proposition's truth.

Is there then for "possibly" something like a deduction metatheorem and a conditionality of consequence principle? Do we have it that for the correctness of ⌐α, β, so, possibly, γ⌐ it is necessary and sufficient that possibly (α, β ⊃ γ)? The answer is no. It gives nothing like possibilistic conclusionality. ⌐Possibly (α, β ⊃ γ)⌐ is much too permissive, true whenever γ also is.

Much the same applies to the conditionality of consequence thesis. Suppose that we have it that {α, β} ⊢ γ if and only if α, β —3 γ. Suppose further that "possibly" is an admissible qualifier of "⊢" and "—3", giving "⊢ᴾ" in the first instance and "—3 ᴾ", in the second. Then couldn't we say that {α, β} ⊢ᴾγ if and only if α, β —3ᴾγ?

I have no idea. The standard logics give no guidance, and my own ear fails me. I don't know what these sentences mean. That is, I don't know what they mean when "possibly" occurs, not as a pragmatic hedge or weakener, but rather as a semantic operator on a relation and in the same way on a conditionality connective. In such uses, "possibly" applied to implication gives a new relation of possible implication, and when applied to "if ... then" gives a new connective of possible conditionality. In which case, possible implication will be a special case of standard implication (which it manifestly isn't) or a species of an implication relation of which standard implication is also a species, but a different one. But I have no idea what this more general implication relation might be, and no idea of conditions that permit possible implication to be a species of it. It is the same way with possible conditionality. If possible conditionality is conditionality of a sort, here, too, I am stumped about what this generic conditionality might be. If, on the other hand, there is no conditionality of which possible conditionality is a sort, I am at a loss to see why we would call it possible conditionality. For, call it what you like, possible conditionality simply falls out of the ambit of the conditionality of consequence principle.

Would it help if we construed ⌐{α, β} ⊢ᴾγ⌐ as ⌐Possibly ({α, β} ⊢ γ))⌐ and ⌐α, β —3 ᴾγ⌐ as ⌐Possibly α, β —3 γ)⌐? If we did, we would have given expression to the idea that possible implication in this world just is actual implication somewhere else, and that possible conditionality in this world just is actual conditionality in some other. Again, my ear is not up to these transpositions. When I hear someone say that this proposition possibly implies that, or that if this one is the case then, possibly, that one is also the case,[290] these paraphrases have an alien ring. I wouldn't know

---

[290] As opposed to, ⌐If α, β then it is possible that γ is the case⌐, in which "it is possible that" could have a purely semantic role.

271

how to judge them. How could I? I haven't been able to grasp the *semantic* import of "possibly" in these attempts at non-pragmatic employment. I haven't been able to hear them as regular.

Proposition 8.3
OBSCURITY: *In its uses as an operator-operator, relation-operator and connective-operator, "possibly" is a semantically obscure adverb, which no mainstream modal logic is appropriately set up to remove. That is, they aren't built for the analysis of these usages.*

Proposition 8.3 should not occasion a rush to judgement, especially as regards possibilistic consequence. I'll come back to this at the tag-end of section 8.9.

8.4 *Probability*

What now of "probably"? Consider the further quartet:

Schema (i)	*Schema (ii)*
α	α
β	β
So, probably, γ	So, probably (γ).

Schema (iii)	*Schema (iv)*
{α, β} probably-implies γ	If α, β then, probably, γ

Let us quickly dispose of a technical point. In the probability calculus, the distinctions at hand are inexpressible: For example, ⎺Pr(α)⎺ is the name of a real number and not a sentence. So "Pr" in ⎺Pr(α)⎺ is not a sentence-operator. But let that pass. It is a technical problem which we know how to fix. Besides, no one denies that there is an established common usage in which "probably" is indeed a sentence-operator. On the suggestion before us, it is also an operator-operator. If this is right, then here too it is well-motivated to ask whether the schemas are equivalent, and, what if any, are the further linkages to (iii) and (iv)?

Here is an echo of the complaint of lines above. To my ear, schema (iii) is out of place. It is the old problem. I don't know what (iii) means in English. Again, by this I mean that, in its use as a relation-modifier, I'm not sure how it goes semantically. Of course, in its other uses things are easier. Sometimes "probably" performs the same function as "as far as I know". So when you say $\ulcorner\{\alpha, \beta\}$ probably implies $\gamma\urcorner$ you're saying $\ulcorner$As far as I know, $\{\alpha, \beta\}$ implies $\gamma\urcorner$ or $\ulcorner$As far as I can make out, $\{\alpha, \beta\}$ implies $\gamma\urcorner$. But this is not its use in schema (iii).

Perhaps this is no more than an idiomatic hiccough. Perhaps in (iii) "probably" is doing duty for "probabilistically". "Probabilistic" is certainly allowable as an adjective of "reasoning". Isn't probabilistic reasoning what most inductive logicians take as a central part of their mandate? If so, wouldn't there be a consequence relation peculiar to probabilistic reasoning? Wouldn't it be natural to call it probabilistic consequence? So why not probabilistic implication? Suppose it were right to say that a deduction metatheorem suffices for these linkages. Then it would seem that in the case of probabilistic reasoning, probabilistic consequence wears the trousers. If we had it that γ is a probabilistic consequence of α, β, wouldn't we also have it that if α, β then, probabilistically, γ? Wouldn't we also have it that when γ is drawn as a probabilistic consequence of α, β this is expressible as $\ulcorner\alpha$, β, so, probably, $\gamma\urcorner$? I'll briefly pick this up again in section 8.7.

8.5 *Defeasibility*

What now of readings in which an adverbial operator is a sentence-operator? In the standard approaches to modal logic, α and $\ulcorner$Possibly $\alpha\urcorner$ are different propositions. Perhaps it will be thought that on the ordinary everyday reading of "probably", the same is true for α and $\ulcorner$Probably $\alpha\urcorner$. If so this would allow for the underlying consequence relation to be both monotonic and, in some cases at least, as classically strong as it gets. Consider, for example,

> Possibly α
> So, necessarily, possibly (α)

and

> Probably α
> So, necessarily, probably (α).

It would appear to be the same way with sentence-operator defeasibility:

Defeasibly α
So, necessarily, defeasibly(α).

If so, the equivalence question for the operator-operator and sentence-operator reading of "defeasibly" is settled – negatively. That is, there are inferences whose premisses and conclusions are defeasible, which are not themselves in the slightest degree defeasible.

The same is not true of the operator-operator sense of defeasibility. Defeasible consequence drawing is robustly nonmonotonic. Accordingly,

Proposition 8.5a
MONOTONIC AND NONMONOTONIC DEFEASIBILITY: *If we admit the sentence-operator use of "defeasibly", and if we allow reasoning to qualify as defeasible when its inputs or outputs or both are defeasible, then there are cases in which the consequence relation of defeasible reasoning are monotonic, and other cases in which it is not.*

Let's turn now to the four-fold schema for "defeasibly".

Operator-operator
α
β
So, defeasibly, γ.

Sentence-operator
α
β
So, defeasibly (γ).

Relation-operator
{α, β} defeasibly-
implies γ.

Conditionality-operator
If α, β then, defeasibly, γ.

In the AI literature, even more than in theoretical jurisprudence, "defeasibly" is a term of art, with respect to whose properties common usage is of scant use. Perhaps, then, the nod should go to its AI-coiners. Since they have pledged themselves to the idea of defeasible consequence, I will do so as well, at least for now. But at this juncture I am not sure about "defeasibly" as an operator-operator. If there were a deduction metatheorem for defeasibility logics, perhaps my hesitation could be lifted. But there isn't. That is, there is no full deduction

274

metatheorem for logics of this sort. The imputed equivalence fails (Morgan, 2000).[291]

How significant is the Morgan result? The answer is "Not as much as we might have supposed." Whether or not there is a deduction metatheorem for defeasible consequence and defeasible conditionality is largely beyond the point of what concerns us here. To see why, consider again the difference between consequence-having and consequence-drawing. When introducing the distinction in chapter 1, I said that consequence-having was wholly a matter of semantic structure. I said that consequence drawing was partly a matter of semantic structure but also a matter of psychological structure. Consequence-having obtains in logical space. Consequence-drawing occurs in a reasoner's mind, in psychological space. What then is the right way to interpret our operator-operator schemas? Up to now, I have been giving them a reading that qualifies them for consideration as the LH-component of a deduction metatheorem. But, on reflection, this is not a reading that captures the consequence-drawing aspect of reasoning from premises. This is most easily seen in the deductive case, which (simplified) is: $\langle \alpha, \beta, \gamma \rangle$. What the deduction metatheorem requires is that, in the technical sense of proof theory, $\langle \alpha, \beta, \gamma \rangle$ be a *deduction*. In that context a deduction is set a sentences – a sequence of them – in which every line arises from a properly anchored opening line by finite application of transformation rules to preceding lines. Correspondingly, γ is deducible from α, β, just in case $\langle \alpha, \beta, \gamma \rangle$ is a deduction. Supposing that γ is indeed deducible from α, β, the question posed by the deduction metatheorem is whether α, β entail γ (and vice versa). There is nothing in this construction of $\langle \alpha, \beta, \gamma \rangle$ that qualifies it as psychological. That γ is deducible in this technical sense from α, β leaves the question of whether γ is something that should be *drawn* unanswered. It leaves this underdetermined precisely because the deduction metatheorem provides that whenever $\langle \alpha, \beta, \gamma \rangle$ is a deduction, γ is a consequence of $\{\alpha, \beta\}$. But this is consequence-having, not consequence-drawing. Accordingly,

---

[291] Morgan (2000) defeats only half of the defeasibility deduction theorem. It leaves it open that whenever we have a true defeasible consequence statement we might also have a valid expression of defeasible conclusionality in the manner of $\langle \alpha$, so defeasibly, $\beta \rangle$. Even so, I daresay that some readers might be of a mind to allow the operator-operator sense of "defeasibly" provided we also acknowledge a defeasible consequence relation and in the absence of information that the defeasibility deduction metatheorem also fails right-to-left.

Proposition 8.5b

DEFEASIBILITY INEQUIVALENCIES: *We are left with the inequivalence of the operator-operator and sentence-operator readings of "defeasibly" (which might or might not be surprising), and the like inequivalence of defeasible consequence-having and defeasible consequence-drawing (which shouldn't be at all surprising).*

Let's briefly renew our bearings. The standard view is that the distinctiveness of defeasible reasoning derives from the distinctiveness of the consequence relation in which it is embedded. When defeasible reasoning is the drawing of conclusions from premises (or data-banks or knowledge-bases), it is also the received view that the conclusion drawn is a consequence of those premises (etc.). The defeasible consequence rule imposes on the logician the necessity to specify the requisite consequence relations and to demonstrate their key properties. By the consequence drawing principle, it also falls to the logician to specify the conditions under which a consequence of premises (etc.) should or might be drawn. Given the conditionality of consequence thesis, yet another obligation looms. This is the *conditionality search requirement*, which bids the logician to find a sense of "if … then" according to which β is a defeasible consequence of α just in case if α then, defeasibly, β. With these things in mind, it would matter whether the conditionality of consequence thesis were actually true of defeasibility. We have decided to approach this task circumspectly, beginning with adverbs of pre-theoretical currency, and working our way towards an understanding of the defeasibility constructions. But let us turn our attention now directly to the conditionality of consequence.

8.6 *Sufficiency*

The conditionality of consequence property is a reflection of Aristotle's original insight that a consequence relation is a kind of necessitation. For if there is a sense in which the truth of α necessitates the truth of β, how could it not be the case that there is a corresponding sense of sufficiency in which the truth of α suffices for the truth of β? And if that is so, how could it not be the case that there is a sense of "if … then" in which if α then β?

A good deal of ink and high feeling has been spilt over the conditionality of consequence thesis. In the early days, there was an instructive battle between Russell and Hugh MacColl over the horseshoe. Russell thought that there was a relation of material consequence and a sense of "if … then" for the conditional sentences ⌐α ⊃ β⌐ that express it. MacColl thought that there was no sense of "if … then" for which ⌐α ⊃ β⌐ is a conditional sentence. So, if there were a relation of material consequence, the conditionality of consequence thesis wouldn't be true.[292] MacColl's objection anticipated a similar one by C.I. Lewis.[293] Whether or not there are conditions under which ⌐α ⊃ β⌐ is a conditional, and whether or not there is a relation of material consequence which that conditional expresses, material consequence isn't strict (i.e. honest-to-goodness) consequence, and "⊃" doesn't express it. Lewis went on to propose that this omission could be rectified by introducing a new conditional symbol "—3   ". In effect, he thought that "⊃" fails the conditionality of consequence principle, and with it the consequence rule, whereas "—3" satisfies them both.

   Was Lewis right about "—3   "? Are the truth conditions he assigned to "—3 "-sentences such as to verify an "if … then" sentence? Putting "◊" for the possibility operator, Lewis defined "—3" as follows:

   α —3 β iff ~◊(α ∧ ~β).

It is interesting to compare this with the classical definition of "⊃":
   α ⊃ β iff ~(α ∧ ~β).

 MacColl's point was that the truth of ⌐~(α ∧ ~β)⌐ was insufficient for the truth of any sentence in the form ⌐If α then β⌐. Anticipating Lewis' definition, MacColl also thought that there were indeed sentences of the form ⌐If α then β⌐ for whose truth the truth of ⌐~◊(α ∧ ~β)⌐ is sufficient.

   Sometimes the move from relation-operator to sentence-operator causes no alarm. On its standard definition "—3   ", "if … then" can be rewritten as "Necessarily, … ⊃ …". Fine as far as it goes, it doesn't establish the conditionality thesis for "necessarily", never mind that the conditionality thesis is true of it. The problem again is that there appears to be *no* sense of "if … then" in English for which if α ⊃ β then if α then β. For suppose otherwise. This would give

---

[292] See MacColl (1908), 151-152, and 453-455 and Russell (1908). A good survey of MacColl's contributions to logic is Rahman and Redmond (2007).
[293] See, for example, (1912), 522-531.

POWERS' PARADOX
1. If α ⊃ β, then, if α then β   assumed for *reductio*
2. If (if α then β) then α ⊃ β.   2, Impl.
3. If α then β iff α ⊃ β   1, 2, Bicond.

But (3) is absurd. It validates the paradoxes of material implication for entailment. One of (1) or (2) will have to go. How could it not be (1)? So let's make it official.

Proposition 8.6a

NOT A CONDITIONAL: *There is no sense of "if ... then" in English for which if α ⊃ β then if α then β.*[294]

Lewis championed the idea that consequence should be conditionally expressible. Although he didn't say so explicitly, it is evidently his view that

Proposition 8.6b

SUFFICIENCY: *No sentence ⌐If α then β⌐ is true unless there is a sense of sufficiency in which the truth of α is sufficient for the truth of β.*[295]

What might these senses be? The obvious candidates include: Logical sufficiency, mathematical sufficiency, metaphysical sufficiency, causal sufficiency and physical sufficiency. Although they can be set down with a confidence that bespeaks their obviousness, no one should think that the similarities and differences among them have been worked out to everyone's satisfaction. I invoke them here to assist in making a small but hardly trivial point. It is that when it comes to an antecedent's sufficiency for the truth of its consequent, there is more than one way to skin a cat, never mind that the complete story has yet to be told. If the sufficiency

---

[294] I owe Powers' Paradox to Lawrence Powers, who told me it drinking coffee one night in Amsterdam in 1994.

[295] Relevant logicians dispute the sufficiency of ⌐~◇(α ∧ ~β)⌐ for the entailment of β by α, and presumably also for the truth of ⌐If α then β⌐. This is because no relevant logician would accept (except ironically) any sentence in the form ⌐If α ∧ ~α, then β⌐ where β is arbitrary. It doesn't matter. Whatever their truth conditions for consequence, their view will also be that for any α and β that satisfy them, α will be sufficient for β and ⌐If α then β⌐ will be a true conditional.

claim is right, the conditionality of consequence thesis falls out rather easily. If there is a sense in which α is sufficient for β then there is a sense in which ⌐If α then β⌐ is true and a sense in which β is a consequence of α (the same sense throughout).

In fact, the story is as hard to tell about sufficiency as it is to tell about necessity. Even in the case of established modal logics such as S5 and S4, specifying the different senses of necessities captured by them is discouragingly difficult. Some logicians identify logical truth by virtue of logical form alone as alethic necessity, and provability by virtue of logical form alone as apodictic necessity. Suppose we decided to follow suit. Then here is an observation from John Burgess:

> ... nothing said so far constitutes even an informal proof' that no formula not a theorem of S4 is correct for apoditic [necessity], or that no formula not a theorem of S5 is correct for alethic [necessity]. And indeed there is no generally accepted informal *argument* for the first claim, though a convincing one can be given for the second claim. (Burgess, 2009, p. 65. Emphasis added to contrast argument with proof.)

What's sauce for the necessities of S5 and S4 is sauce for all the other necessities stirred into contention by modal logic's vigorous and pullulating pluralism. And what's true of all these necessities is at least as true of all those sufficiencies presupposed by Proposition 8.6b.

8.7 *Probabilistic consequence*

I am not quite ready to give up on relations of nonmonotonic consequence. The conditions and principles currently on view hold (well enough) for deductive reasoning. Deductive reasoning is monotonic and defeasible reasoning isn't. It is a large difference. It makes deductive consequence an information-resistant semantic relation and defeasible consequence an information-sensitive epistemic one. Epistemic relations are pragmatic. They hold or not at least partly in virtue of the role played by agents. The agents who interest us here are language-using agents. They are the entering wedge of the pragmatic. This should give us pause. Why would we think that so many of the *types* of conditions and principles that hold for deductive reasoning would also hold for defeasible reasoning? Before settling our minds about this, let's consider an intermediate case. Let's turn to the kinds of statistico-experimental and sample-to-population reasoning which is a central focus for mainstream inductive logicians. In plainer words, before we consider defeasible consequence, let us have a closer look at "probabilistic consequence".

There is a common and long-held view according to which β is a probabilistic consequence of $\{\alpha_1, ..., \alpha_n\}$ if (and on some tellings only if[296]), the conditional probability of β on $\ulcorner \alpha_1 \wedge ... \wedge \alpha_n \urcorner$ is sufficiently high. Perhaps this is right. For a certain large class of inductive reasonings, I think it is right. If it is, it is so notwithstanding that the truth of the α_i are not sufficient for the truth of β, and, correlatively, that the conditional probability at hand is not such as to license $\ulcorner$If $\alpha_1 \wedge ... \wedge \alpha_n$, then $\beta\urcorner$. But, right or wrong, there is a further question to put. It is the question whether the relations of lending support to or being evidence for can obtain between $\ulcorner \alpha_1 \wedge ... \wedge \alpha_n \urcorner$ and β without its also being the case that β is, in the sense at hand, a probabilistic *consequence* of the α_i. Of course, it depends on whether we're prepared to hold probabilistic consequence to a sufficiency condition of its own:

Proposition 8.7
PROBABILISTIC SUFFICIENCY: *β is a probabilistic consequence of Θ only if there is some sense of sufficiency in which the truth of all the α_i in Θ is sufficient for the truth of β.*

Here, too, opinion is divided: One way in which the probabilistic sufficiency principle could fail would be where β is rightly concluded from evidence that supports it, notwithstanding that β is not a consequence of it. Let us suppose so. How would this bear upon the question of whether there is a logic for inductive reasoning? It would bear on it, or not, depending on whether we accept the consequence rule: No consequence relation, no logic. Period.

Not surprisingly, inductive logics brim with attempts to hang on to the idea of probabilistic consequence. Here is Jon Williamson on this point, speaking for logical theories of probability:

> Perhaps the most obvious thing to try first is a generalization of entailment $\vDash$ to partial entailment $\vDash_x$, where a set Θ of sentences

---

[296] I myself am not in the only-if camp. Consider again our old friend the ocelot generalization. I have said why I think this a competent induction, for a realistically broad notion of induction. But judged by, say, Bayesian standards, it is a train wreck.

partially entails sentence Φ to degree x, $\Theta \vDash_x \Phi$, if and only if $p(\Phi \wedge \Theta) = x$. Under such a view classical entailment is the case where $x = 1$. If Θ is empty we get a concept of degree of logical truth which corresponds to unconditional probability.[297]

Of course, partial entailment gives partial consequence, which is a relation prefigured in Keynes (1921) and vigorously worked up in Carnap (1950). My view is that partial consequence is consequence in name only. Partial consequence fails the probabilistic sufficiency principle.

Perhaps there is another way of getting probabilistic consequence back into gainful employment. Suppose, as before, we grounded probabilistic consequence in conditional probability. Suppose that whenever $p(\alpha \wedge \Theta) \geq n$, for some suitable value of n, we would have it that

(1) If Θ, then *probably* (α)

and with it that

(1') α is a probabilistic consequence of Θ.

This provides a key contrast between the partial consequence relation of logical theorists of probability and – as we have it here – probabilistic consequence. The difference is what happens *in the interior* of the corresponding conditional sentences. And, note well, what happens in the interior is an occurrence of "probably" in its sentence-operator sense. So on this analysis, the sentence-operator sense is *not* an outlier in the logic of inductive consequence. It is a full partner in this analysis. This stands in stark contrast to the case of necessary implication. To infer from the fact that α necessarily implies β that if α then it is necessary that β is Sleigh's Fallacy. There is no Sleigh's Fallacy for this sense of probabilistic implication.

In the case of partial consequence the conditional, if there were one, would be

(2) If Θ, then α

---

[297] Gabbay *et al.* (2002), p. 404.

for some suitable sense of "if ... then".

But (2) differs from (1) essentially. The consequent of (1) arises from the consequent of (2) by prefixation of the sentence operator "probably" to (2)'s consequent. If we think (2) false and (1) true in virtue of "probably"'s respective absence and presence, there is a way now of producing the conditional corresponding to partial consequence. Perhaps we could simply *rewrite* (2) as (1).

Of course, there is a problem with this, or at least the appearance of one. A classic problem for probability is the fixing of prior probabilities. There is in (1) a similar problem. Suppose the stand-alone probability of α is n. We don't want (2) to say that Θ makes it the case that the stand alone probability of α is not n. That would be the gambler's fallacy. What in its occurrence there does "probably" mean? None of the going probability theories answers this question directly. Perhaps it doesn't matter. Perhaps we could say, for generality, that α is "evidentially" probable if and only if for true evidence-statements Θ the conditional probability of α on Θ is itself sufficiently high. Why might we accede to this? Because there appears to be a prior meaning of "probably (α)" in English which just might support it.

8.8 *Misgeneralizing the above*

Suppose we found the sufficiency principle to have enough intuitive clarity to justify sticking with it until something better came along. Suppose we found its application to (1) and (1') to be reasonably convincing.[298] Then a natural question would be whether this treatment of probabilistic consequence could be generalized to the other consequence relations variously called to service in analyses of third-way reasoning – defeasible consequence, plausible consequence, and whatever else. If so, these third-way consequence relations would instantiate underlying structural conditions, as follows:

Proposition 8.8a
A PROPOSED TEMPLATE FOR THIRD-WAY CONSEQUENCE: *If the third way reasoning currently at hand is of kind* k, *then the consequent of the conditional invoked by the*

---

[298] Of course, not everyone will think that it *is* reasonably convincing. In which case, I am hard-pressed to imagine how they can hang on to probabilistic consequence without having to jettison the conditionality of consequence principle.

conditionality of consequence thesis would be prefixed by an occurrence of a k-adverb functioning as a sentence-operator.

If the generalization hypothesis held true, then the present conditions would be fulfilled on any interpretation of "*k*" for which it is correct to suppose that *k*-reasoning is a form of third way reasoning.

Let "*k*" denote "defeasible". Then for there to be a relation of defeasible consequence, we must have it that there are inputs Θ and sentences α such that, in conformity with the sufficiency requirements, if Θ, then defeasibly (α), or in equivalent variations, it is defeasibly the case that α.

Recall now our assumption that fallibilism is true; indeed that we make errors, lots of them. Recall that, big though this class is, more numerous still are the beliefs that lie open to the possibility of error, that is to say, the class of beliefs that aren't *infallible*. With exceptions duly noted (whatever they are is fine), this is just about all the beliefs that beings like us will ever hold.

This is problematic for the third-way template if we extend it to defeasibility. For almost no interpretation of Θ and α do we have it that it is the truth of Θ's members that makes it the case that α is open to possible correction. Consider again the Ozzie case. That Ozzie is an ocelot defeasibly implies his four-leggedness. On the conditions presently in view, this couldn't be so unless Ozzie's ocelothood made it the case that it is open to possible correction that he is four-legged. But this it does not do. Ozzie's ocelotness is evidence of the truth of his status as four-leggedness, not of the possibility that he is not. Accordingly,

Proposition 8.8b

THE TEMPLATE FAILS: *The treatment of probabilistic consequence does not generalize to defeasible consequence. Even if the treatment of probabilistic consequence were tenable, or worth serious consideration, a like treatment of defeasible consequence is not.*

COROLLARY: *The adverbs of defeasibility are irregular.*

8.9 Possibilistic consequence

What, then, about possibilistic consequence? Can our treatment of probabilistic consequence be generalized to it? If so, there would be some sufficiency-imputing sense of "if … then" in English for which

(3) If α is a possibilistic consequence of Θ then, if Θ then
possibly (α).

It is clear that what we are looking for here is not the same as modal
proofs in standard logic, in which the consequence relation is as close to
standard deduction as you please and its consequence is a proposition to
the effect that α is not inconsistent. What is wanted is a sense of
"possibly", other than truth in alternative worlds. "Possibly" here is rather
more earthbound. As already mentioned, my own ear gives me little
guidance, beyond the suggestion that when the RH of (3) is asserted in
real-life situations, Θ serves as evidence to keep α in contention. The
evidence in the Spike case is not intended to establish that there is an
alternative world in which Spike did it or that Spike's having actually
done it involves no self-contradiction, but rather that there is reason to
keep Spike on the investigator's list of suspects, or, to use of the language
of the British copper, to keep Spike "in the frame".[299]

Suppose we found ourselves inclined to accept the RH part of
(3). Then there would exist a relation between Θ and ⁻possibly (α)⁻ in
virtue of which that sentence is true. Before we can accept (3) itself, we
would have to satisfy ourselves that that relation must, in some requisite
sense or other, be a relation of logical consequence. My ear tells me that
on the present reading of the possibility attributed to Spike, there is no
particular reason to suppose that this consequence-claim is true, or even
plausible. Perhaps we will have a better handle on this point once we
have reflected a bit more on the fortunes of plausibilistic consequence.

8.10 *Plausibilitistic consequence*

Is there a relation of plausible consequence? If there is and the
conditionality of consequence thesis is true, the conditionality rule
demands that we find conditions under which whenever β is a
plausibilistic consequence of α, there is a true sentence ⁻If α then it is
plausible that β⁻ in which the truth of α is in some requisitely distinctive
sense sufficient for the plausibility of β.

Consider a case. Suppose that "There's been a burglary" is a
plausibilistic consequence of "The door's been left open and the side-
window smashed". Then on present assumptions, there is a sense of
sufficiency and a sense of "plausible" for which the truth of "The door's

---

[299] Although, not in the North American sense of *frame-up*.

been left open and the side-window smashed" is sufficient for the truth of "It is plausible that there's been a robbery".

Suppose that we inclined toward the present construal enough to keep plausibilistic consequence at least provisionally in play. Then it would appear that a like accommodation might be accorded to *evidential consequence*; provided, of course, that there were some counterpart expression playing the same role as "plausibly" played in the treatment of plausibilistic consequence. A natural suggestion is that "evidently" would fit this bill. Thus,

4.   α is an evidential consequence of Θ if and only if there is a sense of "if ... then" and a sense of "evident" according to which

4'.   If Θ then it is evident that α.

Here my ear rebels. ⌐It is evident that α⌐ says something stronger than that there is evidence for it. The same is true of its cognate "evidently", as in. ⌐Evidently α⌐. The same problem besets "makes it evident that", which is a stronger claim than the claim that some evidence exists. This is not good news for imputations of evidential consequence. If so, we have something of interest to reflect upon.

Proposition 8.10
AN ASYMMETRY: *The idioms of evidence do not conform to the template for third way consequence, even if the idioms of plausibility do.*

Why would this be so? Where lies the difference between being made plausible and being made evident, given especially that evidence, each time, is the maker? I'll come back to this below.

Consider again

(1) α is a plausibilistic consequence of Θ.

On the present account, for this to be so there must be a sense of "plausibly" in which if Θ then plausibly (α), where "plausibly" is a sentence-operator on α. By parity of reasoning with the probabativity case, this will be the case if the *conditional plausibility* of α on Θ is sufficiently high. The probability calculus details the conditional probabilities underlying ⌜If Θ, then probably (α)⌝. Is there a like provision for plausibility. Is there a plausibility logic that gives the conditional plausibilities needed for the truth of "if Θ, then plausibly α?" Is there a well-defined notion of conditional plausibility to be found in such logics?[300]

Plausibility is unruly. It behaves very differently from probability. For one thing, it has no stable notion of negation. A given body of evidence can make incompatible propositions equally plausible, with obvious implications for closure under conjunction. Certainly there are logicians ready to say: *No negation, no logic!*[301]

One of conditional plausibility's better treatments is Nir Friedman's and Joseph Halpern's theory of plausibility measures, which is a low-structure generalization of probability.[302] Plausibility is a partially ordered relation subject to a distinguishing axiom that says that a set of sentences must be at least as plausible as any of its subsets. Addition of two further axioms gives the so-called KLM properties for default logic. The first of this pair provides that if (i) A, B and C are pairwise disjoint sets, (ii) the plausibility of A ∪ B exceeds that of C, and (iii) the plausibility of A ∪ C exceeds that of B, then the plausibility of A alone exceeds the plausibility of B ∪ C. The other axiom stipulates that if A and B are both utterly implausible, so is A ∪ B. The KLM properties of a putative default conditional → are set by a reflexivity axiom, and the rules of left logical consequence, right weakening, conjunction, disjunction and cautious monotonicity.[303]

An important feature of plausibility measures is that the spaces they measure are direct generalizations of probability spaces. This gives

---

[300] There is no conditional plausibility in Rescher's logic of plausible reasoning (1976) or Walton's account of plausible argument (1992).

[301] Thus Quine's dismissal of dialethic logic: "They think they are talking about negation, '~', 'not'; but surely the notation ceased to be recognizable as negation ...." (Quine, 1970/1986), p. 81.

[302] Friedman and Halpern (1995).

[303] The KLM properties are named after their proposers: Kraus, Lehmann and Magidor (1990). Reflexivity provides that $\alpha \to \alpha$. Left logical equivalence is: If $\vdash \alpha \Leftrightarrow \alpha'$, then from ⌜$\alpha \to \beta$⌝ infer ⌜$\alpha' \to \beta$⌝. Right weakening is: If $\beta \Rightarrow \beta'$, then from ⌜$\alpha \to \beta$⌝ infer ⌜$\alpha \to \beta'$⌝. Conjunction is: From ⌜$\alpha \to \beta_1$⌝ and ⌜$\alpha \to \beta_2$⌝ infer ⌜$\alpha \to (\beta_1 \wedge \beta_2)$⌝. Disjunction is: From ⌜$\alpha_1 \to \beta$⌝ and ⌜$\alpha_2 \to \beta$⌝ infer ⌜$(\alpha_1 \vee \alpha_2) \to \beta$⌝. Cautious monotonicity, as before, is: From ⌜$\alpha \to \beta_1$⌝ and ⌜$\alpha \to \beta_2$⌝ infer ⌜$(\alpha \wedge \beta_2) \to \beta_1$⌝.

rise to a notion of conditional plausibility analogous to what is required for Bayesian networks. The question is whether conditional plausibility is grounded in the required way. Putting "pl" for "plausibly", does ⌐Pl($\alpha \wedge \Theta$) = x⌐ ground a sufficiency conditional, ⌐If Θ then pl(α)⌐, when x is big enough? If so, wouldn't plausibilistic consequence be back in business? Notwithstanding the weakness of some of the KLM properties,[304] it will seem to some that it might. We won't be able to say so with any conviction until we've found a sense of "plausibly" which allows for the plausibility of α just when its plausibility relative to a true Θ is sufficiently high. Why, then, don't we simply say so? Didn't we do this very thing for "probable". Didn't we say that α is probable just in case its conditional probability on the evidence is sufficiently high? Yes, of course. But we didn't do this blindly. We did it because there is an antecedently available sense of "probably" that seems to enable the connection to be made. There is a meaning of "probable" in English for which it might be true that α is probable if and only if the conditional probability on a true Θ is sufficiently high. The question presently before us is whether likewise there is an available meaning for "plausible" in English supporting the idea that α is plausible if and only if its conditional plausibility is high. Perhaps there is such a sense. Perhaps it is the sense we've already mentioned just pages ago. But, as before, there are costs in calling the relation between Θ and α plausibilistic consequence. We can no longer regard consequence and implication as converses.

8.11 *Myopia?*

The point generalizes. If we accept present attempts to give sense to the purported relations of nonmonotonic consequence, there remains a striking difference between these senses of consequencehood and the kind of consequencehood exhibited by monotonic relations of deduction. If α deductively implies β, then β is the consequent of α. But if α plausibilistically implies β, there is no sense of sufficiency in which the truth of α is sufficient for the truth of β; hence no sense of "if ... then" for which if α is true β also is. The third way template requires that in attributing consequencehood to these nonmonotonic cases, we in effect

---

[304] For example, as mentioned earlier, reflexivity fails intuitively; no proposition is a *default*-consequence of itself. Moreover, if we allow that if ⊢ β then α ⊢ β, then we have the paradox of necessity for →, and likewise for right weakening.

change the subject. When α plausibilistically implies β, it is not β that is α's consequence. It is some adverbially qualified sentence arising from the prefixing to β the appropriate sentence-operator. We can take this course if we like. But if we do, it will cost us. It will cost us another logical commonplace.

Proposition 8.11a
LOSING AN EQUIVALENCE: *If we retain the third world template for probabilistic, plausibilistic and evidential reasoning, the relations of implication and consequence cease to be one another's converses.*

COROLLARY: *Neither is a* consequence *of such a relation the* consequent *of its corresponding conditional.*

COROLLARY: *The adverbs of plausibility are irregular.*

From its very inception, logic has seen reasoning as the drawing of conclusions, and has seen the drawing of conclusions as the drawing of consequences. This makes for a quite determinate way for a person's reasoning to go wrong. It goes wrong if, for the conclusion that he has drawn, there is no consequence relation in whose converse domain that proposition occurs. Thus no consequence relation, no consequences; and no consequences, no conclusions. To the extent that they've been tolerated at all, non-deductive consequence relations have had to beg for their supper at the back doors of deduction. They've had to organize themselves around the properties of deduction not excluded by their differences from it; and they have tried to achieve those differences by adapting or qualifying the deductive conditions whose outright instantiation their differences from deduction also preclude. Think here of the consequence relations of the going $\mathcal{D}$-logics. They are classical consequence at two substantial removes from it. They are its second cousins.

But suppose that the hegemony of deduction rested on false assumptions. Suppose that consequence having were not intrinsic to conclusion drawing. Suppose that the basic relation is the conclusionality relation and that consequence were but a case of it. Then in trying to understand the Spike case and the Ozzie case and the countless others of like stripe, it would no longer be necessary to toil for consequence relations within which to embed those reasonings and against which to assess their rightness or wrongness. No longer would it be necessary to contrive these consequence relations by making them as similar to deductive consequence as their differences might permit.

The Can Do Principle gives procedural advice to theory builders. In the first instance, it tells them to try to build their theories with intellectual resources ready to hand or with circumspect adaptations of them. It counsels the budding theorist to begin *in medias res* with the tried and true. Can Do has a degenerate case. We called it Make Do. In its extreme form, it is the imposition on a research target of an established way of proceeding in the absence of any reason to think it is the way to go *in the circumstances at hand*. ("A bad theory is better than no theory at all.)" The respect for data principle is an attempt to keep Can Do in its place, to keep it from migrating *faute de mieux* to the shabby precincts of Make Do. These reflections give us occasion to consider whether the dominance in logic of the consequence rule isn't a manifestation of Make Do, and an unwitting capitulation to paradigm-creep. So let's make it official.

Proposition 8.11b
MISCONCEIVING NON-CONSEQUENCE: *That α is not in any sense a consequence of Θ does not imply that concluding α from Θ is an error of reasoning. It does not imply that α is not a conclusion to be drawn from Θ.*

Proposition 8.11c
DOWNGRADING CONSEQUENCE: *When a third way premiss-conclusion reasoning relation R is nonmonotonic, the construal of R as consequence makes no load-bearing contribution to the logic of R.*

8.12 *Making*

We said two sections ago that there are propositions α and β for which claims in the form ⌐α makes it plausible that β⌐ sometimes have the ring of truth, indeed could actually be true. It might be of some interest that syntactically similar claims ring less true in some ears than it, e.g. ⌐α plausibly-implies β⌐. Perhaps these are mal-tuned ears, deaf to the relevant similarities. I will not press this further. For present purposes it will be enough to stick with ⌐α makes it plausible that β⌐. Claims of this sort are frequently advanced in contexts of evidence-gathering and evidence-assessment. Something has occurred, reportable in sentences α_1, α_2, ..., α_n. One or other of two questions might be in play in that context. One asks what the α_i are evidence for. The other question is more general. It asks, for some proposition β, how, if at all, the α_i bear on β. There are different ways in which the α_i might bear on β. One is when the α_i make it plausible that β. But, only lines above, have I not allowed that when the α_i are evidence for β it might well be the case that the evidence makes it plausible that β, while frequently enough falling short of making it evident that β? Both plausibility and evidentness are cognitive virtues. Why would it be – what would make it the case – that the very evidence that bestows the one virtue would fail to bestow the other? Perhaps the obvious answer is that evidentness has tougher bestowal standards than plausibility. On the other hand, perhaps the answer lies elsewhere. Perhaps the nub of it all is that plausibility is, and evidentness is not, in the converse domain of the making-relation in question. This would be so if plausibility were a state of mind and evidentness not.

Making is what interests me here. I have a stake in giving it a load-bearing role in the epistemics of reasoning. Making is a challenge. It is a mode of necessitation, and causing is a mode of it. So making is sometimes causing. Making is the stock-and-trade of the CR-approach to knowledge. Knowledge, we said, is well-made belief that meets further conditions. Correct inference, we said, is belief produced by well-made belief-update mechanisms. Making wears the trousers here. Belief lies in its converse-domain. Beliefs are made, not found. Believing that α is a state in which your belief-forming devices put you. Inferring that α is a state in which your inference-making devices have installed you. The causal story of belief and inference, like causal stories elsewhere, is a story of degrees, of saturated end-states, of nearness to saturation, and of causal dispositions thereto.

If the evidentness of a proposition is not a state of mind and is not a causal disposition to be in one, the story of its possession or want of it must lie elsewhere. Perhaps we could say that its story will be a story of propositional relations, hence a story told in logical space. If plausibility is a state of mind or a mental disposition, its story will be told in psychological space. Since a naturalized logic is psychologistic on purpose, this is a space in which its practitioners should feel themselves at home.

I have already mentioned that in *Agenda Relevance* (2003), Gabbay and I causalized the logic of relevance in ways that emphasized the role of making. We defined relevance as a four-place relation on the set $\langle X, I, B, A \rangle$, where X is an individual cognitive agent, I is information, B is his background information, and A his cognitive agenda. Then I is relevant for an X with background information B if and only if in processing I, X is put in a state of mind which advances or closes his agenda. When this happens, the state of mind in which X's processing of I has put him either produces a saturated b-state with respect to A or a b-state to the effect that A, while still open, holds promise of closure, or, finally, in a state which, whether or not he experiences it as so, improves his prospect of attaining closure. Since the human individual is an efficient evader of irrelevant information, it stands to reason that the causal processes that advance the mind in the direction of agenda-closure won't fire if his processing devices fail to screen out irrelevant inputs. So effective irrelevance protection is a natural inhibitor of information processing.

The causal modal relevance can be adapted for the representation of inference or conclusion-drawing. Take the basic case. New information makes it past Harry's processing-censors, and enters Harry's data-base. In so doing, Harry's belief-producing mechanisms start up and place Harry in a state of belief with respect to some further matter. Crudely, being informed by I makes Harry see something that had heretofore escaped his notice. If we were to formulate this simple example in propositional terms, we could say that the semantic content of the input information is proposition α and that the semantic content of the further object is β. In which case, Harry's inference of β from α would be correct when his processing and belief-forming mechanisms were in good working order and operating here as they should. In other words the inference is correct when Harry's information and beliefs are well-produced.

The received view is that whenever someone correctly infers something from something else there exists a relation from the one to the other, something that we have been calling the conclusionality relation. In our discussions here, we have invested a good deal of time and effort in trying to sort out the status of this relation. We have been trying to settle our mind about the kind, if any, of $\mathcal{D}$-consequence relation it is. My view is that we haven't made much headway with this, and I have conjectured that the reason why may be that conclusionality is not $\mathcal{D}$-consequence of any kind. Perhaps this is the right thing to conclude. But it leaves the question of the status of the conclusionality relation. There was a time when logicians thought that relevance was a two-place propositional relation, either a semantic or probabilistic one. On the CR-approach, that was the wrong way to go. Seen the CR-way, relevance is a four-place causal relation. Information is relevant to an individual when, given his background information it is processed in such a way as to advance or close his cognitive agenda. That way of thinking marks quite a departure from the received view and, if true, leaves a large gap between how the relevance relation actually is and what logicians routinely think of it as being. Might not there be a similar gap for conclusion-drawing?

Here is another simplified example. The time is now. Harry, like the rest of us, possesses a b-state inventory, made up of what he occurrently believes and remembers, together with their implicit counterparts in background information and dormant memory. Computer scientists call such inventories "data-bases". Some information I becomes available to Harry. In processing it, there is some further proposition β not in Harry's data-base Δ and not encoded in the newly arrived I, Harry's belief-forming devices fire to produce the belief that β, whereupon, details aside, the update Δ(β) is also produced. The empirical record lends no support to the idea that whenever a being like Harry is presented with new information, he will respond to it, or even try to, in the ways at hand. Harry's success in life depends utterly on the success he has in falling into line with the Clutter Avoidance Maxim. It is not in Harry's interest to process an incoming I in ways that adds it (or its semantic context) to his data-base. Still less is Harry well-served by the mere finding of new things to believe with the help of Δ(I). In a rough and ready way, we don't want Harry's conclusion-making devices to fire when there is no good reason to. Undisciplined firings – even when mechanically sound – are the source of cognitive suffocation. Enquiring minds must be uncluttered minds. Whether or not Harry's devices on this occasion take his information to the lengths of β and Δ(β) will always be a matter conditioned by the agendas in play then, by what his interests are, by what it is in his interest to know. In the present example, and in a loose way of speaking, we say that Harry has drawn β from I as a conclusion. It would more nearly be correct to say that his conclusion was drawn from some interaction of I and Δ. No matter which, it is clear that this is not the story of a two-place relation. It is more likely that

Proposition 8.12a

A CAUSAL FIVESOME (FIRST PASS): *Conclusion-drawing is a five place relation defined on quintuples ⟨X, I, Δ, A, α⟩. Thus in processing I, Harry's devices interact with his data-base in such a way as to produce the belief that α, in advancement or closure of Harry's cognitive agenda at the time. When this happens, we say that Harry concluded α from I or, if we are being more careful, from some item α' in Δ(I). Either way, what we say underdescribes the reality. The conclusion-drawing relation is not binary but quinternary, and it is not intrinsically propositional but causal.*

Propositions enter the picture here, but only as the objects or semantic contents of beliefs. If this were actually so, the haplessness of our search for $\mathcal{D}$-consequence would be readily explained. When Harry wanted to know whether it had snowed overnight, he went to the window and looked down into the street. Sure enough, it had snowed overnight. The information contained in Harry's visual field, produced a belief that closed that agenda. Harry's situation seems to fit the specifications of 8.12a, but it doesn't sound right (to me) that Harry's belief that it snowed last night was achieved by inference, that the proposition that it snowed last night was a conclusion he drew from some or other premiss. Other philosophers are otherwise positioned as to the inferential character of perception. This is another fight that I have no inclination to join. I'll simply say that my reluctance to think of what Harry was up to as conclusion drawing if he added to 8.12a a clause that first occurred in Proposition 3.8. It is a clause that responds to what I take to be a plain fact. It is that if we actually asked Harry what led him to believe that it had snowed overnight, and if this were a question that Harry were disposed to answer, he would be disposed to say something like, "I saw that it had." Dispositions are causal propensities. So is the disposition to cite α' if asked. As developed by logicians over the ages the language of premisses and conclusions have taken on the forensic character that explains their appeal to CC theorists. The last thing a CR theorist should be prepared to do is to give an account in which what we've been calling third way premiss-conclusion reasoning is an intrinsically forensic enterprise. What our present reflections suggest is a more causalist twist for such reasoning. Concluding that α from α' is being disposed to individuate some α' in response to the question "What makes (made) you think that α?" So let's amend 8.12a accordingly. Let D denote the disposition to individuate information in response to this question. This gives

Proposition 8.12b

A CAUSAL SIXSOME: *Conclusion drawing is a six place relation defined on sextuples $\langle X, I, \Delta, A, \alpha, D \rangle$.*

8.13 *Closure*

The closed-worlds assumption raises a question of large importance. As regards most matters of concern to beings like us, the world is never closed. This, in its turn, would appear to mean that we are almost never in a position to adopt a position that closes or even advances a cognitive agenda. That would appear to mean, in further turn, that invocation of the closed world-assumption is hardly ever something that we are justified in doing. Certainly this would be so on intellectually pure readings of the CC-position, and with it would come a scepticistic tsunami.

The CR-position offers promise of relief. On the CR-model, Harry knows that α when his cognitive mechanisms place him in a b-state that stills his impulse to know. His b-state, we said, is an end state of cognitive saturation. B-states close agendas as a matter of fact, and agenda-closure, also as a matter of fact, implements the closed-world assumption. Resistance to the closed-world assumption is a standard way of keeping an agenda from closing. The reverse is also true. Well-produced information that closes an agenda also closes the world. Needless to say, agenda/world closure is necessary but not sufficient for knowledge. But when someone's agenda does close on some given information, if the object α of the b-state is true and well-produced, he does indeed know α. The same is so under suitable adaptation of well-produced conclusions.

When AI theorists write about the closed-world assumption, they are often read as saying that a sentence in the form "There is no unknown defeating evidence" is entered as a premiss of a conditional argument that ends with the drawing of some conclusion α. Then the conclusion is soundly drawn only if the hypothetical premiss is true, and it is reasonably drawn only if the hypothetical premiss is well-justified. This is the wrong way of thinking of these things. The closed-world assumption is not a premiss in a chain of propositions. It is a causal condition in the closure of agendas. It plays this role without propositional representation. Take the Spike case again. There arrives a point in the investigation at which the police come to the realization that Spike was the culprit after all. At that point, the investigators' agenda closed. They wanted to know who did it. Now they do. Spike did it. Of course, the police also have another agenda. They want to know whether they have a case against Spike that meets the criminal law's high standard of proof beyond a reasonable doubt. It would be quite wrong to think that the information that closed their who-did-it agenda would also close their case-making agenda. They are not the same agendas and they typically have different closure conditions. No one should doubt that criminal case-making has its seriously CC moments. But the fact remains that the police won't recommend prosecution unless they are satisfied that the evidence against Spike carries a very high probability of conviction. They won't proceed unless, with respect to *that* proposition, their agenda closes. When the case goes to trial, the police's thoroughness will be critically probed. How hard did you search for alternative possibilities?", they will be asked. "Did you explore exculpatory possibilities?" These are good and necessary questions. But one question that they won't be asked, and which no judge would allow, is "What, apart from your testimony so far, justifies the proposition "In nothing of what remains of our world and of what is yet to be disclosed of it is there any prospect of discouraging the belief that Spike did it?"

Further adjustments lie in wait. Let me announce them rather than making them. A great deal of conclusion-drawing is done down below. A great deal of conclusion-drawing is done in point-and-shoot mode, even some of what occurs up above. A lot of the most interesting conclusion-drawing occurs in manual mode, and must be laboured after with much effort and deliberation. Some conclusion-drawing is a simple causal response to something in the domain of the evidence-for relation. Some conclusion-drawing is the result of case-making inferences from circumstances deemed to have the evidentiary force. There is quite a lot of difference in this network of goings-on. But at each turn there is a common element: The conclusionality relation is epistemically loaded and rupturable each time in the face of new information. One of the tests of any proposed paradigm for epistemology is to account for all this traffic in ways that preserve and illuminate this similarity. The extent to which the CR-paradigm pulls this off is not presently known. That is why, as I write, the natural sciences of cognition are a growth industry. Naturalized logicians should stay in touch.

One of the questions posed in chapter 1 arose from the observation that consequence-having occurs in logical space and consequence-drawing occurs in the mind of the drawer. The question was how, as an occupant of the causal order, a mind gains access to the occupants of logical space. The counterpart question arises for conclusion-drawing. Conclusion-drawing occurs in psychological space, conclusion-having in semantic space. How is access achieved? It is achieved by the devices of belief. True, not everyone likes this answer.

They see in it a recurrence of the very problem it seeks to solve. That is not my purpose here. The question was: "How is access achieved?" not "What shows that it actually can be achieved?" If we allow ourselves the indulgence that belief is indeed the negotiator-in-chief of this sort of cross-worlds intercourse, there are further parallels to draw. No one thinks that in any strict or literal sense that being in a state of belief is a proposition or even that it is a state admitting of propositional residency. Much in the way that a proposition is the semantic content of a bit of information, so a proposition can be taken as the semantic content of believing. No one is much confused when beliefs-states in turn come to be taken as their own semantic contents. They aren't, of course; only in a manner of speaking. It is a way of speaking that facilitates communicational fluency about matters of common interest. Propositions are the formal representatives of the beliefs whose contents they are.

Our most powerful logics have been the logics of formal representations. The elementary logic of propositions is the logic of formal representations of simple declarative sentences of English (or some such) and certain classes of compounds thereof. Modal logicians have made – or purported – considerable headway with formal representations of belief, actually two representations at once. Beliefs are represented as propositions, and propositions are represented by uninterpreted pieces of syntax. Sentences of the form $\ulcorner\beta\alpha\urcorner$ are their fusion. The belief that α is formalized as the scope of the sentence modalized by the belief-operator. Why, then, couldn't we set ourselves the task of adapting the logic of belief to the specific purposes of accommodating those that arise from conclusion-drawing? In the logics of belief the B-logics, are the dominant target is doxastic consequence. In their C-logic adaptations, why wouldn't we give like weight to conclusion-having or, indeed, to *consequence*-having? Wouldn't this be occasion for the reinstatement of consequence?

I don't mind. In for a penny, in for a pound. The technical virtuosity of B-logics is not in doubt. But there are plenty of epistemologists and psychologists for whom no discovery of any note about belief is to be found in these formalizations. Perhaps this is too harsh. I will not press the point here beyond saying that, historically, there has been a twofold purpose that formalizers of a subject-matter of enquiry have sought to fulfill. One is to give a perspicuous summary of what is already known. The other is to do this in a way that opens us to the things not yet known, and which settles antecedently contentious differences of opinion. A case in point is the KK-hypothesis. This is something that theorists of knowledge disagree about. The KK-hypothesis is provable in Hintikka's epistemic logic, which is an extension of S4. Indeed the KK-hypothesis is the epistemic counterpart of S4's distinguishing axiom. Is the fact that KK is provable in Hintikka's logic a fact of probative weight? Suppose you knew someone who denied the KK-hypothesis, and you told him Hintikka's theorem. Would you now expect him to change his mind?

8.14 *The ring of truth*

Probability attracts different interpretations of it. Some theorists see it as an objective property of events and, by extension, of statements about them. Subjective theorists associate it with degrees of an individual's belief. Seen the first way, there is a fact of the matter about the probability of an inductive consequence of Θ which constrains the accuracy of a reasoner's beliefs about it. Seen the second way, there is a relation between a reasoner's attachment to Θ and his prior attachment to α that imposes a coherence condition on his subsequent attachment to it. Either way, the probability that flows to α by way of its high conditional probability on Θ is tied up with what, in the circumstances, the reasoner will or should believe. The more that something is probable, the more that believing it is accurate. The more probable it is, the more that believing it is coherent.

Plausibility doesn't work this way, at least not always. A remark by Peirce bears on this in an interesting way:

> By plausibility, I mean the degree to which a theory ought to recommend itself to our belief independently of any kind of evidence other than our instinct urging us to regard it favorably. (CP, 8. 223)

On first reading it may appear that Peirce does not support the suggestion that plausibility is belief-neutral. Certainly, Peirce is not proposing that a theory with no supporting evidence should be believed. Why, then, would he say of a theory that we shouldn't believe that it "ought to recommend itself to our belief ... if our instinct urg[es] us to regard it favorably"? Let $\mathcal{T}$ be such a theory. What Peirce is proposing is a favourable reception for $\mathcal{T}$ short of belief, based on nothing else than our instinct to favour it thus. However this favour might be bestowed, it is clear that judging a proposition as plausible is to assume a pro-attitude towards it. We might

297

not like the contents of some plausible propositions, but their Peircian plausibility is a cognitive virtue nevertheless. It is a cognitively valuable property for a belief to have, short of there being any reason to believe it.

Plausible propositions are attractive, they are favoured in a certain way; they are cognitive assets even when wholly unevidenced. We might say that when propositions are plausible they have "the ring of truth". What Peirce would be telling us is that propositions that ring true needn't be true, that they recommend themselves to our attention before the evidence is in. If this were right, plausibility is yet another of our filtration devices, screening our hypothesis-selection spaces for candidates worthy of attention. One form of this attention is experimental trial. There are others. They will occupy us two chapters on.

Peircean plausibility makes no claim on belief; it is not belief-presumptive. The implausibility in 1891 of Cantor's theorem was not evidence against it. And, as the evidence mounts in the Spike case, one may even come to believe that Mother Theresa was the one who did it, no matter the implausibility that she would have done so. Suppose we find ourselves thinking that Mother Theresa, not Spike, is the actual perpetrator. Consider the proposition, "This is not something that Mother Theresa would do." But she did. Does this contradict the proposition that it's not something that she would do? My ear tells me "No". "Mother Theresa wouldn't do it" has the feel of a generic claim. Although Mother Theresa did do it, it remains the case that this is not Mother Theresa's thing. It is a truth about her that a true negative instance leaves standing.

Cantor's theorem is also like this. Although some infinite numbers are infinitely larger than other infinite numbers, this is not the way of numbers. This is not what one would have expected of numbers. Of course, we've known about nondenumerability for well over a hundred years, and mathematical sensibilities change. There was a time when "The reals are nondenumerable" lacked the ring of truth. There was a time – perhaps not now – when it lacked the ring of truth even for those who knew it to be true. It is uncharacteristic of numbers to be this way. It is uncharacteristic of saints to act that way. The evidence might be, equally, that Spike did it and, contrariwise, that Mother Theresa did. Even so, that Spike did it has the ring of truth. That Mother Theresa did it lacks the ring of truth. In this same way, although there is evidence that Spike did it and the same evidence that Mother Theresa did it, "Spike did it" is plausible and "Mother Theresa did it" is not.

Perhaps the most interesting feature of a plausible proposition, a proposition that rings true, is that its falsity might be established without laying a glove on its plausibility. The extent to which this is so is the extent to which propositional plausibility is non-probative. But what, if the ring of truth is impervious to demonstrations of falsity, is the good of it? How could it be to our *cognitive* advantage to hear in propositions we robustly disbelieve the ring of truth? To what good end is our habit of favouring theses we have no reason to believe?

The first logician to recognize something like the ring of truth is Aristotle. It is a notion that lies at the heart of his concept of fallacy. The view that Aristotle most frequently advances is that a fallacy is an argument that has the look of a syllogism but is not a syllogism in fact. In some places, Aristotle expands this to say that a fallacy is any argument that looks good in a certain way but isn't in fact good in that way. Thus fallacies have the ring, not of truth, but of *correctness*. Still, Aristotle's false appearances are not quite the same as Peirce's plausibilities. Aristotle advances no claim that, even in the absence of evidence of its goodness, we *should* favour an argument that looks good, notwithstanding the plain fact that this is just what we do. As applied to arguments, that is certainly Peirce's intent, never mind that he, no less than Aristotle, is aware that the good looks of good-looking arguments often disguise their badness.

We could take it that an argument that looked good and for which there were no available contra-indications is one that most people would in fact be inclined to accept, albeit provisionally. Peirce's has a stronger view of the matter. If an argument has the ring of rightness then, absent contrary indications, it ought to be accepted, or anyhow to "recommend itself to our belief". We ought to yield to "our instinct urging us to regard it favourably."

It is evidently Peirce's view that our sensitivity to the plausible plays a large role in the management of matters of which we are ignorant. Not knowing what is the case, not knowing what to do, we are endowed with the capacity to favour some propositions and some courses of action over others. Plausibility plays an enormous role in cutting large spaces of alternative hypotheses and courses of action down to size. Plausibility guides us to favourable consideration well in advance of the evidence. It is necessary to ask why this would be so. Why would we have been built thus? The pedantic answer is that there is an adaptative advantage in it. The plain answer is that we'd be dead if we didn't.

The favouritisms fostered by plausibility, fallible as they surely are, resemble in a certain way our gift for hasty generalizations. Seeing that this, my first ocelot, is four-legged, I am induced to look with favour on the proposition that that's the way ocelots are. Here, too, I can be wrong. Perhaps there is a three-legged species I've not yet heard of. But the striking thing about hasty generalizations is the extent to which, when we actually draw them, we get them right, not wrong. No one should think that we have nearly so good a track record with Peircean plausibilities. That is to say, the correlation of propositions favoured for consideration and propositions subsequently shown true is not the equal of hasty generalizations made and hasty generalizations got right, but even where a favourably alternative hypothesis doesn't pan out, *its* alternatives are easier to spot; and it is with respect to them that the likelihood of accuracy spikes upward. For if *this* isn't the right hypothesis, the one that is has a high probability of occurrence in the space of its alternatives.

Here is one more thing worth repeating:

Proposition 8.14
MORE IRREGULARITY: *The adverbs of plausibility are highly irregular.*

8.15 *Two options*

Modern logicians make up more consequence relations than they can shake a stick at. They are needed, it is said, to undergird all the different kinds of conclusion-drawings made in the course of our reasonings. I have tried to show such multiplicities are actually not needed for the conclusional varieties with which they have been associated and, worse, that in the case of defeasibility consequence the multiplicities purported by the mainstream might not exist. If this is right, conclusion-drawing is not intrinsically a matter of drawing consequences, and the general relation between the concluded and the concluded-from must be sought elsewhere. In earlier chapters, I have suggested a place to look. It lies in the make up of the cognitive ecologies in which we operate, in the causal nexi of the responses of our cognitive equipment to the information that passes our way. Seen the old way, when Harry concludes that α from Θ, there is a consequence relation binding α to Θ which Harry somehow recognizes, and a selection function – or some such thing – that enables Harry to go with α rather than Θ's infinitely many other consequences. Seen the new way, Harry's information-processors operate on Θ in such a way as to engage with Harry when two conditions are met. One is that belief-producing devices form in Harry the belief (or supposition) that α. The other is that the conditions under which this belief is formed induce in Harry some disposition to recognize that it was the information in Θ that got him to see that α or made him think that α.

If my case has legs, the logical orthodoxy as two options to consider. One is to *shrink*; the other is to *expand*. Shrinkage is a way of sticking to your guns. Since its very foundation, logic's dominant and anchoring interest has been the relation of logical consequence. If, contrary to what has been supposed these past fifty years, the consequence relations of third-way reasoning have been either bogus or dubious or at least unnecessary, then third-way reasoning is not the proper business of logic. In the absence of the requisite consequence relations, logics of reasoning, are logics in name only and, in all strictness, unworthy of the name. I am not much fussed by this nomenclatural fastidiousness. This is a conservatism which pretty much guarantees an early end for naturalized logic. If the cost of continuing with the naturalized programme is refusal of the orthodoxy's refusal to lend it the name of logic, that is fine with me.

The expansion option we have already met with. Details are set out late in chapter 2. If there aren't ready to hand the desired relations of defeasible and plausible consequence, who's to say that they can't just be made up, introduced by the theorists' stipulations in much the manner of a model-based scientist's quite general introduction of theoretical constructs?

I said before that I have no principled objection to this. But, as with stipulated entities elsewhere in science and mathematics, the returns had better be good. We should gain more understanding of third-way reasoning by endorsing the stipulation than not. A central thesis of this book is that, for beings like us, most right reasoning is third way reasoning; that is, reasoning which, when right, owes its rightness to the meeting of standards other than deductive validity and inductive strength. Whatever the consequence relations a logician of reasoning may wish to stipulate into existence, their instrumental value will lie in the indispensability of their contributions to the satisfying things the theory manages to say about getting things right (and wrong) when third-way reasoning is under way. It is not a trivial onus. Perhaps this is why there is so little evidence of its having been met (or attempted) in the mainstream logics of the third-way.

Up to a point we can call an object of interest by any name we choose. Names are conventional, and how their nominata actually *are* is not. But here is a further convention: In calling something an 'N', it is usually better to call it what it is rather than what it isn't? Isn't that what the respect for data principle calls for? Besides, in calling something what it isn't, don't we risk giving it properties which the thing it isn't has, but which it itself doesn't have?

CHAPTER NINE

BEING TOLD

"The floating of other men's opinions in our brains makes us not one jot the more knowledge, though they happen to be true. What in them was science is in us but opiniatrety." John Locke

"[T]here is no species of reasoning more common, more useful, and even necessary to human life, than that which is derived from the testimony of men, and the reports of eyewitnesses and spectators." David Hume

"Anyone who has obeyed nature by transmitting a piece of gossip experiences the explosive relief that accompanies the satisfying of a primitive need." Primo Levi

9.1 *Being told things*

The majority of errors made by individual agents are errors of misinformation, not reasoning. There are whole classes of cases in which we have no defences against misinformation. In these situations, we might reasonably say that the information was erroneous, but do we really want to say that in taking it in we ourselves made an error? Made by us or not, they are certainly not errors we are guilty of. Even so, openness to misinformation is an important fact about our cognitive make-up. Since a theory of reasoning requires some acquaintanceship with how the human reasoner is put together cognitively, it is reasonable to give this particular feature of it a degree of attention before moving on.

What is it that makes us so vulnerable to false facts? A short, empirically suggested answer is that

Proposition 9.1a
ADAPTIVITY: *The link between being told something and being got to believe it is infused with adaptive significance.*

COROLLARY: *There is a like propensity to share information.*

This gives us some further facts to take note of, beginning with

Proposition 9.1b
TOLD-KNOWLEDGE: *Very much of what we know we know by having been told it.*[305]

Let α be something that your being told it gets you to believe. Suppose α is true. Then unless your belief was malproduced or the belief-inducing information bad it follows that you now know that α. That is, it follows from the CR notion of knowledge. Any true well-informed, well-produced belief induced in this way is knowledge. It is, as we might say, "told-knowledge".[306]

Here on the ground, in the cut and thrust of real life, the values of α are strikingly many. Not everything we're told is true, but when we consider the totality of our true beliefs, the percentage of those arising from telling is exceedingly high, high enough to lend the cognitive abundance thesis some considerable encouragement.

If Proposition 9.1b is true, so are the following.

Proposition 9.1c
TRUTHFULNESS: *For the most part, informants are sincere in what they tell.*

Proposition 9.1d
RELIABILITY: *To a considerable extent, the information imparted in tellings is reliable, that is, is not misinformation.*[307]

The tellings of Proposition 9.1b are proximate. When someone tells someone that α, there is a direct informational link between her who tells and him who is told. Proximate tellings contrast with what I will call "distal" tellings, about which more in section 9.5.

---

[305] Richard Foley writes to this same effect, understating it: "… belief systems are saturated with the opinions of others." (Foley 1994, p. 61)

[306] There is an already substantial and still growing literature on knowledge arising from testimony. For a sample, see Hintikka (1962), Welbourne (1979), Welbourne (1981), Ross (1986), Hardwig (1991), Coady (1992), Govier (1993), Welbourne (1993), Adler (1996), Audi (1997), Adler (2002), chapter 5. See also online Adler's 2006 article "Epistemological problems of testimony" in the Stanford Encyclopedia of Philosophy.

[307] These propositions jointly are what Adler (2002), p. 144, the "positive bias" position.

We should take pains to avoid a possible confusion. In some approaches to told-knowledge, the claim that a great deal of what we know we know because we were told it (Proposition 9.1b) is an inference that is drawn from propositions 9.1c and 9.1d functioning as default premises. That is to say, the fact and scope of told-knowledge follows from the truthfulness of informants and the reliability of information received, supplemented by the purely factual premiss that a great deal of what people are told they believe. This is not my approach here. I proceed in the reverse direction. I open with the told-knowledge proposition. And I infer from it the general truthfulness of tellers and the proportionate reliability of tellings – proportional that is, to frequency with which they impart knowledge. Given that orientation, Propositions 9.1c and 9.1d pop out quite straightforwardly, not as premisses but as conclusions.

Told-knowledge covers a complex class of cases, linked to one another by family resemblance. Much of this complexity falls outside the province of logic, even a naturalized, empirically sensitive one; a great deal of it awaits further development in the maturing precincts of cognitive science. So the big story of telling will not be told here. There is no knowledge of α unless α is believed or is the object of some saturated end-state.[308] What is it about being told that α that kits you up with the belief that α? The adaptive advantage claim of Proposition 9.1a is the way to go.

Proposition 9.1e

TELLING AS CAUSAL: *For large classes of cases, telling someone something causes him to believe it.*

There is in Proposition 9.1e an unmistakable CR cast. It suggests that someone gets you to know by telling you what he knows. It suggests that much of what you know was transmitted from other knowers to you. Sarah knows that α. She tells it to you. Hey presto! You know it too.

The phenomenon of knowing by having been told somehow is as much a matter of dark knowledge as it is of light, only much more difficult to get a grip on. Tellings that effectuate light knowledge, or knowledge up above, is not without its own theoretical challenges. I have more to say about these cases than I do about the dark. So for the most part, the chapter will deal with the phenomenon of up-above knowledge occasioned by telling.

---

[308] I will revisit the belief-condition in a later chapter . For the present we'll stick with belief, but on sufferance.

9.2 *Transmission*

The phenomenon of epistemic telling raises two important epistemological questions. One is whether knowledge is transmissible. The other is whether it is transmissible by telling.

An affirmative answer to the first question is provided by systems of epistemic logic having the transmissibility of knowledge rule (the TK-rule),

$$K_X K_Y(\alpha) \vdash K_X(\alpha),$$

according to which if there is something that you know that someone else knows, then you know it as well.[309] TK is a rule about knowledge transmission that says nothing about the mode of transmission. But this is precisely what engages our present interest. We want to know the conditions, if any, under which your knowledge that α arises from its being transmitted to you on the *sayso* of another. We should amend the TK-rule to reflect this feature of the case. Accordingly, putting **t** as the "telling" operator, we replace the TK-rule with the condition, TtK, as follows:

TtK: $\mathbf{t}(\alpha) (Y, X) \vdash ((K_X K_Y(\alpha) \rightarrow K_X ((\alpha)).$

Informally: If agent Y tells agent X that α, then if X knows that Y knows that α then X also knows that α.

Suppose that Sarah tells Harry that e = mc². Suppose that Harry doesn't understand the sentence "e = mc²". But Harry might believe and be right in believing that the sentence "e = mc²" expresses a true proposition, and that Sarah knows that proposition to be true. What this tells us is that there are contexts in which not only the TtK rule fails, but also the epistemicized T-schema:

Harry knows that "e = mc²" is true, but he doesn't know that e = mc².

I don't say this with any alarm. The T-schema is an extensional context. But the epistemicized T-schema is intensional contexts, made so by the intensionality of "know". Accordingly, the TtK rules require a comprehension clause, as follows, putting "U" for "understands":

TtK^U: $\mathbf{t}(\alpha) (Y, X) \rightarrow ((K_X K_Y (\alpha) \wedge U_X(\alpha)) \rightarrow K_X (\alpha)).$

---

[309] The TK-rule is to be found in, for example, Hintikka's epistemic adaptation of S4 (Hintikka, 1962).

Informally: If Sarah tells Harry that α, and Harry takes Sarah as telling him that, then Harry's knowing that Sarah knows that α means that, if Harry understands it, he knows it too.

We may take it that TtK^U is a serviceable constraint on conditional transmissibility. We may also take it that TtK^U holds in both the CC and CR-models. If you know that Harry knows that $e = mc^2$ and you yourself understand "$e = mc^2$", then you too know it. One of the problems for a logic of told-knowledge is specifying conditions under which it may fairly be presumed that the antecedent of this condition is true.

Our present reflections assume an affirmative answer to the *credo/intelligo* question, which asks whether understanding is necessary for knowledge. Consider any model of knowledge in which belief is necessary for it. Then our question is complicated by the ambiguity of "believe". If believing that α is a matter of accepting its truth with understanding, then the answer is in the affirmative: Understanding is necessary for knowledge. If, contrariwise, believing α is a matter of accepting its truth without the necessity of understanding it, the answer to our question is in the negative: Understanding is not necessary for knowledge. The acceptance sense of "believe" is captured in St. Anselm's famous declaration, *credo ut intelligam* – "I accept in order that I might come to understand". The opposite view is the converse injunction, *intelligo ut credam* – "I understand so that belief might be made possible." A question for epistemology in general is which, if either, of these Latin tags to favour.

Neither is an option to be dismissed out of hand. Consider again our friend Harry. Thanks to a reasonably satisfactory high school education, albeit a number of years ago, Harry is aware that $e = mc^2$. However, Harry has long forgotten the precise meaning of e, and he has never had an adequate grasp of what mass is. One thing not in doubt is that Harry, like the rest of us, knows that the sentence "$e = mc^2$" expresses a true proposition, indeed a law of physics. In the first case, we can see belief operating as a *de re* modality. In the second it functions *de dicto*. We find in such cases a little-mentioned mingle of the two senses of "believe". In the *de dicto* case, Harry believes (in the psychological sense) that the sentence "$e = mc^2$" expresses a true proposition, and accepts without belief the proposition that $e = mc^2$. Putting 'H' for an agent, 's' as a sentences variable, 'α' as a proposition variable, 'B' for belief with understanding and 'A' for acceptance without understanding, this gives

Proposition 9.2a

THE MODAL-MINGLE SCHEMA: *There is an s with respect to which Harry believes there is a proposition α it expresses, and Harry accepts that α. More formally:* $\exists s \ [(B_H(\exists \alpha \ (s \text{ expresses } \alpha))$ $\wedge \ (\exists x \ (x = \alpha) \wedge A_H \ x)]$.[310]

Harry's situation is far from uncommon. The opposite is true. As the old saying has it, Harry's situation is as common as dirt. Harry's situation triggers the *credo/intelligo* question. Either way, there are consequences. If we deny to Harry the achievement of knowledge, then we deny it to anyone who is unable to give the meaning of terms occurring in sentences expressing propositions they take themselves as knowing, or to characterize in some mature way their denotata. If we take this option, we make ourselves Anselmians about knowledge.

It is possible to reject the Anselmian option for an individual's knowledge and yet, even so, to consider the phenomena schematized by the modal mingle schema as providing a non-trivial cognitive benefit. Take Harry again. His modal mingle with respect to that law and to the sentence that expresses it lifts him considerably above a state of sheer ignorance. Harry is not totally in the dark about this law of physics. His modal mingle gives him, in sentence *s*, an expressive *source* of the knowledge he lacks, and it lays down, albeit schematically, a *discovery* strategy: "Find out what *s* means". His modal mingle gives to Harry a source-discovery map with respect to a proposition he doesn't know. Compared to sheer ignorance, this is a considerable advantage, a positive orientation in the direction of knowledge and a stabilizing factor in the cognitive economy. Knowing where to look and having some idea of what to do to get it may not count as knowledge. Indeed, we might say that when in a modal-mingle state with respect to it, Harry is *quasi-ignorant* about α. But considered *en large,* quasi-ignorance constitutes a significant expansion of humanity's knowledge-potential.

Although it may be too much for some mainstream philosophers to bear, the Anselmian option for knowledge converts these virtuous potentialities of modal-mingle states into the bloom of a generous epistemic actuality. At a certain level of generality nearly everyone

---

[310] For those who distrust the sentence-proposition distinction, there is another way of making the present point. Harry doesn't believe that $e = mc^2$ but he does believe that "$e = mc^2$" is a true sentence. As we might expect of intensional contexts, the T-schema fails here: That Harry believes that "$e = mc^2$" is a true sentence doesn't imply that Harry believes that $e = mc^2$.

knows that nearly everyone knows that $e = mc^2$. Seen in the Anselmian way, this is perfectly true. Seen in the non-Anselmian way, it hasn't a chance.

Given that so many of our true beliefs satisfy the conditions of this example, an affirmative answer to the *credo/intelligo* question would crimp the cognitive abundance thesis. Perhaps this fact alone is inducement to consider a reformation of it. As we currently have it, the abundance thesis asserts that beings like us have lots and lots of knowledge. Perhaps a more congenial wording would be:

Proposition 9.2b
COGNITIVE ABUNDANCE REVISED: *There are lots and lots of propositions which beings like us know to be true.*

COROLLARY: *The things known to be true out-number the known. That is to say, the predicate "knows the proposition expressed by α to be true" has a larger extension than the predicate "knows α".*

COROLLARY: *Both these predicates have abundant extensions.*

This is not a matter on which we should rush to judgement on. If this is the way to handle Harry's case, it is necessary to handle like cases in the same way. Our ordinary, everyday cognitive situations provide copious examples of Harry's situation. But so do our scientific and theoretical practices as well. A case in point is a celebrated experiment in particle physics.

John Hardwig's "Epistemic dependence"[311] discusses a multiauthored contribution to *Physical Review Letters* entitled "Charm photoproduction cross section at 20 GeV". The letter reports experimental results in particle physics, achieved by a widely scattered team of ninety-nine co-authors of varying backgrounds and expertise. Since the case is well-known to philosophers, a detailed description is unnecessary here. Suffice it to say that the experiment was a huge composite event, involving strikingly diverse kinds of specialization, prompting from Hardwig the observation: "Obviously, no one person could have done this experiment – in fact ...no one university or national laboratory could have done it – and many of the authors of an article like

---

[311] (1985), 335-349.

this will not even know how a given number in the article was arrived at." (p. 347) Hardwig adds: "Of course, only a few people actually write the article, but it does not follow that these people are masterminds for the whole procedure or that they completely understand the experiment and the analysis of the data. According to William Bugg[312], although a few persons – 'the persons most actively involved in working on the data and who therefore understand most about it' – wrote up the experiment …, they really only prepared a draft for revisions and corrections by the other authors." (p. 347, n. 10).

This is a fateful disclosure. It virtually guarantees that for each experimenter there was at least one aspect of what the experiment showed, for which there was at least one experimenter didn't understand it. By non-Anselmian lights, we would have it that whereas the team produced the knowledge in question neither of its members were possessors of it. Cases such as this are legion in science and mathematics. To deny knowledge to the people who produced it seems excessive.

My present focus is not the *credo/intelligo* question. It is the transmissibility of knowledge by sayso question. It is a question of absorbing importance no matter where your sympathies lie on *credo/intelligo*. I won't be pressing on with it. Sayso is now front and centre.

9.3 Bootstrapping

Whether justification is a condition on knowledge by sayso is linked to something called *bootstrapping*,[313] which Stewart Cohen sees as a case of the Easy Knowledge Problem and Jonathan Vogel sees as a problem for the teller-reliability assumption. Consider the case of Roxanne. Roxanne is an imaginary driver who keeps a careful eye on her car's fuel gauge. In the beginning, she did this out of an interest in the state of her gas tank, but in time she developed an interest in the state of her gas gauge. On each occasion of her checking the level of gas in her tank, she is on pain of a Moorean blindspot committed to a proposition in the form

(1) On this occasion the gauge reads accurately.

---

[312] William Bugg was professor of physics at the University of Tennessee and a participant in the experiment.
[313] Woods (2003), Cohen (2002) and Vogel (2000, 2008).

She is committed to (1) because it would be self-defeating for her to say "I know that the tank is full because the gauge tells me so, but I doubt whether the gauge is reliable on this occasion."

Since Roxanne is a fastidious and frequent gauge-checker, there is a substantial number of propositions of this same form to which she is committed. Sooner or later, induction has to click in, taking Roxanne to the further proposition

(2) This is a reliable gauge.

Vogel doesn't like this induction. It is, he says, induction by bootstrapping. Bootstrapping is a kind of circularity or non-well-foundedness. It violates the expectation that an agent shouldn't be able to arrive at the knowledge that a knowledge-source is reliable by relying even in part on that source. Vogel regards this as a sound constraint on epistemic justification. Call it the "well-foundedness principle for knowledge", or WFK for short. Vogel's point is that bootstopping is prohibited by WFK.

It is easy to see the connection with told-knowledge. Consider any case in which you believe α because you heard it from some Y (Harry, say, or the BBC). It would be self-defeating for you to say that this is how you've come to know that α and yet that you have real doubts about the reliability of your source.[314] There are a great many cases of this kind, a veritable plethora of them. In some of them, Y is a frequent source of what you know. So Y is subject to the same sort of inductive manoeuvre as in the Roxanne example. If Roxanne's induction violates WFK, so do the present ones. Even where Y is not a recurring source a weak induction applies: "Y has been a reliable source so far". In these situations Vogel thinks that WFK is breached. If Roxanne were bootstrapping then it would appear that told-knowing is also bootstrapping. And if WFK is a sound principle and the bootstrapping of the first case runs foul of it, it is hard to see how the bootstrapping of the

---

[314] We should not confuse Y's situation with his interlocutor's situation. Suppose Y says to X, "I heard on the BBC that α". Suppose that Y distrusts the BBC for its blatant partisanship, especially on α-like matters. Then even if Y adds the clause "but, of course, nobody can trust the BBC on this sort of thing", it is possible that X will nevertheless believe he BBC's report, and not accept his informant's reservations. Accordingly, if as it happens X's belief is well-produced on good information and reported report α is true, then X's knowledge was transmitted by a teller who lacked it.

second case doesn't likewise offend it. In the first instance, the reliability assumption would be the culprit, and in the second told-knowledge.

WFK is a principle of epistemic justification. It disqualifies bootstrapping justifications on grounds of circularity. I am not so sure that the induction that Vogel condemns is actually a genuine instance of bootstrapping. Whether it is or not need not concern us here. If it is bootstrapping, the induction is not well-justified. If it isn't bootstrapping and yet the induction is still defective, it is not so by virtue of its discomportment with WFK. Present considerations converge on an essential question. Can such inductions be accommodated by the right model of knowledge?

For me, knowledge in general is not well-catered for in CC-approaches. So the question that attracts my interest is what the CR-model makes of particularities of situations like Roxanne's. The pivotal fact of the Roxanne case is that she can't with pragmatic consistency take herself as coming to know that α on the basis of her source's telling of it while at the same time thinking that the source is not on that occasion a reliable teller. The claim that Vogel finds offensive is generated by such facts in sufficiently large numbers of cases to constitute a fair sample for the induction at hand.

How does the CR-model handle these pivotal facts? Let us go back to Harry. Harry tells you that the 2014 Winter Olympics are to be held in Sochi. This causes you to believe it. Could we say that if your belief-forming devices were in good working order, you wouldn't believe what Harry tells you on his telling of it unless you were also made to believe concurrently that Harry is a reliable teller? It seems that we could. What, on Vogel's analysis, is a *Moorean* impossibility is, on a CR-analysis, a *causal* impossibility imposed by the workings of your cognitive equipment. Telling couldn't be efficacious if it were believed to be unreliable. A telling couldn't transmit knowledge if it were believed to be knowledge-nontransmitting. Do these causal facts transgress WFK? They do not. WFK is a principle of epistemic justification. These causal facts are facts about a certain way in which our knowledge is produced. It is a way that makes no claim on justification. It is a way that makes WFK an irrelevancy. That is, it makes it an irrelevancy unless, contrary to what I've been saying about the matter, justification is a condition on told-knowledge.

9.4 *The causal tug*

The causal tug of being told things is very strong. Consider the so-called *reason rule*. It isn't a rule strictly speaking. It is the statement of an empirical regularity.

> **Proposition 9.4a**
> THE REASON RULE: *"One party's expressed beliefs and wants are a* prima facie *reason for another party to come to have those beliefs and wants and, thereby, for those beliefs and wants to structure the range of appropriate utterances that party can contribute to the conversation. If a speaker expresses belief X, and the hearer neither believes nor disbelieves X, then the speaker's expressed belief in X is reason for the hearer to believe X and to make his or her contributions conform to that belief."* (Jacobs and Jackson, 1983, p. 57; cf. Jackson, 1996, p. 103).[315]

A lot of knowledge is passed along in conversational contexts. The reason rule reports an empirical commonplace in communities of real-life discussants. Where the rule states that one person's acceptance of a proposition is reason for a second party to accept it, it is clear that 'reason' means 'is taken as reason' by the second party, that is, would be deferred to if challenged. Thus a descriptively adequate theory will observe the Jacobs-Jackson regularities as a matter of empirical fact. This leaves the question of whether anything good can be said for these regularities from a normative perspective.

If the NN-convergence thesis is given standing here, the reason rule is also in good shape normatively, albeit defeasibly so. We said at the beginning that the NN-convergence thesis expresses a default. It stands except in the face of particular reasons not to let it stand. A fair question is whether told-knowledge (or would-be knowledge) is problematic enough to override the default. My answer to this question is No. It is what in this chapter I am trying to establish.

There is an obvious corollary to the reason rule.

> COROLLARY: *Human agents tend to accept without challenge the arguments of others except where they know or think they*

---

[315] Cf. Tyler Burge's Acceptance Principle: "A person is entitled to accept as true something that is presented as true and that is intelligible to him unless there are stronger reasons not to do so." (Burge, 1993, p. 467)

know or suspect that something is amiss, or when not resisting involves some non-trivial cost to themselves.

How does this work? It works in the following way. The regularity reported by the reason rule provides for the default that subjects of tellings accept the conclusions of arguments without challenge. This being so, it renders nugatory any critical examination of the argument's force. It makes the goodness of the argument moot. There are enormous savings in this.

9.5 *Told-knowledge dominance*

There are contexts in which it is advisable not to believe what we are told until various checkpoints have been successfully negotiated. In a criminal trial it is forbidden to convict on the sayso of the prosecutor. Suppose someone tells you that your colleague is a pedophile. Prudence, fairness and epistemic integrity alike require a certain caution. What, you rightly ask, is the *case* for the accusation? Sometimes the case for something that was told you takes the form of a case for the reliability of the teller. So these are two ways in which case-making may influence your belief that a colleague is a pedophile. You can see the documented evidence against him marshalled by the police. And you can be told it by someone whose reliability is not in doubt, someone who is in a position to know. (Perhaps she is a BBC news reader.) What is particularly striking is the dominance of the latter over the former. It is a dominance that reveals itself in two ways. One is that working one's way through the typically massive documentation of the police's case (supposing improbably that we had access to it) would be much more time-consuming and onerous than being told by someone well-placed to do the telling. This we might say is *economic* dominance. The other is that even where the police's case is open to our inspection, most of the facts of the case are themselves the fruit, not of case-making but rather of telling. Call this the dominance of *sayso*. Virtually without exception – apart perhaps from instances of non-representational, non-propositional knowledge – non-testimonial support for a proposition embeds components that depend on sayso. The police case depends on what was said by witnesses and experts. And what an expert knows he knows largely on the basis of what is published by other experts, by the writers of textbooks and the authors of peer-reviewed scientific papers. So we have a further fact to take official notice of:

Proposition 9.5a

SAYSO INELIMINABILITY: *Most of what we know depends ineliminably on something being told to someone.*

Proposition 9.5a introduces the promised notion of *distal telling*. Consider a chain of tellings originating with agent X and passed on recurringly from Y, to Z, and so on and on, eventually coming to you. What makes X's telling distal is his distance from you in the chain. What makes a distal telling of α interesting is that you couldn't have come to have known α without X's distal telling of it, never mind that X didn't tell it to you proximately. Proximately telling then is not a transitive relation. Suppose that you produce a proof of something α that relies on Lindenbaum's Lemma. You've been acquainted with the lemma for a long time, and you have no clear memory of how you first came upon it. Did a teacher tell you it? Was it in one of your textbooks of long ago? You don't know. Suppose you were told it by your teacher. Suppose he gave a nice little proof of it. Unless your teacher was Lindenbaum himself, someone or some textbook in turn will have told it to him. Even in Lindenbaum's own case, most of the premises of his proof he will have learned from someone or somewhere, and at least some of those learnings will have originated in still prior tellings. That is your situation, too. Still, someone told you something distally. But for this to be true you need not have the slightest idea as to the identity of the first member of that telling-chain. So your knowledge is a product of distal, not proximate, telling. If your own proof succeeds, you now know that α notwithstanding that no one told you it. Still, you couldn't possibly have attained your knowledge that α without the distal tellings that your proof relied upon.

In section 9.1, we spoke of proximate tellings, and said that their distinguishing feature is that there is a direct informational link between source and recipient. But in those cases, too, the source's command of what he tells you expressly will nearly always be dependent on what *he* has been told distally, and this in turn will have originated in someone or other being told something or other proximately. In each case the tellings at hand, proximate or distal, are typically the termini of chains of prior tellings, both proximate and distal. So what is distinctive about your being told something distally is that someone told someone else something and the someone else was not you.

9.6 *Multiagency*

The charm reproduction example introduces the idea of multiagency. It is an idea that explodes the myth of the scientist as romantic solipsist, as executor of his genius in performances of solo bravado. It is a silly idea, more attractive than true. All intellectual and creative effort imposes the need for solitude. But it is a transient need, not a systemic one. Much of science is done by teams, and all of it depends on the literatures that record the state of play up to now, the combined efforts of all who went before. Joint involvement is an important aspect of science, but not by any means peculiar to it. It is a feature of cognitive practice across the board. As Quine once noted, we all begin our cognitive lives in *media res*.

Group participation and community involvement are forms of multiagency. The concept of multiagent arises in AI.[316] A multiagent system is a composite of individual agents working interactively according to some operational plan or in fulfillment of some conventional arrangement. When a multiagent's goal is to achieve a knowledge of something, or to reach an improved understanding of it, or to arrive at a decision about what should be done, we may characterize it as a cognitive multiagent. In some cases, multiagency is an additive combination of its separate parts. In others, it is an emergent fusion of subsets of its parts, a cohesion of "the mangle of practice."[317] Plans typically take one or other of two basic forms. In one, the individuals act cooperatively in quest of some overall goal, each providing a partial contribution to the task at hand. In the other, the participating agents operate competitively, either one-by-one or in subgroups, where the finalization of the process lies in the hands of just one of the individuals or some small group of them. Other forms of multiagency involve a blend of the additive and emergent paradigms. Multiagency has also attracted considerable attention from

---

[316] The scope of its research probes is impressive: Distributed problem solving, non-cooperative game theory, multiagent communication and learning, social choice theory, mechanism design, auctions, coalitional game theory, logical theories of knowledge, belief, and other aspects of rational agency, probability theory, classical logic, Markov decision processes, and mathematical programming.

[317] Quoting from Perkins (1985). Recent and important is Shoman and Layton-Brown (2009). See also Floreano and Mattiussi (2008); Eiben and Smith, (2010); and Tettamanzi, Thomassini and Jansen (2010). Also in the works is d'Avila Garcez *et al.* forthcoming.

psychology and learning theory,[318] and comparatively recent developments in epistemic logic have subjected the concept of intelligent multiagency to interesting, and sometimes rather powerful, formal developments.[319]

In their competitive formats, a distinction is often marked between combatants (sometimes called "egos") and superagents ("superegos"). Superagents are not themselves combatants, but rather adjudicants of the interagency contests in which the others are involved.[320] Aside from the large, and growing, multiagency literature which one finds in AI and psychology, valuable insights have been developed by philosophers of science, especially as regards the role of individual agents in multiagent contexts.[321]

Cognitive multiagents execute both knowledge transmission and knowledge discovery agendas. Multiagency of both types resolves into a further pair of contrasts – "facilitative" and "essential". A multiagent facilitates a cognitive outcome when the task at hand is more easily or efficiently performed by individuals working cooperatively or competitively rather than solo. Multiagency is essential to a cognitive outcome when in some contextually appropriate sense of the term it is not possible for the outcome to be achieved at all except for the combined

---

[318] See, for example, Hutchins (1995), and Salomon, (1997). Pacuit, Roy and van Benthem (2011).

[319] See here Vanderschrsaf and Sillari (2007), Meyer and van der Hoek (1995); Fagin, Halpern, van der Hoek and Kooi (1995); Shoham and Leyton-Brown (2009); and van Benthem (2004).

[320] Trial by jury is a good, and well-studied, example of superagency at work. The origin of the competitor notion of multiagency is the Greek notion of *elenchus*, developed by Aristotle in his early logical work, *On Sophistical Refutations*. Closer to home, and more recent, is J.S. Mill's championship in *On Liberty* (1859) of parliament as the marketplace of opposing ideas. Thus, early and late, a theoretical framework for the management of such markets is dialogue logic. Game theoretic logic would come much later. The papers that connected the mathematical theory of games to set theory are Gale and Stewart (1953), and – to logic – Henkin, (1961).

[321] For game theoretic approaches to the epistemology of science see Kitcher (1993) and Zamora-Bonilla (2006). On Kitcher's approach, the relations between collective rationality and the decisions of individual agents are modelled as optimal division of cognitive labour. Kitcher gives conditions under which individual rationality can depart from collective rationality. Zamora-Bonilla models the interplay of individual scientists as persuasion games, and builds a framework which integrates "rationalist" and "constructivist" views on science.

317

contributions of the participants. When this condition is met, let us say that the multiagency in question is "causally essential" in that context. Facilitative multiagency we can call "causally contingent".

The properties of causal essentiality and causal contingency cross the type-token distinction. Some actions are of such a type that every performance of it, hence every token of it, requires a multiagent. Others admit of tokens in which multiagency is a causal contingency.

9.7 *Scientific multiagency*

It is a while since we've heard from our visiting friends from afar. Let's get back to them now. There is among their human subjects the quite common belief that there are critical differences that obtain between the epistemic practices of science and those of Harry. One is that Harry is not positioned to rise to the epistemic standards of science. He lacks the talent and the training. Another is that to a considerable and principled extent Harry depends on sayso for his knowledge of what the scientist achieves by way of proof and demonstration. Since sayso lacks the steadfastness of demonstration, Harry is at risk in a way that the scientist is not. For Harry has a sayso-confirmation burden which the stand-alone rigours of demonstrative science are able to get along without.

When the visitors became aware of these sentiments, they could hardly believe their ears. In short order, they were ringing the changes of the proofs of Fermat's Last Theorem and Poincaré's Conjecture. Here is a much condensed version of what they found.[322] A special case of Fermat's Theorem encompasses what are called Pythagorean triples, whose investigation stems from Babylonian times. The theorem embeds a Diophantine equation, whose investigation stems from the third century. The theorem arises in the context of Fermat's investigation of Diophantus' sum-of-squares problem. Fermat himself proved the theorem for $x^4 + y^4 = z^4$. In the period 1676-1966, no fewer than twenty-six proofs were produced for $n = 4$. From 1637 to1839, the theorem was proved for the three odd prime exponents $p = 3$, 5 and 7. (Indeed, $p = 3$ was first stated by Abu-Mahud Khojandi in the 10th century.) In the interval in question, there were at least sixteen proofs, some of them defective, for $p = 3$, eight for $p = 5$, and seven for $p = 7$. Cases $p = 6$, 10, 14 have been

---

[322] I am told these things by Singh, (1998). See also ⟨http://en.wikipedia.org/ wiki/ Wiles's_proof_of_Fermat's_Last_Theorem⟩.

served by no fewer than fourteen proofs. In the early nineteenth century, Sophie Germain verified the theorem for all odd prime exponents less than 100. Further essential work was done in the decades that followed. Of particular note is Ernst Kummer's verification of the Louis Mordell proposal that Fermat's equation has at most a finite number of principled primitive integer solutions when $n > 2$. This was proved by Gerd Faltings in 1983. In the second half of the twentieth century, Kummer's insights were refined and extended – with the aid of a SWAC computer – for all primes up to 2521, and by Samuel Wagstaff, who stretched the result to all primes smaller than 125,000. In 1993, this was extended to all primes less than four million by Bubler, Crandell, Ernvald and Metsäkylä.

The basic strategy for proving the theorem outright was Gerhard Fey's modularity theorem of 1984, with subsequent improvements by Jean-Pierre Serre and, in turn, Ken Ribet's proof of the so-called epsilon conjecture in 1986. A special case of Fey's theorem was proved by Wiles in 1995. There remained to prove the Taniyama-Shimura conjecture for semistable elliptic curves. Using, first, the resources of Horizontal Iwasawa theory and, subsequently, those of an Euler system developed by Victor Kolyvagin and Maatthius Flack, and with the specialized help of Nick Katz, Wiles eventually arrived at a result which he announced in 1993. The proof was flawed and, notwithstanding the help of Richard Taylor, was abandoned. In 1994 Wiles returned to a redeployed Horizontal Iwasawa theory, and the task was completed. Fermat's Last Theorem was now proved after three hundred and fifty-eight years of sustained mathematical effort.

Much the same story can be told about the solution of Poincaré's Conjecture.[323] Again, Poincaré's Conjecture proposes that every simply connected, closed three-manifold is homeomorphic to the three-sphere. Poincaré raised the question in 1900, but it wasn't seriously taken up until by J.H.C. Whitehead in the 1930s, who advanced and then retracted a proof. Other flawed proofs emerged in the 1950s and 60s, and various attempts to refute the conjecture were also made. A weak version of the conjecture was proved by R.H. Bing in 1958. Further failed attempts to broaden the finding were tried in the 1980s and 90s.

In 1961 Stephan Smale proved the Generalized Poincaré Conjecture for dimensions greater than four. This theorem asserts that a homotopy n-sphere is homeomorphic to the n-sphere. In 1984 Michael Freedman proved it for dimension four. The eventual solution was given

---

[323] I am told these things by Cornell, Silverman and Stevens (1997) and Morgan and Tian (2007). See also ⟨http:en.wikipedia.org/wiki/Poincaré_conjecture⟩.

a considerable boost by a programme of work initiated by Richard Hamilton in 1982. Hamilton showed how the so-called Ricci flow on manifolds can be used to prove special cases of Poincaré's Conjecture. Hamilton's efforts to extend these results to the general case were not successful.

In his three major papers, Grigori Perelman outlined a proof of the conjecture, as well as of a more general one, known as Thurston's geometrization conjecture, and brought to consummation the Ricci flow program launched by Hamilton. Perelman's contributions were sketches. Omitted details were furnished in turn by Kleiner and Lott, Huai-Dong Cao and Xi-Ping Zhu, and in 2008 by John Morgan and Gang Tran. Whereupon 106 years later, Poincaré's Conjecture was proved.

It is impossible to say with any precision which of these antecedent mathematical events – those that succeeded and those that failed in telling ways – figured in the Wiles and the Perelman breakthroughs. It is impossible to say with any precision which of these latter were objects of Wiles' and Perelman's propositional attitudes, as opposed to the subjects of their background presumptions. All the same, two things can be said with some confidence. One is that only a comparative handful of these elements, if any, was renegotiated *ab initio* by the provers of these theorems. Wiles' proof runs to over 100 pages, and draws upon algebraic geometry and number theory, including the category of schemes and Iwasawa theory. It expressly calls upon named contributions from Felix Klein, Robert Fricke, Adolf Hurwitz, Erick Hecke, Barry Manzur, Dirichlet, Dedekind, Robert Laglands, Jerrold B. Tunnell, Jun-Ichi Igusa, Martin Eichler, André Bloch, Tosio Kato, Ernest S. Selmer, John Tate, P. Georges Poitu, Henri Carayol, Emil Artin, Jean-Marc Fontaine, Karl Rubin, Pierre Deligne, Vladimir Drinfelsd and Haruzo Hida. The very idea that Wiles would have re-proved the army of results thus drawn upon is ludicrous. Clearly, they were available to him on some other basis. How could Wiles know the truth of all this borrowed mathematics short of being told it? How could he deploy it here short of re-telling what he was told?

Boethuis once said that it is our lot to stand on the shoulders of giants. Boethius was right. It is a fruitful connection which takes me to a second observation. It is widely conceded that in 1994 hardly any mathematician – never mind anyone else – could follow Wiles' proof even in its essential details. It is true that this small circle is though to have enlarged somewhat with the publication of the proceedings of the Cornell Conference in 1995. But on *any* interpretation of "understanding

Wiles' proof", the probability approaches certainty that the mathematical multiagencies which possess such information are to a nontrivial degree both opaque and causally essential for its provision. That it is to say, it is probable to the point of certainty that, for each of a hefty majority of participants in this multiagency, there are aspects of the proof that are not well-enough understood to count as knowledge, save in the Anselmian sense. Short of the unattractive conclusion that hardly anyone – if anyone at all – knows the Wiles proof, Anselmian knowledge must be given its head, and with it the acknowledgement of the dependence of even the most rigorous parts of mathematical knowledge on sayso.

The visitors were satisfied that this defeats the canard that scientific knowledge differs from ordinary run-of-the-mill knowledge by virtue of science's comparative freedom from and ordinary life's bondage to sayso.

A finishing touch to the dominion of told-knowledge thesis was the necessity of being told things, and of witnessing others being told them too, in the acquisition of language. Accordingly, any knowledge requiring the possession of a language likewise falls under the sway of sayso. If we yield to this pull of Anselminism, we will need to adjust the CR-characterization of knowledge.

Proposition 9.7
ANSELMIAN KNOWLEDGE: *Harry knows that α when α is true, Harry believes that 'α' expresses a true proposition, and his belief is well-produced.*

9.8 *Sayso-manifolds*

The institution of telling bears in an interesting way on the question of multiagency. There are certain epistemic states that cannot be brought about except for the response an agent makes to a telling. Told-knowledge requires a teller of it and a told. Told-knowledge is a shared phenomenon. Examples such as the charm photoproduction case might have inclined us to think that causal essentiality and opacity are peculiar to the large and diverse multiagencies. But simple two-person tellings put the lie to this, at least for those cases in which the function of the telling isn't to remind the addressee but to inform him – to tell him something he didn't know before. By construction of the cases, the telling is causally essential. In the context at hand, Harry wouldn't have known that Barb's birthday was on the 3rd had Sarah not told him so. Indeed, there is no context in which Harry could have acquired that knowledge without someone's (or something's) having told him it. Equally, whenever someone is on the receiving end of a causally efficacious telling of α, there is one or other of two conditions in which the addressee will have been put. He will be made to have a *de re* belief with respect to α, or a *de re* belief with respect to the proposition that α expresses a true proposition, combined with a *de dicto* belief with regard to α itself.

Tellers and tolds constitute epistemic-transfer pairs. Epistemic-transfer pairs are causally essential multiagents. They are causally essential to the production of told-knowledge. Tellings are both proximate and distal. Sarah tells Lou that α to his face, hence proximately. Again, Sarah got it from a book, the author of the book from a teacher, the teacher from a journal, the journal author from a proof some of whose premises he got from another book, all distal tellings for Lou. Since telling is implicated in everything or virtually everything that Lou or anyone else will know, Lou and they are at some remove or other participants in numberless sayso-multiagencies and the various and copious merelogical compositions thereof. When Sarah tells Lou to his face that α, *manifolds* of distal tellings converge on this event. If Sarah passes on to Lou the fact that every simply-connected three-space is homeomorphic to the three-sphere, her proximate telling is lodged in chains of distal tellings that include, among much else, the historical unfolding of the Conjecture's eventual proof, as well as the history of tellings that get Lou in fit intellectual shape to actually be informed by what Sarah tells him. It takes more than Sarah and Lou to make it the case that he now knows that α because she told him it. It takes the histories of tellings that Sarah brings to this exchange, and the history of tellings that makes Lou receptive to it. This is getting to be quite a lot of tellings. In recognition of the point let us say that Sarah and Harry's face-to-face exchange is a fixed point in a *sayso-manifold.*

Proposition 9.8a
SAYSO-MANIFOLDS: *A sayso-manifold with regard to an agent's knowing that α at time t is the spacetime worldline aggregating all the tellings causally implicated in his knowing at t that α.*

In the light of this, the visitors arrived at some further points of agreement.

Proposition 9.8b

THE MULTIAGENT THESIS: *Sayso-manifolds induce cognitive multiagencies made up of the teller-told pairs of the tellings from which the manifold is constructed.*

Proposition 9.8c

THE CROSS-TYPE THESIS: *Sayso manifolds cut across the distinction between scientific and non-scientific multiagency* [324]

Consider everything that Andrew Wiles knows about mathematics. Consider all the prior tellings causally implicated in this knowledge. Call this the Wiles manifold with respect to what he knows of mathematics or the Wiles m-manifold. It is a manifold that comprehends a vast chunk of the history of the teaching of mathematics over the ages. It is a manifold causally implicated in the establishment of Wiles' proof of Fermat. Wiles stood, in the manner of Boethius, on the shoulders of the m-manifold. The multiagent defined by the m-manifold contains Wiles himself. Call this multiagent M^w. Of course, it is true to say that Wiles brought about the proof of Fermat's Theorem. But it is truer to say that M^w did. I said at the beginning that I would concentrate on the cognitive and reasoning practices of the individual agent. As we saw, there is scarcely any part of the exercise of an individual's cognitive agency that is not transacted within a partnering multiagency.

---

[324] Except for some modest mention by formal epistemologists, the formal dynamics of told-knowledge haven't had much of an innings in the philosophical mainstream. Of most direct consequence for epistemology is the theory of telling investigated by public announcement logic (PAL). It originates in Playa's (1989). PAL is just one of many logics that model the dynamics of knowledge and belief in multiagent settings. PAL is a relatively simple extension of epistemic logic (EL) got by the addition of dynamic operators on formulas. PAL is a logic that permits agent's knowledge states to be updated by the public announcement of epistemic formulas. Thus PAL extends multiagent EL so as to models and the communicational consequences of announcements to multiagents. Each formula of PAL can be rewritten as an equivalent formula of EL. An advantage of the reduction is that PAL is intrinsically more succinct than EL. There is a rapidly growing PAL literature. Important papers are, among others, Baltag, Moss and Solecki (1998); van Ditmarsch, van der Hock and Kooi, (2005), van Benthem, van Eijck and Kooi (2006); and Kooi, (2007), and French and van Ditmarsch (2008).

9.9 *What Fermat knew*

We were wondering earlier whether a mathematical theorem could be known in the absence of a proof of it. Let us briefly revisit the question. Our visitors were interested in Fermat. They wanted to determine whether Fermat *knew* that except where $n = 2$ there is no solution for the equation $x^n + y^n = z^n$. Putting aside its *credo/intelligo* complications, the visitors noted that pretty well everyone agrees that Andrew Wiles knows this. He knows it, it is supposed, precisely because he has a proof that shows it. But consider Fermat's situation. In the most famous marginalia in all of mathematics Fermat says that he has a proof of this theorem. Perhaps he did. But for the purposes at hand the visitors were ready to assume not. Fermat was not a silly man – anyhow not a silly mathematician. He was a well-experienced mathematician, a highly talented, indeed a gifted practitioner. He was a mathematical virtuoso. Fermat wrestled with his theorem, and in time became convinced of his truth. When the time came, his belief-forming devices with respect to the proposition expressed by his theorem actually fired. There is no reason to suppose that they were other than in good working order, and none other to fear that some hostile externality was in play. So, on the CR-model, Fermat knew to be true the proposition his theorem embodied, and did so notwithstanding his inability to show it so. This, anyhow, was the visiting investigators' conclusion.

Notice that if Fermat did know the facts forwarded by his theorem, it is not the case (or not at all likely) that he knew it in the sense in which Harry knows that $e = mc^2$. That is, Fermat's ken was not restricted to his *de dicto* knowledge that here is a formula that expresses a true proposition of number theory. Unlike Harry in relation to "$e = mc^2$", Fermat full-well understood the sentence, "There is no positive solution of $x^n + y^n = z^n$ for values of $n > 2$". If the CR-model is right, then Fermat had *de re* knowledge of that very proposition. He knew it if it is true (as it indeed is), he believed it (as indeed he did), his belief-forming devices were in good order (as indeed they were) and not in this instance mal-produced. It is not in the least unlikely that for someone like Fermat all these conditions could be met in the absence of proof. If so, a rather large consequence emerges as a kind of bonus. The multiagent known loosely as "modern day number theory" does *itself* know Fermat's theorem in the absence of a proof it cannot wholly (or even substantially) understand.

The same is true for a good many of us, not number theorists but sufficiently numerate to understand what the theorem says.[325]

This is a significant development. It reveals a confusion about the showing-condition routinely associated with the technical arcana of scientific knowledge. If in such cases showing is restricted to proving, then the community of knowers of such matters shrinks alarmingly, to the point of big-box scientific scepticism. This the visitors were prepared to treat as a *reductio* of the present assumption, in which case, showing that α is not in general a condition on knowing it, even if showing it *is* possible.[326]

9.10 *Dependency elimination?*

It is sometimes pleaded, especially in the CC-community, that these attestative dependencies are removable *in principle*, that if we went back far enough (and tried really, really hard) we could come upon all the truths of, say, pathology, or of experimental psychology, without recourse to anyone's sayso. It takes little reflection to see that any such notion is Enlightment confabulation of the first order. Claims about what could be done "in principle" are a deep and natural harbour of confusion. They embody tricky ambiguities. Suppose there is some language that you can't read – Aramaic, say. But you *could* read it. That is, it lies within the

---

[325] The visiting team was not unmindful of metaphysical reservations about mathematical knowledge in general. They acknowledged the difficulty of how epistemic contact with abstract objects could be negotiated by concrete beings like humans. But this, they reckoned, was not a relevant consideration in the present case. For if the abstractness of its objects preclude Fermat's knowledge-without-proof of his theorem, still less could he have known it with the aid of the proof.

[326] The visitors also reflected on what they came to know as Benacerraf's Dilemma. If "our best" theory of truth tells us that mathematical sentences are made true by the properties and relations of abstract entities called numbers (or perhaps sets), and if "our best" theory of knowledge tells us that when someone knows of some x some fact about it, his knowledge arises from a kind of (possibly distal) causal content with x, how are these two theories to be made compatible with one another? While not making light of the dilemma, the visitors weren't much troubled by it. Certainly the CR-model doesn't require causal contact with the subjects of our knowledge as a general condition. It requires a causal contact between our beliefs and our belief-forming devices. But it was never their view that having a belief about x necessarily required causal contact with x.

reach of your actual capacities to learn it. It may not, even so, be possible for you to learn it, what with the heavy claims that already exist on your time and cognitive assets. So you can't learn Aramaic. That is to say, you can't learn Aramaic in fact, but you could learn if your circumstances were different. It is also true that you can't make all the observations and can't perform all the experiments which, successfully made and performed, would produce all of what is presently known of pathology or experimental psychology. It goes without saying, given your other commitments and the other claims on your cognitive resources, learning all of pathology or experimental psychology from the ground up is not on. Still, there is a difference. Neither is it within your capacity to learn it, your wholly unfettered capacity. It is not that you are too busy to learn it, or too tired. If you weren't busy in the slightest, and if you were bristling with energy and bursting with talent – with genius, even –it still couldn't be done. You won't live long enough, you won't remember enough, you won't have computational capacity enough, and so on. Learning all of experimental psychology from the ground up in the absence of proximate or distal telling of any kind, exceeds your constitutional limitations; it is denied you by how nature has put you together as a cognitive being. So, in the sense of the Aramaic example, you couldn't learn all of experimental psychology just by yourself *even in principle*.

9.11 *The* ad verecundiam

The epistemology of telling is helpful for a sensible handling of the *ad verecundiam* fallacy. As traditionally conceived of, the *ad verecundiam* fallacy is a mistake of reasoning. But, as I have been saying, most of the errors arising from misinformation are errors of belief not errors of reasoning. The *ad verecundiam* arises in a different and more natural way. When someone tells you something and you believe it, there might be a challenger lurking by. "How do you know?" she asks. "Because he told me", you reply. In so saying, you have pledged yourself to an argument in the following form, which for ease of reference I'll call the "Sayso-Schema":

1. Source s told it that α

The traditional view is that the *ad verecundiam* fallacy is committed under precisely those conditions on which the Sayso-Schema would be a bad argument.

In Locke's treatment, an *ad verecundiam* argument is one of four strategies for rebutting an opponent in an argument over some disputed matter. So understood, the *ad verecundiam* is a premises-introduction manoeuvre. In executing it, one party proposes to the other that there is a proposition α which the two of them should accept, and that what makes it so is that it has been attested to by a person of "eminency", that is, a person whose superior learning places him in a position to vouchsafe it. Typically, α is a premiss favourable to the *ad verecundiam*-maker's side of the dispute, and the point of his *ad verecundiam* move is to break down his interlocutor's resistance to it. Locke formulates the move in an interesting way. I paraphrase as follows:

3. Eminent person X attests to α.
4. It would be *immodest* or *impudent* of us not to accept X's attestation.
5. So *modesty* requires that we accept α.[327]

The Latin "*verecundia*" means modesty.[328] It is something of an embarrassment that fallacy theorists persist in translating it as "authority". Still, it is clear that an *argumentum ad verecundiam* is indeed an argument from authority, from the authority of persons of superior knowledge, from the authority of experts – "men of parts".

It is significant that Locke's discussion lends no support to the modern habit of treating *ad verecundiam* arguments as fallacies. The conception of *ad verecundiam* that has come down to us is not the Lockean conception. It specifies conditions whose non-fulfilment would make the Sayso-Schema a bad argument. Of particular importance are the following two, whose formulation I borrow from Tindale:

---

[327] Here are Locke's words: One way of arguing "is to allege the opinions of men whose parts, learning, eminency, power, or some other cause has gained a name and settled their reputation in the common esteem with some kind of authority. When men are established in any kind of dignity, it is thought a breach of modesty for others to derogate any way from it, and question the authority of men who are in possession of it. This is apt to be censored as carrying with it too much of pride, when a man does not readily yield to the determination of approved authors which is wont to be received with respect and submission by others; and it is looked upon as insolence for a man to se up and adhere to his own opinion against the current stream of antiquity, or to put it in the balance against that of some learned doctor of otherwise approved writer. Whoever backs his tenets with such authorities thinks he ought to thereby carry the cause, and is ready to style it imprudence in anyone who shall stand out against them. This I think may be called *argumentum ad verecundiam*." (Locke, 1690/1961, volume 2, pp. 278-279)
[328] Also shame and shyness.

Proposition 9.11a

THE RECEIVED CONCEPTION: *The Sayso-Schema avoids the* ad verecundiam *fallacy when (a) source "is identified and ...[has] a track record that increases the reliability of the statements [attested to] over related statements from sources that do not possess the expertise .... Appeals to unidentified experts with unknown or weak track records can be judged fallacious"; and (b) "direct knowledge [of the matter attested to] could be acquired by the [recipient of the testimony], at least in principle. That is, there must be some way of verifying the expert's claims.* (Tindale, 2007, pp.135-136))[329]

Tindale's words emphasize the role of experts in sayso-knowledge transmissions. It has long been recognized that the Lockean man-of-parts conception of tellers generalizes in a natural way to cases in which a person's testimony is sanctioned by his position to know, not necessarily his learnedness. (The woman who tells you the whereabouts of Central Station needs no PhD in urban geography.)[330] This being so, it is clear that a satisfactorily comprehensive treatment of the modern concept of it requires that expertise be generalized in this way. There is a further reason. All expert tellings conceived of in the narrow man-of-parts way, depend on tellings both proximate and distal, very many of which arise from tellings that are not in this same way expertly sourced.

Other logicians anticipate Tindale, both before and after Hamblin. Salmon (1963) judges it necessary to ascertain that the "vast majority of statements made by [the source] concerning [the matter at hand] are true." (p. 64) The requirement of a reliability check as a condition of one's epistemic response to sayso is not limited to straight-up acceptance or straight-up belief. Rescher (1976) also makes it a condition on what he calls plausible reasoning, where the plausibility value of a piece of information has a principled connection to the reliability of its source.[331]

---

[329] I note in passing the absence of the *ad verecundiam* in the latest pragma-dialectical treatment of the fallacies. See van Eemeren *et al.* (2009).

[330] Woods and Walton (1974). Reprinted in Woods and Walton (1989/2007).

[331] Rescher (1976) is a pioneering contribution to the logic of plausible reasoning, but he is as wrong as the fallacy theorists are about the across-the-board performability of reliability checks. For a critical examination of Rescher's plausibility logic, see again Gabbay and Woods (2005), pp. 222-238.

It is instructive to compare actual cases of the exercise of due diligence with the conditions Tindale imposes. Consider a case. On its six o'clock news programme the BBC has just reported that Ace Petroleum has drilled a significant find off the Alaska Panhandle. You reach for the phone with a buy-order for your broker, but then you pause. The BBC is an institution on which you have relied for this information and for the course of action it prompts you to take. But have you established the *bona fides* of the BBC? Have you, in the spirit of Salmon, determined that the vast majority of statements made by the BBC concerning the matters it reports on are true? Have you, in the spirit of Tindale, confirmed that the BBC has a track record that increases the reliability of its statements over comparable statements from sources without the BBC's expertise (Sky TV, perhaps, or Ariana Huffington's blog)? How would you go about doing this?

There is little point in belabouring the obvious beyond observing that you wouldn't do it and, even if you wanted to or thought you should.

Proposition 9.11b

DUE-DILIGENCE EMBUBBLEMENT: *The whole structure of your due-diligence determinations would have the same epistemic character as the assurances you are trying to validate.*

How would you know that the BBC is a reliable reporter of the news? Could the answer be that it lies in the ethos of the media's news organizations to be reliable and trustworthy? Yes; but how do you know this? Perhaps you have a producer-friend. Perhaps your friend attests to this ethos. Perhaps he assures you of the BBC's integrity. But do you know – and if not, could you come to know what Salmon requires you to know – that the vast majority of statements made by your friend about matters of interest to you are true? What is the sample on which this generalization is based?

Considerations of due diligence raise two interesting questions. One is the question of what triggers, or should trigger, a due-diligence search. The other is what connection, if any, obtains between a due-diligence requirement and a justification requirement for sayso-acceptance? Let us begin with triggers.

9.12 *Triggers*

Suppose someone has told you something and that you neither believe nor accept what you've been told. It is neither necessary nor sufficient to the triggering of a   search that the telling failed to get you to believe α or to accept it. If what your informant told you is something you already knew, his telling didn't get you to believe it, and yet he might be a fully reliable teller of like things, and you might know it. Equally, he might be wholly out of his depth with respect to these matters, and you might know that too. Or you might already know what he tells you is false. When, for this reason, a contrary telling fails to induce the belief, this needn't be accompanied by the belief, or even the suspicion, that your informant lacks the *bona fides* for reliable pronouncement about things of this sort. No one thinks that reliable tellers are infallible. Even the person who knows what he is talking about makes mistakes. There is plenty of room in the present kind of case for you to think that your informant is a genuine expert about α-like things who made a mistake with respect to this particular α.

There are also cases in which it is a plain fact that an addressee will accept what a teller tells him while concurrently harbouring doubts about the teller's *bona fides*. A quite common linguistic indicator of this sort of cognitive dissonance is the addressee's seeking assurance from the very person whose *bona fides* are in doubt. ("Are you really sure that α?"; "How could you possibly know that?"). This is a factor of critical importance for a correct understanding of the modern version of the *ad hominem,* a point I'll take up in due course.

It is extremely difficult to get a grip on the triggering conditions for due diligence exercises. Not only is so much of what we know been told us or our distal chains, but most of the tellers are either anonymous or persons whose track records are unknown to us. It is perfectly true that there are more or less stateable constraints on when sayso would be sought out and when a negative judgment on sayso would be arrived at. But these are not our questions here. What we are after is an understanding of the conditions under which due diligence exercises are *occasioned*, something that precedes both the seeking of the requisite sayso and a negative judgement of it (or any). When lost in Amsterdam, you ask a passerby for the whereabouts of Central Station, you are multiply-cued by contextual factors. You are also guided by an observation loosely in the form "She has the look of someone who would know". "In her bearing and in how she replied, she has the look of someone who knows" is a default, it embodies a generic generalization. It decomposes into something like "She has the look of a local" and "Locals know the whereabouts of things like railway stations". Besides, the very fact that she would answer implicates that she knows it.

Checking an expert's *bona fides* is not to be confused with the quite separate matter of identifying a source as an expert. It is one thing to seek the advice of a medical expert or an income tax expert or an encyclopaedia. It is another thing to demonstrate that the expertise a source claims for himself is indeed the expertise he actually has. In the large majority of such cases, confirmation that the source is an expert is *self-attested*, a vivid example of sayso enchainment. We identify Dr. Zed as a qualified rheumatologist on Dr. Zed's own sayso, whether his listing in the Yellow Pages or in the Directory of the Medical Building, or on the sayso of our general practitioner, who in turn has had it in these same ways from Dr. Zed. If we were sceptical (or nosy), we might search Dr. Zed's walls for duly framed diplomas and degrees, notwithstanding that the assurances they convey are themselves subject to self-attestation. A diploma is a document that says of itself that it is a diploma. We should not be made nervous or embarrassed by our utter openness to these commonplace indications of expertise. They are not in the least stand alone instruments of reassurance, but do their work within a rich network of relevant defaults: "Medical schools don't give false assurances", "People wouldn't risk legal redress by misrepresenting themselves in this way", "The Yellow Pages wouldn't publish fraudulent information", and so on. Besides, that they would publish these attestations implicates their authenticity.

Confirming a source's *bona fides* is like running a credit check on an applicant for a mortgage, or executing a title search in a house-buying exercise. Perhaps closer to home is the job interview, in which, if properly conducted, a good deal of due diligence is performed. But here too the sought for assurances are themselves deeply embedded in prior sayso, albeit not exhaustively. True, a "job-talk" for an academic appointment is a display of expertise rather than a claim to it, but it is nevertheless replete with assertions offered to the audience on the basis of the speaker's sayso. The more important point however is that

Proposition 9.12

TRIGGER RARITIES: *As a matter of empirical fact most by far of our actual acceptances of sayso are not induced by reliability confirmation triggers. And even if it were otherwise, the performance requirements for such enquiries would be largely beyond our ken.*

COROLLARY: *There is a considerable economy in this.*

Identification and verification have significantly different structures. It suffices for the *identification* of Dr. Zed as an expert rheumatologist that I hear it said on the BBC. But it does not suffice for a *demonstration* of his expertise that I hear it said on the BBC that he is an expert. Identifying him as an expert requires little more than having good reason to believe that he possesses the right formal qualifications. Demonstrating possession of the expertise which those credentials attest to is a different matter, and a more difficult thing to bring off. Whatever the details, it involves coming to know "of your own knowledge" (as the lawyers say) the nature and quality of his medical gifts. Nobody thinks that *this* can be done by the mere determination of credentials, and nobody thinks it can be done much short of acquiring something much closer to a first-hand acquaintance with Dr. Zed's methods. If this last is what is meant by the call for an independent confirmation of expertise, it falls well beyond what most people could conceivably manage. And, even for those who could bring it off – say by the BBC's own medical expert – the process of confirmation would be replete with sayso-dependency even so. Sayso-chains cannot be severed. We operate as cognitive beings by virtue of the sayso-manifolds of which we are undetachable constituents, by virtue of the multiagencies of which we are *epistemic particles.*

9.13 *Confirmational impediments*

The defaults that guide us in the ways in which we structure our reliance on others draw upon common knowledge. It is typical of what is known as a matter of common knowledge that it is accepted and acted on independently of a due diligence with regard to its *bona fides,* but also in the absence of an identifiable teller. It is true that often we get to know that something is a matter of common knowledge by being told so in some particular case. But it is not to the telling that the commonness of common knowledge is due. A detailed etiology of common knowledge would be at least as interesting to have as it would be difficult to provide. But what counts here, in both its purveyance and its acceptance, is the radicalness of our alienation from the routines of the independent check. It turns out, then, that the confirmational impediments of teller-identified told-knowledge are but a particular case of this more general alienation.

An especially poignant instance of confirmational incapability is afforded by the CR-model. I say that Sarah knows that α if α is true, she really thinks that α, and her thinking so is brought about by cognitive devices operating as they should on good information and in the absence of hostile externalities. The debate between the CC and CR-models of knowledge further inclines me to think that it is not in general true that the conditions under which a person's cognitive devices are firing as they should are conditions that satisfy the knowledge-endowing criteria of the CC-model. If this is right, then independent direct confirmation of the across-the-board reliability of those devices is beyond the reach of even the most perfervidly utopian epistemologist. Consider again the case of perception. We say with a certain justified confidence that our perceptions are accurate when they are caused by perception-forming devices functioning as they should on normal inputs. But hardly anyone knows anything much about what those devices are, still less the conditions for their proper operation – their final cause, so to speak. There is nothing in the least surprising about Everyman's ignorance of the mechanics of perception. Proper functionalism with respect to perceptual knowledge could be true (and is), notwithstanding its attendant confirmational impotence. Proper functionalism with respect to knowledge more generally counsels a similar lesson for knowledge across the board. Having it depends on having cognitive devices that work properly. It does not depend on knowing that they do work properly or the conditions thanks to which this is so.

Confirmation is itself a common sense notion with a bristling technical side. In speaking just now of our confirmational incapabilities as regards the good workings of our cognitive mechanisms, care should be taken not to confuse the ordinary and technical meanings of "confirmation". If confirmation is understood in the sense of contemporary confirmation theory, then for the individual agent on the ground there is little that he can confirm, never mind that his cognitive engineering is in apple-pie order. In the less technical sense, things are not quite as bleak. At a certain level of generality, it seems right to say that, like your perceptual equipment, your cognitive engineering is working right when you are *healthy* and *injury-free*. There are lots of instances in which the state of your health is beyond your ken. Your blood-pressure can soar without your knowing it and, up to a point, your reflexes can dissipate unawares. But you are far from being totally in the dark about the state of your health. No doctor would ask how you are feeling if he thought that you couldn't tell him. The question, "How are your eyes, since you were in last?", or "Are you still tasting your food?" are answerable questions. So too are: "How's your memory been?" "Are you having difficulty with the monthly cheque-book balancing?" Even the general question – say after a head injury – "Are you still thinking straight?" is something to whose answer you can usually make a nontrivial contribution. Lying behind these competencies is an important assumption:

Proposition 9.13a
SYMPTOMS: *To a considerable extent an agent's equipment-breakdowns and other deficiencies produce symptoms discernible to the agent himself.*

Proposition 9.13a generates the further default that

COROLLARY: *It is reasonable for an untutored individual agent to presume that his cognitive gear is in good working order in the absence of symptomatic indications to the contrary.*

What the individual cannot do is run a technical check on his equipment. He hasn't the time; he hasn't the wherewithal, and he hasn't the know-how. But this is not to say that he lacks all sense of how he is doing cognitively, far from it.

But surely, it will be insisted, a good part of our ability to determine when a belief-forming device is not working as it should is when it starts producing lots of false beliefs. So isn't there something to be said for the truth-tracking conception of the good-orderedness and proper functioning of a cognitive device? No. Again, the error abundance thesis casts doubt on the suggestion. Since we make lots of errors, and false belief heads the charts of error-commission, we have lots of false beliefs. But having lots of false beliefs doesn't indicate the breakdown of cognitive equipment. It is rather the *kinds* of false beliefs and the circumstances of their commissions that carry diagnostic significance. The general idea is easy enough to state, but details are elusive and difficult to knit together. The general idea is that of Harry's cognitive equipment were disordered, the rest of us – anyone privy to Harry's behaviour – would know it. This, of course, is an autoepistemic argument which, like all such arguments, pivots on the truth of its conditional premiss, "If Harry's cognitive devices were misperforming on the scale embedded in the error abundance thesis, people would know it (including Harry himself)." Getting at the truth conditions of such conditionals is no trifling matter, made so in part by the fact that they are underscored generic propositions, not universally quantified ones. Autoepistemic reasoning is the business of the chapter after this.

If we persist with the traditional conception of fallacy it might come back to us that the *ad verecundiam* really is a fallacy after all. But is it? A fallacy is an error that people in general have a strong disposition to commit with a notable frequency. A case can be made for its general appeal, but its status as an *error* is precisely what is in doubt here. It is an error if it is an error to subscribe to the Sayso-Schema in the absence of due diligence and sayso-elimination in principle. If what I have been saying here can be made to stand, both these conditions are failed on a *massive* scale. If this is the source of error, most of what we think we know we don't, and most of our knowledge-acquiring behaviour is a failure. Either way, the cognitive abundance thesis is ransacked, and the concurrence of the normal and the normative is gutted. Not even the most puritan of the purveyors of the traditional approach to the fallacies intended the near-wholesale wrecking of knowledge. But knowledge *is* lost if these *ad verecundiam* practices are errors. So I conclude, *modus tollens tollendo,* that they are not errors. Whereupon, another of the eighteen slips out of the embrace of the traditional concept of fallacy.

Proposition 9.13b
A NON-FALLACY: *The* ad verecundiam *is not a fallacy.*

9.14 *Exceptions to told-knowledge*

An important class of exceptions to told-knowledge are evaluative tellings, especially those involving matters of moral purport. The existence of this asymmetry is not hard to miss and should not be minimized. It is that concerning "matters of fact", the default position is that, if someone purports to be in a position to pronounce authoritatively, he is. However, concerning "matters of value", there is no such presumption. When it comes to ethics there is not much backing for the idea that people who tell you what to do know better than you yourself. Here, too, there are exceptions. Young children are still within the ambit of parental inculcation, and religiously observant adults as a matter of theology accede to the importunities of the clergy, who in turn as a matter of theology are importuned by God (who actually does know more about it than they).

There is a natural-seeming, if rather formal, generalization of this exception. It is that what a person knows passes to you by his telling you it except when his telling you it doesn't get you to believe it, when his telling you it does not move you from non-belief to belief; that is, when it fails *causally*. Although true, this is not of much help. It doesn't identify exceptions by type, and it pays no mind to the causal particularities of this non-responsiveness. These are matters to which we shall shortly return. For the present it is desirable that we take note of one of the benefits of generalizing in this way.

Other exceptions include perceptual beliefs and beliefs arising from "personal experience". You look out the window and exclaim, "Look, an ocelot!" Someone might reply, and you might believe him, "Not an ocelot; it's an orangutan" In so doing, he is not correcting your perception, he is not doubting that you see a hairy, four-legged primate there; he is correcting your belief as to what to call it. Or you say, "Barb has a sunny disposition" (after all, you have known Barb since college days), and someone – her husband perhaps – says that it's all a front. It is not that this would not change your mind. Perhaps it would, but it would probably not do without some resistance.

Also often excluded are theoretical beliefs in which their holders have significant stake, for whose demonstration they may have worked rather hard. Nearby, and partly overlapping, are beliefs which, in one way or another, have claimed our "doxastic loyalties",[332] some of which, in turn, may be rooted in faith. Then, too, there are the causal provocations of commercial advertising and the entrapments of propaganda, indeed, of all forms of belief-inculcation, including the innocent ones.

Other exclusions are less sensitive to considerations of belief-type. If Harry tells you that Barb's birthday is on the 3rd, you may well believe him, but if it were Lou who told you this, you might not. So it is teller-identity that is relevant here.

All these exclusions, and the others we haven't the time to chronicle here, provide the would-be theorist with manifold explanatory challenges. Much of the present-day literature on testimony is a response to just part of that challenge, and a contentious response at that.[333] Nearly

---

[332] Woods (2004), chapter 8, section 5.

[333] Of particular interest from the CR perspective is simulation theory's analysis of the attribution of mental states to others. See here Goldman and Sripada (2005), Goldman (2006) and, for psychological findings, Rolls and Scott (1994), Adolphs *et al.* (1994), Sprengelmayer *et al.* (1999), Small *et al.* (1999), Calder *et al.* (2000), Lawrence *et al.* (2002), Wicker *et al.* (2003) and Lawrence and Calder (2004). Goldman's version of mind reading is formulated as a serious rival of the theory-theory and modular approaches. The version of Nichols, Stich and Weinberg (2003) is more accommodating and integrative, and there are modularists and theory-theoryists who acknowledge that simulation plays an

all of these issues outrun our preoccupations here. I am interested in errors of conclusion drawing, and I am interested in a certain kind of error of premiss selection. By and large, they are errors occasioned by not using our heads, as the saying goes. We have already said that a fact of major importance for a theory of error is that an agent's belief-errors significantly outnumber his reasoning errors, and that a good part of the explanation of this fact, is the dominant presence of sayso in the production of human belief. This is the question that will occupy us in the next two sections.

9.15 *Sayso-filters*

For all its abundance, told-knowledge is subject to these exclusions. In a rough and ready way, we could say that told-knowledge, voluminous as it is, is a *filtration* of even more numerous tellings. They are filtrations of tellings to believings. It would be handy to have a name for these devices. So let us call them *sayso-filters*. Sayso-filters are part of a larger class of such things, mentioned earlier, paraconsistency filters, relevance filters and plausibility-filters being three of them. Paraconsistency filters minimize inconsistency even for a consequence relation that tolerates it. Relevance filters screen out information irrelevant to thought and action, enabling us to fulfill most of Harman's Clutter Avoidance Maxim. Plausibility-filters separates the plausible from the implausible, and in so doing play a central role in hypothesis selection. Filters are solutions of Cut-Down Problems. They cut quite

---

important though not exclusive role in mind reading, for example, Botterill and Carruthers (1999). Both theory-theory and simulationism agree that our ability to grasp states of mind is a biological inheritance. An important demurrer is the view that our folk psychological abilities have a sociocultural rather than biological basis, a view captured by the Narrative Practice Hypothesis (Hutto, 2008). As mentioned earlier, suggestions that NPH the only way in which children can acquire the capacity for mental state recognition and attribution is by engaging in narrative practices. Given that telling is a kind of narrative, NPH is something epistemologists of told-knowledge should pay attention to. Hutto's philosophical narratology is anticipated to some extent by the script theories of AI (Johnson and Reeder, 1997), and even more so by early 20th century developments in sociocultural psychology occasioned by the work of Lev Vygotsky (1962). Also of interest is the sociometric approach to signal-reading advanced by the organizational engineer Sandy Pentland, in which the idea of "reading" the social networks that attend signal-sending and signal-receiving plays a central role (Pentland, 2008).

large information spaces down to size. Size is important. It is tightly keyed to economies necessitated by an individual agent's resource-limitations. A further such filter is the consistency filter, which – at a certain level of abstraction – operates stepwise. In the first instance, it cuts down massively larger inconsistent information spaces to subsets which, though still large, are consistent. In step two, it cuts consistent big spaces down to locally manageable consistent subsets, in response again to the individual agent's need for economy. Consistency-filters tell us something important about filters in general. Consider the step from an agent's total beliefs and commitments, TBC, to any locally manageable consistent subset. It is natural to think of the filter as running a consistency-check on TBC. After all, if in the first instance the required task is to find the largest consistent subset of them, then how could a consistency check over them all not be made? But, as mentioned before, it is well-known in complexity theoretic circles that the consistency-check problem is at least NP-hard. NP-hard problems are intractable. They cannot be solved by beings like us, even "in principle".

This creates a problem for any account in which individual agents filter consistent subsets from inconsistent supersets, or tolerably inconsistent subsets from intolerably inconsistent supersets. However it may operate in fine, the filter solves a computationally intractable decision problem. The same can be said for the other filters – for the ones that cut big information spaces down to relevant subsets, and for the ones that cut those same spaces down to plausible subsets. A similar problem afflicts the sayso filter, but in an interestingly particular way. On any given occasion of its deployment, the problem solved by the sayso filter is not intractable. It is not in general necessary to examine the huge space of what might have been said in order to filter the now believed α from what was previously unbelieved. The intractability issue achieves its purchase in contexts of hoped for sayso-elimination. If, as Salmon avers, a condition of your taking the BBC's word for α, is that you satisfy yourself that everything or at least the majority of what the BBC reports as true is in fact true, and that, as a condition of doing everything the BBC reports as true on the basis of anything that depends on the truth of someone's report to someone, then intractability again looms. Checking the reliability of all that would be like running a consistency-check. But consistency-checks can't be run.

The computational discouragements of these filtrations face us with the question of how their intractability is overcome. It is a vexing question, concerning which there isn't at present much light to shed. Still, one thing is clear. It is that the CC-model has a terrible time with these filtrations. However they may be achieved, they are not in the general case achieved by way of the CC-modalities. Even if we were to suppose that there exist true theories, T, T', T'' of relevance, plausibility and told-knowledge, on the CC model the filtrations in question are brought about by the agent's own intellectual efforts, the free deployment of procedural routines subject to his command and control. They are procedures that enable him to implement the requisite theories, with T's rules giving him relevant information, T''s giving him plausible information, and T'''s giving him knowledge. There is not a CC-theorist alive who thinks of these executions as within the reach of actual agents in the general case.

9.16 *CR-relief*

CR-orientations are more amenable to filtration-problems, but they are far from problem-free. One clear advantage is that the burdens of filtration solutions are transferred from the individual knower to his devices. With it comes the thought that the filtrations that actually obtain are the causal upshots of how we are built, of how *Natura sive Deus* has put us together as cognitive beings. The reliabilist part of CR further provides that filtrations will have been well-produced when our filtration devices were working as they should. The causalist part of CR provides that the filtrations that actually obtain are embedded in lawlike correlations. Jointly, filtrations made in the circumstances in which they actually show up are filtrations which in those circumstances, had to have been made. Throw in Enough Already, and we have it that in the general case the filtrations that had to have been made are the filtrations that serve us well.

A prime datum for theories of told-knowledge is the extent and sheer complexity of the exclusions transacted by sayso filtrations when working as they should. Notwithstanding the causal tug exerted by being told things, exceptions are numerous and not especially easy to classify. They range, as we saw, from false belief inducements for which we are blameless, to those that we should have seen coming a mile away, and the tellings of which in turn range from innocent fact-sharing to the tendentious assertions of advocacy, to the murderous lies of Dear Leader propaganda.

Much has been written of these matters, some of it quite important, and nearly all of it beyond our range here. What matters for us is the empirically evident fact that our sayso-filters are a good deal less efficacious than our consistency, relevance and plausibility-filters. It matters for the logic of reasoning errors because it bears on the issue of premiss-selection. Of central importance is the problem of sorting out why we are less good at screening out bad tellings than we are at avoiding bad reasonings. Let us call this the "asymmetry problem". I will say a word about that now.

9.17 *Surfaces*

Premiss-selection and conclusion-drawing alike can be interpreted as belief-change. Belief-changes are affected in contexts appropriate to their kinds. In its primary sense, premiss-selection is a matter of taking in new information. It is the endpoint of a communication. Communication is the medium of premiss-selection. Conclusion drawing operates on premisses at hand – old or new, it doesn't matter. Conclusion-drawing is a matter of exploiting what's already there. This is done by inference. Inference is the medium of conclusion-drawing.[334] In the case of third-way reasoning, the medium of conclusion drawing is defeasible inference.

Let us now introduce the idea of the *surface* of a medium of belief-change. Surface is a function of the length and complexity of the causal nexus between origin and endpoint. Information communicated by sayso is in the general case an element of a sayso chain, which for large classes of cases, are multi-stranded and long. Let us say, then, that sayso communication is a *large-surface* medium of belief change. In contrast, the causal chain between the start-up and finish of your inference from α_1, ..., α_n to β is short and sweet, although certainly not devoid of its only complexities. The route from the beginning to the end of the completeness proof for first order logic is a good deal less short and sweet than the journey from "This ocelot is four-legged" to "Ocelots are four-legged". But by comparison with sayso communications, the surface of inference is small. Reasoning, then, is a *small-surface* medium of belief change.

We are now in a position to advance a hypothesis about the asymmetry problem. It is that

---

[334] Conclusions, once drawn, also serve as potential premisses for subsequent inferences. We may say, then, that conclusion-drawing abets premiss-selection in a secondary sense and, accordingly, that conclusion-drawing is the communication of premisses in a secondary sense.

Proposition 9.17

ERROR AND SURFACE SIZE: *The larger the surface size of a medium of belief change, the greater the likelihood of error.*[335]

Proposition 9.17 is reflected in the common assumption that repetition is an accuracy-degrader, as with the joke about the sailor and the parrot, which is hardly recognizable after a dozen or so (serial) tellings. It takes little reflection to see that, common though it may be, the assumption is untrue for a considerable mix of cases. One can only imagine the millions of tellings of the year of Columbus' footfall in the Americas, or of Pythagoras' theorem, but there's not much to be said for the idea that, as we have them now, they are pretty well dead-wrong. Even so, there is something right about the surface-size prediction. The larger the surface-size of the communication chain from days of yore to today's telling, the greater the number of the tellings in between and of the number of tellers and persons told. Large surfaces are made so by the number of the cognitive processes at the links of embedded causal chains. Fallibilism provides that each of us – that is, all individual agents – will in the course of transacting our cognitive tasks, make errors with a notable frequency. So it just stands to reason that the greater the number of links, the greater the medium's size, the greater the likelihood of aggregate error.

The surface-size hypothesis is something worth testing empirically. In the end, it may come to nothing, but I doubt it. For the present, its appeal is largely abductive. If the surface-size hypothesis were true then it would be a matter of course that we are less successful at screening out bad telling than we are at keeping our reasonings in good order.

If the asymmetry problem is of special importance to a logic of error, of even more central importance is an issue to which we now turn. As will become evident in due course, not only is it an interesting problem in its own right, but it also bears directly on the purported fallacies of begging the question and *argumentum ad hominem*.

9.18 *Knowledge by contradiction*

Consider a quite common kind of case. Let us lay it out in simple dialogue form.

---

[335] If we felt in a joking mood, we could call this the "surface-to-err" principle. Counting against it, apart from considerations of taste, is that "err" is pronounced "urr", not "air".

341

Harry: "Tomorrow is Barb's birthday".

Sarah: "No it's April 3rd."

Harry: I'd bet my bottom dollar that it's tomorrow".

Sarah: "Well, it isn't".

Harry: "No kidding! I thought it was tomorrow. Good. This give time to get her a present"

The phenomenon of told-knowledge gives rise to a number of difficult problems. But the most difficult of them by far, and the *central* task of a theory of told-knowledge, is figuring out the present case. Considered in CC-terms the case is hardly intelligible. You are so sure of Barb's birthday that you'd "bet your bottom dollar" on it. True, that's just an expression, but it attests nevertheless to a certain solidity of conviction. Even so, four seconds from now you are separated from that conviction, by two simple utterances that have all the look of begging the question, even more so in its stripped-down form:

Harry: "Tomorrow is Barb's birthday".

Sarah: "No it isn't".

Harry: "Oh, I see".

Cases such as these are critically important. I regard them as a primary datum for an account of told-knowledge. Left undealt with they are theory-breakers. They are cases in which belief change is occasioned by the simple denial of what is now believed. Harry believes that α. Sarah contradicts α. So now Harry doesn't believe α. He knows that ~α. This is *knowledge-by-contradiction*, an astonishing defection from a prior confidence, with no CC-reason to do so. But these cases are legion.

If the CC perspective is the right perspective and if the CC-perspective embraces the traditional notion of the fallacy of begging the question, then in *all* these cases we have lost our reason, felled each time by the starkest kind of question-begging. On the traditional conception, not only is begging the question a failure of reasoning for the beggar, it is like a failure for him who accepts it. In chapter 12 I will say why I think that begging the question doesn't in fact instantiate the traditional notion of fallacy. Even if I'm not right about this, there is still something right about knowledge-by-contradiction. And it falls in turn to any theory that knows this to be to explain why knowledge-by-contradiction should be given a generally clean bill of health.

CHAPTER TEN

NOT BEING TOLD

"The little I know I owe to my ignorance."

Sacha Guitry

"Ignorance is the necessary condition of life itself. If we knew everything, we could not endure existence for a single hour."
Anatole France

"… it occurs to me that no publication either before or since the invention of printing, no theological treatise or scientific creed, has ever been as narrowly dogmatic or as offensively arbitrary in its prejudices as a railway timetable."
Nero Wolfe

10.1 *The good of ignorance*

Because the acquisition or enlargement of knowledge is one of reasoning's main goals, it is natural to think of ignorance as a defeater of the first and a despoiler of the second. A lack of knowledge often plays a role in motivating a person to reason in the first place. Harry wants to know, but doesn't, whether he has a positive balance in his current account. He starts balancing his cheque book. After awhile he knows. He's in overdraft. This we might think of as the motivational value of Harry's not knowing in the first place. It prompted him to take steps to convert not knowing into knowing. If ignorance can be a motivational good for reasoning – a goad to its own elimination – we might think that it makes no contribution on the positive side. Harry also wants to know why the dog didn't bark in the night and, to that end, engages in some reasoning. Once the reasoning has ended, he still doesn't know. If we allow for degrees of epistemic success, it could be that Harry's reasoning has made no advance in the extent to which he now knows what he set out to know. His reckoning has achieved no reduction in the ignorance that triggered the reasoning in the first place. Ignorance is present in this reasoning, but no positive good comes of it, or so it would seem. When someone set out to know something, and there is a salient element of ignorance which his reasoning failed to subdue, hasn't his reasoning been, if not defective, unproductive?

As naturally as it may strike the ear, it may not be true that the good of ignorance is at best motivational. There are two interesting kinds of case which, on the face of it at least, defeat that impression. In each of them ignorance seems to play a positive and cognitively benign role. One is reasoning *from* ignorance. The other is reasoning that *preserves* ignorance. Reasoning of the first sort is *ad ignorantiam*. Reasoning of the second abductive. I'll discuss *ad ignorantiam* reasoning in this chapter, and reserve abduction for the next.

What makes for the specialness of *ad ignorantiam* and abductive reasoning? As a first approximation, we might venture the idea that

Proposition 10.1

IGNORANCE AS INFERENTIALLY PRODUCTIVE: *It is a defining condition of both* ad ignorantiam *and abductive reasoning that they embed true premises in the form ⌐α is not known⌐ which function there with some kind of cognitively advantageous inferential force.*

10.2. *Ad ignorantiam*

In most of today's treatments, *ad ignorantiam* reasoning is always a fallacy, that is, a fallacy intrinsically so. According to one recent account, the

> *ad ignorantiam* fallacy amounts to the following way of concluding: If it cannot be proven that x is the case, it is therefore proven that x is not the case (or, alternatively, if it cannot be proven that x is not the case, it is therefore proven that x is the case).[336]

For any reader disposed to accept chapter 9's findings on told-knowledge, this approach will be an epistemic discouragement. There is for most α neither a proof that α nor a proof that not-α. But there being no proof of α is by this definition itself proof of ⌐not-α⌐; and there being no proof of ⌐not-α⌐ is likewise a proof that α. Accordingly, for all these α – for much of what we take ourselves as knowing – it follows that they and their negations are proved. It is a hefty consequence, needless to say, excessive even by the lights of dialethic logic.

---

[336] van Eemeren *et al.* (2009); p. 193.

Christopher Tindale's 2007 treatment captures something of the spirit of the present conception without generating the excesses occasioned by its overstatement. Seen Tindale's way,

> [b]asically, the argumentative strategy in question involves drawing a conclusion on the basis of the absence of evidence against that conclusion – at least, that is the simplest form of the *ad ignorantiam*. (Tindale, 2007, p. 117)

Tindale allows for cases in which reasoning of roughly or approximately this form may not be fallacious and, in so doing, countenances cases in which the absence of evidence might actually be of some probative significance. It is a welcome concession, echoing an early claim to the same effect.[337]

Still, for at least large classes of cases traditionally minded fallacy theorists are inclined to condemn the *ad ignorantiam* as an outright error. My own view is that here again the traditional distrust is misplaced. I am encouraged in this by the substantial empirical fact that in actual practice *ad ignorantiam* reasoners quite often instantiate a cognitive manoeuvre known in the AI literature as *autoepistemic reasoning*, in which there isn't the slightest suggestion of fallacy.

10.3 *Autoepistemic logics*

The best known formal treatments of autoepistemic reasoning arise from logics in the manner of Moore (1983, 1988), whose basic language is the modal propositional language $\pounds_M$ of McDermott and Doyle (1980). These logics do well on the score of mathematical virtuosity. Moore (1984) gives an elegant semantics along the lines of Kripke's treatment of S5. Soundness and completeness are proved, and further applications have been developed for a more general logic of belief (Levesque, 1987 and Lakemeyer and Levesque, 2000).[338] Autoepistemic logics are a major force in the efforts of AI researchers to achieve an understanding of defeasible reasoning. Defeasible reasoning occupied us in chapters 7 and 8. Now, two chapters later, autoepistemic consideration is a kind of home-coming. In this instance, I will take as

---

[337] Woods and Walton (1978). Reprinted in Woods and Walton (1989/2007) as chapter 11.

[338] A good overview is Meyer and van der Hoek (1995). Gabbay *et al.* (1994) is the classic general survey.

given the formal success of these logics. I want to focus instead on matters of more conceptual and empirical interest.

Typical examples of an autoepistemic argument are:

1. If there were a department meeting today we would know it.
2. But we know no such thing.
3. So, presumably, there won't be a meeting today.

and

a. If there were a late night Tuesday flight from Vancouver to London it would be announced on the Departures Board [The Departures Board would make this known to us.]
b. But no such flight is listed [The Board gives us no knowledge of it.]
c. So, presumably, there is no such flight.

The factor of ignorance is expressly introduced by the respectively second premisses. These I will call the "*ignoratio* premiss". Since those premisses are inferentially productive probative – that is they support the *presumptions* which their respective conclusions embody – the reasoning that issues in these conclusions satisfies the conditions of Proposition 10.1 and thus is reasoning *ad ignorantiam,* reasoning in which ignorance has an inferentially load-bearing role. The inferential adequacy of the connection between the profession of ignorance and the conclusions *ad ignorantiam* from which they are derived is furnished by the truth of the respective conditional premisses, whose consequents are the negations of the *ignoratio* premisses. Accordingly, the conditional premisses have an expressly epistemic character; indeed, when the reference of the premiss is to its own utterer, its epistemic character is *auto.* Call these the "autoepistemic premiss".

It is commonly said of cases such as these that the conclusion is a default defeasence from an autoepistemic conditional, itself backed by a proposition of appropriate generality. In the first example, the tacit generalization is said to be something like "Department meetings require proper notice to members." In the second example it is something like "Airports and airlines don't schedule unposted flights." In the absence of our coming to know differently, we may take it that there is no such meeting and no such flight. The "no" in these conclusions is thought to express negation as failure. "There is no such meeting" means "We have failed to determine that such a meeting will occur". The "no" of negation as failure reflects a failure to know. In these contexts, "no" is an epistemic term. This, it is said, establishes the link between negation as failure reasoning and *ad ignorantiam* reasoning.

A good many people seem to think that to get the hang of autoepistemic reasoning it is necessary to stress the "auto" part at least as much as the "epistemic" part. In fact, autoepistemic reasoning is slightly misnamed. Deriving from the Greek, "auto" means *self*; and it is true that in lots of cases the self is the subject of a first person pronoun. In our present examples, the "we"/"us" of the premisses is the speaker or maker of the inference. Accordingly, what makes that inference an auto-inference is that the person making the inference is in the denotation of the pronominal expressions contained in his own premisses. But it is not necessary to have this linkage between the premissary pronouns and the person drawing the inference. For consider,

    i.   If there were a department meeting today, Harry's Dad, Professor Zed, would have been notified of it.

    ii.   But he hasn't been.

    iii.   So, presumably, there won't be a meeting

in which premiss (i) is pronoun-free and the pronoun of premiss (ii) is of the third-person variety, and the speaker and the unknower are not one and the same. Still, for the time being, I'll continue to call the conditional premisses "autoepistemic" and the second premisses "*ignoratio*".

Autoepistemic inference make ineliminably significant claims about the actual states of knowledge possessed by contextually indicated persons. In its purely auto form, the speaker is either the sole referent of the singular terms of the premisses or is a member of the reference class of their pluralized forms. In its other than auto form, the persons cited in the premisses must be one and the same, but the speaker may be some third person. Accordingly, we have a first pass at a general schema for autoepistemic reasoning:

Proposition 10.3
AUTOEPISTEMIC   SCHEMA: *Reasoning   counts   as autoepistemic when: There is a premiss stating that were α the case, then some specified or contextually indicated persons would know it. There is a premiss stating that, as it happens, these selfsame persons do not know α. The final line concludes that it is reasonable to presume that α is not the case.*

Or in simplified form:

1. *If α were the case then X would know it* (The autoepistemic premiss)
2. *X does not know it* (The *ignoratio* premiss)
3. *So, presumably, α is not the case.* (The conclusion *ad ignorantiam*)

10.4 *Support vs. inference*

Suppose that Harry were privy to the information displayed in the first two lines of the flight example. This information lends autoepistemic support to the schema's conclusion that there is no late night flight to London on Tuesday. Some will think that when an agent is in possession of information that lends autoepistemic support to some conclusion in which he has an interest, he will draw that conclusion from that information by autoepistemic defeasence. Intuitive though the supposition may be, there are cases that discredit it. Here is one of them.

Harry at present doesn't have a headache. If he had, he would have known it, but he doesn't. (How could he?) So he hasn't got a headache. Of course, Harry knows that he hasn't got a headache, but he doesn't know it on the basis of the twin facts that had he had one he'd have known it and he doesn't know it. What Harry knows he knows directly, not by default from some autoepistemic conditional.

Such cases are far from rare. Harry awakes early one dark January morning, apprehensive about a predicted overnight snowfall. He pads over to the window and look down into the well-lit street. The street is clear, dry as a bone. Driving to work will be no problem. It is true that had there been a dump of snow overnight, you would have known it. You would have seen it with your own eyes. But that it didn't snow overnight. So it can't be that you know that it did, and do know that it didn't. If an autoepistemic logician were lurking about, he might want to impose his organizational signature on these facts. He might make a sequence of them in the order of their reportage here. Having done so, suppose our logician were to see the third proposition as autoepistemically derivable from the prior pair. Suppose that he now puts it to you that that's how you came to know that there had been no overnight snow. You knew it by autoepistemic inference. Would you believe him? It is true that the facts of the situation lend autoepistemic inducement to your thinking that last night was snow-free. That's what it was that made you think so. But

349

nothing in the facts of the case affords any encouragement to the idea that this is the manner in which your knowledge was actually achieved. The autoepistemic conditional is true but, in the circumstances in which you actually found yourself, it played no probative role.

10.5 *Downgrading the autoepistemic*

It is 10:15 in the evening and, on impulse, Harry wants to know whether he can catch the late show at his local cinema. He checks the listings in the paper and sees that the last screening was at 10:00. A few blocks away Sarah has had the same impulse. Lacking a paper, she phones the box-office and is informed that the last show started at 10:00. In the literature it is customary to see Harry's type of case as instantiating the autoepistemic schema. Had there been a later showing he would have known it. He would have known it because it would have been listed in the showtimes section of the paper. But, not being listed there, Harry doesn't know that there is a later showing than the nine o'clock one. So there isn't. (Presumably.)

There might be little enthusiasm for construing Sarah's situation in like manner. But why not? Had there been a later showing, she would have known it. She would have known it because the box-office clerk would have told her so. But he didn't and she doesn't. If Harry knows that 10:00 is the last show by autoepistemic reasoning, isn't the same true of Sarah as well?

Perhaps the perceived tension between the two cases lies in differences between the propositions that back their respective autoepistemic premises. It is generally accepted that the autoepistemic fact about Harry is backed by the generalization that cinemas announce their screenings. This lends itself to the following schema:

1. The paper informed Harry that there was no showing after 10:00.
2. So Harry now knows that there is no showing after 10:00.

But doesn't Sarah know this too? Why wouldn't Sarah's situation schematize like Harry's?

1'. The box office informed Sarah that there was no showing after 10:00.
2'. So Sarah now knows that there is no showing after 10:00.

What is it that distinguishes the two cases?

There are, in fact, two versions of the Sarah case, one supporting the indistinguishability thesis, and the other not. In the first, her interlocutor at the box office reads off the list of show times, in which 10:00 is the last item. In the second, he utters the words," 10:00 is our last showing". In the latter case Sarah *was told* the time of the last showing, and the former Sarah herself was able *to tell* the time of the last showing from what she was told. Is it then really the case, then, that Harry's paper told him that there was no showing after 10:00? True, it contains no inscription in the form, "There is no showing after 10:00"; it contains no list, "Times when there are no showings" on which we find the inscription "10:20"; and it contains no admonition, "These are all the showings we have". What the paper does contain is a heading, "Show times", followed by a list of times. Our question is whether "Show times", together with the list that follows it, tells Harry that there is no screening after 10:00? It would now appear not. What Harry was told enabled *him* to tell when the last show starts.

Here is a further point of interest. If we examine lines (1) and (2) and (1′) and (2′), it is easy to see that they provide a natural context for inclusion of an autoepistemic premiss in the form "Had there been a late showing Harry/Sarah would have known it" and an *ignoratio* premiss in the form "But Harry/Sarah didn't know it." But (1) and (2) and (1′) and (2′) are good inferences without the need of these further premisses. In those arguments the premisses are redundant. So the reasoning they typify is, on present definitions, neither autoepistemic nor *ad ignorantiam.* Compare this with the meeting example of Prof. Zed. A key difference is that Zed expected to be told if there were a meeting today and took no steps to discover on his own whether or not there would. Here the autoepistemic premiss and the *ignoratio* premiss are not redundant.

10.6 *Telling vs. knowing*

What enabled Harry to tell from what the paper told him when the last show started? If Harry is anything like the rest of us, he will know that it is not in the theatre's interest to withhold from customers when its showtimes are. He will be familiar with *standard operating procedure.* There is here a kind of closed world assumption at work. Given the nature of the movie business and the importance of matching audiences with showings, communications of this sort are subject to a convention, roughly as follows:

Proposition 10.6a
NO TELLING, NO FACT: *Unreported facts don't exist.*

The provenance of Proposition 10.6a is remarkably varied. In its own way, it echoes the wide range of told-knowledge, the vast and multiplex enterprise of informing people of things. Of course, there are clear cases in which the no telling-no fact principle doesn't apply, and there are many areas in which it applies only partially or selectively. When the police were questioning Spike about his involvement with the crime, even if we assumed that his answers were true, it would be naive to expect a complete accounting. Why would it be, given that it might not be in Spike's interests to give one? Similarly, when the BBC reported the riots in Britain in the summer of 2011, we may take it that it didn't report whether, concurrently with those disturbances, the Duke of Westminster was having a hair-cut. But reported or not, he was or he wasn't. Another thing the BBC didn't report was the razing to the ground of St. Paul's Cathedral. Given that it is the interest of the BBC to report any facts lawfully in its possession and in accordance with its importance and its interest to listeners, why wouldn't we conclude that the fact that the BBC reported no damage to St. Paul's tells us that St. Paul's was undamaged?

Inferentialists are a tough lot. They see tangles of inference where others see the less cluttered economies of direct apprehension and the other forms of causal attachment. Perhaps there isn't much likelihood of resolving differences of opinion so confidently held. But it would repay us to note that it matters which model of knowledge these disputants find themselves most at home with. Any version of the CC-model in which evidentialism is favoured will be one in which knowledge acquisition rather generally – not excluding perception itself, some would say – is the natural and necessary habitat of inference. Things are dramatically otherwise on the CR-model of knowledge whose proponents accuse CCers of over intellectualizing the causal. Suppose that it is beyond question that Harry has correctly inferred some proposition γ from the true and rightly held beliefs α and β. Then, on the causal model, the belief-forming devices that brought about Harry's belief that γ were caused to fire by his states of belief regarding α and β. To some ears, this won't sound like inference at all. My own ears are more generously attuned. They hear this as bona fide inference. When Harry has a disposition to specify α and β when asked "What makes you think that γ?" But they are alert to salient differences. Let Harry believe α and β, as in the first case. Suppose that here too he believes γ. Suppose further that propositions α, β lend some probative support to γ. But, once again, as the headache and snow-fall cases remind us – and I would say the showtime and St. Paul's cases as well – it is possible that Harry's belief that γ hasn't been causally induced by his belief that α and β, that the causal inducements lay elsewhere. Again, what we have here is belief-update in the presence of probatively adequate inferential support, which is causally inert in the context at hand.

The no telling-no fact principle is a closed-worlds assumption for information-systems. When it holds, if you haven't been informed of the fact that α, the reason is that there is no such fact. When this is so, it is also the case that had α been the case you would have been told it is, or your senses would have advised you. None of this means that every true proposition which, in your present circumstances, is governed by the no telling-no fact assumption is one you would *know*. What is true is that you would have been *told* it. You would have had it *represented* to you as known. Of course, telling is very often knowledge-transmitting. This was a principal thesis of chapter 9. So in the case before us, had you been told it, there's a statistically favourable likelihood that this would have brought about your knowing it.

Being told things and knowing them are not unconnected phenomena. It is an important connection, but not the one that matters most here. Here the primary link is between something happening and its being *reported*. Thus there is a second respect in which the name "autoepistemic reasoning" is misleading. The "auto"-part is misleading for reasons sketched some pages ago. The "epistemic" part, as we now see, is also not quite right. Of course, if the reporting turned out to be knowledge-inducing for its recipients, that would be a welcome byproduct, a nonlogically contingent collateral effect. But the fact of the matter is that in cases such as these the description "autoepistemic" overstates the general character of the reasoning. Reasoning that ~α in default of one's not having been told that α is intrinsically autoepistemic. So let's make it official.

Proposition 10.6b
MAKING TOO MUCH OF THE AUTOEPISTEMIC: *The epistemic character of no telling-no fact reasoning is a frequent but not invariable contingent byproduct of particular circumstances.*

Proposition 10.6c
NOT AD IGNORANTIAM: *Autoepistemic reasoning is not,* as such ad ignorantiam *reasoning.*

This changes the character of the autoepistemic schema. Since not *being told* is the key factor, the autoepistemic conditional premiss should be rewritten accordingly. Instead of ⌐If α were the case I would have known it⌐, we should have

1. If α were the case I would have been informed of it;

and instead of calling it the autoepistemic premiss a better name would be "the told-it premiss." Equally, the *ignoratio* premiss ⌐I don't know that α⌐ should be replaced by

2. I haven't been informed that α

and re-named "the not-informed premiss". The same adjustment is required for the name of the conclusion, no longer a conclusion *ad ignorantiam* but rather a not-been-informed conclusion. Thus the revised schema – the told-it schema – would be

1′. If α. If α were the case I would have been informed of it (The told-it premiss).
2′. I haven't been so informed. (The not-informed premiss)
3′. So α is not the case (The conclusion from not-being-informed).

It is well to note, however, that

Proposition 10.6d
NEGATION AS FAILURE RETAINED: *The present de-autoepistemiized schema preserves that negation as failure interpretation of "not" in the schema's conclusion.*

It is necessary to emphasize that the original autoepistemic schema isn't dead wrong. In fact it is close to dead right, made so by the following pair of considerations: First, as was proposed a chapter ago, there is a statistically favourable correlation between been told that α and coming to believe that α. Second, although there are lots of contexts in

354

which tellings are causally inefficacious, not-being-informed contexts are not to be counted among them. The very fact that not being told that α makes you believe or inclines you to believe, that ~α is a strong indication that had you been told it would would have believed it. Since those tellings would be sanctioned by a standard operating procedure of which you are at least implicitly aware, your believing that α had you been told it would, with a high likelihood, have put you in a state of knowledge with regard to α.

10.7 *Public truths*

Let α be any arbitrary truth. Most values of α are unknown. Most of what is true is unknown. That is, most values of α are unreported. In contrast, consider any proposition β fulfilling the condition:

If true, then told.

The values of α transfinitely outnumber the values of β. Notwithstanding, β's values play a very large role in the cognitive economies of beings like us. The values of β constitute a set $\mathcal{P}$ of what I will call *public truths*, or truths "with a public". In a good many cases, for α to be a public truth two conditions must be met:

1. α is a proposition whose truth can't in the normal run of things go unnoticed, and
2. There exists a quasi-nomological regularity between being noticed and being reported.

The public for a proposition in $\mathcal{P}$ is the class of persons it is reported to when it fulfills condition (2).[339]

---

[339] Except for some modest mention by formal epistemologists, the formal dynamics of told-knowledge haven't had much of an innings in the philosophical mainstream. Of most direct consequence for epistemology is the theory of telling investigated by public announcement logic (PAL). It originates in Playa (1989). PAL is just one of many logics that model the dynamics of knowledge and belief in multiagent settings. PAL is a relatively simple extension of epistemic logic (EL) got by the addition of dynamic operators on formulas. PAL is a logic that permits agent's knowledge states to be updated by the public announcement of epistemic formulas. Thus PAL extends multiagent EL so as to model the communicational consequences of announcements to multiagents. Each formula

355

Had St. Paul's been razed to the ground, it would have been noticed; and had it been noticed, it would have been reported. The BBC would have told it to its viewers. The world's media would have told it to the world. The razing of St. Paul's would have been a public truth, and its public would have been anyone anywhere who watches or reads the news. The department meeting is a different case. Its public is smaller, and the quasi-nomological regularity is different. It is a matter of policy (and in some cases of law) that department meetings be announced in advance. Not only is there a *de facto* regularity between announcings and havings, it is brought about by the normative fact that policy requires it. The BBC also has a policy to report matters of interest, the more shocking the better. Had the BBC failed to report the collapse of St. Paul's it would have been a professional dereliction. But there is a more fundamental regularity at work having nothing to do with professional policy. News of St. Paul's would have been passed on with lightning speed by word of mouth by the social, not professional, media. BlackBerrys would have been alight the world over. The story would have spread like wildfire, not as a consequence of professional policy but in response to a deeper regularity: There are some things that people can't keep to themselves. Therein lies the causal foundation of the institution of gossip. [340]

Our two cases exhibit a further difference. With St. Paul's, noticing precedes telling. With the department meeting, telling precedes noticing. Accordingly, some slight adjustment of the definition of a public truth needs to be made, requiring in turn a distinction between tellings before the fact and tellings after the fact. Conditions (4) and (5) above fit the class of tellings after the fact. The department meeting example typifies the class of tellings before the fact. St. Paul's typifies no noticing without subsequent telling. The department meeting typifies no noticing without advanced telling.

When a public truth fails to reach its public and this is subsequently discovered, a further indication of the quasi-nomological character of the embedded regularity is that the discovery is met with a kind of normative double-take – the most general sense of which is:

---

of PAL can be rewritten as an equivalent formula of EL. An advantage of the reduction is that PAL is intrinsically more succinct than EL. There is a rapidly growing PAL literature. Important papers are, among others, Baltag *et al.* (1998), Gebrandy (1999), van Ditmarsch *et al.* (2005), van Benthem *et al.* (2006), Kooi (2007), and French and van Ditmarsch (2008).

[340] See here Magnani (2009), pp. 410-411, and Magnani (2007).

"Something's not right here", or "This is not how things go." Policies are breached, professional responsibilities are shirked, rational self-interest is comprised, human nature defected from. The nomological intensity of these reactions varies significantly. One can think of these intensities in modal terms. It is procedurally necessary to announce meetings. It is self-interestedly necessary to publicize what one wishes to sell. It is causally impossible to suppress public disasters.[341]

Other cases are trickier, in which the role of telling is less clear-cut than in the examples we've been discussing. Conventions are like this. In some places, everyone knows that pedestrians walk of the right of the sidewalk and that the left is reserved for passing. Defections are met with resistance: This is not the way it's done. "Walkers keep to the right" is a public truth in these places. It is possible that some members of this public would have received instruction about walking, instead of learning it on their own. But most people learn this convention without instruction. Sometimes they learn it by bumping into people. But mainly they learn it by noticing things. They learn it by hasty generalization.

10.8 *Surprise*

If all this is right, we may have found a hospitable context for AI's notion of default. Staying with the movie example, Sarah's default position is that there is no showing after 10:00. What this means is that should she become aware of a showing after 10:00, this would be met with one of those responses, "Something's amiss" or "This is not how it goes, this is not a matter of course." We need a name for this kind of response. I propose that we adopt a usage of Peirce. Let's call them *surprises*.[342] The word "surprise" is ambiguous, referring alike to the event and to the response the event evokes. But on my proposal surprises are reactions to defeated defaults. They react to them as dislocations.[343]

---

[341] It is not causally impossible or even all that difficult to deny a normally public event any witness of it. If you don't want it known that you've hanged Robin Hood, then hang him privately and murder the hangman and the guards. But if you hang him in the public square, the whole country will know it sooner than later. Sooner or later, movies will be made of it.

[342] *CP*, 5.189.

[343] "The time is out of joint". (*Hamlet,* Act I, scene 5, 188)

Proposition 10.8

DEFAULTS (FIRST PASS): α *is an agent's default position if he believes that α and information to the contrary would occasion his Peircean surprise.*

We should take care not to allow the idea of surprise to get away from us. Sometimes a surprise is an event which is not only unexpected but unattended by any particular reason to think it would or might occur. When Harry was a boy, his parents took steps to keep him from knowing what Santa would bring. They wanted it to be a surprise. Most years, Harry had a pretty good idea of the alternatives: roller-skates or ice-skates or a catcher's mitt, something of that sort. Suppose that on the big day it was the Boy's Mechanical Encyclopaedia that appeared. Have Mom and Dad succeeded in keeping this a surprise? Yes, but not in the Peircean sense. Similarly, when Sarah first met her boss at work, she was pleasantly surprised that he was taller than she expected. He is not all that tall, and Sarah had no particular reason to suppose that he would be less tall than that. Still, it was a surprise, but not in the Peircean sense. In each of these cases, the missing Peircean element is the originating default position. Harry had a pretty good idea that the skates or a mitt might be what Santa would bring. When the encyclopaedia appeared, Harry's response was not that something had gone wrong. That it would be skates or a mitt was not a default position for Harry. Equally, Sarah's surprise in seeing how tall her boss was was not expressible as a response in the form that "This is not how it's supposed to go". She was not in the appropriate default position.

We have already taken note of the varied usages of the idioms of default in the AI literature. A default could be the instantiation of a generic proposition or of an adverbially qualified non-quantification. It could be the generic proposition itself, or the adverbially qualified non-quantification. It could be a rule instructing (or permitting) a reasoner to hold a proposition in the absence of information to the contrary. In still other contexts "default" describes a putative consequence relation whose inputs are default-propositions in the first sense and whose outputs are default-propositions in the second sense, and whose corresponding consequence-drawing rule is a default in the third sense. A still broader usage is one in which a default is any proposition provisionally held until such time as we may come to know better. Slightly more broadly still, defaulting is the appropriate cognitive response to defeasibility in all its forms. In its most general form, defaulting is what fallibilists do. Their motto is: Don't box yourself in with assertion or belief. The better course is to hedge.

After some discussion, we decided not to rein in these several uses in favour of a tidier and more semantically uniform idiom. I'm not much minded to urge the reversal of that latitude here. Slovenly as they may be, these various meanings are entrenched in the literature. Still, I admit to a certain fondness for the conditions of Proposition 10.8, not least because they arise from the context in which the AI notion of default made a canonical appearance, the logic of autoepistemic reasoning. True, Proposition 10.8 doesn't define defaults expressly for autoepistemic contexts. It defines it for predecessor contexts, the contexts of public propositions.

Had there been a screening after 10:00, and had Sarah and Harry subsequently learned of it, each would have exhibited Peircean surprise. They would have done so even though Sarah believed that there was no screening after 10:00 because she was told it, and Harry believed that there was no screening after 10:00 because he could tell it is so from what he was told. Each defaulted to "There is no showing after 10:00" on the basis of the same general proposition," The last show is at 10:00", never mind that Sarah was told it and Harry wasn't. Professor Zed's case is different. No one told him that there wasn't a department meeting today. Nor was this something that *he* could tell from the things told him that day or on preceding days. If there actually were a meeting today and Prof. Zed got to hear of it, it would have been met with Peircian surprise (and possibly a good deal of umbrage). But the proposition that there wasn't to be a meeting today is not made to be a default by Zed's surprise on learning otherwise. Our definition requires Zed's antecedent *belief* that there wouldn't be a meeting today. But, on the face of it, this is precisely what Zed did not have. He never gave the matter a second's thought one way or the other. So how could he be said to have believed it?

This gives us two options we could consider. One is to brazen it out, insisting that the very fact of his subsequent Peircean surprise establishes the prior existence of its occasioning belief. The other a more modest alternative. It is to think of something else to call it. Truth to tell, I don't quite know what to call it. But let's give "implicit belief" a try.

10.9 *Implicit belief*

For a very large class of cases, it would appear that when someone believes that α, there is a state of mind that he is in. By a wide margin, the dominant view is that he has a propositional attitude whose content is the semantic object α.[344] Propositional attitudes include what in earlier chapters we called k-states and b-states. Being in a k-state or a b-state with respect to α is a mode of propositional engagement. Whatever the details, one couldn't hold the belief that α without grasping a propositional object. Suppose this person is an intelligent and maturely centred individual. By and large, he knows his own mind. So, by and large, when he believes that α, α is something that he has on his mind. When these conditions are met, it is customary to say that his is a state of *occurrent* belief.

In our department meeting example, much of what Zed believes he believes implicitly, not occurrently. It may well be that Zed has never had express occasion to engage or entertain the proposition that it is standard operating procedure for departments to announce their meetings. It would appear, even so, that this wouldn't have been a fact entirely lost on him. It seems right to say that this is something of which Zed is aware, if not expressly so. What, one wonders, would Zed's state of mind be when he believes non-occurrently that it is standard operating procedure for departments to announce their meetings? Here, too, there are alternatives to consider.

One is that Zed fulfills the conditions on occurrent belief, subject to further qualifications. That is, here too Zed is in a state of belief – a b-state – with respect to the department's standard operating procedures, but fulfills the further conditions that render this doxastic state one of implicit belief that there won't be a meeting today. For this to be so, it must be possible for the engagement of this propositional object itself to

---

[344] Quine is a notable dissenter. He is famous for his hostility to propositional attitudes. Quine is the honorary father of naturalized epistemology. Why wouldn't anyone recommending the naturalization of logic have the same reservations? Quine dislikes propositional attitudes for two reasons. He finds it difficult to see how they could be given an adequate home in physics; and he sees that they give rise to intensional contexts, that is, that they are semantic objects. The first offends his fondness for physicalism. The second requires the failure of a law of logic, the intersubstitutivity of extensional equivalents. We saw earlier that Quine's naturalism is of the replacement variety. But even replacement naturalism does not of itself require an attachment to physicalism or an intolerance of intensional contexts. My naturalism is of the cooperative variety. My naturalism is entirely open to propositional attitudes until such time as the mature sciences of cognition make a convincing case against them. My view is that we shouldn't hold our breath.

be implicit. Critics will see little progress in explaining implicit belief via implicit engagement. They will see it as circular.

A second possibility is that the condition that Zed is in when possessed of the implicit awareness of the department's meeting policy is not one of having a propositional attitude to it, implicitly or otherwise. Zed's awareness that departments don't have unannounced meetings involves neither engagement with or entertainment of any proposition to this effect. He is not in a doxastic state with respect to it. Instead, the state he is in is one in which he has a causal disposition to meet defections from standard practice in some or other particular way. One of these ways is Peircian surprise. There are others.

Here is one of them. Lou has a young daughter whom Harry has not met. One Saturday morning Harry bumps into Lou at the hardware store. Lou is helping a little girl with her mitts. "Harry, I'd like you to meet Olivia". Harry returns the greeting and thinks to himself" "Oh, not 'Audrey'." Harry has never in his life put his mind to the proposition that the name of Lou's little girl is not 'Audrey". But he does so now. That proposition is now the object of a b-state. Ten minutes ago, indeed at all prior times, it was no such thing. There was no such state. It seems right to say that prior to now Harry believed implicitly that it isn't the case that the little girl wasn't named 'Audrey'. But it wouldn't be at all right to characterize Harry's reaction to what Lou has told him as Peircian surprise. Olivia's not being named 'Audrey' violates no quasi-nomological regularity. There is nothing in it to "put the time out of joint". Harry's situation is utterly common. We implicitly believe the negations of the contraries of what we currently believe. So let us say that

Proposition 10.9
IMPLICIT BELIEF: *Someone implicitly believes α at t if at no t′< t has α been the content of his b-states, and also at t he is causally disposed to put his mind to α on becoming aware of information that falsifies its negation.*

It is important to distinguish the cases satisfying Proposition 10.9 from run of the mill joggings of the memory. Imagine a slight variation of the present example. Lou introduced his daughter, and Harry thinks "Oh, yes, the name is 'Olivia'." Harry had once known this occurrently and in time it passed into the far reaches of his memory. This is what psychologists call dormancy. Harry has ceased putting his mind to the semantic propositional object that the name of Lou's daughter is "Olivia". Dormant

361

memories are subject to reinvigoration given the right stimulus. Here the stimulus is having been reminded of it.

Contrast this situation with a third. Harry doesn't know and never did what the girl's name is. Harry has entertained no beliefs as to its identity. All that Harry knows is that Lou has a little girl. Had Lou introduced her as "Vomit" there would be no doubting Harry's Peircian surprise. But what Lou said was that the little girl's name was "Olivia". In being told this, no previously implicit belief was now called to mind, and no memories were stirred. All the same, let's not lose sight of the fact that the implicity caught by Proposition 10.9 is not dormancy, even if dormancy is implicity of a kind.

It is possible that once upon a time someone induced in Zed the occurrent belief that department meetings don't occur unannounced. The dynamics of the reminiscential economy favour the timely and short-term conversion of the dormant to the occurrent. Dormant beliefs are beliefs that don't clutter up the mind; they are beliefs that are ready to hand for occurrent redeployment, as needs must. They are a large part of the just-in-time inventories of human cognition.

Of course, it is also possible that Zed has never had occasion to make an occurrent belief out of his awareness of the department's policy, to say nothing of his belief that there will be no meeting today. We should take care not to make more of Zed's situation than is in it to make. We might find ourselves wondering how it is possible to stock the mind with beliefs one has never had and never will. It is a good question, but this is not quite the right way to ask it. The better way is, how would it be possible for Zed to believe that there is no meeting on any day for which it was not announced, if never once in his lengthy and distinguished career did he have occasion to engage with and entertain a proposition to the effect that meetings must be announced? That was our earlier question. How is it possible for it to be Zed's default position that there is no meeting today if this is a proposition to which Zed has given no mind?

In the cognitive economies of beings like us, implicit belief is a major work-saver. Like the inventories of modern manufacturing, explicit beliefs are items that arrive "just in time", that is, ready as needed. When needed, they are causal dispositions stimulated into action. Having one's beliefs implicitly is one good way of dealing with Harman's clutter avoidance admonition: "Do not clutter up your mind with trivialities." Trivialities, here, are factors inhibiting the quality of cognitive performance . It is true that ocelots are four-legged, but knowing so, even implicitly, plays no role in Harry's quest for showtimes at the Roxy.

Perhaps a more natural way of making Harman's triviality point is that the ocelot proposition is irrelevant to Harry's agenda. That unannounced screenings are contrary to policy is not irrelevant to Harry's agenda. Harry wants to know whether he's in time for the last show. Even so, the theatre's listing policy is not something which Harry must expressly engage in attaining the saturated endpoint that closes that agenda. Here, too, triviality is not quite the *mot juste.* It is better to think of express propositional engagement not as trivial, but as unnecessary for reaching the agenda's saturated endpoint. Wrapping your mind around the theatre's policy would be redundant. To be influenced by the policy doesn't require that you call it to mind. Calling things to mind is expensive. Implicit beliefs don't clutter up the mind. They have no residency in minds.

10.10 *Virtuality*

Let's return to an earlier point. As we have told it here, it is possible for Zed to have defaulted to "There won't be a meeting today" without ever having occurrently believed either "It is a departmental requirement to give notice of its meetings" or "There won't be a meeting today." The same is true of "No meeting has been announced for today". Yet (I say) it is true today that Zed implicitly believes those propositions and that this state of affairs is construed by a causal disposition to react in a particular way to tomorrow's information that there was a meeting yesterday ("Why, we missed you at the meeting, Zed.") Cases of this sort are legion. Cases of this sort might start us to wonder whether "implicit" is quite the word for them.

As long as "implicit" retains a cognate tie to "implied", it would seem not to be *le mot juste.* The same reservation applies to "tacit". "Tacit" means "There or present, but unspoken." It means "silent." But in the cases that presently concern us, there is nothing there to *be* unspoken. However, an even better candidate for suspicion is "belief". Isn't its application to causal dispositions a bit of a semantic stretch? How can it in any sense be believing α now just because, although you have never yet turned your mind to it, you are so constituted – when circumstances are right – to turn your mind to it later. When you are thus positioned you are clearly in the described state. It is an important state to be in. It should have a name. "Implicit" seems wrong, and "belief" seems wrong, and so do the two together. Perhaps "virtual" will work, even in apposition to belief. Thus virtual belief is not belief of a kind. It is a state that precedes belief and, in lots of cases, makes it possible.

Let me say again that Zed's situation is very common. Zed and all the rest of us owe our survival and prosperity to our success in learning the world, in sussing out how things go. To a considerable extent, learning how things go is learning how things are supposed to go. This couldn't happen without a ready and early facility with hasty generalization and a ready and early sensitivity to quasi-nomological regularities. "Learning the world" is something of an overstatement, needless to say. No one comes close to learning it all. What is striking about this expression is not so much its exaggeration as the comparative modesty of its extent. The human animal owes his survival and prosperity to the ready and early mastery of enough of the world to make these achievements routinely possible.

What the human agent doesn't learn by hasty generalization, much of it he learns from being told things, or from not being told things, or from such more direct encounters with the world's particularities as he is able to notice and remember. Most beliefs induced by hasty generalization are both achieved and retained virtually, even when contrary instances are taken note of. Beliefs occasioned by tellings exhibit a greater likelihood of occurrency at the point of their telling, but quickly recede to virtuality, as do occurrently held beliefs, as soon as occurrency ceases to be an advantage. Beliefs, occasioned by not telling are, I now propose, beliefs more in name than substance. They are virtual.

Virtuality is occurrency in-waiting. It is occurrency a circumstance may demand. People who know that ocelots are four-legged rarely, if ever, have occasion to propositionalize their belief, to wrap their minds around the semantic object that occurrency requires. Still, there are exceptions. If you're interested in establishing whether the proposition expressed by "Ocelots are four-legged" has a quantificational structure, you have no recourse but to wrap you mind around that semantic object.

There is a tendency among philosophers to favour the occurrent. They seem to think that the optimal state for belief is occurrency and that the best place for it is in the front of its possessor's mind. I am contrarily inclined. The optimal state for belief is virtuality. The best place for belief is in the causal dispositions that structure the believer's down-below. If there were a slogan about belief, it would be "Keep it virtual, stupid". Occurrent beliefs are expensive, made affordable mainly by the shortness of their careers, by their readiness to revert to the virtual. This tells us something about learning the world. To the extent possible, we store our acquaintance with singular events as generalizations. When you spot your first ocelot, four legs and all, you might store it as, "That ocelot had four legs". You would be better served if you stored it as "Ocelots are four-legged". How is this done? It is done by acquiring the disposition to respond in a particular way if quocelots were to make the scene.

One of the hardest questions to ask of a theory of belief is how virtual belief is made occurrent – how the passage from down there to up here is effected. A similar question is how a memory of something is retrieved from storage. Notwithstanding the impressive effort by psychologists and others, much conjecture but little solid headway has been made with these matters. This is not to say that we are wholly in the dark. There are regularities to be noted. One is that the transition from the virtual to the occurrent is governed by context-sensitive relevance constraints.[345] By and large, beliefs are occurrentized as needed. By and large, they dissipate when the need passes. By and large, the frequency and numbers of occurrent beliefs aren't too much for a mind to handle. The admonitory tone of Harman's rule, "Don't clutter up your mind with trivialities", is unnecessary. We don't clutter up our mind with non-trivial occurrent beliefs. We couldn't even if we tried.

Virtual belief now gives us occasion to revise Proposition 10.8:

Proposition 10.10
DEFAULTS (SECOND PASS): *α is an agent's default position if and only if he (occurrently or virtually) believes that α and information to the contrary would occasion Peircian surprise.*

COROLLARY: *Defaults are but a particular kind of virtual beliefs.*

10.11 *Virtual knowledge*

Needless to say, if belief is retained as an unqualified condition on knowledge, most of what we know we know virtually. In comparative terms, hardly any of what we know will have been the result of propositional engagement. If virtual belief is not a propositional attitude, neither would virtual knowledge be if we allowed virtual belief to fulfill the belief-condition on knowledge. Some will not like this implication. They will resist the very notion of a person's knowing that α but having no grasp of it. They will see in this too little to qualify as genuine knowledge.

---

[345] For some formal modelling of this and related matters, see Gabbay, Rodrigues and Woods (2002), (2004).

This is not an unreasonable reaction. The favouritism which philosophy extends to occurrency is deeply dug in, and there is no knock down argument guaranteed to over-run it. But there are costs that can hardly be denied. If virtual belief doesn't satisfy the belief condition on knowledge, then most of what we virtually believe, even when true and produced by cognitive devices in good operating order and working as they should without external interference, isn't knowledge after all. The cost of this is twofold. We must give up the cognitive abundance thesis. We must give up, in turn, the value of knowledge thesis, the thesis that we owe our survival and prosperity to the truth of the cognitive abundance thesis. If so, we must re-think our resolve not to define knowledge out of existence.

The idea of having no mental grasp of so much of what we know will take some getting used to. But there are already signs that resistance may be less stiff than we might suppose. "When is Barb's birthday?", Harry asks. "April the 3rd", replies Sarah. "Ah, the third! Of course, I knew it all along." The moral of the example is clear. As long as we are prepared to accept that much of what we know lies in dormant memory, there should be no shock left in the idea of virtual knowledge.

In earlier chapters, I gave up on the J-condition for knowledge. It is, I said, an over-strict requirement, making for a good deal less knowledge that we actually have. The difficulty by quite a wide margin is that it is harder to justify things than to know them; that for much of what we know, there is no access to a justification of it. Justificationists have answers to this. They say that justifications are always available in principle. They say that to have a justification it is not necessary that you know how to give it voice. I said at the time that these are phantom justifications, recourse to which is little more than smoke-blowing. It might now be objected that in so saying I am hoist on my own *pétard*. Isn't it the case that what these J-defenders were claiming is that, for those cases of knowledge in which no justification seems to be present, it is in fact present and operational, not occurrently but virtually? And how, it will be asked, can I be such a ready virtualist about knowledge while refusing the same recognition to justification? Isn't the J-condition now back in business? Doesn't this restore to serious contention the CC-model of knowledge? If virtual justification is just an epistemologist's phantom, how can virtual belief not be the same?

I have two things to say about this. One is that if we liken the virtuality of justification to the virtuality of belief, then we make of justification – and knowledge too – the disposition to respond to informational stimuli in causally appropriate ways. Details aside, this is not an agreeable fit with the CC-model, which has little liking for the idea that knowledge is mainly a state that you've been put into, rather than a state that you've put yourself into. The other, more importantly, is that in causally appropriate circumstances virtual belief is convertible to occurrent belief. This is also sometimes true of justification. I may not know my justification of α until you demand it of me. Lo and behold, it now comes forth and attains occurrency. However, there is no reason to think that the success rates of such conversions are anywhere close to saving the J-condition as a congenial contributor to knowledge rather than an inhibitor of it. Virtual belief is a great deal more easily occurrentized than virtual justification. Occurentization of virtual belief is a commonplace of the everyday dynamics of belief-update. New information arrives, and adjustments are made. And, rather routinely, beliefs are occurentized. For any well-made and properly functioning human individual, such transformations are a matter of course, for which no particular talent or tutelage is required. Occurentization is a regular byproduct of information-flow. It is notably otherwise with justification. Justifications have to be paused over. They have to be sought out. Some people have little talent for them. In a rough and ready way, occurentize belief is something that just happens. It happens all the time. We can't say this of justification. Comparatively speaking, the occurentization of justification hardly ever happens. This matters, if true. If the J-condition is to be persisted with, knowledge hardly ever happens either. I don't like this result. It allows epistemology to be too hard on knowledge.

10.12 *Background information*

The economies achieved by virtuality are nowhere more evident than in the inferential uses of virtual belief, of the role it plays as the implicit premisses of reasoning. The importance of this for a theory of reasoning is unmistakable. Since implicit belief is belief *sans* propositional engagement, the reasoning in which it is implicated retains this feature. Reasoning that incorporates virtual belief is itself virtual to the extent that it does. It is to that same extent reasoning in the absence of propositional objects.

A central challenge for the philosophy of science is to achieve an orderly understanding of the role of background information in a scientific theory. The idea of background information also overlaps the notion of common knowledge (where it is understood that the "knowledge" of common knowledge is sometimes an honorific – not knowledge, in fact, but belief.) A good deal of the idea of background information is captured by the virtuality of the beliefs that make it up. Virtuality is what supplies the background. It also is clear that common "knowledge" intersects with background information in significant ways. Whatever its other properties, a good deal of common "knowledge" is virtual belief with a public face.

We saw in earlier chapters that one of the occasions of error is the phenomenal inapparency in the here and now of k-states. K-states are the engagers *par excellence* of propositional objects. K-states are states of mind at their least virtual. Virtuality is also major occasion of error. Virtual belief is belief uncalled to mind. Virtual belief hovers in the background. Virtuality is potency. Yet virtual belief is causally productive of knowledge and constructively fruitful for reasoning. When this is so, it is not propositionally recognizable as such. So when errors are committed, it only stands to reason that they will go unspotted. And when they are discovered, Peircean surprise often plays a major role.

10.13 *Presuming*

We are now at a good point to say something further about presumption. Presumption is less an occurrent state of mind than a causal disposition to be in one. Presumption is a form of virtual belief. Taken *en large* our presumptions embed aggregated unconceptualized expectations of how things go in life, one-by-one. The AI literature routinely jumbles up the idioms of presumption, defeasibility and defaults. I have a mind to keep them apart. Ozzie will help us here. There are a great many people who have the virtual belief that ocelots are four-legged. When the conditions are right, this is a belief that amounts to virtual knowledge. Let Harry be one of those people. Harry believes, without having seized upon it as a propositional object, that ocelots are four-legged. He presumes that *Ozzie* will likewise be of the four-legged kind. But he hasn't presumed that *ocelots* are four-legged. By construction of the case, he *knows* it. When it turned out that Ozzie wasn't four-legged after all, Harry's presumption was defeated, but his virtual knowledge was unmolested.

When Zed virtually believes that department meetings don't occur unless announced, we find it natural to say that Zed presumes that there will be no meeting today. Suppose that Zed, like Harry, also has a clear and confident memory that Barb's birthday is tomorrow. Of course, as in most things, he might be wrong and he might know that he might. Is it then Zed's *default* position that tomorrow is Barb's birthday? Is the state that Zed is in one of presuming that tomorrow is Barb's birthday? I think not. Suppose that Zed found out that Barb's birthday is on the 3rd, not tomorrow. How likely is Zed to think that Barb's actual birthday is contrary to the flow of things? Zed is a pretty confident fellow, but here not even he will be met with Peircean surprise. He will simply see that he was mistaken about when the birthday fell.

In actual speech, the idioms of default and presumption have yet to settle into a comfortably established usage. But if we did extend this latitude to the birthday belief there would be consequences. One is that most of our occurent beliefs would be defaults. Another is that when such beliefs did indeed turn out to be mistaken, they would hardly ever be met with Peircean surprise.

This risks the confusion of the fallible and the presumptive. It presses us to consider whether it is a distinction with theoretical legs, that is, a distinction that is worth the trouble of thinking through. AI theorists have taken the view that this is indeed a matter to make something of and to find a place for in our scheme of things. In much of the literature the proferred home is autoepistemic reasoning. If our reflections of recent pages have merit, the autoepistemicity of this context is overstated. But the underlying insight is deeply right.

Most of what we know and need to know we experience and/or store as generalizations, as what we've been calling quasi-nomological regularities. By the necessities of design and the requirements of manageable economies, most of what we know is virtual, stored as causal dispositions to respond in circumstantially particular ways. Most of our cognitive health is realized down below, and the little of it that is achieved up above is linked in turn, and indispensably so, to the occurrences further down.

Autoepistemic logicians were right to notice that a good many facts are facts with a public face, supported by generalities in the form ⌐Were α the case, its public would have been told it.⌐ Counterinstances are met with Peircean surprise. But Peircean surprise is not by any means restricted to contrary instances of public-face generalizations. When you presume that Ozzie will be four-legged, it is on account of his ocelothood that you do, not – even if it is so – because you haven't been told the contrary. You've gone to the zoo to visit Ozzie the ocelot, and now that you're there you see that he has three legs, not four. This you greet with Peircean surprise. This is not the way of ocelots. But there is nothing autoepistemic about your reaction. Neither are you in the catchment of

369

the proposition – even if true – "Had Ozzie not been four-legged, X would have been told so." Your Peircian surprise doesn't require that you be a value of X.

Defaults, again, are a cheap way of avoiding occurrent belief. For most ocelots, O, hardly anyone who knows that ocelots are four-legged will have the occasion to wrap his mind around the semantic object expressed by "O is four-legged." Defaults enable us to avoid this cost even at the point of the discovery that occasions Peircean surprise. You can discover that Ozzie hasn't four legs without ever having formed the belief that he had four of them.

In a good deal of the AI literature, defaults and presumptions are made for one another. Just as a default is the disposition to treat a negative instance of a quasi-nomological regularity with Peircean surprise, presumption is the means whereby that disposition is effected. It is effected by virtual recognition of how things are supposed to go.

From time to time I have cast a gimlet eye on a pair of what I take to be philosophically dubious practices. One – briefly revisited here – is the bad practice of intellectualizing the causal. The other is the bad practice of semanticizing the pragmatic. Presumption is a case in point.

Like modal expressions and the idioms of defeasibility and plausibility, the notion of presumption is available in adverbial form as both an operator-operator and a sentence-operator. As I have been trying to emphasize, these uses are pragmatically significant. They serve the pragmatic purpose of hedging. ⁻Presumably α⁻ is an assertion-weakener. ⁻{α, β} so, presumably, γ⁻ is a conclusional-force weakener. It is interesting to speculate on the semantic connections, or want of them, between these uses, and the further uses of presumption-attribution: ⁻I presume that α⁻, ⁻That's only my presumption⁻, and so on. Whatever our findings turn out to be, it is a mistake to suppose straightaway that they translate into semantic conditions on *presumption*. Here again is a common example of this mistake. Since "presumably" is adverbially affixable to conclusion-indicator words such as "so", "hence," and so on, there must exist between the premises and conclusions of such arguments a consequence relation of a type that reflects the adverb's semantic features. Suppose that we decided to call this relation the relation of *presumptive consequence*. Then the logician's task would be to define for it a well-made distinction between presumptive consequence-having and presumptive consequence-drawing. His further task would be to ground a semantics of presumptive inference in what his theory had to say about presumptive-consequence drawing. I have already had my say about logic's over-fondness for rendering conclusionality as consequence. Others will think otherwise. I leave it to them to find a tellable story about presumptive consequence.

10.14 *Taking for granted*

With these things said, it takes little further reflection to see in presumption a load-bearing role for closed world reasoning, not all of it perhaps, but a nicely representative chunk of it all the same. When Harry first drew the inference that Sarah drives an American car, it is more strictly correct to say that he thought she *still* drives an American car. He is right, and he is wrong. Sarah is still an American-car driver but, contrary to what Harry thinks, it's not true that she still drives a Buick. Harry doesn't know it, but she now drives a Cadillac. Harry knows that people don't change their cars with the frequency with which they change their shirts, say. Car-replacement behaviour is subject to certain norms, subject of course to local conditions. Some people like to hold onto their cars. Others are happier with a brisker turnover. Many people stay with the brand that they are now driving, or at least with the company that made it. Others favour a more varied plot. Money is a factor, and so is taste. Even so, we have it by construction of the Gettier case that Harry's inference involves that a normal-practice generalization is still in force.

One day Lou drove by Sarah's house and saw that her car wasn't in the driveway. He took it that Sarah had yet to come home. This was backed by a true typicality statement. Let's suppose that Sarah's car was in the shop for minor repairs, and that she had come home by taxi. So Lou is wrong in particular even though he is right in general. His conclusion was false even though the backing generalization was true and remained so. Other examples abound. For generality, let α be the thing drawn and N be the thing drawn from. Perhaps the most interesting feature about the N-α link is that α could be false without damage to N. If N were a universally quantified conditional, α's falsity would have put it out of business. In all these cases we may take it that the agent in question has the virtual belief that N is still true, and that his attachment to α reflects that belief. The closed world question enters our story in an interesting way. In all three cases, the agent's virtual N-belief is a world-closing move. If he believes that it is still true that N he believes that the world at present has no defeaters of it. Thinking so is a mistake when N is a universally quantified conditional, but no mistake at all in the other cases. Ozzie's three-leggedness puts paid to α but lays no glove on N.

371

When N is a universally quantified proposition, α's falsity makes the N-belief a closed world mistake. In the other cases, α is a mistake and N is true; so there is no closed world mistake here.

It would be handy if there were some settled ways to capture the peculiarities of generically impervious reasoning. In the AI and philosophical literatures there is a strong inclination to be influenced by the *hedging* uses of "presumably" and "default". There are uses designed to weaken the speaker's commitments, and so is a kind of damage-control manoeuvre in advance of any damage that might actually ensue. Fair enough. Such idioms have their uses, but not as descriptors or indicators of what is characteristic of third way reasoning. When Harry says that he presumes that α, it may sound that he is buying some insurance against failure, that he is flagging the weakness of his position. But when *I* say that α is Harry's default position, I impute no such weakness to α, or any such intention to Harry. Suppose that Lou interprets N as a universal quantification. Lou is in less safe hands than Harry. Lou is at risk of two mistakes, an instantial mistake about Ozzie and a closed world mistake with respect to ocelots. It is much the same way with presuming which, in some uses, is indistinguishable. These are misuages. Assuming something is supposing it. Presuming something is taking it for granted, taking it as *given*. This is an ancient difference, found in the Latin predecessors *assumēre* and *praesumēre*.

If in our theoretical speculations we wanted to make something of these lexical facts, we could propose that

Proposition 10.14a
EXPLOITING BRITTLENESS: *The more a generalization is elastic the more its instantiations are subject to presumption, the readier they are to be defaulted to.*

COROLLARY: *In the family of instantial inferences, default inference – that is presumption – has a higher than usual degree of epistemic security, made so in part by a lesser susceptibility to closed world error.*

It greatly matters that most of what we need to know is experienced or stored as quasi-nomological generalizations, expressible as generics, or typicals or non-universal quantifications N. It would not be far wrong to think of the world as the net aggregation of its N-states, and it would be dead right to say that an individual's survival depends utterly on his readiness to attune himself to these arrangements. Take Harry again. The heart and soul of Harry's sussing out of the world is the totality of Ns with which he achieves a working congress, no matter how virtual. Even for a man who lives so routine a life as Harry, that's a great many Ns, well beyond the reach of exhaustive specification. This tells us something interesting about presumption. Let α denote a state of affairs recognizable to Harry as an instantion of a given N. For each N, there are a great many such αs. All in all, there are a great many more αs than there are Ns, of which there are also very many. When Harry is causally disposed to react to the news of α's falsity with Peircian surprise, we say that α is a default for Harry and, equivalently, that α was something that Harry takes for granted. Accordingly,

Proposition 10.14b
THE DOMINANCE OF PRESUMPTION: *Presumption massively exceeds occurrent belief.*

COROLLARY: *Presumption is the workhorse of the human individual's cognitive economy.*

COROLLARY: *Most of what makes for a properly functioning cognitive life is most of the time nowhere in the sight of its possessor.*

CHAPTER ELEVEN

ABDUCING

"The chicken you say pecks by instinct. But if you are going to think every poor chicken endowed with an innate tendency toward a positive truth, why should you think that to man alone the gift is denied?" Charles Sanders Peirce

"It is sometimes said that the highest philosophical gift is to invent important new philosophical problems. If so, Peirce is a major star [in] the firmament of philosophy. By thrusting the notion of abduction to the forefront of philosophers' consciousness he created a problem which – I will argue – is the central one in contemporary epistemology." Jaakko Hintikka

11.1 *Abduction*

The initial focus of the last chapter lay in charting the course of right reasoning when not knowing something makes a positive contribution. In trying to bring this off, we generalized the notion of not knowing to the broader idea of not being told. Of course, the two phenomena aren't unlinked. Consider those cases in which being told that α would cause you to know that α. Then in each such case, not being told it might well preclude that knowledge. But now the focus I want shifts back to not knowing, to the contribution to good reasoning effected by your ignorance of something you *desire* to know. Abduction provides us with a context in which to explore this further.

In earlier chapters I proposed that two of an individual's most basic and successful scant-resource adjustment strategies are hasty generalization and hearsay. Most of the generalizations we actually draw and are right to draw come from small samples, and for most of what we know, arises from sayso manifolds in cognitive multiagencies. To these we can now add a third. Much of what we know we know virtually, owing to the absence of reports to the contrary. Much else, also virtual, arises from presumption. A further strategy is abduction.

Although there are stirrings of it in Aristotle's notion of *apagogē*, we owe the modern idea of abduction to Peirce. The idea is encapsulated in the *Peircean abduction schema*, as follows:

The surprising fact C *is observed.*
But if A *were* true, C *would be a matter of course.*
Hence there is reason to suspect that A *is true.* (CP, 5.189)

Peirce's schema raises some obvious questions. One is how central to abduction is the factor of *surprise*. Another is the issue of how we are to construe the element of suspicion. A third concerns what we should make of the idea of something's being a *matter of course*. Together these three – surprise, suspicion and matters of course – form what we could call the Peircean Triad for abduction. I will take it as given that anyone selling an elucidation of the Peircean notion will give the Triad due attention

Like so many of his better ideas and deeper insights, Peirce has nothing like a fully developed account of abduction. Even so, the record contains some important ideas, seven of which I'll mention here.

P1. Abduction is triggered by surprise. (CP, 5.189)
P2. Abduction is a form of guessing. (CP, 5.171, Peirce, 1992, p. 128)
P3. A successful abduction provides no grounds for believing the abduced proposition to be true. (Peirce, 1992, p. 178)
P4. Rather than believing them, the proper thing to do with abduced hypotheses is to send them off to experimental trial. (CP, 5. 599, 6. 469-6. 473, 7. 202-219)
P5. The connection between the truth of the abduced hypothesis and the observed fact is subjunctive. (CP, 5. 189)
P6. The inference that the abduction licenses is not to the proposition *A*, but rather that *A*'s truth is something that might plausibly be conjectured. (CP, 5. 189)
P7. The "hence" of the Peircean conclusion is ventured defeasibly. (CP, 5. 189)

We have already discussed the factor of surprise. Surprise is an agent's reaction to the defeater of a default and is recognition, virtual or otherwise, that something is contrary to a quasi-nomological regularity, is out of step with matters of course. In abductive contexts this flavour of surprise is retained, and a further feature is emphasized. Abductive surprises are agenda-motivating cognitive irritations. They are irritations prompting an interest in determining what, if anything, would "lift" the now surprising event to a regularity of course. This is not to say surprises are intrinsically agenda-setting. Ozzie's three-leggedness may induce

Peircean surprise, but it is hardly likely that it would motivate a search for conditions, if any, under which three-leggedness would be a matter of course for ocelots.

As regards (P4), let us note a possible confusion between what I mean by cognitive economics and what Peirce means by the economics of research. Peirce advances the interesting suggestion (which is nicely discussed in Rescher ,1976b) that a decision whether to send an hypothesis to experimental trial is rightly affected by the effort of so doing, not excluding the financial costs involved. For me, a cognitive economy is one in which some practical equilibrium is sought between cognitive aspirations and the cognitive resources available for their attainment. In an obvious but not useless caricature, we might say that Peircean economics is white-coated and grant-sensitive and, mine is everyday.

In much of the AI literature, abduction is any form of backwards chaining. This is a usage more loose than principled and, as such, there is no harm in it. Historical origins aside, we could, if we liked, regard the backwards chaining formulation as fixing a generic notion of abduction, with the Peircean idea a conceptual species of it. What matters here is the importance of the Peircean idea for the cognitive economics of individual agents.

Nearly all abductive logicians drop the name of Peirce. Anything less would be discourteous. But when they turn their efforts to exposing the logical structure of abductive reasoning, all these seven features of abduction are rarely present. This is easily seen by inspection of *the standard schema for abduction.*[346]

Suppose we have it that E is a sentence reporting some event or state of affairs, K is a knowledge-base, $\rightarrow$ a consequence relation, H a hypothesis, and $K(H)$ the revision of K upon the addition of H. Assume that the following facts obtain:

1. E
2. $K \nrightarrow E$
3. $H \nrightarrow E$

---

[346] Notwithstanding some variation in formulation, essentially this schema is the one to be found in Aliseda, (2006), Flach and Kakas (2000), Kowalski (1979), Kuipers (1999), Kakas *et al.* (1995), Magnani (2001), parts of Magnani (2009), and Meheus, *et al.* (2002) among others. In Gabbay and Woods (2005) we called it "the AKM schema", after the letters that begin the surnames of some of the important writers who espouse it.

Then an abduction is the derivation of H from three further facts:

    4. $K(H)$ is consistent.
    5. $K(H)$ is minimal.
    6. $K(H) \rightarrow E$.

Accordingly,

    7. H.

The Peircean elements triadic and otherwise, are not expressly present here. The standard schema imposes no requirement that E be surprising, or that successful abduction be non-probative or evidentially inert, or that the sentence $\ulcorner K(H) \rightarrow E \urcorner$ be in the subjunctive mood, or that the conclusion of an abduction be simply that H can plausibly be conjectured, or that the schema's conclusional operator mark the inference as defeasible. Neither is there any reflection on what it is for something to be a matter of course. It is desirable to have a schema in which these omissions are repaired. To that end, let us introduce the idea of an *ignorance-problem.*

11.2 *Ignorance-problems*

Abductions are responses to ignorance-problems. An agent has an ignorance-problem in relation to an epistemic target that cannot be reached by the cognitive resources presently at his command, or within easy and timely reach of it. If with respect to some fact or state of affairs there is a question Q and some proposition α that would answer it, you have an ignorance-problem with respect to Q if α is not presently in your knowledge-base. You have a problem in the form "What do I need to know to answer Q. What is the missing α?" Two of the most common responses to an ignorance-problem – neither of which is abduction – are (1) *subduance* and (2) *surrender.*

In the first case, one's ignorance is removed by new knowledge, and an altered position is arrived at which may serve as a positive basis for new action. In the second case, one's ignorance is fully preserved, and is so in a way that cannot serve as a positive basis for new action. (New action is action whose decision to perform is lodged in reasons that would have been afforded by that knowledge, once acquired.) For example, suppose that you are writing a letter to a friend and that you can't remember how to spell "accommodate". Has it got two *m*s or one? You don't know. If you consult a dictionary or go online or ask your officemate, you will come to know in short order. This is subduance. You would be so acting as to eliminate your ignorance. But if you are alone in a canoe in the middle of a lake north of Sioux Lookout, and if you want to finish your letter then and there, you'd be better advised to give up on "accommodate" and make do with "take care of". This is surrender.[347]

There is a third response that is sometimes available. It is a response that splits the difference between the prior two. It is abduction. Like surrender, abduction is ignorance-preserving, and like subduance, it offers the agent a positive basis for new action. With subduance, the agent overcomes his ignorance. With surrender, his ignorance overcomes him. With abduction, his ignorance remains, but he is not overcome by it. It is a response that offers the agent a reasoned basis for new action in the presence of that ignorance. Of course, no one should think that the goal of abduction is to *keep* oneself in ignorance. The goal is to make the best of the ignorance that one chances to be in.

11.3 *The G-W schema*[348]

The basic features of abduction can be described informally. You want to know whether something α is the case. But you don't know and aren't in a position here and now to get to know. However, you observe that if some further proposition H *were* true, then it together with what you already know *would* enable you to answer your question with regard to α. Then, on the basis of this subjunctive connection, you entertain the suspicion that H *is* true and, on *that* basis, you release it provisionally for subsequent inferential work in the relevant contexts.

More formally, let T be an agent's epistemic target at a time, and K his knowledge-base at that time. Let K^* be an immediate successor-base of K that lies within the agent's means to produce in a timely way. Let R be an attainment relation for T and let $\leadsto$ denote the subjunctive

---

[347] Other responses are hybrids. At Victoria Station you catch a glimpse of a man in a crowd, who looks like your brother, Harry. It can hardly be, for isn't your brother in Shanghai? But wasn't the man in the crowd a dead-ringer for Harry? The crowd is slowly approaching. So you decide to wait and see. When he gets close enough, you'll know for sure. This is not quite subduance. You don't go out and get new knowledge; you wait for new knowledge to get to you. Nor is it surrender. You haven't given up your interest in knowing whether the man is Harry. You have put it on hold, waiting for him to get closer. You don't mutter, "Who cares!", and stalk out of the station.
[348] After Gabbay and Woods (2005).

conditional connective. *K(H)* is the revision of *K* upon the addition of *H*. *C(H)* denotes the conjecture of *H* and H^c its activation. Let T! $Q(\alpha)$ denote the setting of T as an epistemic target with respect to an unanswered question to which, if known, α would be the answer. Accordingly, the general structure of abduction is as follows.

1. *T!* $Q(\alpha)$
2. ~*(R(K, T)*   [fact]
3. ~*(R(K*, T)*   [fact]
4. *H ⊬ K*   [fact]
5. *H ⊬ K**   [fact]
6. ~*R(H, T)*   [fact]
7. ~*R(K(H), T)*   [fact]
8. *H ⤳ R(K (H), T)*   [fact]
9. *H* meets further conditions S1, ...S_n [fact][349]
10. Therefore, *C(H)*   [sub-conclusion, 1-7]
11. Therefore, H^c [conclusion, 1-8][350]

---

[349] Of course, the devil is in the details. Specifying the S_i is perhaps the hardest open problem for abductive logic. Happily, it is not necessary for present purposes to solve this problem. In most mainstream treatments, the S_i include the consistency and minimality constraints at lines 4 and 5 of the standard schema. (I have my doubts, which I'll expand upon later in the note just below). Since for many abduction problems there are unboundedly many candidate hypotheses, one of the jobs of the abducer is to shrink this space from typically many to hopefully one, that is, by producing a poset at least, and preferably one that lodges the best choice in a unit set. We have already seen the difficulty in seeing how this is done, both formally and empirically. See also Gabbay and Woods (2005b) for discussion of cut-down filters.

[350] For a more detailed formal treatment see Gabbay and Woods (2006b). I might add that although part of the AKM consensus in his early writings (see n. 1, 2 pages ago), Magnani's views have evolved. (Magnani, 2009) develops an ambitious and sweeping taxonomy, in which G-W abduction is but a part. Also included are the following abductive types: creative diagnostic, explanationist, habitual, instinctual, instrumental, manipulative, model based, morphodynamical, multimodal, narrative, neural, and visual. Let's note, however, that Magnani thinks, and I too, that the ignorance-preservation property cuts fairly deeply across most of these distinctions. For an excellent survey of Peirce, see Park (forthcoming).

It is easy to see that the distinctive epistemic feature of abduction is captured by the schema. It is a given that H is not in the agent's knowledge-base. Nor is it in its immediate successor. Since H is not in K, then the revision of K by H is not a knowledge-base successor set of K. Even so, $H \rightsquigarrow R(K(H), T)$. So we have ignorance preservation, as required.

At this point, it is advisable to guard against some misconceptions. When I say that an abduction involves the activation of a hypothesis in a state of ignorance, it is not at all necessary, or frequent, that the abducer be wholly in the dark, that his ignorance be total. It need not be the case, and typically isn't, that the abducer's choice of a hypothesis is a blind guess, or that nothing positive can be said of it beyond the role it plays in the subjunctive attainment of the abducer's original target (although sometimes this is precisely so). Abduction isn't mysticism. In particular, it is not foreclosed that there might be evidence that lends a hypothesis a positive degree of likelihood. But when the evidence is insufficient for activation, sometimes explanatory force is the requisite "top-up". Abduction is often a deal-closer (albeit provisionally) for what induction cannot bring off on its own.

Relatedly, we should also note that Gabbay-Woods schema caters for the *degree* of the ignorance that an abduction preserves. Let α be a proposition with respect to which you have an ignorance-problem. This means that your knowledge doesn't meet the epistemic standard embedded in your cognitive target. But there are epistemic standards and epistemic standards. Let S be the standard that you are not able to meet (e.g., that of mathematical proof). It is possible that there is a lesser

---

Coming back briefly to the consistency and minimality conditions, my main worry is that they lack a convincing motivation. Most logicians who insist on consistency have a bad track record in managing inconsistency. That's not a good enough reason to suppress the fact that in real life some of the most fruitful hypotheses weren't consistent with what was known at the time of their engagement. Minimality is differently problematic. If minimality means "of least complexity", this might lead to smoother search-algorithms. If it means "most similar", it provides a much too strong constraint. A reasonable constraint would be to favour hypotheses of least overall complexity consistent with an overall good result. However, as my undergraduate student Frank Hong has pointed out, perhaps the minimality and consistency constraints could be reinstated by a maximal similar worlds semantics for the subjunctive conditional at line (8) of the G-W schema.

epistemic standard S' (e.g., having reason to believe) that you do meet. What is preserved is your ignorance relative to the first standard S, not the second S'. Suppose now that there is an H such that $K(H)$ is a successful abduction with respect to this target. That this is so is no reason to have a *higher* degree of belief in H than you might have pre-abductively. So it bears repeating that abducers needn't be wholly in the dark about the matters that concern them.

There are lots of cases in which abduction stops at line 10, that is, with the conjecture of the hypothesis in question but not its activation. When this happens, the reasoning that generates the conjecture does not constitute a compellingly positive basis for new action, that is, for acting *on* that hypothesis. Call these abductions *partial* as opposed to full. Peirce has drawn our attention to an important subclass of partial abductions. These are cases in which the conjecture of H is followed by a decision to submit it to experimental test. Now, to be sure, doing this is an action. It is an action *involving* H but it is not a case of *acting on* it. In a full abduction, H is activated by being released for inferential work in the domain of enquiry within which the ignorance-problem arose in the first place. In the Peircean cases, what counts is that H is withheld from such work. Of course, if H goes on to test favourably, it may then be released for subsequent inferential engagement. But that is not abduction. It is induction.

Epistemologists of risk-averse bent might be drawn to the idea that what I am calling partial abduction is as good as abduction ever gets and that complete abduction, inference-activation and all, is a mistake that leaves any action prompted by it without an adequate rationale. This is not an unserious objection. Suffice it to say that there are real life contexts of reasoning in which such conservatism is given short shrift, in fact is ignored altogether. One of these contexts is the criminal trial at common law.[351] Another is various kinds of common sense reasoning.

The Gabbay-Woods construal exposes abduction as a very common form of practical reasoning. It tells you how to act in the absence of knowing what to do. It provides the benefits of action without the cost of knowledge. It provides this guidance not on the basis of supporting evidence but in the absence of it. It provides the agent with an answer not to an epistemic question but rather a prudential one. It tells the agent that it is worth the risk of putting the conjectured proposition to provisional inferential use, and of acting in ways consonant with its

---

[351] See, for example, Woods (2007b, 2008a).

provisionality. It is an action-guiding practice of immense and indispensable economy. It is a scant-resource adjustment strategy.

The Peircean element of surprise (P1) is preserved in the Gabbay-Woods model as the cognitive irritant inherent in an ignorance-problem. Since abduction is ignorance preserving, and since it produces, at most defeasibly (P7), a warrant for conjecturing H, and since the link between $K(H)$ and T is merely that of subjunctive attainment (P4), the Peircean elements of non-probabitivity (P3), conjecture-only (P6) are clearly present. What there may be some doubt about is (P4). Peirce's insistence that the only way in which it is justified to act on the conjecture of H's truth is to send it to the on-going research programme for experimental test. This is partial abduction by the lights of the G-W schema. Full abduction may seem to take things further than Peirce was prepared to go. I am not so sure. Certainly one way to act on $C(H)$ is to send it to the lab and suppress all premissory use of it (no matter how provisional) until the tests have been run. But not everything plausibly conjecturable is experimentally testable. One way to test an experimentally untenable conjecture is precisely by putting it to provisional premissory work in the domain of discourse in which the abduction's ignorance-problem arose. (Consider for example the action-at-distance hypothesis.) So I conclude that in this respect the G-W model's full abduction is not seriously at variance with Peirce's abductive conservatism. But what of the element of guess-work (P2)? Where is it to be found in the G-W framework, and is it really missing from the standard schema? The purport of the G-W model is that it is reasonable to detach for provisional action a proposition that doesn't repair your ignorance, and that a central part of this rationale is that the *subjunctive* fact that if true it would play a role in hitting the desired target. If this isn't guessing – admittedly not blind guessing – I don't know what would be. Perhaps the guessing element is indicated by the standard schema, but a good deal less vividly, I should say, given the schema's suppression of the essentially subjunctive character of abduction.

Back briefly to (P7), defeasibility of abduction has a twofold presence in abduction. The "therefore" of the G-W schema's line (10) exhibits the defeasibility of links in which the schema's premises are linked nonmonotonically to its conclusion. But line (10) also embodies the defeasibility of hedges, a defeasibility which is also carried over to line (11).

There remains to say something about the triadic notion of matters of course. As we have it from the previous chapters, α is a default for X when indications to the contrary would come as a Peircean surprise to him. Accordingly, let us say that

Proposition 11.3
MATTERS OF COURSE: *An event or state of affairs reported by a proposition α is a matter of course for X if it is a default for X.*

This would be a good place to correct a possible misconception. When I say that the standard schema fails to capture Peirce's concept of abduction, that it misses out on Peirce's Triad, I don't mean to impugn its contributions outright. Far from it. True, Peirce's abduction is a good deal more different an inferential enterprise than the reasonings charted by the standard schema. But it might well be thought that going with the broader notion offers the promise of a smoother and more unified theoretical treatment. Perhaps this is so. My own view is that the specialness of the Peircean idea is worth the trouble. What makes it special are the three properties of the Triad – surprise, matter of courseness and suspicion – each of which has been neglected by philosophers and each of which has a huge provenance in the cognitive affairs of the reasoning individual.

11.4 *Explanation*

My claim so far has been that abduction is a basic scant-resource adjustment strategy and that it is its Peircean features that make it so in a quite peculiar way. To the extent that the standard model leaves these features unrepresented, it generates some unclarity about how nonmonotonic reasoning operates in the cognitive economy. Even so – and notwithstanding these differences – there is something that the two models have in common. They accommodate – or appear to – the widely held view that abduction is typically, or at least frequently, inference to the best explanation.[352] Here is Hintikka on this point:

> Most people who speak of "inferences to the best explanation" seem to imagine that they know what explanation is. In reality, the nature of explanation is scarcely any clearer than the nature of abduction. (Hintikka, 2007 p. 40)

---

[352] Harman (1965) and Lipton (1991/2004). See also Thagard (2007).

383

Well, then, what *is* explanation and how is abduction related to it? Hintikka himself advances a well-recognized answer to the first of these questions:

> ... explaining an explanandum E is to derive it from [a] ... background theory T plus a number of contingent truths A that are relative to E and that have to be found in order for an attempt to explain E is to succeed. (Hintikka, 2007 p. 41)

We see that Hintikka's background theory T plays the role of the standardist's knowledge-base K, that Hintikka's A_i play the role of the standardist's hypotheses H_i, and that Hintikka's derivation is the standardist's "→". Similar pairings are obvious for the G-W model, except that Hintikka's derivation is the Gabbay-Woods attainment relation R. It is obvious upon inspection that Hintikka's analysis of explanation, or anything like it, makes it impossible that abduction, whether taken in the standardist's way or the G-W way, is inference to an explanation. In the explanation schema Hintikka's A_i are contingent truths. But in the abduction schemata, the A_i are hypotheses and would be disqualified as hypotheses were their abducer to advance them as contingent truths. We see, then, that on this quite common philosophical understanding of explanation, abduction cannot be inference to an explanation, even though sometimes it is inference *from* a would-be (or subjunctive) explanation. This critical feature is precisely what the standard schema leaves out of account and what the old version the Gabbay-Woods schema was not sufficiently clear about.

There is a further difficulty posed by linking abduction to explanation. Let E describe some phenomenon and suppose that someone X wants to know what explains this phenomenon or what, if true, would explain it. This is an ignorance-problem for X. Details aside, suppose that X settles upon H. H explains E or, if true, would explain it. This solves X's ignorance-problem. It tells her what she wanted to know. But this is not abduction. With reference to the standard schema, X stops at line (6). With reference to the G-W schema, it stops at line (8). It does so each time in a way that furnishes a *premiss* for a possible abduction. But it produces no abductive *conclusion*.

Proposition 11.4
AN ABDUCTION-EXPLANATION RIVALRY: *Let the R of the Gabbay-Woods schema be interpreted as explanation, and let the schema's subjunctive conditional premiss be* $^{\neg}H \rightsquigarrow K(H)$

explains T⁻. *Then for any sense of "explain" for which explanation is evidentially probative with respect to* H, *the schematized inference is a failed abduction.*

It is notable that Peirce's schema makes no express mention of explanation. It provides that if A were the case then C would be a matter of course. I can see in this nothing to suggest any intrinsic tie to explanatory force. It is true that, beyond the schema, there is in Peirce's scattered remarks about abduction nothing that we could call a comprehensive account of it.[353] But as far as I can tell, it is not simply the case that Peirce's abduction is, as such, inference to the best explanation, even on a conception according to which explanation is evidentially and probatively inert. Consider, for example, the day in 1900 when Max Planck announced to his son a discovery that would rival Newton. Planck was troubled by what the physics of the day had to say about the mathematical structure of the laws of black body radiation.[354] The mathematics for high frequencies were strikingly different than those for low frequency. Granted that high frequencies differ from low, Planck thought that the gap between their respective laws was larger than those differences called for. The black body situation was a Peircean surprise for Planck. The physics of the day was out of joint. Of course, we could on Planck's behalf ask for an explanation of this dislocation. Planck himself was not in much doubt about the answer. The explanation is that physics of 1900 wasn't quite up to the task of handling black body radiation as it should. Planck had an ignorance problem, rightly enough. But it wasn't the problem of not knowing what explains the Peircean

---

[353] But see again Park (forthcoming).

[354] The question was how to describe the spectal radiance of electro-magnetic radiation from a black body at a given temperature. The problem was that experimental results were thought to suggest two laws, the Rayleigh-Jeans Radiation Law for very low frequencies, and Wein's Radiation Law for very high frequencies. Let v be a frequency, c the speed of light, k the Boltzman constant, T the temperature in kelvins, and B a black body. Then the Rayleigh-Jeans Law is:

$$B_v(T) = \frac{2v^2 \, kT}{C^2}$$

Let α is a constant giving the numerical solution of the maximization equation and h the Planck constant. Then Wein's Law is:

$$v_{max} = \frac{\alpha kT}{h}$$

surprise created by physics' contemporary provision for the black body asymmetries. Planck wanted to know what it would take to alter physics in a way that would make those asymmetries go away. That was the ignorance-problem with which he was faced. The great breakthrough was achieved when he saw (or thought he did) that if the quantum hypothesis were true it would enable a reformation of physics in which the black body dislocation is set and healed. It is a tribute to Planck's genius that, as things turned out, not only did a quantumized physics set and heal the black body dislocation, but that the new physics was subsequently graced with empirical confirmation on a scale that would have astonished its founder. Still, in 1900, it never occurred to Planck that the quantum hypothesis had explanatory force. How could he have done? In 1900 he thought that the quantum hypothesis lacked physical meaning.[355]

Still, explanationism persists in the present-day literature. It is not difficult to see why. Explanationists prefer to override the distinction between pragmatic, instrumental, accommodative, and other forms of nonexplanationist abduction. There is a virtue of sorts in abstracting away from differences. Again, there are gains in the form of generality and unificational ease. Readers who favour this way of proceeding would do well to organize their thinking around Lipton and Thagard. Those of contrary disposition will have to tough it out with Magnani (2009).

If abduction is taken in Peirce's way, if it possesses its triadic features, it follows at once that a good deal of hypothesis-selection is non-abductive. Consider a case. It is a dark late afternoon in January. You peer out the window and are able to see nothing but the sidewalk under the street lamp in front of your window. You look down and see that the pavement is wet. "Damn", you think, "it's been raining". Of course, it might not have been raining. Perhaps some mischief-maker spilled a barrelful of gyclerine in front of your house. Had it been raining, the wetness below would be a matter of course. But there is no surprise here. How could wet sidewalks in January (in London or Vancouver) surprise anyone? By Peircean lights, this is not abduction. This is induction.

---

[355] Other such examples are Gell-Mann's postulation of quarks, and Newton's of action at a distance. (When Newton declared that he did not dain to fain hypotheses, he wasn't denying that action at a distance is a hypothesis, but rather a hypothesis with explanatory force). Magnani (2009), pages 72-76, ascribes a non-explanative abductive character to Gödel's reflections in (1944) and (1947) on the ontology of mathematics. Gödel's non-explanative abductivism, and Russell's too, is also discussed in Gabbay and Woods (2005), section 5.6.

11.5 *Is abduction ignorance- preserving?*

If the Gabbay-Woods analysis stands up, there is an attractive and straightforward way of characterizing three of the main modes of inference. Deductive inference is truth-preserving. Inductive inference is likelihood-enhancing. Abductive inference is ignorance-preserving. In fact, however, a little further reflection shows that even if the first two of the three are satisfactorily characterized, the third isn't. At least it isn't if the CR model of knowledge is the way to go. How so?

Peirce insists that the abductions supporting conclusions $C(H)$ and H^c lend no evidential support to H. That $C(H)$ and H^c are abductively well-grounded is no reason to believe H. Peirce's recommended course is a stout agnosticism towards H. Yet the empirical record is awash in defections from Peirce's prescription. In lots and lots of real life cases abducers of $C(H)$ and H^c will be led to believe H. Call the Hs that discomport with Peirce's recommended agnosticism $B(H)$s. A $B(H)$ is a believed abductively produced H. It can now be shown that when a successfully abduced H is also a $B(H)$, H is something the abducer may have attained a knowledge of.

For concreteness, let's go back to an earlier example. It is 4:30 on that same chilly Tuesday afternoon in January. You have parked your car in the garage at the rear of your house, and have started to approach the back door, which is your habitual point of entry. You see that the door is ajar and the side-window is broken. Something is clearly out of joint. You and your wife are the sole residents, and the only other regular visitor is the house-cleaner, whose inflexible schedule puts her on the scene Thursdays only. Not Tuesdays. Besides, your wife has never yet come home from her work this early. She's hooked on after-hours palates at Fitness World. Of course, today might be an exception, or the cleaner might actually have shifted from Thursday to today. Neither can it be entirely ruled out that when you left the house this morning, you neglected not only to lock the back door but even to close it. And the window might have been broken by the ball of the boy next door, witheredly and wistfully practicing his winterly swing. All the same, you are also seized of the possibility that there's been a burglary, indeed that it might now be a burglary in progress. Nothing short of your wife popping out of the door with garbage bag in tow would induce you to set foot in the place. You believe the burglary-hypothesis H. Your H is a $B(H)$. You dial 911.

Let us stipulate that a burglary is indeed in progress. Your $B(H)$

is true. The question now is whether $B(H)$ was well-produced. This is to ask whether your hypothesis-forming devices and your belief-forming devices have fallen into an accord in which each partner was in good working order, and operating properly on good information. It is important to emphasize that an affirmative answer may leave it open whether the facts of the situation are evidence for H, but does not require it. The question isn't whether your information justifies $B(H)$, but whether it stimulates your devices to produce $B(H)$ and the belief was well-produced. I see nothing in the facts of the case to suggest that these conditions must be failed. So we may conclude that on the CR-model of knowledge,

Proposition 11.5a
IGNORANCE-PRESERVATION REVISITED: *Although successful abductions are frequently ignorance-preserving, it is* not *intrinsic to abduction that this be so.*

Proposition 11.5b
KNOWLEDGE-STIMULATION: *When successful abductions of* C(H) *and* H *discomport with Peirce's recommended agnosticism, there are ranges of cases in which they stimulate the knowledge that* H.

Not everyone will like this turn of events. Either way, there appears to be something amiss.

Proposition 11.5c
HOBSON'S CHOICE: *If we retain the CC-model of knowledge, it is difficult to see how to avoid granting to successful abductions an* intrinsically *ignorance-preserving character. On the other hand, we can recover an ignorance-removing potential for abduction if we adopt the CR-model of knowledge. But if we do that, we make room for* unevidenced *knowledge both in science and cognitive life in general.*

Some will say that the first view makes abduction too grudging towards knowledge. Others will say that the second view makes abduction too casual about knowledge. I see no way of diffusing this tension, short of settling the more general CC-CR rivalry which it instantiates.

11.6 *What Church and Turing knew*

This returns us to a question we considered in chapter 9, section 9.9, "What Fermat knew." It is the question of what it takes to have a good mathematical mind, of what it is that mathematicians have that the rest of us lack. Most of us would agree that good mathematicians have good mathematical instincts. Not perfect, but good; indeed good beyond the common norm of goodness. Good mathematicians have a talent for good mathematical ideas. The rest of us, in comparison, have hardly any. It lies in the nature of a mathematical intelligence that good ideas will be thought up with a notable frequency. Still, it would be difficult to credit that a good mathematician's good ideas wouldn't outrun the ones that are true. In this regard, mathematicians are like everyone else. All of us have well-produced beliefs that significantly outrun the true ones. It lies in the general nature of well-produced beliefs that lots of them will be false. It strains credulity that the same wouldn't be true of well-produced mathematical belief. Mathematicians, like the rest of us, have more good ideas than true. Mathematicians have a facility for catching the ring of truth, never mind that what rings true often turns out to be false. Without this facility nothing would be accomplished in mathematics. The heart and soul of mathematics is the disciplining of good ideas. This could not be done were there no good ideas to test. Proof is the preferred manner of mathematical testing. But it would be foolish to overlook the frequency with which a mathematical idea wins acceptance by its indispensability to proofs whose conclusions are antecedently known or taken for granted. The ideas of the creative mathematician are hypotheses. Hypotheses circulate in mathematical circles respectably and fruitfully, at a rate that exceeds their eventual demonstration.

One of mathematics' most hard-worked conjectures is the Church-Turing thesis, a fusion of a good idea of Alonzo Church and an equivalent idea of Alan Turing.[356] Church's idea was that the Turing computable functions coincide exactly with those that are computable in an intuitive sense or, as is said, effectively calculable. Turing computable functions coincide precisely with recursive functions, and also with those that are computable by register machines. The Church-Turing thesis is a conjecture, not a theorem.[357] It is almost universally believed not to admit of proof. Yet it has foundational impact on computability and recursion theory has been enormous.

It would be hard to imagine what modern-day mathematics would look like had the Church-Turing idea not been thought of, or, once thought of, so willingly embraced by the algorithmics community. A rather impressive abduction is discernible in so broad and favourable

---

[356] Church (1936) and Turing (1936-1937).

[357] It is accepted as a theorem in some branches of constructivist mathematics.

adoption of an unprovable idea. We could schematize it as follows: Let us suppose that it is 1936. Let T! Q (α) be the target constituted by the desire to know what proposition α would answer the question, "What is the precise definition of effective calculability?"

1. T! Q (α)   [setting the target]
2. What is presently known doesn't meet the target, that is, doesn't provide an answer Q.
3. Nothing else that lies within our easy grasp would answer it either.
4. The Church-Turing thesis is not an object of knowledge; it is only a hypothesis.
5. Nor is it something whose knowledge can be attained in an easily effected update of our knowledge-base.
6. So the Church-Turing thesis doesn't close our epistemic agenda.
7. Neither does the addition of the thesis to what we actually do know.
8. But if the Church-Turing thesis were true, we'd have a knockdown answer to what we wanted to know. If the thesis were true, Church-Turing computability would fully capture the meaning of effective calculability.
9. Hence there is reason to suspect that the thesis is true.
10. So let's release it for provisional premissory work in the general theory of recursive and computable functions.

It is interesting that a sizeable majority of present-day practitioners would be inclined to reject premises (4), hence (2), (6), (7) and to regard (8), (9) and (10) as otiose. At least, that would be their virtual position, or what they would take for granted. Perhaps they would change their minds if asked. If we replaced in the effected lines all occurrences of "known" with "proved", those reservations would evaporate. But, as we saw in chapter 9, to make provability a condition on mathematical knowledge is a decision of striking epistemological hostility. It makes ignoramuses of even our most gifted mathematicians on a scale that doesn't bear thinking about. The expression "Known but not proved" should not be an oxymoron.

Suppose that the Church-Turing abduction is sound. In the period from 1936 until now, the fruitfulness of the Church-Turing thesis has become steadily apparent. The fact that modern mathematics would be considerably dishevelled by the conjecture's suppression is revealing. The conjecture is a piece of good mathematics that has withstood

390

more than eighty years of high-stakes wear and tear. That alone is reason to suspect that the thesis is true. If so, it is important for the CC-CR rivalry. The question before us is whether the Church-Turing is knowledge. For people of a CR-persuasion, the answer is Yes, provided the thesis is true and its belief well-produced. For people of a CC-persuasion – certainly a JTB-persuasion – the answer is also Yes, provided that the thesis is true and the belief well-justified. *Anyone* who agrees that a conjecture's cumulatively good service over eighty years of high-stakes wear and tear *is* good reason to believe it is an epistemologically secure position. If he is a CR-ist, he can claim that, if true, his well-produced belief amounts to knowledge. If he is a CC-ist, he can claim the same thing on the basis of what both parties agree is good reason to believe. From which we have it that

Proposition 11.6a

DOWNGRADING PROVABILITY: *Given that taken together the CC and CR-models are most of what passes for contemporary epistemology, the requirement of proof as a condition on mathematical knowledge is a decidedly minority position.*[358]

---

[358] This is not to overlook the longstanding influence of the Hilbert Programme in the foundations of mathematics (Hilbert, 1904) and the consequences for it engendered by the Gödel incompleteness proofs (Gödel, (1931). Put informally, Hilbert's position is that in matters mathematics there is nothing more to a proposition's truth than its provability in the system that expresses it. What Gödel established is that, under Hilbertian assumptions, there is a way of formalizing arithmetic in which there is a true sentence that is provably unprovable. Call that sentence the Gödel sentence. We now have two readings of this result. One is that, contrary to Hilbert, there is indeed something more to mathematical truth than provability. For isn't the Gödel sentence true but unprovable? The other is that, in conformity with the Hilbert equation, the Gödel sentence is a true propositions proving its own *falsity*. In which case the Gödel sentence is paradoxical in the way that the Liar sentence is, and arithmetic when formalized in Hilbert's way is inconsistent. This, to be sure, is a minority opinion among philosophers of mathematics, but it is dignified by at least one celebrated proponent. (See Wittgenstein, 1964, pp. 180-188 and Wittgenstein, 1956, pp. 500, 116a, and 174e.) We needn't decide this matter here. If Wittgenstein's rendition is true, and classical logic prevails, then all of mathematics is out of business. But no *epistemologist* seriously believes that all of mathematics is out of business, no matter the details of his resistance. See again the end of chapter 1. I will let the point rest here.

On the other hand,

Proposition 11.6b
GIVING PROOF ITS DUE: *When proof of α is an agent's cognitive target, then by definition it is required that a proof of α be found.*

We remarked in chapter 9, the importance of proof as a condition of peer-reviewed publication or other forms of public attestation.

So far as I can see, there is nothing that requires CC and CR-theorists to divide on the question of whether good mathematicians have well-produced mathematical ideas. Where they will differ is on the question of whether well-produced true belief should count as knowledge. I have already had my say about this. So I shan't press the issue any further here.

11.7 *Lexical undetermination*

I want now to return to abduction, or at least the appearance of it. The granddaddy of the formal logical fallacies (FLFs) is the *non sequitur.* At a certain level of generality, it is the only formal deductive fallacy there is. Closer to the ground are the FLFs of affirming the consequent and denying the antecedent. We may say that a *non sequitur* is an error of reasoning in any context in which an agent's cognitive target embeds the attainment standard of deductive validity. For this to count as a fallacy in the traditional sense, it would have to be the case that getting *intendedly* valid arguments wrong is something to which virtually all of us are especially drawn, that we foul up with a notable frequency, and that we don't improve much even after glitches have been pointed out to us. Empirically speaking, there is nothing to support the view that we are hapless deducers by and large. It bears on this that most of our reasoning isn't governed by the validity standard. But even when it is, there is no particular reason to charge ourselves with inferential misperformance of sufficient scope, gravity and tenacity to qualify the *non sequitur* as a fallacy in the traditional sense.

Might affirming the consequent and denying the antecedent be a different story? Yes, they are a different story. Take affirming the consequent (*AC*). When you construct an argument, or engage in a piece of reasoning, in which one of the premisses affirms the consequent of another, and if you then draw as your conclusion the first premiss's antecedent, you have done something whose form at a certain level of generality is this:

1. If α then β
2. β
3. So α.

If the context in which you advance this form is one that requires you to meet the validity standard, then you have made a mistake. You have argued, or reasoned, invalidly. One of the central claims of this book is that in the situations of real life, it is comparatively rare for the individual reasoner to be saddled with the validity standard. Another of our central claims is that errors are made only in relation to attainment standards currently and rightly in play. This gives us the means to say something intelligent about *AC*. One is that when you correctly characterize a piece of reasoning as invalid, it is undetermined as to whether it is an error. Another is the likelihood is that it was not an error. A third is that if indeed it was not an error, then the logical form of affirming the consequent underdetermines – does not capture – what makes the reasoning good. Since interpretation precedes assessment, the default position with respect to reasonings in this gross form is to find a reading of them under which they come out all right. One such is an abductive reading:

1′. If α were the case, β would be a matter of course.
2′. β.
3′. So, defeasibly, it is reasonable to conjecture that α.

The present case provides another example of the lexical undetermination of speech act identity. Here is another example to ponder, also having to do with *AC*.

It is well-known to logicians and linguists that the English particle "if … then" has numbers of uses in which conditionality is not expressed by them. "Beer-conditionals" are a case in point. "There is beer in the fridge if you're thirsty", expresses no conditional at all. If such uses of "if … then" are non-conditional, other uses are what we might call *hyperconditional*. Consider an example.

a. If Harry is a bachelor he's an unmarried man.
b. Harry is an unmarried man.
c. So, Harry's a bachelor.

This has the gross form of an *AC* argument, hence is an argument with an invalid form. But as has long been known, invalid forms do not necessarily invalidate the arguments having them, as witness the materially valid argument before us.[359] So we must amend our earlier claim. Arguments having the *AC* form of affirmations have invalid forms, but they are not necessarily invalid arguments.

Our example raises an interesting question. Why if the argument is valid, would we give formally invalid voice to it? This too is the central question of enthymemes. Why, if enthymematic arguments are (in some sense – perhaps materially) valid, would we express them invalidly? Enthymemes are interesting in their own right. They are not our concern here, beyond noting that formally invalid enthymemes are an economical way of making materially valid arguments.[360] It is cheaper to give informally invalid voice to materially valid arguments than to shuffle about one's propositional space for premises that will remove the informal invalidity in a nontrivial way. It is much the same with materially valid *AC*s. For one thing, sometimes missing premises are hidden in virtual belief.

The lexicon is our concern here. It is possible to *say* "if … then" and not *mean* "if … then". In uttering "If you're thirsty, then there's beer in the fridge", your host is not saying that the beer's presence in the fridge is in any way conditional on the state of your thirst. It is, as we said, a non-conditional use of "if … then". The bachelor example, just above, is both similar and different. It illustrates what we are calling a hyperconditional use of "if … then", a use which expresses not a conditional relation between bachelors and unmarried men, but rather a biconditional relation. In plainer words, it is not infrequently the case that when we say "if … then" we mean what we would mean if we said "if and only if". If this is right, it would go some way towards getting non-abductive instances of *AC* off the hook. That alone is something that counts in favour of it. It counts favourably because the default position of argument interpretation is not to read an argument as bound to the validity standard, and the default position of argument assessment is not to take an argument's invalidity as grounds to think badly of it. It would be reassuring if we could find additional reasons to support the present suggestion. The question is, can we? Yes.

---

[359] Also called semantic validity, material validity is discussed in Brandom (2000) and Woods *et al.* (2002).
[360] See again Paglieri and Woods (2011).

If we examine the untutored lexical performance of native speakers, it is easy to see that a command of "if ... then" is a fairly early achievement. It is not hard to see why. A human individual who lacks the concept of consequence (especially causal consequence) is sorely at risk. So consequence spotting is part of the early endowment of such beings. A similar point can be made about identity, and the related requirements of individuation and re-identification. Our capacity for these also bristles with adaptive significance, and the human animal is not long in getting a grip on it. What, then, of equivalence? We might see equivalence as occupying a point midway between consequence and identity. It is more robust than consequence and less robust than identity. The conjectures to which we find ourselves drawn are that the concept of equivalence is, if at all, a later and tutored achievement, that likewise the idiom "if and only if" has to be taught, and that we can go through life with an abundant linguistic competence without ever giving utterance to "if and only if" with equivalence-expressing intent. It is true that "if and only if" has some currency in English conversational usage. When Harry says "I will attend the reception if and only if Sarah does" this is more often than not to lend emphasis to the alternative "I will attend the reception only if Sarah does". Thus "If and only if" here has the force of "I will attend the reception only if Sarah does, and I am *not* kidding about this!".

Another possibility is that when ⌐If α then β⌐ is made true by the biconditional fact that α if and only if β, any argument in *AC* form is materially valid. And, as before, not only is material validity often more cheaply achieved than formal validity, achieving material validity also makes it unnecessary to achieve formal validity. In a quite general way, formal validity is expensive; and in comparison, material validity is cheap. So it is not for nothing that formally invalid *AC* arguments underdescribe materially valid ones. Whatever we might in the end think of these speculations, they apply with suitable re-adjustment to the companion fallacy of denying the antecedent (*DA*). I leave it to the interested reader to work through the details.

Most of what we find in the psychological literature is at variance with the views proposed here.[361] My view is that, like mainstream logic, mainstream psychology has got *AC* dead wrong. This is not to say that there aren't valuable deviations from this orthodoxy.[362] Of particular

---

[361] Again, and famously, Wason (1966) as well as Marcus and Rips (1979). Stein (1996), p. 2, is a good recent summary of the received view.
[362] Consider, for example, the reservations of Griggs and Cox (1982) about Wason (1966).

interest are suggestions developed in Verschueren *et al.* (2001). One is that reasoners confuse necessity with probability. The other is that reasoners confuse "if ... then" with "if and only if". Aside from the mention of "confusion", which I think misplaced, I am struck by the affinity of these suggestions to our own conjectures here. One way in which to make a "probabilistic" argument in "necessitarian" form is to use an *AC* to make an abductive argument. One way in which to use "if ... then" in our "if and only if" way, is in making an *AC* argument that is materially valid. So I welcome these suggestions from Verschueren and his colleagues.

Further recent encouragement comes from a philosopher. Floridi (2008) argues that *AC* and its dual *DA* are degraded versions of Bayes' theorem – and, as such, are "informational shortcuts, which may provide a quick and dirty way of extracting useful information from the environment." (p. 317) I am much drawn to the resource-bounded character of this analysis, notwithstanding my inclination to think that Bayesian degradation is not an essential part of the story.

11.8 *Ad populum*

Before quitting the discussion of abduction, we should say a brief word about the *argumentum ad populum*. As traditionally conceived of, the fallacy is one in which the tenability of a view is confused with its popularity. Proponents of this view hold that this is a confusion that invalidates the inference of tenability from the fact of popularity, and it is there that the fallacy is committed. We could schematize the fallacy as follows:

1. It is widely believed that α.
2. So α is true.

There are lots of cases in which nothing succeeds like popularity. This is especially evident in the rise and fall of fads, in the world of fashion, popular music and popular culture more generally. The full story of the attractions of the popular would include a chapter on the tug of the popular on belief. Part of that story would have to give weight to our claim in chapter 9 that in lots and lots of cases it is typical of our being told things that we believe them. However, legion as such cases may be, people who are caught up on the sweep of a popular idea do not typically represent its tenability as *following from* its popularity. So most of those cases are of little interest, or importance, to the logical mainstream. I am differently disposed. I think that there are aspects of popular belief other than its fallaciousness that would rightly engage the interest of any naturalized logician intent on producing an empirically sensitive and epistemologically responsive account of the reasoning practices of beings like us. It strikes me, therefore, that before the issue of *ad populum* fallacious ones can profitably be taken up, there are two prior questions that call for attention.

One is how popular beliefs are formed. The other is how we become aware of a belief's popularity. Consider an imaginary case. Harry is out of town on business. When he returns to his hotel early one evening, his friend Sarah has left a message on his voice-mail: "Harry, something dreadful has happened. The Prime Minister was assassinated an hour ago. I'm too upset to say anything more." Harry is shocked by the news and impatient to learn more. But he believes what his friend has told him. By the provisions of chapter 9, if indeed the P.M. has been assassinated and Harry's believing it has been well-produced, his belief amounts to knowledge. Harry knows then and there that the Prime Minister has been murdered. By the provisions of chapter 11, Harry also knows something else, albeit virtually. He knows that the *whole world* knows of the P.M.'s murder, and hence that the proposition "The P.M. has been murdered" is a widely popular belief – widely popular not in the sense of widely approved of, but rather in the sense of being widely *held*. (Although you never know with Prime Ministers.). A few years later, when Harry was recounting the sad event to his small children, one of them asks, "How did you learn of this, Dad?" Harry replies that his friend Sarah had told him it. If some philosopher were lurking nearby he might have asked how Harry came to know, then and there, that the whole world knew it too. Harry might not have had a ready answer. So I will answer for him. Harry presumed – took it for granted – that the P.M.'s murder was a fact that instantiates the no report – no fact regularity (actually, its contrapositive). How could the whole world *not* have known this? Don't prime ministerial murders have a public?

Suppose now that the report of the Prime Minister's assassination was later discovered to have been a hoax. This, too, would be a fact with a public face. The whole world would know of it in a trice, and there would be widespread relief. Of course, as things actually are, the default position is that the P.M. has not been assassinated. In our imaginary example, the default position is that it remains an uncontested fact that the assassination did occur. In each case, information to the contrary would generate Peircean surprise. In each case the new information, if true, would be out of joint with respect to a quasi-nomological regularity. They are different regularities: the prohibition, and sheer difficulty, of prime ministerial murder in the first instance, and the autoepistemic regularity enshrined in the contrapositive of the no report – no fact principle in the second.

More generally, we might say that a belief is popular in a population P to the extent that everyone or nearly everyone holds it (if only virtually). These cases break down into subcases, each turning in some necessary way on features commonly distributed in P, factors such as occasion and capacity. If P is made up of beings like us, it can be said with some assurance that pretty well everyone will believe that crows are black. There is ample opportunity to spot an occasional crow and ample capacity to generalize elastically and hastily. Those who haven't had the pleasure of a personal encounter will have been told it. There will hardly be anyone in the group who hadn't learned this in middle school biology. On the other hand, if the belief in question is that the capital of Bangladesh is Dhaka and P is made up of people like me, it is likely that we have all been told it on some shared occasion by an informant with an uncommonly wide reach – the BBC for example. Of course, there are lots of people in P who knew this geopolitical fact already. But it wasn't a belief held by all until the BBC had occasion to tell it to everyone more or less at once. Such occasions include the slaughter of hundreds of Dhaka's inhabitants in the fiery collapse of a clothing factory. Some facts, including some of the awful ones, instantiate the no report – no fact regularity with full force. Some facts cannot occur without everyone knowing them.

Several chapters ago, we noticed an ambiguity in the question "How do you know?" It would repay us here to give it some further attention. If someone asks, "How do you know that Dhaka is the capital of Bangladesh?" you might say that it's been all over the television lately. For one of its meanings, this is a right and proper answer. But if the question was otherwise intended, if it called for a justification of your belief, the same answer might pass muster. If it did it would be because your interlocutor, like everyone else, knows that this is the sort of thing the press doesn't get wrong, that on this matter your source of information is authoritative. It's what the BBC is in business to do.

Other cases are different. You are on the bus and you overhear the following conversation:

Passenger #1: "Fred's got a pretty bad drinking problem."
Passenger #2: "No kidding! Are you sure?"
Passenger #1: "Well, that's what everyone says."
Passenger #2: "Poor Fred. What a shame!"

The case is not well enough described to make it a safely determinable question as to whether you yourself know that someone named Fred is having trouble with the bottle. But I daresay that there are plenty of such occasions on which overhearing what one stranger tells another about a third does indeed occasion your belief that what the first said of the third is true.

Let's now change the example. Retain the dialogue but replace each occurrence of "Fred" with an occurrence of "Tiger Woods". In each case, you overheard a rumour. In the Fred-case it is not likely to have any circulation beyond Fred's own circle. In the Tiger-case, it has a good chance of "going viral". If it did, it wouldn't be long before the whole world knew that it had widely been rumoured that the world's greatest golfer was again in trouble. In due course, the gutter press would pick it up. Everyone would be talking about it. Everyone would know that everyone is talking about it.

I don't myself think that at this stage the proposition that Tiger is a drunkard would be a well-produced belief. Too much counts against it. Alcoholism is professional death for athletes of this caliber. Moreover, nearly everyone loves to receive and to distribute juicy gossip, the more scandalous the better. They love to see the high and mighty brought low and isn't it the job of the gutter press to smear the big shots? Suppose that the BBC's sports editor instructs his golfing team to start looking into this. Suppose he tells them to pull out all the stops, that he wants to get to the bottom of the Tiger story once and for all. Would this show bad professional judgement? Would it show that the editor's hypotheses-forming mechanisms weren't working as they should?

Some days have now passed. You notice that there hasn't been a word of rebuttal from the Tiger camp, nor has Tiger himself been anywhere in sight. You begin to think that had there been anything to say against the rumour, it would have been said by now, and that had Tiger not gone to ground, this too would have been known, and if known, reported. Then, two days later, the BBC claims that according to an unidentified source at the PGA, there is considerable concern that the story might have legs. You now find yourself thinking that Tiger really does have a drinking problem. Would that have been an ill-produced belief? Would its production show that there was something not quite right with your head? Perhaps we could agree that it would be entirely out of place to circulate your belief as a matter of fact. For one thing, it might not be a fact. For another, fact or not, it would be a nasty thing to do. But that isn't the question. The question is whether, if you *believed* it, your belief-making equipment would have been defective?

We are now at the nub of things. The more widely a belief is held – or proposition entertained or taken note of – the greater the likelihood of correction or rebuttal, if indeed it is something susceptible of such. The greater a public's interest in such a proposition, the greater its interest in its correction or rebuttal and the greater likelihood of their being reported if made. Needless to say, the Tiger-case is a contrary-to-fact possibility. It is only a made-up example. But there are lots of real-life cases to which it generalizes. Clearly, that Tiger is a drunkard is a "proposition of interest". It is, true or false, a proposition that will catch and hold attention, and have a large circulation. Let α be any such proposition, and let $C(\alpha)$ be its "contradiction-space", that is, the space of occasions to deny it, or rebut it or demonstrate its opposite. Let me now propose the following:

Proposition 11.8a
CONTRADICTION-SPACE SIZE: *The more widespread the distribution of α, the larger its $C(\alpha)$.*

Proposition 11.8b
REPORTS OF CONTRADICTIONS: *Let α be a widely distributed proposition of interest and c the purported realization of a member of $C(\alpha)$. Then the distribution of the report of c's occurrence varies proportionally with the popularity and interest of α.*

Proposition 11.8c
INFERENCE *AD POPULUM*: *Let α be a widely distributed proposition of interest and $C(\alpha)$ its correspondingly large contradiction-space. Suppose also that no c of α has been reported. Then if α were actually true, the present circumstances would be a matter of course. Hence, there is reason to suspect that α is true.*

COROLLARY: *In particular, uncontradicted widely-distributed bad news has a way of being true.*

It would be ill-judged to grudge this inference a passing grade. True, it makes essential use of factors *ad populum*, and it is certainly not rock-solid reasoning. But it doesn't need to be rock-solid to pass muster. Hardly any good reasoning is rock-solid. Often such reasoning is good. It is good if, as in this case, it is *abductively* good. Of course, it would be quite wrong to over-value the goodness of this kind of goodness. It is not the goodness of the rock-solid. But it would be at least as wrong – and very stupid – to undervalue its goodness as well.

In the conditions of human life as lived, I don't think we should overlook or make light of the frequency of utterances in the form (or the ease with which find ready lips): "How could α be so widely believed if it weren't true?" I have already conceded the traditionalist his dismissal of the schema: For all values of α,

1. α is widely believed
2. So α is true.

I reject any inference in which it is purported that (2) is a consequence of (1), that (1)'s truth is in some sense sufficient for (2)'s own. Here, too, interpretation precedes assessment. Who in his right mind would make this benighted schema the default interpretation of these not uncommon remarks. The schema is important. By this I mean that it is negatively important. It is in the general case the wrong interpretation, and further illustration of lexical underdetermination. Yet it is precisely this schema that attracts the traditionalist's verdict of fallacy. This is consequential. Even on those occasions on which we think or say, "How could something as widely believed not be true?", we are hardly ever invoking the schema. So we are hardly ever committing the schema's mistake. We have it, then, that

Proposition 11.8d
NOT A FALLACY: *The traditionalist's* ad populum *doesn't occur with the notable occasioned frequency demanded by his own conception of fallacy. It is not a fallacy in the traditional sense of that term.*

It hardly needs saying that it is not my position that received opinion is sacrosanct. Anyone drawn to the views that I am espousing in this book must be at some variance with logic's received wisdom. Anyone holding the concept-list misalignment thesis has entirely

401

separated himself from the received views of the fallacies tradition. The import of our present position should not be misconceived. It amply allows for defection from received opinion. It is paid for when you show cause to give it up.

11.9 *Common knowledge*

Beliefs having the character of the Tiger-belief are common. That is to say, they are widely distributed (in P) and virtually everyone (in P) believes this to be so. Common beliefs play a large role in Aristotle's logic of dialectical argument. They are what he calls *endoxa*. They are the beliefs that everyone holds, or that most people hold, or that are held on the sayso of experts. They are the beliefs in common currency in the Athens of the time. They are received opinion there. Much of it is also received opinion here and now.

We can generalize the notion of common belief to get a conception of common knowledge.

Proposition 11.9a
COMMON KNOWLEDGE: *It is common knowledge (in P) that α if and only if α is a true common belief (in P).*

There is a situation in which all of us have found ourselves. We've remarked to someone that α and have been asked in turn, "How do you know that?" After a bit of mental foot-shuffling we reply, "It's common knowledge that α". What is so striking about this kind of situation is the number of values of α for which it obtains. We have in this the makings of a problem. On the face of it, it is an embarrassingly weak answer, whose underlying reasoning is certainly faulty. But it is not an infrequently given answer. Indeed it seems to be the default answer whenever we find ourselves not knowing what else to say. It is a situation that we all have been in. It is a situation that applies to a great many of our beliefs *ad populum* – that is, beliefs that we ourselves hold and believe that everyone else (in our P) believes as well. And, when α is challenged, the common-knowledge answer is the answer that everyone always thinks of when they can't think of what else to say. Given that the reasoning behind the answer is bad, don't we have here an error of reasoning that occurs with a notable occasional frequency? Wouldn't that take us most of the way to having exposed a fallacy in the traditional sense?

I have my doubts. Consider a case. Harry remarks that α, Sarah challenges with "How do you know?", and Harry replies that it's common knowledge. Now if it *is* common knowledge that α, Harry's original assertion of it is true. We may take it that Harry believes that α is commonly believed. Indeed, by construction of the case, if he believes it he believes that virtually everyone else believes it too. Common beliefs are widely distributed propositions of interest, widely believed to have been uncontested. Harry's position is that had a $c \varepsilon C(\alpha)$ been realized – that is, had α been contested, denied or refuted – he and everyone else would have got wind of it. News of it would not have failed to reach them. So – negation by failure – it may be supposed that there is no such c. For anyone disposed to take a CR-view of such things, the question is whether Harry's beliefs here are well-produced. I see nothing in the nature of the case to support a generally negative answer. Chapters ago, we considered the phenomenon of dialectical erasure. We considered situations just like the present one, except for Harry's quite different response. In the earlier example, Harry didn't know what to say. So he just folded. He folded in a quite particular way. His previously confident belief was now eroded, if not entirely erased. Harry's dialectical impotence caused the original belief to shut down. This is a striking difference from what we have here. Here Harry doesn't fold. He answers the challenge. He tells Sarah that everyone knows that α, and in so doing implicates that Sarah's not knowing it is somehow out of joint. How could Sarah *not* know it?

Does Harry have any call to say that something is common knowledge? In the case before us, his belief that α is common knowledge is well-produced. If α is indeed common knowledge, Harry knows that α is common knowledge and knows that this is something that Sarah should know too. It matters a great deal that people don't usually resort to the common-knowledge answer for α when they think that it is not common knowledge after all. This is not just a matter of conversational sincerity. More basically, it wouldn't usually *occur* to us to make the common-knowledge move in the absence of thinking that α is indeed something that everyone knows. There are well-known regularities about this: One is

Proposition 11.9b
WELL-PRODUCED (1): *Except where there are indications to the contrary, the belief that α is a common belief is a well-produced belief.*

Another is

Proposition 11.9c
WELL-PRODUCED (2): *Except for indications to the contrary, the belief that a common belief α has not been contested (etc.) is a well-produced belief.*

Accordingly,

Proposition 11.9d
A MYSTERY: *If these things weren't so, it would be a mystery as to how common knowledge could come to be.*

Of course, that's just what the nattering nabobs say: Common knowledge is hardly ever knowledge. Common knowledge is what passes for knowledge among the *hoi ploi*. The common man is an ignoramus. So say the nabobs. Rather irritating, say I

CHAPTER TWELVE

ASKING

"How can a philosophical enquiry be conducted without a perpetual *petitio principii*?   Frank Ramsey

"[I]t is only within the framework of a theory of argumentation that one can take an accusation of *petitio principii* into consideration and examine whether or not the implied criticism is legitimate".   Perelman and Olbrechts-Tyteca

12.1 *The modern conception of question-begging*

Harry bumps into Barb at Starbucks. "Don't you have a birthday coming up, Barb?" "Yes", she replies, "it's on the 3rd of next month." Harry's was a question that prompted an answer that gave him the knowledge he sought. Situations like this are utterly common. Asking-telling pairs are senior partners in the sayso-manifolds of cognitive multiagency. Of the two, fallacy theorists have been more interested in asking than telling. True, there are logics of assertion, but in comparison with the booming trade in interrogative and dialectical logics, they are rather small beer. Chapters ago we touched on the idea that the fallacies on the traditional list are inherently dialectical. The answer proposed there was that, with the possible exception of, e.g., begging the question and many questions, they are not. The aim of the present chapter is to cancel the suggestion by showing that begging the question and many questions are not dialectical fallacies after all. The reason for this is not that question-begging and many questions aren't (by and large) dialectical *practices*.

Let us begin with the dialectical case. That begging the question, BQ for short, is a fallacy is an idea which originates with Aristotle. Given logic's already long history, it should not be surprising that Aristotle's views of these matters have in some ways been superseded. But the traditional view retains the original connection between conception and instantiation. Whereas BQ in Aristotle's sense is said to be a fallacy in Aristotle's sense, so too is BQ in the modern sense said to be a fallacy in

405

the modern sense.[363] As currently conceived of, BQ fallacies can be characterized in the following way:

> *The BQ schema*: Let τ be a thesis advanced by Harry. Let α be a proposition forwarded by Sarah as counting against τ. Then Sarah begs the question against Harry's thesis τ if
>
> (1). α's truth is damaging to τ;
> (2). α's truth is not conceded by Harry, does not follow from propositions already conceded by Harry; and
> (3).   is not otherwise ascribable to Harry as a reasonable presumption or a default (for example, the belief that water is H_2O or that London is the capital city of England).[364]

As before, a fallacy is anything that instantiates the traditional concept. It is an error of reasoning that satisfies the other conditions: attractiveness, universality, incorrigibility (and badness).

Not only is it not obvious that the modern BQ falls under the traditional concept of fallacy, it would seem that it actually does not. Question-begging is an *attribution*-error. Sarah's advancement of α against Harry's τ is tantamount to the assertion that Harry already accedes to, or at least is bound to it. If, in so saying, Sarah begs the question against Harry, it is clear that her error is one of false ascription. If this happens in the context of dispute about some issue τ, in which Sarah is attempting to discredit Harry's attachment to τ, Sarah's move can also be seen as a premiss selection error. In her attack on Harry Sarah appropriates as a premiss *a* proposition α that Harry does not accept and is not committed to accepting. Does this perhaps bring us a little closer to the idea that BQ is a mistake of reasoning? I have already made the point that premiss-selection errors are not typically errors of reasoning, hence not fallacies in the traditional sense. But I've conceded two exceptions to this. In selecting a premiss, you commit an error of reasoning if its acceptance is either an affront to reason (as in the case of monstrous belief) or a case-making defect. No one seriously suspects any intrinsic tie between question-begging premisses and affrontery to reason, but

---

[363] See, for example, Johnstone (1967), Stanford (1972), Barker (1976), Woods and Walton (1975, 1977, 1978), Mackenzie (1978), Walton (1991).

[364] BQ is characterized here for single propositions α. It is easily generalized to sets of propositions and its theories.

premisses that damage one's own case are another matter. In the present example, two things are clear: Sarah begged the question against Harry and, in so doing, she certainly didn't advance her case against him. But does her case-making ineffectiveness on this occasion deserve the name of error of reasoning? And – a separate matter – do we really want to call it a fallacy?

Certainly there will be cases – quite a number of them if the empirical record is to count for anything – in which the faulty attribution of belief to another is *inadvertent,* that is, made in good faith and in the mistaken belief that one's addressee actually holds it, or is presently committed to it. When a question is begged through ignorance of what one's vis-à-vis actually believes, there isn't in the general case the slightest reason to convict the beggar of a breakdown of reasonability. Nor is there much to be said for the suggestion that, even so, she has wrecked her own case. In real-life exchanges, the begged upon party has only to acquaint the beggar with her error for the attribution to be withdrawn. If there had been case-making damage, it would have been momentary and reparable. Such is the value of feedback.

There is little doubt that situations such as these occur with some frequency, but not typically in dialogue logics. In the idealized environments of those logics, they are prevented by fiat. It is a requirement of these systems that the parties' concessions be explicitly listed, and that the list be subject to a negation as failure condition. So constrained, you may attribute to your vis-à-vis only those concessions that occur on the list, and nothing gets on the list except by mutual acquiescence. Let us say that in formal systems of dialogue logic parties are bound by a closed world assumption with respect to one another's concessions. In closed world logics, misattribution errors are both knowing and advertent. But closed world logics are idealizations. Nothing in life comes close to rallying to their constraints. There is a reason for it. It is more economical to make an attribution which, even if wrong, could be repaired on the spot than to draw up a list of everything the parties have already said and everything else that this commits them to. Recall here the distinction between errors of accuracy and errors of aptness, and the related distinction between mistakes whose effects are easily reversible and those whose effects we're somehow just stuck with. As with the stem + "ed" rule for the past tense, it is often preferable to favour the apt to the accurate for economic reasons. Repair is usually cheaper than avoidance.

Everything that is known about the empirical record suggests that by a wide margin premissory misattributions of this kind arise from nonculpable ignorance. This gives us grounds to say that

Proposition 12.1

BEGGING THE QUESTION AS GOOD ECONOMICS: *By and large, question-begging in the sense of premissory misattribution achieves the economical advantage of aptness. They are more easily corrected after the fact than avoided in the first place.*

Let us also ask even in the idealized contexts of dialogue logics where question-begging is on-purpose and knowingly bad. Who is to say that it couldn't be resorted to in hopes that the injured party might fail to notice it? To be sure, this is a sneaky debating trick, but do we really want to say that it is a reasoning error? Some argumentation theorists are of the view – Walton is a notable example – that fallacies liberally include, efforts by one party to put something over on the other party, to slip past his attention some dialectical or logical impropriety.[365] Doing so is not unusual; perhaps it even meets a universality requirement. Dialectical slipperiness is something to dislike in any context of argument for which forthrightness, impartiality and cooperativeness are prerequisites.[366] But it comes nowhere close to instantiating the traditional concept. So, interesting as it certainly is, I shall have nothing further to say of it here.

Perhaps the basic way in which to bamboozle a dialectical adversary is simply to lie to him. In actual practice, as opposed to a logician's dialogue games, it is entirely common for an antagonist to add an α to her adversary's stock of beliefs by telling it to him. Suppose this happens. Sarah tells Harry that α, and Harry believes her. But, as it happens, α is damaging to Harry's thesis, and α is false and Sarah knows it. In this pure form Sarah's invocation is not even an attribution error. It is a lie. Who in his right mind would be drawn to the suggestion that the

---

[365] "To say that an argument is fallacious is a strong charge, entailing more than just the claim that the argument is weak, or has been insufficiently supported by good evidence. A fallacy is an underlying, systematic kind of error or deceptive tactic of argument used to deceptively get the best of a speech partner." (Walton, 2000, p. 25)

[366] But aren't rational arguments of all kinds subject to these same requirements? No. Legal arguments at the common law criminal bar are not only not bound by them, but are incompatible with them. The same is true of negotiation-arguments in collective bargaining.

failure to spot such a lie is in general an error of reasoning or, indeed, an error of any kind? And who would be drawn to the idea that telling such a lie is an error of reasoning?

12.2 *Metadialogues*

There is a special class of non-culpable premissory-attribution inaccuracies that requires comment. This is the class of attributions concerning whose accuracy the parties are landed in a sincere disagreement. Take the case in which Sarah's attribution of α to Harry is grounded in the disputed belief that α is a consequence of concessions that Harry has already made. Or consider the case in which Sarah grounds her ascription of α to Harry in the belief that α is common knowledge or that there would have to be something wrong with Harry were he to reject it, and yet Harry does reject it. What makes these cases interesting is that they themselves are occasion for question begging. Sarah selects α as a premiss. Harry objects that Sarah has no business doing so. Sarah replies that α is implied by what Harry has already said. Harry denies this. Now what?

There is a kind of second-level disagreement involving what Eric Krabbe calls a *metadialogue*,[367] that is, a dialogue about a dialogue. In our example, Sarah and Harry are not arguing about α. They are arguing about something else, β, to which Sarah thinks α would be relevant. But now they are arguing about α. More precisely, they are arguing about whether Harry is committed to α. And they are having *that* argument because Sarah intends to use α in her attack upon β, and Harry wishes to show that doing so is defective case-making. (Harry: "You can't make hay with what I don't believe, Sarah. It's not allowed.") So there is point in characterizing this dispute about α as a dialogue about the dispute about β – a metadialogue about that first-level dialogue. When discussants move from an originating dialogue to a metadialogue we might say that they have been party to a *dialogic ascent*.[368] An important questions for dialogue logic is whether there exists a principled and natural upper bound to dialogic ascent.

Let us characterize a "deep disagreement" between parties as one that strains the resources required for its non-question begging

---

[367] Krabbe (2003). See also *Argumentation,* 21, no. 3 (2007) which is a special issue devoted to metadialogues. Finocchiaro (2013) is also important.
[368] Woods (2007c).

resolution.[369] At their deepest, deep disagreements *cannot* be dialectically proceeded with without begging the question. We might take this as a definition of disagreements that are "maximally deep", concerning which consider again a case well known to philosophical logicians. Dialethic logicians claim that there is a small class of semantically distinctive propositions, made so by the fact that they are concurrently classically true and classically false; that is, they are true contradictions. Everyone else insists that no contradiction is ever true. There is a story told about an exchange between the dialethist Richard Routley and the anti-dialethist David Lewis, in which Lewis claims that the Law of Non-contradiction (LNC) is a true law that brooks no exceptions. Routley challenged Lewis to prove this. Lewis refused to be drawn. He thought the demand absurd. LNC is a "first principle", he said (echoing Aristotle). Its truth neither requires nor admits of demonstration. There are no resources to prove the law's exceptionless truth. Its unqualified truth, said Lewis, is beyond the reach of dispute-resolution. But consider this: If LNC is a first principle, then it is a truth whose truth is unprovable. For generality, let α be a proposition which the parties agree to be a first principle. List this agreement as premiss (1) of the following argument:

1. α is a first principle   (mutually accepted as fact)
2. α is true   (from 1, by def. of "first principle")
3. α is unprovable.   (from 1, by def. of "first principle")

But $\langle (1), (2) \rangle$ *is* a proof of α; α's truth follows from its first principleship. And (3) *contradicts* the fact that $\langle (1), (2) \rangle$ is a proof of α. Thus (1), the claim that α is a first principle, generates a contradiction. So we have it by *reductio* that there are no first principles.

History doesn't record whether these further points were actually considered by Lewis and Routley. But for some people certainly, $\langle (1), (2) \rangle$ is a question begging argument, whereas for others $\langle (1), (2), (3) \rangle$ is a valid dismantling of *any* claim of first principleship, including LNC. In the first camp would be those for whom valid single-premissed arguments are question-begging (De Morgan, for example). The other camp has a slighter membership, but is not empty, as witness Woods (2005). Of course, people in the first camp might be mollified by

---

[369] The name "deep disagreement" arises as a technical term in Fogelin (1985). Much the same concept is discussed by Woods (1996, 2000) under the name "standoffs of force five".

extending the proof of α's truth to a doubly premissed argument as follows: ⟨α is a first principle, if α is a first principle α is true, α is true⟩. Since this expanded argument is a proof of α, we can use the same device to show that it isn't: ⟨α is a first principle, if α is a first principle it is unprovable, α is unprovable⟩. The first argument proves α. The second argument proves it unprovable. Contradiction. It is hard to see how disagreements of this sort could be anything other than paralyzing.

It is interesting to notice how quickly our present discussion rose on the wings of dialogic ascent. It started over the issue of whether LNC is true. In no time at all it became an issue of whether LNC was provable, and thereupon whether the proof that it was provable was a good proof and whether the proof that it wasn't provable (because false) was a good proof. It would bear on our interest in BQ if the following turned out to be true:

Proposition 12.2a
ASCENT TO PARALYSIS: *The greater the level of dialogic ascent the greater the likelihood of deep disagreement, hence the greater the occasion for dialectical paralysis and question-begging.*

Sarah and Harry are having it out about something. They can't agree about doctor-assisted suicide. They are now well-launched into the flights of metadialogue, and they are now squabbling, about the definition of following-from. Before they know it, they are locked on the question of the validity or otherwise of the disjunctive syllogism rule. Lou verbalises this stalemate: "Just listen to you! What in the world does disjunctive syllogism have to do with doctors helping people kill themselves?"

Proposition 12.2b
ASCENT TO IRRELEVANCE: *The greater the level of dialogic*

But, of course, irrelevance is not question-begging. It would not serve our present purposes to shape the assisted suicide discussion around the claim lodged by Proposition 12.2b.[370] This is not to say that BQ and irrelevance have nothing in common. They are *both* responses to deep disagreement. But they are responses that tug in opposite directions. They are complementary vectors or polar opposites. We might even say that

---

[370] The connection of irrelevance to dialogic ascent is discussed in further detail in Woods (2007c).

Proposition 12.2c

POLAR OPPOSITES: *In contexts of dialogue ascent, when a disagreement is very deep, irrelevance and question begging are related in the following way: Each is at risk in any effort to avoid the other.*

If you are determined not to beg the question against your opponent, the greater the likelihood that your further contributions will be irrelevant even if true. If you are determined to deep your contributions strictly relevant, the greater the likelihood that they will be question-begging. These polarities serve as a rough working definition of deep disagreement. Even so, since irrelevance is the opposite of BQ, it cannot be directly implicated in the analysis of question-begging and so I will let it be.[371]

12.3 *Theses and thesis-holders*

Readers will have noticed that the BQ schema offers only a sufficient condition of begging the question. It underdescribes it. It might strike us that this is not a good position to be in. It risks that what the account of BQ leaves unsaid might be decisive for the question of its fallaciousness. Would it not behoove us to try for a fuller description?

The present characterization tells us, rightly, that question-begging often occurs in argumentative contexts in which the beggar is the would-be refuter of a thesis forwarded by her interlocutor and defended by him. On the face of it, however, it is by no means necessary that to refute a thesis you might be in any kind of dialogical contact with its holder. It is possible to marshall a rebuttal of the doctrine of false consciousness without having to have tea with Marx. It ought also to be possible to refute *Marx* on false consciousness without refuting false consciousness. Although the originator of it and the doctrine is true, Marx might have got his own version of it wrong.

This helps us see the importance of distinguishing between launching a refutation of a thesis and launching a refutation of someone who holds it. If true, it also helps us see that the BQ schema doesn't yield a necessary condition on question begging. The latter requires the presence, or the involvement, of the person challenged and demands in

---

[371] The interplay of question-begging and irrelevance is discussed at greater length in Woods (2003).

turn that the thesis in question actually be held by him.[372] Neither of these requirements applies in the former case. You can build your case against the doctrine of false consciousness in the absence of Marx or any Marxist sympathizer. This should matter for what we say about BQ. In the latter case, when you are intent on refuting my subscription to the doctrine, you beg the question against me when the conditions mentioned in the BQ schema are fulfilled. In the former case, you beg the question against the doctrine when there is a thesis-damaging α meeting certain conditions. But what conditions? It is not at all easy to say.

Here is a candidate to consider. For a refutation of τ itself, as opposed to a holder of it, perhaps we could require that a τ-damaging α abide by the condition that τ not itself contain or imply any proposition β incompatible with α. Or, invoking yet a further distinction between τ and C, the *case* for τ, we might decree that α outright contradict nothing said or implied by τ and nothing contained in C or implied by it. Why would we be drawn to such a suggestion? We would be drawn to it if we were to allow the concepts of commitment and concession a certain *non-dialogical* latitude. What if we were to allow that a thesis, in this non-dialogical sense is, "committed ND" to what it says and implies and that it is also committedND to what is said or implied by the case which grounds it. Let us say further that, like commitment, commitmentND has a negation rule. That is, if β is a commitmentND of τ then $\ulcorner\sim\beta\urcorner$ is not a commitmentND of τ, hence not a concessionND of it either. Then an α that contradicts any such commitment of τ, implies a proposition to which τ is not committedND and which therefore is unconcededND by τ. And isn't this begging the question against τ? (Or, if you prefer, begging the questionND against it?)

Let's suppose for the moment that it is. How would a successful refutation of τ go? What would be the conditions of its non-questioning-begging success against τ? Consider how the same sort of question plays out for τ-*holders* – for Harry, say. A refutation succeeds against Harry in relation to τ when Harry is shown to have commitments that contradict or conflict with τ's or τ's case's commitments. In other words, refutation of Harry is possible if and only if in relation to τ Harry has contradicted *himself*. In this case, the refuter is permitted to violate the constraints on the refutation of theses, because no self-consistent thesis is capable of

---

[372] It is interesting to note an asymmetry with respect to misattribution of premises and misattribution of the thesis against which those premises are launched. While the former is BQ, the latter is *ignoratio elenchi* in Aristotle's narrow sense (or straw man in the modern sense).

contradicting itself, whereas a holder of a self-inconsistent thesis is entirely free to do so.

We see in this the wisdom of Aristotle's, and later Locke's, insistence that refutations are arguments *ad hominem*. When they come off, the damage is done to Harry, not to Harry's thesis. The damage done to Harry is the disclosure that his commitments are inconsistent, that in his effort to justify τ Harry has fallen into contradiction.[373] We cannot succeed against Harry in relation to τ unless we bring to the surface the contradiction that Harry is now in. Sound refutations demand sound measures *ad hominem*, which, in neither Aristotle's nor Locke's senses, are fallacious. On the other hand, as we have it now, the anti-BQ constraint on the refutation of theses is so tough as to make theses irrefutable. Why? Because, although we have agreed to extend the notions of a speaker's commitments and concessions to a thesis's commitmentsND and concessionsND and a case's commitmentsND and concessionsND, there is no room for a like extension to *other* commitmentsND and *other* concessionsND that a thesis or a case might be revealed to have. That is, for theses and cases alike, there is no general sense to the idea that they have fallen into contradiction with respect to themselves. True, a thesis might turn out to be self-inconsistent and a case for a thesis might embody an internal inconsistency, but this is not the case in general. So the present point retains some considerable force.

Proposition 12.3

THE IRREFUTABILITY OF THESES: *No self-consistent thesis for which there is a internally consistent case is strictly speaking subject to non-question-begging refutation. So if BQ were a refutation-wrecker, theses would be irrefutable.*

But isn't this a result shocking enough for self-disqualification?

12.4 Falsification

We needn't worry. It is a false alarm. The irrefutability of theses affords them no *carte blanche*. Theses are still subject to falsification, and

---

[373] In both Aristotle's and the logical tradition's discussions of the refutation of τ-holders, a refutation succeeds when it exposes the τ-holder's inconsistency, which is a deductive matter. Actual practice suggests the presence of a more general notion, which also allows for a refutation to succeed when the refuter brings it to light that the holder is committed to propositions that weigh against his thesis non-deductively – probabilistically, for example.

although in everyday speech "falsification" and "refutation" can be used interchangeably, we now see that there are good theoretical grounds for distinguishing them with some care; but not before noticing a still further distinction which the word "falsification" sometimes blurs. This is the distinction between *making* false and *showing* false. Consider a proposition β and a proposition α incompatible with it. Then if α is true it makes β false, that is, the state of affairs represented by β falsifies α. The *world* makes α false. Consider now the same α and β. Then someone shows that β is false, if *he* advances α with requisite probative force. To do this it is unnecessary that holders of α believe that β is true or that β's truth is suitably grounded. This keeps open the possibility that an original thinker might have shown something which no one (else) at the time believed. There is a story that circulates about Galileo. It recounts Galileo's anticipation of Dedekind infinity, which he is said to have suppressed out of theological scruple. If the story is true, Galileo showed-false the received view that actual infinities are inconceivable. He showed it notwithstanding that no one was shown it.[374] We have it, then, that

Proposition 12.4a
NON REFUTING FALSIFICATION: *A successful showing-false of α needn't succeed as a holder-refutation of it.*

If freedom from BQ made the refutation of theses impossible, the falsification of theses would make BQ virtually impossible. Refutation and falsification – like irrelevance and question begging – would themselves be polar opposites or complementary vectors. There is a useful lesson in this tension. It is that attacks on theses are better left to the resources of falsification and that attacks on thesis-holders call for the defter hand of refutation. But let us be in no doubt. If you hold a thesis τ, I can move to falsify τ without having to bother with you. I can do so, even if the moves I am making against τ are moves that would be question-begging if made against you as τ's holder. If my target is τ rather than you, I can move against τ and I might succeed against it. Have I then succeeded against you? No. I have shown that a thesis you hold is false. But I haven't, just so, shown *you* this. I haven't refuted you,

---

[374] Here, too, we might note that any impulse we might have had to generalize the notion of thesis-refutation – in application to holders refutation – to those that turn on nondeductive or probabilistic considerations, would be dealt with by a suitable adjustment of falsification.

because I haven't shown that you are inconsistent with respect to τ. This discloses a point of importance:

Proposition 12.4b
BURDENS OF PROOF: *In some respects, the burden of proof for refutation is much higher than the burden of proof for falsification.*

Would this matter for BQ? It appears that it would. Refutation is often a tricky thing to bring off. It calls to mind the old saw that it "takes two to tango". We saw earlier that deep disagreement is a natural abettor of question-begging. Why wouldn't the same be true of the tough sledding that refutation makes necessary?

12.5 Proof

We are still not quite where we want to be. The distinction between thesis-refutation and falsification is critical. In its absence, we also appear to have the irrefutability of theses problem. With it still in play, we appear to have the problem of BQ-free falsification. Seen the first way BQ is unstoppable. Seen the second way BQ is impossible. Clearly, we would be better served by seeking a further way of conceiving of these things, a way that preserves the intuition that there are forms of argument for which BQ is both a definable property and an occasionally instantiated one. While we're at it, would it not also be advisable to look for a conception of argument which is neutral as between hostility towards a proposition and support of it? After all, if it is possible to beg the question in attacking a proposition, shouldn't it also be possible to beg the question in supporting a proposition?

We have what we need in the notion of proof. Showing this will require us to mark a three-way distinction between "object language proof", "metalanguage proof" and what I will call "the standard notion of proof". I will explain these distinctions as we go along, beginning with object language proof. In the usual approaches to proof theory, a proof is a construction of the object language. It is any finite sequence of sentences each of whose members satisfies certain conditions. For example, in axiomatic proof theory, each line of the proof must be an axiom or must arise from prior lines by finitely many applications of the system's transformation rules. So, for example, in any system in which ⌐α ∨ ~α⌐ were an axiom ⟨α ∨ ~α, ~(~α ∧ ~~α)⟩ would be a proof. So would

416

$\langle \alpha \vee \sim\alpha \rangle$ be, and also $\langle \alpha \vee \sim\alpha, \alpha \vee \sim\alpha \rangle$. Proofs in the proof theoretic sense are object language sequences unfettered by a noncircularity requirement. Historically, as well as contemporaneously, circularity and BQ are linked in ways that we have yet to investigate (but see just below). Let us say for now that what the present examples suggest is that BQ is not a definable notion for proofs of this object language, or proof theoretic, sort.

If our sought-for notion of proof weren't the object language entity of axiomatic or natural deduction proof theory, consider now metalanguage constructions of a type also common to systematic logic of all stripes. To give them a name, let us say that they are metalanguage proofs. Metalanguage proofs are proofs that certain object language constructions are (or are not) object language proofs of a given system of logic. Consider a stylized example

1. α theorem
2. $\alpha \supset \beta$ theorem
3. β 1, 2 MP

in which formulas in the numbered lines constitute a proof of β in the object language sense. It is not, as I say, subject to a BQ prohibition. The numbered lines *together with* the remarks entered on their right constitute a proof the proofhood of $\langle \alpha, \alpha \supset \beta, \beta \rangle$, hence a metaproof. It is clear that the metalanguage proof answers to conditions not applicable to the object language proof. One is that the proof be constructed in $\langle \alpha, \alpha \supset \beta, \beta \rangle$'s metalanguage, which in the present case is English. Another is that a metalanguage proof cite the conditions under which $\langle \alpha, \alpha \supset \beta, \beta \rangle$ is an object language proof. The object language proof must satisfy these conditions, but it needn't – in fact it cannot – *state* them.

It is interesting to note how little difference there is between the present conception of metalanguage proof and the dialogician's conception of refutation. Earlier we said that in systems of dialogue logic, parties are subject to a closed world assumption with respect to one another's concessions. Something of the same sort is also true of metaproof. To see how, consider a dispute that might arise between Harry and Sarah about our present example.

The dispute arises as follows. Harry complains that α is *not* a theorem, that Sarah has no business intruding it into the proof. Of course, it is or it isn't a theorem. By the settled conventions of the logic of proof, this is a question governed by its own closed world assumption. One may attribute theoremhood to a sentence in a proof such as this one only if it identifiably occurs in a list of sentences already established to be theorems. Thus, irrespective of whether α is or is not a theorem in fact (it might be a distant theorem not yet proved), if it is not on that list it is not eligible for citation. The resemblance to dialogical setups is rather impressive. In a formal dialogue, attributing an unlisted proposition to a party violates the closed world assumption. In metaproofs of the sort under review, citing a theorem not on the list also violates a closed world assumption. If we can count the first violation as begging the question, we can surely do the same for the second. Accordingly,

Proposition 12.5a
BEGGING THE QUESTION IN PROOFS: *If in a purported metalanguage proof the closed world assumption is abused, it gives rise to BQ of a sort resembling faulty attribution errors in refutatory contexts.*

Needless to say, the formal distinction between object language and metalanguage proof is too narrow for general service. Most arguments occur in contexts in which the object-meta distinction is somewhat forced if applicable at all. But we can save the lessons now learned by generalizing to a more common distinction. It is the distinction between the full version of one's argument for α, and the stripped down version made up by its core sentences. The full version is the narrow or core version supplemented by the requisite justificatory remarks, step-by-step.

The lessons now learned are valuable, but they still don't quite take us to where we want to be, not at any rate until we introduce a further feature of metalanguage proofs. With this I mind, consider the following proof, *Pr*, in which "rep" names a repetition rule.

1. α_1 theorem
2. α_2 theorem
3. α_1 1, rep.

That $\langle \alpha_1, \alpha_2, \alpha_1 \rangle$ is a proof theoretic proof even though it sanctions the application of a rule that makes $\langle \alpha_1, \alpha_2, \alpha_1 \rangle$ circular. This is of no mind. Proof theoretic proofs tolerate circularity perfectly well. It is, however, a point of some importance that the metalanguage proof that $\langle \alpha_1, \alpha_2, \alpha_1 \rangle$ is an object language proof doesn't itself make any move that makes *it* circular. In allowing the repetition rule, it allows its application to $\langle \alpha_1, \alpha_2, \alpha_1 \rangle$, but in so doing does not itself instantiate it. The point generalizes. Metalanguage proofs do not *themselves* tolerate conclusion-rendering applications of the rep. rule or of any other move that renders them circular in that way. Accordingly, we may now say that metalanguage proofs instantiate a more general model of proof. It is the third member of our troika. It is what we were calling proof "in the standard sense" or *S-proof*, a defining characteristic of which is that S-proofs are defeated by circularity.

12.6 *Circularity*

When is circularity a case of question-begging? The answer is

Proposition 12.6a
QUESTION-BEGGING CIRCULARITY: *Circularity in a metalanguage proof or any other form of S-proof is question-begging.*

Where do we now find ourselves? There is no commonly practised kind of proof with respect to which the closed world assumption holds strict sway. We have already touched on the reason for this. It is that in the ups and downs of real life – of life beyond the leisured rigours of nonmonotonic formal systems – it is very expensive to honour the closed world assumption. In a suitably generalized form, the present notion of full-version proof is akin to the showing-false sense of falsification under two adjustments. One is that "showing" is now what I will call a "quasi-success" term. The other is that the focus is expanded from showing false to showing true or false, as the case may be.

The quasi-success requirement on showing is matched with a *quasi-closed world* assumption on ascription. In each case, it is a condition on showing people things that this be done in a way that facilitates being shown it. Proof is now the full-version S-notion of showing things to be what they are: true, false, inconsistent, unsupported, well-evidenced, silly, and so on. The quasi-closed world assumption is that an S-prover's attributions will stand some reasonable chance of being seen as acceptable by an arbitrary inspector of the argument. It is difficult to say with precision what this "reasonable chance" comes down to. But it is something like this. An S-proof of τ is aimed at an assumed constituency of addressees. It might be the readers of a philosopher's book. It might, for matters of general interest, be the population at large. It might be a first-year class of microbiologists. Call this community $\mathcal{K}$. It is further assumed that if α is cited in the construction of a full-version

S-proof of τ aimed at $\mathcal{K}$, then one or other of two default conditions must be met for its use to meet the quasi-closed world assumption:

Proposition 12.6b
THE QUASI-CLOSED WORLD ASSUMPTION: *In the absence of particular indications to the contrary, α is something that an arbitrary member of $\mathcal{K}$ already accepts. Or in the absence of indications to the contrary, α is something an arbitrary member of $\mathcal{K}$ would be prepared to accept on the prover's sayso.*

Proposition 12.6c
COMMON KNOWLEDGE: *Any α governed by the quasi-closed world assumption with respect to K is common knowledge in $\mathcal{K}$.*[375]

It is important to see that if Proposition 12.6c is true, then most of the α involved in premissory misattribution fall outside the reach of the quasi-closed world assumption. Also important is what a member of $\mathcal{K}$ would make of an argument all of whose premises pass the quasi-closed world test. There is also the matter of the argument's "justificatory" aspect. Consider the case of an argument aimed at a community, with respect to which all the premises pass muster, but the conclusion arises from those premises only by means of mathematical induction, or only with the backing of some recondite principle of the probability calculus, and that these are the justifications actually proferred in support of the proof's final line. For most communities of real life showees, this is that point at which the argument fails to show its conclusion, never mind that it follows from premises already shown. So clearly something else is required – a further condition on quasi-showability. It is that the conclusional backing claimed by the arguer has the same kind of reasonable chance of being accepted by an arbitrary member of $\mathcal{K}$.

We now see the force of the qualification "quasi". If $\mathcal{K}$ is any real life community of addressees, it is a virtual given that there will be members of $\mathcal{K}$ for whom the showing of α does not show α to *them*.

---

[375] There is clearly a link between common knowledge in this sense and background knowledge in the sense of chapter 11, effected principally by the factors of default and presumption, and the fact that the quasi-closed word assumption embeds a – so to speak – quasi-negation-as-failure assumption.

These are defections that don't defeat the claim that in that population the argument for α did indeed show that α. Their status as non-defeaters is that the conditions on quasi-showing are genericity-implying defaults, and that these defections are negative instances of the kind that don't overturn generalization at hand. This is expressible in a compact way. If an individual is known to be such a defector, he may not play the role of the arbitrary $\mathcal{K}$-member cited in the quasi-showing rules.

It remains to determine whether, as we conjectured above, violations of the quasi-success condition qualify as beggings of the question. That is, do we now have it that the citation of α in a $\mathcal{K}$-directed S-proof of τ begs the question in $\mathcal{K}$ if τ is a proposition which an arbitrary member of $\mathcal{K}$ would not accept? The answer is that it does if the arguer himself is a member of $\mathcal{K}$.

This is an interesting condition. It gets us to see that anyone who fails the quasi-showing conditions in $\mathcal{K}$ is herself an aberrant member of $\mathcal{K}$, that is, aberrant in the sense that, for the purposes of those rules, she may not play the role of the cited arbitrary $\mathcal{K}$-member.

There is a crude way of summing up. The maker of a quasi-showing-argument makes a BQ error when she invokes a premiss which as a member of $\mathcal{K}$, she should have known contradicts what members of $\mathcal{K}$ are presumed to accept.

What now of the invocation of a conclusion-backer, that is not presumed applicable in $\mathcal{K}$? There are two cases to consider:

1. Conclusion-backers which members of $\mathcal{K}$ regard as defective as such.
2. Conclusion-backers considered inappropriately applied in the present case.

With standard practice as our guide, difficulties of this second sort don't attract the complaint of BQ. A critic would say simply that the arguer's conclusion did not follow in the requisite way from her premisses. But perhaps the first case is different enough to pause over. Let us come back to *ex falso quodlibet,* the classical theorem asserting that a contradiction entails everything whatever. Correspondingly, there is something that we could call the *ex falso* rule: From inconsistent premisses derive any arbitrary sentence. Suppose that you are attempting to show that τ. Suppose that the $\mathcal{K}$ to which you are directing your argument is the community of present day philosophical logicians. Imagine that at a certain stage of your argument for τ, you introduce a new premiss by subproof and that the subproof instantiates the *ex falso* rule. There are large (and talented) groupings within $\mathcal{K}$ for whom *ex falso* is a bad rule. If you yourself are also a member of $\mathcal{K}$, then by the present suggestion you have begged the question in that subgroup. Moreover, if the class of defectors within $\mathcal{K}$ are numerous and smart enough, it becomes difficult to say that *ex falso* is a rule presumed good in $\mathcal{K}$. In that case, your use of it would be questioning-begging in $\mathcal{K}$ as well. If so, there would *be* no $\mathcal{K}$ for *ex falso*. It is interesting to observe that on the *ex falso* issue there is indeed at present no $\mathcal{K}$ for it. This, in a nutshell, is the meaning of pluralism in logic.[376]

So we have what we want. Apart from the refutation of thesis holders, we now have a concept of full-version S-proof for which begging the question is a definable property yet only a contingently instantiated one. The question that remains is whether these beggings are fallacies.

A further complication is that in lots of cases there is a key distinction between a refutation's target – whether τ or τ's holder – and the refutation's *addressee* – say the people reading a book on the subject of τ. When in one of your articles or books you falsify a claim τ, you launch a blind showing-false of τ. It might end up as a refutation of a person who unknown to you, subscribes to τ. Beyond a certain point, whether what you advance is a refutation or not is not down to you. It is down to your reader. In such cases, it is trivial to see that the launching of a damaging α might well beg the question against Harry, but not against a reader. Equally, falsifications admit any α causing damage to τ or to Harry that the falsifier actually holds and has grounds for holding. Whether Marx or Richard Routley would accept α, *you* might accept α, and on that basis launch it with refutatory force, in hopes of its reaching a reader whose commitments include α. Suppose, on the other hand, that there are readers committed to ⌐~α⌐. It is interesting to imagine what their response to your move against τ might be. Would they say that, because you have made use of an α that they don't accept, you have begged the question against them? I'm inclined to think not. I'm inclined to think that the general reaction of such readers to your purported falsification of τ would be that since it pivots on a false α, the falsification fails. If this is

---

[376] Pluralism, as we remarked, is a hot ticket these days. See here Beall and Restall (2006), Woods (2003) and (2011).

right, it tells us something else of importance for the analysis of BQ. It tells us that a reader may take himself to have been refuted by a falsification he accepts, yet does not take a falsification that fails as a failed – still less, question-begging – refutation of him. There is an exception to this. If your book's falsification is directed at some person *by name*, then his failure to accept it – say, he denies α – renders your argument a failed refutation of him. Yet it is not at all clear that this also qualifies as a would-be refutation that begs the question against him.

12.7 *Occasions*

The universality of fallacy-making is an occasioned universality. Fallacies are committed only when there are occasions to commit them. As fruitful as the notion of occasion surely is, it bears repeating that it does not smoothly lend itself to precise definition. It presents us with one of those situations that commonly arise in philosophy. The idea is clearer than its application conditions. It seems clear enough that you don't commit the fallacy of hasty generalization (ignoring for now that it is not a fallacy) unless there is something that you take to be a sample, and that you don't commit the fallacy of post *hoc, ergo, propter hoc* (assuming for now that it is one) unless there is something you take for a temporarally ordered positive correlation. But what of begging the question? Of course, you don't commit the fallacy of begging the question unless there is a question to be begged. What are the conditions under which this is so? This, too, is far from easy to answer. Suppose for the moment that these occasions turned out to be conditioned by the presence of deep disagreement. Then – dialogic ascent aside – question begging would be a fallacy by the definition of the occasion for it. If deep disagreement raises the likelihood of begging the question in some substantial way – say, by making it a practical certainty – and if deep disagreement constitutes an occasion for it, then begging the question trivially meets the occasioned frequency condition on universality. More generally, if being in the occasion of error is like being in the occasion of sin, being in a situation which by definition carries a very high likelihood of committing it, why wouldn't this be just what the doctor ordered for attractiveness and incorrigibility, as well as universality? This gives us two options to consider. Either the traditional idea is right, according to which begging the question is a fallacy by definition. Or we've got the wrong definition of "occasion". If the concept-list misalignment thesis is to be given its head, we must make good on this second option.

Upon reflection there is something right about this way of conceiving of occasions, yet something wrong with it too. It is right in a general sort of way, but it is wrong in the particular and somewhat narrow form in which we have it now. What is right about it is this: If the notion of fallacy is undefined for non-occasions, then it is trivial that the rate of fallacy commission is greater when there are more occasions for it than not. What's wrong about it is that having the occasion for it does not in the general case generate the degree of frequency required for the universality condition. Intuitively, this is right. No one seriously supposes that being in a situation that makes fallacy commission likelier is routinely correlated with a fallacy actually being made. Think again of the *post hoc* error. You can't commit a *post hoc* error unless there is something you recognize as a temporally ordered positive correlation. But it is not true empirically that the recognition or assumption of such correlations is routinely attended by the ascription of causal import. So the ratio of false causal attribution to occasions for it isn't high enough to meet the frequency condition on occasioned universality. Of course, even if these observations are sound, we haven't shown that all occasions for commission are free from a definitional link to erroneous commission, and we certainly haven't shown this in the case of begging the question. So we re-press one of the chapter's core question: "What are the occasions for begging the question?" In light of our discussion to date, I think we might propose the following:

Proposition 12.7a
OCCASIONS FOR BEGGING THE QUESTION: *One has occasion to beg a question just in case one is engaged in a holder refutation of some thesis or in a S-proof of some proposition.*

On the universality condition, something is a fallacy only when it is widely committed with a notable frequency under conditions that constitute an occasion for it. In the present case, the occasions for begging the question are contexts of real life holder-refutation and generic S-proof. In each of these cases, occasions are relativitized to communities $\mathcal{K}$. They are relativized to $\mathcal{K}$ in such a way that a person who begs the question in $\mathcal{K}$ is disqualified from playing the role of an arbitrarily selected $\mathcal{K}$-member in the definition of the quasi-closed world assumption. This is a significant constraint. It provides that by and large people who participate in $\mathcal{K}$-directed holder-refutations and generic proof don't in so doing beg questions in $\mathcal{K}$. So we have it that

Proposition 12.7b
BEGGING THE QUESTION IS NOT UNIVERSAL: *Begging the question fails the condition of occasioned universality. It is not a fallacy in the traditional sense.*

COROLLARY: *Note that the concept-list misalignment tally now stands at ten out of eighteen.*

12.8 *Circularity redux*

In the standard contemporary writings, there is a link between circularity and question begging. It is also a link that received Aristotle's early notice. Aristotle considers five ways of begging the question. The first is one in which the rebutting syllogism is syntactically circular, i.e., is something in the form $\langle \alpha, \sim\tau, \sim\tau \rangle$. The second way involves a premiss that immediately implies the negation of the opponent's thesis, exemplified by an argument $\langle \alpha, \beta, \sim\tau \rangle$ in which β is tightly equivalent to $\ulcorner\sim\tau\urcorner$. Such arguments are, as we might say semantically circular. The third way is, by modern lights, the fallacy of hasty generalization, exemplified by $\langle$"Socrates is a Greek who loves to argue" "All Greeks love to argue"$\rangle$. Aristotle's (confused) point is that the conclusion cannot be true unless the premiss is also true. So asserting the conclusion is a way of asserting part of the premiss. Accordingly, the third way reduces to the second. The fourth way turns on a technicality that has no modern resonance and need not concern us here. The fifth way is one in which the conclusion is immediately equivalent to a premiss, and which, therefore, reduces to the second way. Aristotle's fallacies are arguments that appear to be syllogisms but are not syllogisms in fact.[377]

Syllogisms are not to be confused with valid arguments. A syllogism is a valid argument that meets certain further conditions. One is the premiss irredundancy condition, which says that removal of a premiss from a syllogism invalidates the resulting argument. Another is the non-circularity condition, which says the conclusion of a syllogism may not repeat a premiss and may not be immediately equivalent to a premiss. All this has a bearing on how a challenger, Sarah, might proceed against a holder and defender, Harry, of a thesis τ. Consider the case in which, in

---

[377] Doubtless this is the source of the inapparent invalidity conception, said by Hamblin to be the "standard definition" of fallacy. Against this, Hansen (2002) convincingly shows that the empirical record does not substantiate this claim.

425

its occurrence in the following argument, α is *immediately equivalent* to $\ulcorner \sim\tau \urcorner$.

    1. α
    2. $\therefore \sim\tau$.

In its modern sense, we have in $\langle \alpha, \sim\tau \rangle$ an argument that begs the question against Harry if the attribution of α to Harry is unjustified. However, by Aristotle's lights, it is also a fallacy in his technical sense. It is a fallacy because $\langle \alpha, \sim\tau \rangle$, violates the noncircularity condition on syllogisms. Seen Aristotle's way, it is a matter of no consequence as to whether $\langle \alpha, \sim\tau \rangle$ is a fallacy that it begs the question in the modern sense.[378] In the case we are presently examining, we could have it that Harry *accepts* the attribution of α, thus removing any possibility of having had the question begged against him, in the modern sense, and, still, *Aristotle* would find that Sarah had committed it. Why? Because one way of committing Aristotle's BQ fallacy is by taking a valid circular argument for a syllogism. In Aristotle's case, unlike the modern case, it may appear that what makes Sarah's selection of α a fallacy is a matter quite apart from whether it is falsely attributed to Harry. It has everything to do with its syllogism-precluding presence in $\langle \alpha, \sim\tau \rangle$. For those interested in the history of logic, this is an interesting development, as we saw. Aristotle is the originator of the concept of fallacy, but Aristotle's idea is not the traditional conception of it as we have it today.

Aristotle doesn't regard question begging in the modern sense a fallacy. But this in no way diminishes the low regard he has for it. The modern BQ is a violation of a dialectical rule. It is not the mistaking of a non-syllogism for a syllogism. The rule that it violates requires that, in the construction of the syllogism whose conclusion is the contradictory of the opponent's thesis, a would-be refuter may use as premisses only those propositions already conceded by the opponent. Thus Aristotle, too, sees these violations as premiss selection errors and, relatedly, as case-making

---

[378] Let me say again that Aristotle is not indifferent to the problem of faulty attribution. Indeed he thinks that it is a fallacy. But it is not the fallacy of begging the question. It is the fallacy of *ignoratio elenchi* (in the narrow sense), which in Aristotle's hands is the mistake of either misdeducing a consequence of something one's opponent holds or correctly deducing a consequence of a thesis that one's opponent does not hold. As we observed in footnote 10, in its latter form *ignoratio elenchi* is the forbear of the modern straw man fallacy. Aristotle's *ignoratio elenchi* in its wide sense covers all the fallacies, since each is a mistake that damages a refutation.

errors. They are bad ways of going about one's refutatory business. They instantiate – in Aristotle's wide sense of the term – the *ignoratio elenchi*. That is, they are misconceptions of what it takes to achieve a refutation. Fallacious as they certainly are in the modern tradition, they are not so for Aristotle.

Consider now the case in which Sarah's attribution of α to Harry is question-begging in the modern sense. Since Sarah falsely attributes α to Harry and α immediately delivers the goods against Harry's own thesis τ, one might think that Harry has ready occasion to resist the attribution and to call Sarah for her premiss attribution error. However, if one accepts Aristotle's conception of the fallacy, it would be the sheerest folly for Harry to reject the attribution. Better that he leave the attribution uncontested so that Sarah might have unfettered occasion to construct for herself an instance of Aristotle's fallacy. If Harry gives to Sarah that for which she asks ("begs"), he gives her α, and α destroys all chance of $\langle\alpha, \sim\tau\rangle$, attaining the status of syllogism.

It is clear in any case that virtually all that Aristotle has to say about begging the question is linked to syntactically or semantically circular arguments. The same is true of logicians as historically distant from one another as Sextus Empiricus and DeMorgan.[379] It may be that the reasons that logicians have stressed this connection is their desire to make good on the idea that begging the question, both in Aristotle's and the modern sense, is an error of reasoning, hence, as it turns out, a fallacy in the modern sense. But, as we have seen, the connection is contingent at best. Question begging, we said, is premiss misattribution or conclusion backing invocation in contexts of holder-refutation or S-proof in which a quasi-closed world assumption is violated. Clearly, then, there will be plenty of situations in which circularity is not question begging. We turn to these two sections hence but not before a further point about what we will call "correction by contradiction", close kin of knowledge by contradiction.

12.9 *Correction by contradiction*

The inclination of logicians to link circularity to question begging lends itself to the following picture: That where τ is some thesis advanced by Harry, any utterance by Sarah of an α that immediately implies the negation of τ is question begging. There are two things to say about this

---

[379] Woods (2012b).

picture. First, the move it documents is empirically widespread in human conversational practice. Second, in the general case it is neither circular nor question begging. The two most conspicuous cases of this behaviour are "correction by contradiction" and "counterexampling", the first of which we take up now, and the second in the section to follow.

Earlier we saw that the most common situation in which a proposition is used against an opponent and implies in one step the negation of what the opponent holds, is correction by contradiction. This lies at the heart of what we have been calling the primary datum for a logic of told-knowledge. Let us re-visit an example.

1. *Harry* : "Tomorrow is Barb's birthday".
   Sarah: "No, it's the day after."
   Harry: "Oh, I see."

Here are two more:

2. *Harry*: "Bryson is a paraconsistent logician who espouses *ex falso.*"
   Sarah: "Paraconsistency is defined in such a way that that can't be true." *Harry*: "Oh, I didn't know"[380]

3. *Harry*: "Some ravens aren't black."
   Sarah: "Oh no, all ravens are black."
   Harry: "I must have been thinking of swans."

It is easy to see that in each case Sarah contradicts Harry by uttering a sentence which in one fell swoop delivers the negation of what Harry says. In two respects this kind of case differs from those we have been considering. One is that when you contradict someone by uttering an α that implies the negation of what he says, you needn't be attributing to him the belief that α. In some cases, as we saw, you are *informing* him of something of which he appears to be unaware. In others you are *reminding* him of something he seems to have forgotten. The other difference is that a context in which you contradict someone in this way needn't be one in which he is actually defending τ. In fact, when it succeeds, correction by contradiction pre-empts defence; that is to say, it obviates the need for it. For both these reasons, it is easy to see that

---

[380] In fact, Brown (2007) questions the blockage of *ex falso* as a necessary condition of paraconsistency. That is something we needn't go into here.

Proposition 12.9a
CORRECTION BY CONTRADICTION IS NOT QUESTION-
BEGGING: *Correction by contradiction fails to qualify as a BQ-
occasion, hence is a practice which is trivially BQ-free.*

Of course, correction-by-contradiction exchanges sometimes do
not terminate with Harry's withdrawal of his claim. There are situations
in which Harry won't accept the contradicting claim of Sarah. If this
happened in such a way as to move Sarah and Harry into the holder-
refutation mode, then of course questions might end up being begged. But
this would no longer be a case of correction by contradiction.

Closely related to correction by contradiction is the use of
counterexamples as a critical device. As it has evolved in philosophical
practice, β is a successful counterexample of α only if β immediately
implies $\ulcorner \sim \alpha \urcorner$, and typically one of three further conditions is met.

1. α is a generalization and β is a true negative instance of it, *or*

2. α is a definition and β is a proposition that instantiates its
      definiens and fails to instantiate its *definiendum, or*

3. α is an implication statement and β is a true conjunction of
      its antecedent and a contrary of its consequent.

We might note an apparent asymmetry between producing a
counterexample and begging a question. It is an asymmetry of which both
parts pivot on the factor of presumed obviousness. Accordingly, whereas
"*a* is a F that is not G" is, if true, a successful counterexample of "All F
are G", "All F are G", even if true, has the look of a question-begging
move against "*a* is a F that is not G". Similarly, $\ulcorner \alpha$ and $\sim \beta \urcorner$, if true, is a
successful counterexample against $\ulcorner \alpha$ implies $\beta \urcorner$, but $\ulcorner \alpha$ implies $\beta \urcorner$ has
the feel of a question begged against $\ulcorner \alpha$ and $\sim \beta \urcorner$. Examples such as these
draw us to conjecture that

Proposition 12.9b
THE CONVERSE THESIS: *In their typical forms, successful
counterexamples are the converses of apparent question-
beggings.*

Proposition 12.9c
NON-CLOSURE: *Neither counterexampling nor question-
begging is closed under the converse-of relation.*

At the heart of these claims is the factor of obviousness. When β is a successful counterexample of a generalization or definition or entailment statement α, it is taken that β obviously contradicts α and – in many cases – that *once it is pointed out* – β is obviously true. In other words, successful counterexamples embed something that closely resembles correction by contradiction.

Harry has known for a long time when Barb's birthday is. But Harry has no head for dates. It often slips his mind that Barb's date is April the 3rd. Harry knows that the big day is in the early spring and, year by year, as time closes in on the third, Harry knows that the birthday is ready to hand. Today he thinks the birthday is tomorrow. He thinks with confidence enough to prompt a regret for not having sent Barb a proper present. Sarah contradicts Harry. In so doing reinstated the belief that had *slipped his mind*

Counterexampling, or counterexemplification, is not an occasion for question-begging, although it could easily enough become one. Consider:

1. *Harry*: All F are G.
2. *Sarah*: Well, a is F but not G.
3. *Harry*: No, it isn't.
4. *Sarah*: Why ever not?
5. *Harry*: Because all F are G and so no F is not G.
6. *Sarah*: But, look, it's an empirical fact that a is F and not G.
7. *Harry*: No, it's an empirical fact that all F are G.

No one should think that exchanges of this sort don't occur in real life. I will have a brief word to say about them immediately below. For the present, it suffices to say that the situation in which Sarah and Harry now find themselves may have arisen from a counterexemplication situation, it is now no such thing. It is now a matter of holder-refutation, for which BQ is both definable and, in the present instance, vividly discernible. But the main point of this section is that

Proposition 12.9d
COUNTEREXAMPLING IS NOT QUESTION-BEGGING: *Since counterexemplification is not an occasion for it, it is a practice that is BQ-free.*

12.10 *Babbling*

Deep disagreements are dialectically paralyzing. We said that a mark of such paralysis is irrelevance, that another is question-begging, and that the two constitute a kind of complementary polarity. There is a further response to dialectical paralysis which we should pause to consider. It is *babbling*. In its present-day meaning, babbling is foolish, excited or confused talk. For Aristotle, however, it has a quite different meaning which I shall now briefly explain. Suppose again that Sarah has placed Harry's thesis τ under challenge. Suppose further that, until now, Harry has never been challenged to defend τ, that, until now, Harry presumed that τ is a proposition that everyone takes as obviously true. (For concreteness, let τ be the proposition that doctor-assisted suicide is morally wrong.) It is easy to see the difficulty that Sarah has placed Harry in. For, while it cannot be ruled out that Harry has a perfectly satisfactory case to make for τ, very often in just this kind of situation, he has no case to make. If, up until now, no case has been demanded, if up until now, no case has appeared necessary, it is not surprising that Harry may lack the resources to mount a defence of τ there and then.[381] In such cases, Harry is faced with two options, both of them unattractive. One is to admit that τ is a proposition which he is unable to defend. The other is to stand mute. The first is unattractive since it stirs the presumption that challenges that draw no defence require the surrender of the thesis in question. The other is also unattractive. It convicts Harry of unresponsiveness. It is not uncommon in such cases for Harry simply to reassert τ, often in other words. In so doing, he performs the minimally necessary task of avoiding the prior options, each of which would be presumed to imply capitulation. Merely re-asserting a proposition that is under attack is not babbling in Aristotle's sense. But re-asserting it *as its own defence* is. Anyone who pays attention to how disputed matters play out in real life will be quick to see how pervasive babbling is in the lives of beings like us. If babbling were a fallacy, it would meet the condition of occasioned frequency. Apparently it does meet the condition of occasioned frequency. The question is: Is it a fallacy?

Babbling appears to be a kind of question-begging. It most nearly resembles a defence of τ in the form "Why, τ is self-evident". It calls to mind Lewis' mocking answer to Routley's challenge to prove LNC:

---

[381] The dialectical vulnerabilities of this kind of case are discussed in greater detail in Woods (2000).

"Why, it's too obvious for proof!" It is a form of question begging which reverses the roles of Harry and Sarah. In the modern sense, Sarah, the challenger, begs the question against Harry by attributing to Harry a proposition he doesn't concede. In the present case, it is Harry – the party under attack – who begs the question against Sarah. The question, again, is whether it is a fallacy. The answer is that it is not, either on Aristotle's conception of fallacy or on the traditional conception. For, in the first place, no one who knew what a syllogism is could ever think that $\langle \tau, \tau \rangle$ is a syllogism; and, in the second, no one could miss that attributing to Sarah's acceptance of the very proposition that she judges that Harry has no right to hold, is a wholly blatant premiss attribution error. It is an error, therefore, that fails the attractiveness condition. Accordingly,

Proposition 12.10a
BABBLING IS NOT ATTRACTIVE: *Since babbling is so obviously a bad premiss attribution move, it cannot meet the attractiveness condition of appearing to be a good move.*

There is point in emphasizing that babbling happens. It happens abundantly, and does so under conditions of the kind that are presently in view. This might incline us to cast a baleful eye at Proposition 12.10a. It might strike us that Proposition 12.10a passes muster only on the technicality that since $\langle \alpha, \alpha \rangle$ is so obviously a bad move it couldn't possible be one that attracts a favourable verdict, hence could be attractive in ways required by the traditional conception of fallacy. After all, if it happens frequently, and if, in holder refutation contexts, it is question begging, surely that is attraction enough to make *some* claim upon the attractiveness condition. I have some sympathy with this line of reasoning. Proposition 12.10a leaves something important out of account. It is that the point at which parties to a holder refutation start to beg the question in this way, that is, by way of babbling, they have undergone another dialectical shift – this time one that takes them from an originating holder refutation to a subsequent babbling. Babbling marks the *exit* from holder refutation. It is the paralysis of holder-refutation. It is what happens to the parties when the refutatory process falls apart. Accordingly, the more basic point to make is that

Proposition 12.10b
BABBLING IS NOT A BQ OCCASION: *Since babbling is not a BQ-occasion, it is not the traditional fallacy of question begging.*

12.11 *The paradox of analysis*

I have been saying that the form in which it is most attractive to logicians, question-begging is reasoning in the form $\langle \alpha, \sim\tau \rangle$, in which α is unconceded by the defender and immediately contradicts his thesis τ. In this section and the one to follow, I want to touch on features of circularity that tend to attract the special attention of philosophers. One is circular definition. The other is the paradox of analysis. Let us deal with these in reverse order. In making these transitions, we shift from a dialectical to a non-dialectical orientation.

The paradox of analysis is one of a family of problems having to do with triviality. Its modern version was introduced by C.H. Langford (1942), although there are clear anticipations of it in antiquity, most notably in Plato's *Meno* and Aristotle's *Posterior Analytics*. In Langford's version, the paradox arises as follows. Suppose we have a concept A for which we seek a conceptual analysis. Suppose that someone proposes that something is an A if and only if it is a BC. If the putative analysis is correct then 'A' and 'BC' will carry the same information. If that is so, the analysis is trivial. On the other hand, if it is not trivial to characterize an A as a BC, the analysis is faulty.

Another source of the same problem is the complaint of Sextus Empiricus that all syllogisms beg the question or, in its modern variant, that all valid arguments beg the question.[382] Here, too, the concept of information lies at the heart of the problem. For if an argument is valid, all the information contained in its conclusion is contained in its premisses. Let us call this version the paradox of validity.

The connection with circularity is hard not to miss. As we saw, circularity may be understood in at least two ways. In one, it is a certain kind of linking of one and the same syntactic item ($\ulcorner$Since $\alpha, \alpha \urcorner$). In a second, it is a certain kind of linking of semantically equivalent items ("John is a bachelor because John is a man who has never married"). The paradoxes of analysis and validity introduce a third conception. Circularity is a certain kind of linking of expressions having the same information content.

The conclusion of a valid argument produces no new information, that is, no information not already contained in its premisses. Central to this claim is the notion of information-measure, introduced into the literature by Shannon and Weaver (1963). In this technical sense,

---

[382] Sextus Empiricus (1933), chapter 17.

not only is Sextus' claim confirmed, but so too is the classical theorem – *ex falso quodlibet* – according to which an argument with inconsistent premisses is valid for any conclusion.[383] Let α and $\ulcorner\sim\alpha\urcorner$ be premisses. Then since $\{\alpha, \ulcorner\sim\alpha\urcorner\}$ contains all information, there is no conclusion β which contains information not contained in $\{\alpha, \ulcorner\sim\alpha\urcorner\}$. We may say, then, that if, in the claim that an argument is valid if and only if its conclusion contains no new information, the embedded notion of information-quantity is that of Shannon and Weaver, then, valid arguments are always circular (not that hardly anyone would ever notice it). On the other hand, if information is taken as semantically grasped

---

[383] In chapter 1, I argued against *ex falso* on grounds that if it were true then inconsistent theories (intuitive set theory, for example) could teach us nothing whatever about their subject matters. For how could we learn anything about sets if there is no set theoretic sentence that is true rather than false and no set theoretic sentence that is false rather than true? In fact, however, there is a great deal to learn about sets in intuitive set theory, never mind its inconsistency. The fault logicians find with intuitive set theory is not that it tells us nothing about sets, but rather that what it tells us about sets couldn't possibly be true, that is, true as opposed to false. Like that argument or not, it may appear that the present concession that *ex falso* is valid when taken in the sense of information-impartation counts against it. How, it might be asked, can it be true that intuitive set theory contains all information, yet not also be true that its every sentence is true (as opposed to false) or false (as opposed to true)? The answer is that the Shannon-Weaver sense of information has a peculiar feature. We might express it this way: Let α be a sentence you understand and let I be all and only the Shannon-Weaver information contained in α. Suppose now that your and α's circumstances are such that α informs you of something and that you understand α fully. It is *not* a condition on Shannon-Weaver information that what you have thus been informed of is I. What you have been informed of is mediated by α's propositional content. So chapter 1's argument against *ex falso* still stands. A contradiction does not have (or deliver) the propositional content of every sentence. Even if it did, that could not change the fact that none of it is true as opposed to false, or false as opposed to true. Let us again take note that the use of information as an epistemological notion has an interesting and contentious history. Beyond what was said in chapter 6, this is not the place to go into these difficulties at length, but we can give the flavour of some the associated philosophical trouble by noting that on some accounts (e.g. Dretske (1981), the likelihood that message α informs Harry of the fact that α is at least increased by the quantity of Shannon-Weaver information contained in α. The problem, of course, is *ex falso*. If a contradiction β contains all Shannon-Weaver information, how could it not inform Harry of the fact that λ for all λ? For a good overview see Dretske (2008). See again Adriaans and van Benthem (2008).

content, then neither Sextus' claim nor *ex falso quodlibet* stands up to scrutiny.

Seen the first way, a valid argument can't be informative. Seen the second way, it can be. Seen the first way, a correct conceptual analysis of something can't be informative. Seen the second way, it can be. Accordingly, a correct conceptual analysis can be circular (in the Shannon and Weaver sense) without appearing to be (since it may not in fact be circular in the semantically grasped content sense). This being so,

Proposition 12.11
ANALYSIS AS CIRCULAR AND NOT: *While there is a sense in which a correct conceptual analysis is circular, it is not a sense in which circularity necessitates question begging.*

Some would see this as hinging on a pair of contrasts that aren't sufficiently exhaustive to handle the cases at hand. It might be argued that there is a third contract that cuts across the prior distinction. While it is true that there is a sense of information content in which no consequence of a sentence contains more information than it, and a different sense in which it could and typically does, there is also a contrast between the information a sentence contains and the information that it imparts to a conclusion-drawer. Most people who, when first shown the propositions that imply Pythagoras' theorem, wouldn't see that they do. That's why we have proofs. A proof of the theorem contains commentary that links the premises in a way that leads us to *see* the theorem's consequential attachment to them. When this happens, and the proof is understood, we draw the consequence that those premises have. This is what we should do if we are good reasoners. What we don't do is draw the conclusion that the proof is circular. This is not what good reasoning requires of us. It does not require from us even if the proof *is* circular, even if, the sentence that expresses the conclusion contains all the information contained in the premises.

All of this plays on the problem of determining the closure conditions for belief sets. Let $\langle \alpha, \beta \rangle$ be a valid argument, and let α be in Harry's belief set Σ. Is β in the deductive closure of Σ? On the Shannon-Weaver model, all the information in β is already in α. If believing β is just a matter of the information imparted by β being *contained* in what one believes, then in believing α, Harry believes β. So β is in the deductive closure of Σ. On the other hand, if believing β involves an affirmative understanding of it, it is wholly implausible to put β in the deductive closure of Σ as a general principle. Accordingly, classical systems of belief dynamics, such as AGM,[384] in which belief is closed under deduction, are right in so saying if belief is taken in the first way, but quite wrong if belief is taken in the second way.

12.12 *Definitions*

Perhaps the most common form of definition is the lexical or dictionary definition, which is a kind of meaning-teacher. I teach you what 'X' means by telling you the definition of 'X'. Suppose you don't know the meaning of "yclept". I tell you: "yclept" means "also known as" or "alias", as in the scriptural description of Simon, yclept Peter. It has been widely and rightly recognized that a meaning revealing definition is spoilt if its definiens contains an essential occurrence of its definendum. Here "spoilt" means "doesn't teach the meaning in question". Let us say that definitions having this character are "elementarily circular definitions". Giving an elementarily circular definition is a hopeless way of revealing the meaning of a definiendum. But it is a bit of a stretch to say that giving a circular definition of something whose meaning you intend your interlocutor to learn is a mistake of reasoning, stupid though it surely is as a piece of pedagogy. So we would not want to say that elementarily circular definitions are fallacies.

Perhaps there are occasions on which *case-making* embeds a definition that is circular in this way. If that were so, it is important to note that unless the definition were attractive in ways that appear to lend credence to the *case*, even here it will have a hard time qualifying as a fallacy. Since, as we saw, the badness of elementary circularity is a hard thing to disguise, it is correspondingly difficult to make a circularity-containing case to look good.

Definitions whose definiens contains the definendum cannot succeed as meaning-teachers. A recursive definitions repeat the definendum in its definiens. So recursive definitions cannot be meaning-elucidators. What then, are they good for?

Consider the elementary example of the definition of *sentence* in a formal propositional language $\mathcal{L}$. We assume that the grammar of $\mathcal{L}$ has already specified a well-individuated, arbitrarily large stock of *atomic sentences* and a finite collection of connectives, say $\sim$ for negation and $\supset$

---

[384] Alchourron *et al.* (1985).

for material conditionality. Then the definition of sentence is given by the formation rules of $\mathcal{L}$

1. If α is an atomic sentence of $\mathcal{L}$, it is a sentence of $\mathcal{L}$.
2. If α is a sentence of $\mathcal{L}$ so is ⌐~α⌐ a sentence of $\mathcal{L}$.
3. If α, β are sentences of $\mathcal{L}$, so is ⌐$\alpha \supset \beta$⌐ a sentence of $\mathcal{L}$.
4. Nothing else is a sentence of $\mathcal{L}$.

The role of our recursive definition is not to reveal the lexical or dictionary meaning of "sentence" – or, more carefully, of "sentence of $\mathcal{L}$". The lexical meaning of "sentence of $\mathcal{L}$ is "formula of $\mathcal{L}$ generated by its formation rules". The role of formation rules is to provide an algorithmic means for identifying the extension of "sentence of $\mathcal{L}$", irrespective of whether there was some lexical meaning the term already possessed. "(Natural) number" is a helpful example. It has a lexical or dictionary meaning and it has a recursive definition (with '' denoting successorhood):

1. 0 is a number.
2. If n is a number, so is n'.
3. Nothing else is a natural number.

The definition fixes the extension of "(natural) number". It does not give its lexical meaning. It does not and is not intended to be meaning-revealing.

The example of recursive definitions tells us something important about circularity. It is that circularity reflects no discredit, as such, upon the reasonings that instantiate it. The same is true of recursive proofs or proofs by mathematical induction. These proofs are fuelled by induction hypotheses. The proofs are intended to prove that some property holds of all members of an infinitely large set. Consider the property of being a sentence of $\mathcal{L}$. We then proceed to construct the proof in ways that exploit the complexity of formulas. So assuming α to be a sentence of $\mathcal{L}$, so would ⌐~α⌐, and assuming α, β to be sentences of $\mathcal{L}$, so to would ⌐$\alpha \supset \beta$⌐ be. Then the principle of induction says that with respect to an infinite class of formulas if from the fact that a formula of least complexity has the property in question then the formula of next greatest complexity also has it, it also follows that all the elements in this class also have the property. Here, too, the aim of the proof is not to dissuade some feckless sceptic who happens to deny that ⌐~α⌐ is a sentence of $\mathcal{L}$ even if α is, or to get some laggard to see the lexical meaning of "sentence" or "sentence of $\mathcal{L}$". The object of the proof is to show that the finite sentence-rules for $\mathcal{L}$ generates the infinite output of those rules mechanically, infallibly, and in finite time. So there is nothing circular or question-begging as such about proof by mathematical induction.

Proposition 12.12
CIRCULARITY IS NOT A FALLACY: *In relation to the frequency of occasions for it erroneous circularity lacks the frequency to satisfy the EAUI conception of fallacy.*

The circularity of recursive definitions bears a certain resemblance to *self-referential definition*, whose circularity is perhaps not so easily brushed aside.[385] In the interest of space, let us make do with some brief reflections about non-wellfounded sets, or sets that contain themselves as members. Non-wellfounded sets are disallowed in classical systems such as ZF, the standard treatment of sets, by the so-called foundation axiom, FA. It is obvious that if non-wellfounded sets are admissible to mathematics, they create the basis for well known puzzles and paradoxes.[386] For a long time, the dominant position in 20th century mathematics was that non-wellfounded sets are trouble enough to be excluded by fiat. Although this remains the standard view, alternative approaches have been considered and have, in time, been a considerable spur to new and interesting developments in mathematics.

A turning point was the formalization in 1983 of what has come to be known as the anti-formation axiom AFA.[387] The set theory that is built around AFA has proved a fruitful development. It provides a set of tools for studying the phenomenon of circularity in mathematics in a deep way. In this respect, non-wellfounded set theory plays a role not dissimilar to that of dialethic logic. As we have seen, dialethic logic is a logic that lets certain contradictions be true. Standard logic simply banishes them. But dialethic logic makes a real effort to disclose what might be called the deep structure of inconsistency, much as Cantor and

---

[385] Here again, is the self-referential definition of N, the class of natural numbers: x is a member of N if and only if x is a member of every inductive class. (A class is inductive when it contains 0 and is closed under successorhood.)
[386] For example, the Russell paradoxes precludes the universal set which, if it existed, would be self-containing. Consider also the (not quite convincing) definition of "liar": x is a liar if and only if everything asserted by x is false.
[387] Forti and Honsell (1983), Aczel (1988).

his colleagues explored the deep structure of (actual) infinities.[388] Whether in the end we accept that some contradictions are indeed true, dialethic probing will have offered a deeper understanding of the nature of inconsistency than can be found in Lewis' (and in some respects Aristotle's) dismissals of it. Similarly, non-wellfounded set theory aims to probe circularity, not dismiss it. It aims to plumb its structure, not render it inadmissible by axiomatic fiat.

The mathematics in which these investigations are made have spawned new developments beyond the exclusive focus of circularity. Coalgebra, which defines the concepts of corecursion and coinduction, is arguably the leading example[389] but there are also fruitful connections to the universal Harsunyi spaces of game theory, to situations semantics and fractals[390]. What one finds in this literature is a certain open-mindedness about set theoretic circularity. Here, too, the object is not to condemn it outright, but rather to determine how it works and why it is so attractive to mathematicians. The programme in coalgebra is a work in progress. Its final provisions have yet to be fashioned. But there is an unmistakable interim message. Circular reasoning in mathematics is hardly dismissible without theoretically weighty cause.

12.13 *Formal semantics*

Just now I mentioned a connection between circularity and paradox. As the discussion of non-wellfounded mathematics makes clear, it is possible to have the one without the other. What I want now briefly to consider is a case in which both paradox and circularity crop up, but in nothing like the straightforward way suggested by the conditional "If circular, then paradoxical". The case in point is the Tarskian hierarchical definition of predicates "$true_i$", which includes, in part, for the object language sentence α the first level metalanguage biconditional $\ulcorner\alpha$ is $true_1$ $\equiv \alpha\urcorner$ in which "$true_1$" is the truth predicate defined for metalanguage 1. Notice that the connective $\equiv$ is the material equivalence sign, defined in the usual way: $\ulcorner\alpha \equiv \beta\urcorner$ if and only if α and β are both true or α and β are both false. Thus the Tarski scheme is circular, not in the elementary sense in which the definiens contain the definiendum but in the extended sense in which the definiens occurs essentially in the specification of relation

---

[388] I owe this comparison of dialethists with transfinitists to Andrew Irvine.

[389] Paulson (1999), Kurz (2006), Moss and Viglezzo (2006).

[390] Harsanyi (1967), Barwise and Moss (1996), Edalat (1995).

that obtains between the biconditional's relata. Let us call such definitions *embeddedly* circular. Tarski's theory of truth (or truth$_i$) has dominated formal semantics for seventy years, and hardly anyone has worried about the circularity it embeds at each stage beyond the object level.[391] This is so notwithstanding Tarski's express recognition of the problem and his efforts to solve it (Tarski, 1944). Might this suggest that the received view is that the embedded circularity is harmless, that just as the mathematics of non-wellfounded sets can flourish around circularity, so too can formal semantics?

The answer is No. The Tarski's schema is seriously compromised by essential occurrences of ≡ when defined as Tarski intended. The reason for this is not circularity but inconsistency. In the definition, "true" occurs without index, that is, as a predicate of natural language. But Tarski thinks that the liar paradox establishes the inconsistency of this predicate. Tarski is a classicist about inconsistency. He thinks that inconsistent terms are non-instantiable. He thinks that inconsistent predicates have null extensions. Although he himself is not always clear about the point, he is committed to the view that there are no true sentences and, given that negation is classical, no false ones either. This being so, Tarski's biconditional schemata for the indexed predicates "true$_1$", "true$_2$". .... "true$_n$", ... are fatally compromised. Let ⌐α ≡ β⌐ be any biconditional. Then ⌐α ≡ β⌐ is true iff and only if α and β are true or α and β ⌐α⌐ and ⌐~β⌐ are true. Since nothing is true, neither of these alternatives obtains, and no material biconditional whatever is true. The consequences of this for Tarski's theory are immediate. The biconditionals that regulate the hierarchy's truth predicates aren't true. Tarski's semantic theory of truth isn't true.

Tarski considers three ways out of the problem

1. If the objection poised by the purported problem were correct there couldn't be a semantic theory of truth. So *modus tollendo tollens,* the objection fails.

2. The definition of the system's biconditionals lie outside the formal development of the theory. So it doesn't deface the theory.

3. Truth functions don't really matter, since the formal logic of deduction is really about provability (in the syntactic sense).

---

[391] An exception is Jacquette (2010).

It is not to my purpose to evaluate the strengths and weaknesses of these solutions, beyond saying that they are inadequate on their face.[392] What matters here is that none of the statements of these solutions is true. How could they be, given that "true" is an inconsistent predicate, a predicate with a null extension? Could we have it, then, that they are $true_i$, for the requisite value of i? Let us leave to one side the obvious difficulty posed by the fact that none of (1)-(3) is a sentence of a formalized language. Suffice it for the present that since none of them ascribes a truth predicate, we might consider them zero level sentences. If so, then what we want to know is whether (1)-(3) might be $true_1$. The answer is certainly in the negative unless we have it that for any such sentence e.g. (1), "(1) is $true_1 \equiv$ (1)". But now our problem recurs. If $\equiv$ is definable via $truth_1$, the biconditional is circular. If it is definable via $truth_j$ ($j > 0, 1$) then what the biconditional says is that "(1) is $true_1$" and "(1)" are both true j or not true j. But this is impossible. The hierarchical structure of $truth_i$ precludes the ascription of $truth_j$ to sentences appearing on different levels of the hierarchy. Accordingly, there is no true solution of the problem and no $true_j$ solution of the problem, for any i.

The circularity occasioned by the presence of $\equiv$ in the biconditionals that regulate the truth predicates of Tarski's hierarchy is of a particularly lethal kind, wrecking the semantic theory of truth. But, as we have seen, there is circularity and there is circularity. Not all circularity is all that lethal, and sometimes it is quite benign.

Could we help Tarski out of this difficulty by interpreting the truth predicate of the definition of material equivalence in some suitably indexed way? Consider the options. In the clause presently in question, the truth predicate which defines $\equiv$ has an index higher than, the same as, or lower than the index of the truth predicate which the biconditional sentence defines for that level. But this is hopeless. There is no truth predicate at a lower level than 1, in which case the definition of $\equiv$ doesn't compute. If $\equiv$'s truth predicate has a higher index than 1, it is undefined at level one, and so, again the definition of $\equiv$ doesn't compute. If $\equiv$'s truth predicate is given the same index as the truth predicate that $\equiv$ defines at level one, then the definition at level one is viciously circular, in much the same way that elementarily circular lexical definitions are.

---

[392] In this I agree with Dale Jacquette. For a detailed assessment of them see Jacquette (2010).

12.14 *Is anything a fallacy?*

I mean, of course, is anything a fallacy in the traditional sense? Is anything an EAUI fallacy? An objection that could be urged against the concept-list misalignment thesis is that, if true, there is something wrong with the concept or something wrong with the list. As we will see in chapter 15, dialectically-minded argumentation theorists are inclined to retain the traditional list and give up on the concept. I myself have recommended the reverse course. What the reverse course proposes is that the traditional concept is interesting and important enough to justify our investment in a seriously-minded positive theory, and it further proposes that, wherever else their interest may lie, it is not a requirement of a good theory of the traditional concept that the eighteen themselves be in its extension.

Still, there are objections. Let us grant that the eighteen aren't in the extension of the traditional concept of fallacy. That alone would not make it a theoretically uninteresting concept, or an undeserving target of logical theory. But suppose that we found not only that the eighteen aren't in the extension of the concept, but that nothing else is either. Wouldn't that be more than enough to blow the cover of the traditional idea? Two good questions present themselves.

1. Is the traditional concept of fallacy non-empty?
2. If not, what falls under it if not the eighteen?

I don't think that the traditional concept *is* empty. I think that there is a way of answering (2) that makes this clear. But before saying what that is, I want to enter a plea for the non-triviality of the traditional concept *should* the answer to (1) be No. If it were indeed empty, then we would have it that, given the interests, capacities and resources of beings like us, there is no such thing as an error we all tend to make with noticeable frequency, to which we all, by and large, are drawn and which we all, by and large, find it difficult to recover from. If this were actually so, it would be a momentous revelation for the investigation of the cognitive track record of beings like us. And it would put additional weight on the error abundance thesis. If beings like us make errors, lots of them, how this comes to pass will have to be explained. If they are errors, they aren't attractive, they aren't universal, and they aren't incorrigible. Although we all commit errors, lots of them, there is an important sense in which they are very different errors – from occasion to occasion, era to era, culture to culture, and so on. That will, if true, not be an easy thing to account for.

But, of course, it isn't true.   The   traditional   concept   is   not empty. Take our present reflections on Tarski's definitions of "true$_i$". If what we have been saying here is so, then the Tarski-schemata for truth in formal languages is either inconsistent or non-computing or viciously circular. The received opinion is that Tarski on truth-in-L is very much the way to go, a major breakthrough in formal semantics. Let K be the community of formally minded truth theorists, then the Tarski-schemata are attractive and universal in K. They are also an error. But, unless I am mistaken, the K-community is not likely to have a wholesale change of heart about the semantics of truth, even if they concede the difficulties alleged against it. If so, it would be an incorrigible error. So the traditional concept of fallacy is not empty after all. In this case, it is a fallacy where it has occasion to be. It is a fallacy in K.

Begging the question is perhaps the granddaddy of the questioning fallacies (actually, "fallacies"). But also important is the fallacy (or is it "fallacy"?) of many questions. I close the chapter with some remarks about it.

12.15 *Many questions*

This need not detain us for long. The difficulties occasioned by many questions are entirely the product of dialectical artificialities. Suppose that Sarah and Harry are having a conversation which is tightly bound by the Yes-No rule. Suppose, that is, that all contributions from Sarah must be questions answerable Yes or No, and that Harry's contributions must be those answers. Suppose Sarah wants to know whether Harry is a dog beater. She asks, "Do you still beat your dog?" If Harry answers Yes, Sarah has what she was after. But what if, as might well be expected in the general case, Harry answered No. What does this get Sarah? Since Harry's answer is the negation of Sarah's question, and since Sarah's original question was more than one question at once (many questions), Harry's answer is the negation of her many-questions, as follows. Putting U for "used to beat my dog", and N for "beat him now",

?(U and N).

Harry's answer is

Not (U and N)

or equivalently,

Not U or Not N,

which is consistent with but does not entail

Not U and Not N.

Sarah wanted to know whether N. Given the question she asked him, Harry cannot give her what she wants with a No answer. So Sarah asked the wrong question.

Contrary, to what informal logicians seem to think, Sarah did Harry no logical disservice in asking him this question. In answering it as he did, Harry did himself no logical disservice either. Sarah's fault was not to have unfairly placed Harry in a compromising position. Sarah's fault is that she did not succeed in putting Harry in a compromising position. Given her interests, she asked a stupid question.

It is true that, where a question contains an embarrassing or otherwise harmful presupposition P, asking whether it is still the case that P or whether you have stopped Ping or whether you intend to keep on Ping will, if answered No, generate a proposition whose logical form may be misunderstood, whether by the parties themselves or by a third party. Using present schemas, this would be the mistake of interpreting "Not (U and N)" as "(U $\wedge$ not N) $\vee$ (Not U $\wedge$ N)". It is possible, of course, that there are people for whom the inequivalence of these two formulas is inapparent. If so, we might say that not the asking of the complex question but the misinterpretation of its answer is a fallacy in something like the traditional sense. But there is a better answer, a Gricean answer. In his "Logic of conversation" (1975), Grice espouses a quantity maxim. It counsels that when engaged in a conversation with someone, you shape your contributions in such a way that they say no more and no less than what the situation calls for. This "speaking up to one's maximum" is what is missing in Harry's answer to Sarah's complex question. This leaves the following problem for Harry. Assuming that Harry in answering "Not (U and N)" Harry was speaking up to his maximum, he has no defence to offer against the presupposition of Sarah's question. Accordingly, if the best he can say against that question, and against its presupposition, is something less than that the presupposition is false, then there is room for the abduction that the presumption isn't false after all.

It won't do. The Grician answer overlooks the artificiality of the requirement that Harry may make no response to Sarah other than Yes or No. In real life conversations there is no such requirement. If someone asks you a loaded question in the form "Do you still P?", you need only unload it; you need only say "I never did P and I don't now". True, there are real-life contexts in which something like the Yes-No rule is officially in effect. Think for example, of cross-examinations at the criminal bar in common law jurisdictions. In practice, however, where P is a damaging claim not yet attested to by direct evidence, no witness will give "Do you still P?" a Yes-No answer (unless it's Yes), and no judge will permit him to do so.

The point of substance that arises from these reflections is that fallacy-making is not so much a matter of error-making as error-making whose damages *linger*. We saw at the beginning that in contexts of holder-refutation, premiss-attribution errors commonly arise from nonculpable ignorance and are efficiently correctable. The same is true of presupposition-misattribution. If I ask whether you still live in Ann Arbor when in fact you never did, all you need do is say so. Here too, we often have it that an accuracy error is an aptness non-error and that – largely for reasons of economy – aptness trumps accuracy. For, again, repair is often cheaper than avoidance. What this tells us is that

Proposition 12.15
MANY QUESTIONS IS NOT A FALLACY: *By and large, many questions doesn't meet the badness condition of the traditional notion of fallacy.*

So we need say nothing further about many questions.

COROLLARY: *The misalignment tally now stands at twelve out of eighteen, or thirteen out of nineteen if we count circularity as separate from BQ.*

CHAPTER THIRTEEN

GETTING PERSONAL

"Multi famam, conscientiam pauci verentur"   Pliny

"All great truths begin as blasphemy."   George Bernard Shaw

"Same to you, fella!"   Bob Newhart

13.1 *Slanging*

Second-person utterances are about someone other than their utterers. The founder of logic took early note of a small proper subset of them. They are second-person remarks in which the addressee is also their subject, and in which the thing said of the addressee is that his commitments don't sit well – in Locke's later words – with "consequences drawn from his own principles or concessions." (Locke 1961, p. 279) For Aristotle and, following him, Locke, these are remarks *ad hominem* – to the man. Clearly no fallacy as they stand, they are precisely what Aristotle's holder-refutations require of us. *Ad hominem*s have taken on a different face in the modern tradition. It is to this new face that I want now to turn, but not before a brief word about slanging.

There are two impulses for the *ad hominem*. One is broadly rhetorical. The other is broadly logical. *Ad hominem* remarks are the heart and soul of slanging. Slanging is a rhetorical device, as old as the hills. Its objective is to expose, embarrass, infuriate, ridicule, mock, calumniate or humiliate one's opponent, often with the intent of rattling him dialectically. One of the more impressive slangers of the western intellectual tradition was Rabelais. Reflecting on how his ideal Academy should be constituted, Rabelais indulged in some rather aggressive exclusions. He was especially hard on lawyers and, as we now call them, professors of management.

> Enter not base pinching Usurers,
> Pelf-lickers, everlasting gatherers.
> Gold-graspers, coin-grippers, gulpers of mists:
> Niggish deformed sots, who, though your chests
> Vast summes of money should to you afford,

Would ne'ertheless adde more unto that horde,
And yet not be content, your clunchfist dastards,
Insatiable fiends, Plutoes bastards,
Greedie devourers, chichi sneakbill rogues,
Hell-mastiffs gnaw your bones, you rav'nous dogs.[393]

Slanging tells us something important about how differences of opinion engage us. Rabelais didn't want lawyers, accountants and professors of business. His reason is that they are greedy and crooked. If this were true, it would be a pretty good reason for their exclusion. But Rabelais hasn't in these remarks established that these Plutoes bastards are indeed greedy and crooked. Perhaps he was directing these remarks to the supporters or current members of the Academy. Perhaps it was a rallying cry, designed to consolidate their anti-lawyer policy. Whatever its audience, it contains some rather aggressive remarks *ad hominem.* For an exclusionary case to work at, it would have to be established that the would-be excludeds are greedy and crooked. Rabelais doesn't establish that here. He *asserts* it. So his case, if that's what it is, is defective. Rabelais also heaped abuse upon the would-be excludings. Did that contribute to the weakness of his case? It did nothing of the sort. The case was weak independently of the abuse.

It may on reflection strike us that, rather than making the case against lawyers and professors of management, Rabelais is simply announcing a decision and, in so doing, availing himself of the chance to slang them. What matters for the soundness of the policy is whether the excluded are indeed greedy and crooked, not whether "clunchfist dastards" is a polite way of saying that they are.

Awhile ago, Walton and I drew attention to the phenomenon of the *dialectical shift.*[394] Imagine two parties having a refutation argument in which Sarah is the questioner and Harry the respondent. Dialectical shifts can happen in two ways, intra-argumentatively and inter-argumentatively. It happens intra-argumentatively in Sarah's and Harry's refutation if at some point Sarah and Harry exchange dialectical roles, with Harry saying "But now, let me ask you a question, Sarah". A dialectical shift occurs inter-argumentatively when – still staying with the present example – Sarah and Harry get stuck, and a refutation argument metamorphoses into a quarrel (perhaps preceded by babbling). Dialectical shift is a useful concept for the understanding of arguments. It reminds us

---

[393] Quoted from Davies (1979), p. 8.
[394] Woods and Walton (1982), chapter 8.

that, like a certain president of the United States, it is possible to do two things at once – walk and chew gum. It tells us that we can even mount a competent refutation against someone whom we loathe and against whom we're prepared to fulminate, whether for strategic purposes or just to see him squirm. Theorists of argument don't like to see arguments get personal in these ways. They think that impartiality is compromised.

Fulmination is one way to proceed against people you loathe. Irony is another. No one can know the pain of true rhetorical slaughter until he has been dismembered by an upper class English undergraduate of impeccable nasality who is able at will to say how interesting your argument is and mean that it is utter rubbish.

Arguments are what people have together. Disputations are arguments about what the disputants disagree about. It is not generally recognized by argument theorists – certainly it is something not emphasized in the leading theories – that in the grand scheme of things differences of opinion are more unremarked upon than not, and even when spoken of, are so with a noticeable diffidence, well short of argument. In comparison with the sheer abundance of their occurrence, discussions about differences of opinion almost never happen. It matters who the parties are whether strangers, occasional acquaintances, office mates, chums, lovers, shop stewards, statesmen, presidents of the National Hockey League, and so on. In a rough and ready way, a difference of belief will be commented on or not depending on the relationship between the parties, the subject matter of the opinion over which they differ, and upon whether there exists, expressly or by convention, some element of invitation to pursue.

That's what it takes to get a difference of opinion *spoken of*. It is quite another matter of what is required to get it *argued about*. Even here, arguing about something admits of a critical difference between

 a) giving your side of the story
 b) defending your position and/or attacking his position.

In the give-and-take of everyday life, it is often the case that all that the intimacies of the situation will allow is an explanation of your side of the issue or of why you feel bound to adopt it. Consider: *Sarah*: "Aren't you thrilled with Obama's lifting of the prohibition against embryonic stem cell research?" *Leila*: "I'm crestfallen by the news!" *Sarah*: "But why?" *Leila*: "Well, I'm a Muslim." *Sarah*: "Oh, I see. Right, then."

Attempting to talk an interlocutor out of a belief by showing that the belief is false or at least ill-evidenced, much less by showing that her own defence of it is inconsistent,   requires a latitude that the social

448

niceties almost always discourage. The conditions, both personal, interpersonal, topical and circumstantial, under which proving an opponent wrong could even be tried are hard to come by. The chances are that if Sarah and Leila were to have made the cut for an attack-defend go-around, the attendant intimacies would almost certainly free them from the requirements of sunny (or even icy) *politesse*. So we might accept an argument that is by turns aggressive, resentful, upsetting, dotted with cheap shots. This dialectical shiftiness with a vengeance. It is doing more things than one. Remarks *ad hominem* are the life's blood of deeply felt disagreements among intimates. It is not polite to call your opponent a God-damned hypocrite, even if she is one. But it would be a rush to judgement to condemn it outright as a refutation-buster.

It is possible to do two things at once – *rebut* and *fulminate*. Even when engagement is necessitated by professional obligations – think here of the legal aid lawyer landed by the rotation system with the defence of the odious Spike McGurk – there are usually exit conditions triggered by the improper behaviour of one's opponent. Although in the general case, conditions for engagement are noticeably freer than conditions for disengagement – walking away in disgust, for example – there exists very wide latitude going in, and very wide latitude walking out. At a certain level of abstraction, arguments are suspendable agreements to try it on for size. It is here that we meet with an important asymmetry. What justifies a disputant in walking out on an opponent is radically less onerous to achieve than making the case that he presses against him. If you don't like how the game is going, you can always take your ball and go home. The sheer weakness of the conditions on quitting extends to disputants a perfectly natural freedom to rough one another up, that is, to stray from the procedural niceties demanded by argument theorists of the Goody Two-Shoes school.[395] And since it *is* possible to do two things at once, it

---

[395] Is that too cheeky a description? Well, sorry. Goody Two-Shoes theorists model argument as a conflict resolution device in which parties are fair, respectful, open-minded and truthful. A typical example of such arguments is what van Eemeren and Grootendorst (1984) call the "critical discussion". I myself have no problem with people being nice to one another. My reservation about the suitability of the Goody Two-Shoes approach as a general theoretical framework for argument is empirical. In relation to the totality of conversationally embodied differences of opinion, arguments of that sort hardly ever happen. They are links to those 'urricanes in 'Artford, 'Eresford and 'Ampshire. They aren't in the slightest degree typical. Not even diplomats argue this way.

is possible that your slanging of an opponent is no despoiler of the case you make against him. Think here of master cross examinations at the criminal bar whose principal, and legally mandated, intent is to *impeach* the witness, *impugn* him.

Hard-knocking argument is hard work. It is much cheaper just to state your respective positions and then go for a beer. Bringing about one another's dialectical destruction requires a talent which, for the most part, has to be learned, and these sorties require for their persistence, as for their initiation, the acquiescence of parties whose involvement nearly always is discretionary.

There is no blanket requirement either of courtesy or rationality to subject differences of opinion to the cut-and-thrust of dialectical annihilation. Indeed, it is the reverse that is nearly always true. When they arise spontaneously, as opposed to being necessitated by professional duty, we have them at their most intimate, and intimacy is the natural currency of passionate engagement and *ad hominem* provocation.[396]

Getting the rhetorical textures of verbal abuse right is a trickier task than might be supposed, but one thing can be said without much ado. There is no intrinsic tie between slanging and the negative assessment of either party's position or case. Slanging isn't argument assessment. Since, in the logical tradition, this latter is precisely what the *ad hominem* purports to be, the logical tradition has no interest, as such, in slanging. What logicians want to know is whether the *ad hominem* is ever a legitimate part of argument assessment. In particular, they want to know whether it is a fallacy.

Most treatments of the *ad hominem* follow Aristotle's and Locke's lead in having the subject of the remarks and the addressees of them be the same person. In actual practice, an *ad hominem* is often addressed to a third party, and in lots of cases it is left to the third party to take its measure and to furnish whatever answer may be possible for it. In still further cases, as in a scientific or scholarly article or op-ed piece for the *Daily Telegraph,* the subject of a remark *ad hominem* might be Harry, but its addressees are the piece's readers, making up what Trudy Govier calls the "non-interactive audience". While some responses are possible by way of the Replies section of the academic journal or a letter to the

---

[396] "Dialectical annihilation": See section 7.8 of Magnani (2009) entitled "Fallacies as distributed 'military' intelligence", and Magnani (2013), entitled "The non-fictional nature of scientific models: Model-based science as epistemic warfare".

editor, it is not typical of this situation that *ad hominem* challenges are responded to. In what follows, I confine my remark to cases in which the subject and addressee of an *ad hominem* retort are one and the same person. Since most of what is said here about these cases generalizes in a quite natural way to the others, nothing for the most part is lost by imposing this restriction. For an exception, however, see two sections hence.

Consider now the absent party case. It matters here that the rules of case making and *ad hominem* making are at their freest. The empirical record discloses that by far the majority of case-making episodes are directed to an absent party – hence not really a "party" at all. (Is St. Anselm a party to John Mackie's fulminations against the ontological argument?) This means that by and large case making arguments are not transacted *dialogically.* Think here of the whole extant case-making literature in philosophy. How much of it was face-to-face? How much of it is of the replies and rejoinders sort? There are reasons for this. One, as we have been saying, is that dialogical contestation is only comparatively rarely a social permissible thing to engage in. The other is that the rigours and complexities of case-making and counterattack are too much for the discussive limitations of *conversation* to bear. It takes too long, and there is too much to remember, for the short shelf-lives of chit-chats on the ground. The reason that even Plato's short dialogues ring so false as dialogues is their length and their complexity, even for the leisured intelligentsia of Athens. Accordingly, the model for case making is the full service S-proof and for counterattack is the falsification argument, each dealt with a chapter ago.

We glean something of importance from this. It is that dialogically based theories of case making argument will not serve the general case. They will serve best, if at all, in those comparatively limited ranges of cases in which case making and counterattack are of necessity transacted dialogically – as in the courts of law, the mediators' offices, and the collective bargaining table. Nor should we forget the breakfast table, the neighbourhood bar and the office water-cooler. But these are the contexts in which, if there were to be a shift from a position-explaining conversations, it would be much likelier to turn into a bickering or bantering or quarrelsome argument than a case making or counterattacking one, and thus one that provides a natural habitat for ugly provocations *ad hominem.*

It would be wrong to say that *homineming* is an exclusively dialogical phenomenon. But it wouldn't be far wrong to say so. Consider some exceptions. Suppose that you are constructing an attack on Anselm's ontological argument. (Suppose you're writing it down). If your case against St. Anselm included the point that as a Catholic priest, Anselm was precommitted to God's existence, and that, therefore, he wouldn't be sufficiently open to contrary indications, wouldn't that have an been an *ad hominem*, a remark "to the man", and wouldn't Anselm have been the man, never mind that he couldn't have been your addressee? Or, in a slight variation, suppose that you publish your attack in a learned journal. Perhaps, now, we could say that while St. Anselm is the object of your *ad hominem*, your readers are its addressees. Yes, I suppose that these things are possible. But let us not lose sight of the central point of fact. It is that if we examined the historical record, we would be hard pressed to find instances in which the *case* against the ontological argument included, to say nothing of turned on, any such reference. Why would this be? One could say that it is because serious case makers know that the reference would be fruitless. One could also say that the reference is not made, not because it has no relevance to the issue at hand, but because it can't be *answered.* I shall return to this point below.

13.2 *Antipathy*

In a general election, there are lots of voters who would never cast a ballot for someone they disliked or disrespected or outright hated.[397] There are lots of voters who would arrive at these electoral rejections without the slightest regard for the merits of the candidate's policies or competence. When this happens, an interesting ambiguity arises. Suppose that a candidate for high office is revealed to be a lifelong philanderer and betrayer of his long-suffering wife, and that it is on this account alone that he has aroused your ire and lost your support. Then it would be accurate to say that the importance of the candidate's policies and competencies have been overridden by your low opinion of his bad behaviour. It would be correct to say that, for you, his repellent conduct trumps the virtues of his policies and the clear indications of his legislative competence. Suppose you say so in a letter to a friend: "Smith's loathsome behaviour trumps his policies, and his record of

---

[397] Hence the title of this section. "Antipathy" means "having a strongly felt aversion to", but its original Greek meaning is truer to what I have in mind here. *Antipathei* is the fusion of *anti* (= against) + *pathos* (feeling), carrying the clear suggestion of ill-will or hostile regard.

competence." Would it be entirely surprising if your friend wrote back reproaching you as follows: "Smith's moral defectiveness doesn't show his policies to be defective or his competence to be non-existent. You've committed the dread fallacy of abusive *ad hominem*!"

Of course, that's not what you said. What you said, or thought you did, is that the candidate's behaviour triggered your *decision* not to support him, never mind the cogency of his policies and the rock-solidness of his competence.

Campaigns Inc. is an American political consulting firm, established in 1933, said to be the world's first. It was a successful venture. Out of its first 75 tries, 70 of Campaign Inc's 75 clients won election. In a recent piece in the *New York Times,* Jill Lepore summarizes the rules of procedure. "Make it personal: candidates are easier to sell than issues .... Pretend that you are the Voice of the People .... Attack, attack, attack .... Never explain .... Say the same thing over and over again .... Simplify, simplify, simplify .... You can put on a fight, or you can put on a show." I quote these remarks from Andrew Coynes's *National Post* column of 22 September, 2012, p. A 1. There is nothing new in this kind of campaign strategizing. As Coyne points out, Cicero (the one who is identical to Tully) has similar advice for his brother Marcus, in the run up to the consular election of 64 BC. In his letter "Commentariolum Petitionis", Cicero wrote, "it is better to have a few people in the Forum disappointed when you let them down that have a mob outside your home when you refuse to promise them what they want .... Remember how [another candidate] was expelled from the Senate after a careful examination by the censors? .... Stick to vague generalities."

The efficacy of slanging in political campaigns is beyond doubt. There are some questions this raises, I mean seriously raises, and some it does not. It does not raise any serious question about whether the creators and implementers of these campaigns are defective reasoners. It does not raise any serious question about whether the voters influenced by these imprecations have lost control of their reasoning faculties. Virtually everyone thus stimulated knows that it is not his premiss-conclusion reasoning that is calling the shots here. Perhaps it should. But if it should, it is the voter's prudence that is called into question, not his capacity to see what follows from what.

This last remark takes us to a question that seriously does matter. It is how countries whose governments are elected in such unsavoury ways can prosper and flourish in the way that the Western democracies have done for as long as they have done? All I will say about this here is that the truth of the question's presupposition – its predicate, as lawyers say – counts with some favour on the prudence of voters' susceptibility to the modern methods of electoral politics.

This is not to say that voters are quite straight with one another, or themselves, about their responsiveness to political harangue. Rarely do we see it explicitly acknowledged in a negative "character" campaign that the object of the exercise is to strengthen the causal tie between personal antipathy and a decision to withhold one's vote. This is attested to when the candidate in the opposing camp, the intended beneficiary of the causal inducement, is challenged in quite predictable ways. Someone asks, "Why do you think that Smith's personal immorality about matters having nothing to do with his competence to govern should disqualify him from doing so?" Someone else asks, "Why don't you just come clean and admit that all you're really up to is pushing the electors' emotional buttons?" The correct answer to the first question is that Smith's philandering alone *is* no reason to doubt his capacity to govern effectively. The answer to the second is that button-pushing serves the objective of winning this election at all lawfully permitted costs. These answers are never given, never mind that they are widely known to be true and that this too is also meta-known by every campaign team bestriding the earth from Cicero onwards. Instead, these questions are replied to in dissembling and self-serving ways: The misbehaving candidate is not fit for office because he is unreliable. If he'll betray his wife, who's to say that he wouldn't betray his country? He is a person of bad character in marital matters. So who is to say that he wouldn't be a person of bad character in national security matters? Common to all these responses is the implication that facts about his marital life are evidentially probative – and negatively so – for the question of his fitness to hold the office he seeks. As the empirical record makes clear, whatever may be the strength of the correlation between an arbitrarily selected politician and a record of political incompetence in office, that relation is not elevated to the level of statistical significance between an arbitrarily selected philandering politician and an incompetent political performance in office. So the purported evidential tie is insufficient to bear the weight of any such inference.[398]

---

[398] Of course, there is a philosophical view that traces back to antiquity, in which it is impossible for a bad person to be wise. For all I know, there may be some people among the refusniks presently in view who ground their rejection of a bad person's candidacy in this metaethical precept. It doesn't matter. The empirical

Up to now I have centred this discussion on the purveyors of negative "character" advertising in political campaigns. It is interesting to compare these cases with the reaction of voters on the receiving end of the attacks. On the one hand, hardly anyone wants to admit that his decision at the polls was the result of causal manipulation by experts adept at the manufacture of affect. On the other, there were legions of Americans whose vote-casting decisions were constituted by their loathing for Richard Nixon and nothing else, and lots of them had no scruple in saying so. The difference between the two turns on the fact, or the strongly apparent fact, that people often don't mind admitting to emotion-driven decisions, as long as they leave the appearance of having been freely arrived at. It would be useful if these cases had names that reflected this difference. In the first instance, where a voter withholds his vote for reasons of probative import, let's call the decision a "probative" one. In the second instance, where the decision to vote against is rooted in the desire to deny the disliked candidate a goal he is desirous of having – as a penalty for his bad behaviour – let us call the decision a "punitive" one.

When a vote against is rooted in a probative decision, the probative significance it purports is nearly always nonexistent.[399] For anyone whose interest in such cases is borne by his involvement with fallacy theory, the central question is not whether probative decisions embed a faulty inference; for the answer in the general case is obviously that they do. The central question is whether voting inspired by personal dislike is typically or mainly grounded in probative decisions or punitive decisions. My inclination is to favour the second alternative, and to think that ascription of the first very often pivots on the fact that, if challenged about the probative legitimacy of one's decision, a punitive decider – if he answered the challenge at all – would display some likelihood of characterizing his decision as probative. That is, there would be some likelihood that he would *mischaracterize* the decision he actually took. In such cases, two things have gone wrong. The argument he advances is defective; and the argument he advances misdescribes his actual decision. His decision to reject the candidate was punitive, not probative. In

---

record leads it no support and attests to an absence of support by the population at large.

[399] An exception, perhaps, is a candidate for election to the Office of County Treasurer, who is revealed to have had numerous convictions for influence peddling, fraud, and misappropriation of funds; or has been discovered to have an IQ of 50.

reaching it, his intention was to punish the candidate. Perhaps some reasoning was involved in arriving at this decision. But, if so, it was not probatively evidential reasoning. It was means-to-an-end reasoning. If a fallacy is an error of reasoning, there is nothing in this kind of decision that qualifies as an error of reasoning. Accordingly,

Proposition 13.2
THE ABUSIVE AD HOMINEM: *Abusively* ad hominem *decisions are not fallacious.*
COROLLARY: *Further confirmation of the misalignment thesis. The tally now stands at twelve out of eighteen.*

This would be a good place to make a quick return to a feature previously discussed under the heading of "dialectical erasure". As earlier noted, sometimes it happens that there is some proposition that someone now knows – that is, knows according to the CR-account of knowledge. Then the very fact that he has been challenged to show that he knows it may occasion an erosion of belief to the extent that it no longer fulfills the belief condition on knowledge. In which case, whereas a minute ago he knew that α, a minute later he does not. This we called dialectical erasure. Erasure is occasioned by the knower's inability to provide the justification demanded of him by his challenger. As we now see, something of the same sort can also occur when the maker of a punitive decision is challenged to show that it is a probatively justified decision. In the knowledge case, the knower's correct response is to reject the challenge, to point out that justification is not a condition on knowledge. In the punitive decision case the correct response is to reject the challenge, by pointing out that your decision was made without probative intent. In each case, if the challenge is taken up, trouble ensues. In the knowledge case, the proferred justification (if any) is manifestly no good. In the decision case, the proferred case is manifestly no good. In each case, the defence lawyer's rule holds sway: *Take no hostile questions.*

Punitive *ad hominems* are instructive in another way. They help make the point that voting is discretionary. It is yours, free for the casting however you may please. It may or may not be reasonable to vote your loathing of a candidate of great talent, it may or may not be a lapse of civic duty to get personal in these ways, but if the decision is punitive, there is no case you need answer in a charge of fallacious thinking. This moral extends to any behavioural context in which your participation is discretionary. This is no less true of arguing than of voting. It is no less

456

true of capitulating to an opponent's case than shopping at Fortnum's. True, there are conventions of courtesy and fair practice. If you are arguing with someone about embryonic stem cells, it may not be entirely courteous to call your opponent a loathsome Muslim bigot (and mean it), and it certainly isn't likely to advance you in the cause of winning her assent. But the fact remains that when you make yourself available to another person's efforts to attack positions to which you're strongly enough committed, your submission is discretionary and can be terminated without notice or cause. This, too, is important. It is often underdetermined by surface behaviour whether an arguer's getting personal signals argument-withdrawal,[400] rather than faulty inferences about the incapacity of Muslims to hold correct opinions when they are opinions contrary to your own.

It bears repeating that on matters of any complexity, or where disagreement is deeply embedded, case making and counterattack are very difficult to bring off. The importance of this fact was evident to the precursors and founders of logic. A good deal of Aristotle's *Topics* is devoted to the development of the case-making and thesis-defending skills of the citizens of Athens. Aristotle had a low opinion of the argument-making virtues of Athenians, of both its callow youth and the greatest of its great men.[401] Since Aristotle also tended to the position that arguing is reasoning out loud, he was likewise inclined to think badly of the reasoning skills of his compatriots. I myself go part of the way with Aristotle. They weren't very good at arguing (no less we). But they (and we) are quite good at reasoning. In other words, reasoning may be implicated in reasoning, but arguing is not just reasoning out loud. Arguing is more complex than reasoning and, on that account alone, more difficult.

13.3 *Damaging information*

David Hitchcock has recently done us the service of demonstrating that the *ad hominem* has been a fallacy only since the mid-nineteenth century.[402] De Morgan appears to have been the first to

---

[400] Or argument shift, as from case making and counterattack to quarrelsome argument.

[401] Socrates, too, as the *Symposium* cringingly attests.

[402] Diehards may think this a trifle question begging. They will think that the *ad hominem* has been a fallacy all along, and that its discovery had to wait for 2200 years or so.

propose it as a logical error – indeed, as De Morgan sees it, it commits the fallacy of *ignoratio elenchi* (1847, pp. 308-309), which he characterizes "as answering to the wrong point" (Hitchcock , 2007). *Argumenta ad hominem* are discussed by Aristotle in, among other place, book Gamma of the *Metaphysics*. They also captured the attention of mediaeval thinkers and were part of the standard fare of post-Renaissance logic until the 1800s. In all that time, they were interesting to logicians, but not because they were fallacies.[403] Hitchcock's view is that the by now received understanding of these arguments as fallacies is in fact a common misconception, hence a fallacy in its own right. I think that Hitchcock is right about this.

I have a particular and twofold purpose in this section. I want to produce reasons in addition to those put forward by Hitchcock for a less negative assessment of *ad hominem* manoeuvres I will argue that

Proposition 13.3a
Ad hominem *arguments are not fallacies in the traditional sense.*

I want also to suggest a positive thesis about them:

Proposition 13.3b
Ad hominem *remarks play a natural and dialectically benign role in the give-and-take of real life argumentation.*

Propositions 13.3a and 13.3b are instantiations of the more general providence of Propositions 1.3b and 1.3c, the misalignment and cognitive virtue theses. As we proceed, it will become clear that my support of these theses will pivot on various of the features of told-knowledge discussed in chapter 9. My own view is that a proper appreciation of *ad hominem*ing cannot be achieved except that the role of telling in human cognition be given its due.

As we saw, one of the virtues of Aristotle's approach is that disputes that honour the refutation rules are hermetically sealed against begging the question. When a refutation succeeds, the refuted party has only himself to blame. While the avoidance of question-begging is clearly a benefit, it also carries a cost. The cost is that Aristotelian refutation arguments are unnatural to work with and subject to significant inefficiencies. They are costly things to bother with.

---

[403] For a survey, see Nuchelmans 1993).

In contrast, in the adjudication of the contested issues of real life, parties routinely supply their own premisses. Such premisses typically arise in one or other of two common situations. In one, the premiss is a proposition in the common knowledge of the parties, hence is one that doesn't need to be "asked for". In the other, the premiss furnishes the reply to a question which, had it been put to him, the party under attack would not have been able to answer. Here the antagonist (or the critic) plays the role of *informant*. This, I think, is a crucial feature of real life contentions, made so in no small measure because of what they reveal of the causal tug of being told things. When a liar tells you that he is not lying now and you accept it, who could gainsay the power of the telling? We may take it as a fundamental fact about real life refutation arguments that

Proposition 13.3c
COOPERATION AND ANTAGONISM: *Even in holder-refutation arguments in which the parties are antagonists, they are in the absence of indications to the contrary, mutually cooperative in the matter of transmitting and receiving* new information.

Proposition 13.3d
TIE TO TOLD-KNOWLEDGE: *Even in antagonistic contexts of holder refutation, the dynamics of told-knowledge are not by and large significantly interfered with.*

Proposition 13.3e
BENEFITS: *Again, the benefits of such cooperation are dominantly economic. What it gains economically is far from cancelled out by attendant inaccuracies.*

It is at this point that Aristotle's dialectical rules dig in deeply. In chapter 8 of the *Topics,* Aristotle requires that the refuting party never play the role of informant, and that when his interlocutor is unable to answer a question, his ignorance must be removed by supplementary questioning. The classical paradigm of the dialectical removal of ignorance is the slave boy argument of Plato's *Meno.* Everyone familiar with this celebrated text will be aware that Socrates' interrogation is both argumentatively artificial, lengthy and time consuming.

Ad hominem retorts are a way of alerting us to the danger of relying on one's interlocutor for information intended to be used to one's disadvantage. Imagine the following kind of case. Suppose that Sarah is trying to convince Harry to end his friendship with Lou. This is a serious dispute. Harry is Lou's friend and wants not to see the friendship ended. Sarah is determined that it will end. Suppose that Sarah's principal complaint is that Lou is unfaithful to his wife, and that, in support of that claim, she announces that Susie saw Lou at a restaurant last night cosying it up with another woman. Harry is a Lou-loyalist. He is a steadfast friend. He doesn't like what Sarah is telling him. Suppose Harry says (and means it), "No, I saw him in Cambridge last night." This is the last thing that Sarah wants to hear. Her information places Lou in London. Harry's "Well, I saw him last night" is decisive against Sarah's restaurant-charge if she accepts what Harry has told her over what Susie told her. Certainly she would have no good reason to accept this claim if she thought that Harry was outright lying or that in the heat of the moment he was being economical with the truth by misrepresentation or omission or confusion. But otherwise, this would be reason to accept what he says and, with it, the defeat of her claim about Lou's restaurant shenanigans. There is an edgy tension here – or at least the prospect of it. In telling this to Sarah, Harry's intention is to damage her *contrary case* with it. Each party has an interest. Harry's interest lies in getting Sarah to accept information that undermines her position, and Sarah's interest lies in resisting such information. We may say that Sarah has a stake in playing by something like Aristotle's rigid rules. The trouble is that, in real life, Sarah's defensiveness would be seen as unrealistic and pig-headed.

Any form of argument that permits an aggressor to introduce damaging information on his own sayso is at risk for the contaminations of misrepresentation, omission and outright falsehood. When an argument is a serious one about a matter that touches the vital interests of the parties, it matters considerably that the parties correctly assess the informational *bona fides* of their interlocutors. Accordingly, on the face of it, the safer tactic is to refuse the sayso of an antagonist whenever the proferred information is damaging to the defendant's position on a matter that is vital to her. More particularly, when Sarah has reason to think that Harry is a liar, she has reason to refuse anything he tells her if it proves damaging to the position she is defending [404] In fact, however, this is often what does *not* happen.

---

[404] In the spirit, perhaps of the response of a former Chancellor of the Exchequer to the then Prime Minister of Britain: "There is nothing you could ever say to me now that I could ever believe." Quoted in a review of Peston (2006) by Peter Clarke in the *Times Literary Supplement,* June 30, 2006, p. 12.

Proposition 13.3f

LYING DOWN WITH THE ENEMY: *Interlocutor-directed admonitions* ad hominem *aren't typically rejections of the proferred information. What is more, they often invite the informant's testamentary reassurance.*

This, I will say, constitute the primary datum for any theory of the interlocutor-directed *ad hominem*.

Bearing on this is the quite general fact of our dependency on the sayso of others for so much of what we know and for the bases on which most of our decisions to act are taken. So entrenched is the disposition to accept what we are told, that it is *comparatively* difficult to refuse the information that informants direct our way. A measure of this entrenchment is the frequency of our mollification by liars. This creates a default. Unless otherwise indicated, it is better to accept rather than reject the sayso of others. Beings like us have a drive to accept the sayso of others. So whether it is indeed better to do so in the general case or not, we don't have much of an option. It is rather striking, then, that information on the sayso of an opponent that damages one's own position does not, just so, override that default.

The empirical record amply attests that the burden of proof falls on Sarah to show cause for not accepting Harry's damaging information. This burden is certainly not discharged by observing that Harry's information has a destructive motivation. If this is right, then remarks *ad hominem* are often the addressee's sole means of meeting this burden. If Sarah knows, or has reason to believe, that Harry is a liar, may she not properly resist his information on that basis, and may she not tell him so? If, contrary to what I have just suggested, she did reject his information on that basis, this would be an abusive *ad hominem* in the standard taxonomy. It would not be a mistake unless she concluded in some quite strict way from the fact that Harry is a liar that Harry's present claim must or is likely to be false. Either way, no fallacy is committed. In the first instance this is because no error is committed. In the second instance, this is because the error is so egregious that hardly anyone would ever be drawn to it. It is an error that lacks the frequency of commission required to make it a fallacy in the traditional sense.

Similarly, if Sarah's *ad hominem* retort were that Harry carries a bias against her position, this would be circumstantial *ad hominem* in the standard classification. Unlike the previous case, in which Sarah expresses doubt as to whether she can trust Harry, in the present kind of case Sarah's worry is often also about whether Harry can trust himself. For, in one of its common manifestations, bias "closes minds."[405] Someone caught in the grip of a bias can be adversely affected in two main ways. He can assign an undue weight to the evidence for his own position and less than due weight to counterevidence. The other is that he may fail to take proper note of the opponent's case. In the first instance, in effect he tampers with the evidence. In the second he omits to hear it. Of course, in each case, this often happens unawares.

Here too it would be a blatant mistake if Sarah were to conclude from the fact of his bias that Leila's position on embryonic stem cells must be false. But this still gives her lots of room for manoeuvre. In particular, she has reason to suspect that Leila's *case* against it is defective, and is made so by the distortions occasioned by bias. At the heart of this sort of worry is the foundational fact that real life arguments nearly always constitute themselves in nonmonotonic contexts. Accordingly, a case for α is well-made to the extent to which there are sound reasons for α *and* the reasons for α trump the reasons against it, as may be. A case for a claim that does not include some consideration of the case against it lies exposed to the risk that the unconsidered evidence will trump the considered evidence. Someone who is hostage to bias may be excessively preoccupied with favourable evidence and not sufficiently alert to the possibility – and the strength – of counterevidence. When Sarah voices her concern about Leila's bias, she enters what in Johnson (2000) is called the "dialectical tier". Sarah may have it in mind that Leila's is offering an over-narrow case for her position. If Sarah is disposed to accept an opponent's sayso when it comes to information that damages her position, she is surely entitled to expect her opponent also to acknowledge information that might support it.[406]

A further instance of the circumstantial *ad hominem* is one in which an interlocutor's precommitment is challenged. If Sarah knows that Leila is a devout Muslim, she may be troubled by the fact that her opposition to embryonic cell stem research is an obligation of faith, hence not free in Mill's sense. If, thus positioned, Leila is *parti pris*, how likely is it that he will be open to Sarah's defence? Equally, if Lou is communications director for the Liberal Democratic party, the same difficulties are also present. In each case, the views presented by someone

---

[405] The dialectical role of mind-closure is investigated in the eponymous dialogue game, MindClosed, developed in Gabbay and Woods (2001a), (2001b).

[406] Cf. Mill on "free discussions", which are characterized as *free* and *equal,* and *fair* and *thorough* (Mill, 1859, pp. 9, 10, 20, 24-26, 30, 44).

who is *parti pris* are views to which he may be pledged irrespective of the merits of the case. This, too, constitutes a condition of bias, and places in some doubt the party's sincerity (blurring, we may note in passing, the distinction between circumstantial and abusive variants).

The role of bias as an affective intrusion into belief-formation cannot be gainsaid, with obvious consequences for premiss-conclusion belief adjustment. Part of what I was saying just pages ago was, in effect, that negative bias is a spoiler of inference only when probatively intended, and even then not always and yet that the frequency of such intentions to such havings is a good deal smaller than the fallacies tradition seems to think. Still, the fact remains that when probatively intended, the prospects for reasoning are usually pretty bad.[407] But neither should it be overlooked that bias has a huge provenance in human affairs, a good deal of which is *benign*, and much of that it turn a positive good, and a necessity. If we likened frequencies of occurrence to the sizes of television screens, probatively intended negative bias would be a twelve-inch and benignly standard would be at least a seventy. Bias is preconception. Preconception is what you already know or truthfully think that you know. If all knowledge is acquired in *media res,* it is unavoidable that most of what we know at a time we know thanks to no effort of ours. Every empirically or normatively intended theory begins with pretheoretical data. From the theory's point of view, its motivating data are preconceptions. But preconceptions are data. Preconception is everywhere. It is the lifeblood of presumption and common knowledge, whose principal sources of nourishment are observation sayso and memory, and such as belief-revision adjustments as attend them. People who hold a JTB-view of knowledge won't be wholly at ease with the sheer abundance of preconception and the reach of the benignity. But CR-theorists can lighten up a bit. Indeed, they can lighten up *considerably.*[408]

Suppose now that Harry is pressing the anti-smoking position on Sarah, and that Sarah's complaint is that Harry himself smokes, landing himself in a kind of pragmatic inconsistency.[409] In the old classification,

---

[407] See here Blair (2012), chapter 3.

[408] For more on preconception and what I call "doxastic loyalty", see Woods (2004), chapter 8. For insights into the neurological correlates of position bias, see Lauwereyns (2010).

[409] There are two different conceptions of pragmatic inconsistency discernible in the current literature. One is a kind of behavioural incompatibility between what a person practices and what he advocates. This is the conception invoked in this

this would be the *tu quoque* variety of the *ad hominem*. Sarah is accusing Harry of a kind of incontinence. In voicing to her doubts, a number of different motivations may be present, of which three are of particular importance. One is that in pointing out Harry's behavioural defection from his own policy, Sarah might be expressing a doubt as to whether Harry's actual position is the same as the one to which he has given voice. Although Harry has said, "No one should smoke", perhaps what he really means is that no one should smoke to excess, or that young people shouldn't smoke.[410] Another possibility is that Sarah is questioning an unspoken presupposition of Harry's position, namely, that the case he launches against smoking is sufficient to motivate a reasonable person to adjust his behaviour to it. If Harry fails to honour his own policy, then Harry himself is either signalling the motivational inadequacy of his argument or disclosing his own unreasonableness. But if Harry is unreasonable in the way he handles the smoking issue, why should one suppose that the case he has assembled against it is sufficient to compel compliance? If Harry will not *act* reasonably about smoking, why should we think that he is *reasoning* reasonably about it? A third possibility is that Harry's defection signals a kind of insincerity. This would matter in those cases in which the dispute is aimed at getting to the truth of the matter about smoking, rather than a contest in which the parties merely "score off" one another. If there is reason to think that Harry doesn't believe his own propaganda, why should Sarah accept his assurances? Douglas Walton points out that sometimes the intent of *ad hominem* remarks is to raise questions about an informant's reliability. They may be used to raise six critical questions about what he calls "source-based evidence". The questions concern the consistency, honesty, sincerity, reliability, moral character and judgement of the source.[411] A difference between this account and my own is that Walton envisages the interlocutor and the subject of the *ad hominem* to be different persons. In the cases I consider, they are one and the same. This is one of the few respects in which the generalization from such cases to two-person cases is not entirely straightforward.

---

chapter. The other is a blindspot-creator (Sorensen (1988), occasioned by a conflict between utterance-conditions and truth-conditions (e.g., "*P*, but I don't believe it" or "I can't speak a word of English")).
[410] See Woods (1993), reprinted in Woods (2004) as chapter 6.
[411] Walton (1992) pp. 197-198.

In the cases we have examined so far, it might again be insisted that the difficulties that trigger Sarah's *ad hominem* remarks constitute a sufficient ground to quit the argument at hand. If Sarah has reason to think that Harry is dishonest, biased, unreasonable or insincere, who would fault her for breaking it off and going home for a beer? Yet it bears repeating that perhaps the most remarkable feature of *ad hominem* retorts is that they mark not the complainant's decision to withdraw but rather her readiness to *continue* (at least for awhile). This is far and away the most important feature of our circumstantial *ad hominem* behaviour, yet one that has received virtually no attention from theorists. On the face of it, this is amazing. How is it to be explained?

One possibility takes note of this massive dependency on the sayso of others and the deeply entrenched disposition to trust their assurances. As we see, much of the time the point of the *ad hominem* challenge is to prompt the reassurance of the other party. It is a remarkable fact about the dialectico-epistemic lives of beings like us that we are so ready to accept the reassurances of people whom we have reason to think might not be in a good position to offer them. If we think that Harry is a liar, what good is his insistence that he is not lying now? If we have reason to believe that Harry is biased, why isn't his reassurance to the contrary undone by it? And so on. Upon reflection, it seems that reassurance of the kind under review has the function of transforming a presumption into an explicit declaration, of a *causal disposition* into a *propositional attitude.*

Proposition 13.4a

REASSURANCE: *The presumption is that our interlocutors are by and large up to the task of giving their view of the matter at hand a fair run. Reassurance re-issues this presumption as a direct declaration by the party in question. What* ad hominem *contexts suggest is that the hesitations against this presumption embedded in* ad hominem *retorts are subject to removal or mitigation by simple assertion to the contrary.*

This may strike us as decidedly odd, but there can be little doubt as to the presence of this disposition in the empirical record. Of course, all this will be lost on the nattering nabobs of negativism.

A second possibility is that by and large liars lie when it is to their advantage to do so. A truthful person honours the truth before he honours his interests. Liars reverse the ranking. In one of its core senses, lying is the transmission of bad information with misinformational intent.[412] There are contexts galore in which it is easy to see the advantages that flow to the liar from his lying. Equally, there are lots of contexts in which lying would compromise his interests. A case in point is this dispute between Harry and Sarah. If Harry's objective is simply to get Sarah to climb down, then if Harry is a liar, it may well be in his interest to lie now. But if his objective is to get Sarah to see the issue that divides them as Harry knows it to be, then lying would be counterproductive. She might assent to his claim, but she could not be said to have a solid reason for holding it. In raising the flag of his notable dishonesty, Sarah issues a challenge to Harry. But she also creates a problem for *herself*. The problem is that of determining whether in the particular circumstances of the dispute now in progress it would be to Harry's advantage to lie and, if so, whether that advantage is weighty enough to make it likely that that is what he would in fact do.

In these contexts, the object is to get the other person to grasp the fact of the matter concerning the issue in dispute. Since each has this intention with regard to the other, each has a stake in one another performing as well as he or she can.[413] So if Harry is a liar, and proffers information that, if reliable, damages Sarah's case, Sarah has a stake in determining the admissibility of that information. For the protection of her turf, the course of prudence would have Sarah reject it. But given her larger objective of getting to the truth of the matter, prudence counsels that she not be precipitate or knee-jerk in her dismissals. This gives rise to what I've been calling the most important (and most neglected) feature of *ad hominem* challenges.

Proposition 13.4b

UNEVIDENCED PROCEEDINGS: *Sarah must perform this balancing act without recourse to independent evidence, one way or the other, as to whether Harry is lying now (or whether the balance of his mind is disturbed, and so on). In this regard, her situation bears some resemblance to the removal of an answerer's ignorance problem in Aristotelian refutation-*

---

[412] In a more basic sense, its purpose is *denial.*

[413] Although listed by Aristotle separately from refutation arguments, elements of refutation can occur quite naturally in the transaction of real life examination arguments.

dialogues. In each case, the problem has to be solved with internal resources.[414]

13.5 *Plausibility of manner*

Whether we are able to judge that a liar is lying now flows from the facility we have in assessing the *plausibility* of informants. A plausible witness is one who endows his assertions with "the ring of truth" in the absence of independent evidence for them – and, at times, in the face of evidence to the contrary. This is what earlier on I called the plausibility of *manner*. It is a property of persons rather than of the propositions they attest to. Juries, for example, are frequently faced with the task of determining whether a witness's manner affects to any degree the propositional plausibility or implausibility of what is asserted in his evidence. When this happens, we say that the witness's manner lends plausibility (or implausibility) to what he testifies to.

Proposition 13.5
PLAUSIBILITY OF MANNER: *In pure cases, where informant and recipient are strangers and have no knowledge of one another's track record, a witness's plausibility is betokened behaviourally, by tone of voice, the cadence, pacing and emphasis of replies, facial expression, and body language generally – all in the context of background information. Such signs are typically processed sublinguistically and subconsciously, which helps explain why linguistically-minded logicians have paid no attention to them – even those who occupy themselves with the logic of testimony.*

It may strike some readers that the present suggestion misconceives the idea of plausibility. Perhaps this is the case, but I am not one who thinks so. If I have indeed misconceived the notion, it was not something that happened through inadvertence. This is not the place to try to settle the matter decisively. Perhaps it is enough to point out that my plausibility of manner resembles a conception of credibility, a conception in which an essential point is preserved. Credible sources can lend the ring of truth to perfectly implausible claims. If it is agreed that this is so, a problem clearly arises. In dealing with strangers, of whom nothing in particular is known or bruited, how is credibility established?

---

[414] See again *Topics*, chapter 8, concerning what has come to be called the Socratic method in ignorance removal.

Some will say that it cannot be established, except on the strength of the credibility of what he attests to. I say that this is unrealistically narrow, and that in actual practice we consult faces and read body language.

It is a fact of considerable importance that our facility with the recognition of the plausibility of informants is on the whole fairly reliable, never mind that there are clear exceptions. To the extent that this is so, an *ad hominem*er's provisional and qualified acceptance of the reassurances of her adversary can be supposed to be rooted in a plausibility recognizing capacity.

The role of the plausibility of manner is also little discussed – indeed little recognized – by epistemologists and dialecticians. Among psychologists it has generated an already significant literature. There are clinical and experimental studies, according to which the attribution of an emotion to another person – say Sarah's attribution of fear to Harry – is a matter of her responding to Harry's facial expression in ways that stimulate Sarah's amygdala, which is the part of the brain that supports her own experience of fear. On this view, the attribution of fear to another involves the simulation of it in oneself, a simulation cued by the other party's facial expression[415] This, the so-called "simulation theory" has successfully been applied to the attribution of disgust[416] It also appears to explain the attribution of anger.[417]

It would be an interesting extension of the simulation theory of emotion-attribution if it could likewise explain the attribution to others of mental states in general, including belief. Precisely this generalization is proposed by Goldman (2005). This is not the place to assess Goldman's theory of mind-reading in any detail. However, we can note that the theory has two main parts, one of which is more controversial than the other. The comparatively uncontroversial part holds that a person's mental states can be discerned defeasibly in his facial expression and body language, including the character of his linguistic behaviour. The controversial part holds that such readings trigger the brain of the reader to simulate the state in question. Fortunately for my purposes here, it suffices to invoke the uncontroversial part. To put it as simply as possible, if Harry's believing that α is discernible in his facial and vocal expressions and his body language, if knowing that he believes α is in

---

[415] Adolphs *et al.* (1994), Sprengelmeyer *et al.* (1999), Sripada and Goldman (2005).

[416] Rolls and Scott (1994), Small *et al.* (1999), Calder *et al.* (2000), Wicker, *et al.* (2003).

[417] Lawrence *et. al.* (2002), Lawrence and Calder (2004).

part a function of face-based and body-based manifestations of sincerity, then Sarah is in some position to determine whether, in this particular case, Harry is misrepresenting his beliefs. Perhaps a skilful liar would fool Sarah. But most liars are not that skilful. Or perhaps Sarah is not as skilful as the rest of us in seeing what Harry's up to. But the fact remains. There are lots of cases in which it is not unreasonable to accept a liar's assurances that he's not lying now.

We have in these reflections a general answer to the question of how Sarah is to determine whether Harry the liar is lying now. She does this by activating her belief attribution mechanisms and hoping for the best. It is, as mentioned, a defeasible procedure. But it is, I suggest, the way in which the plausibility of manner is reckoned.

It can hardly be denied that one of the functions of Sarah's *ad hominem* retort is to seek reassurance from Harry. But it also serves to put Harry on notice, to the following effect: that, if their discussion is to continue, Harry's case should take on a strength and a weight sufficient to subdue or circumvent Sarah's reservations. We may say, then, that the net effect of the challenger's *ad hominem* is to raise the bar of case making as it applies to the subject of her challenge. What is normally taken as a dialectical presumption – that parties are sincere and competent – is now subject to this weak sense of showing cause that the contrary is not presently the case. So it would appear to be the *ad hominem*'s position that if the elevated bar is not scaled, then, at a minimum, she reserves the right to withdraw for cause.

Before bringing this section to a close, it would be useful to consider a brief more general word about the dialectical role of assurance. Consider some cases. Sarah says to Harry, "I think that you might be lying". Harry replies, "I promise you that I'm not". Or Sarah says to Lou, "Someone I know says you were with that woman last night", and Lou replies "On my word of honour, your friend has made a terrible mistake". Or Sarah says, "I fear that bias has compromised your objectivity". Lou replies, "You need have no worry on that score". On the face of it, these are paradigm cases of dialectical impotence. They are made so by the utter absence of supporting reasons in the respondent's replies. On the face of it, there is something wrong with Sarah's granting them any degree of acceptance, however provisional and qualified. Since Sarah's position is that of all of us, it is interesting to ponder what might we say in Sarah's favour. On thinking it over, I've already said it. Sometimes such assurances are neither unreasonably given or unreasonably accepted.

13.6 *Reprise*

This would be a good point at which to entertain an objection that some readers might wish to press. It is that by and large the quality of an argument is independent of the qualities of the person who advances it. Accordingly the job of the assessor is to take the measure of his interlocutor's argument, rather than the arguer himself. True, there may be exceptions to this, as in eye-witness testimony, the appeal to authority and sayso information, but we shouldn't allow exceptions to be canonical for the thing itself. I concede that this is by far the standard explanation of the fallaciousness of the *ad hominem*, but I don't agree that it is the correct explanation. Let us allow that when an arguer's situation is irrelevant to the case he has advanced, making it a factor in the assessment of his argument is a mistake. But it is not a fallacy unless it is a mistake committed with a requisitely occasioned frequency. It seems to me quite clear that *ad hominem*s of irrelevance don't meet this condition. Of course, sometimes disputants slang one another. But, as I said at the beginning, the object of slanging is not to discredit an opponent's argument, but rather to embarrass, mortify or infuriate its maker. Slanging isn't argument assessment. It is also true that sometimes disputes are mere debates, indulged in for the fun of it or for some more serious ulterior purpose, such as getting Lou an acquittal. In these cases, it is known by all that each party is engaged in *one-sided* argument designed to produce assent (or silence) rather than get at the truth of the matter. In such cases, parties ride these respective biases to the hilt. If this were occasion to challenge one another's *bona fides* or simply to quit the field, then we would see this well-evidenced in practice. But we don't see it in practice with anything like the frequency required by the traditional concept of fallacy. The dominant empirical fact is that, slanging apart, people make *ad hominem* challenges when they think that doing so is of some relevance to the worth of the addressee's case. And, as I have said, it is a further fact of our actual practice that raisers of these doubts are by and large competent judges of when they have relevance and when they don't. If this is right, the practice cannot be fallacious in the traditional sense.

Consider briefly another trio of cases. There are issues galore of such complexity, and people galore of such gormlessness, that it would be folly to listen to, never mind taking the pains to assess, arguments about, e.g., quantum non-locality forwarded by a person of immense stupidity. In that case, the sensible course is not to bother with him.

Certainly, if one did bother with him, the chances of his getting quantum nonlocality right are practically nil; and this is so because he is so stupid. Perhaps one can see the point of *saying* this to another party (perhaps someone who doesn't realize that quantum nonlocality is simply beyond this stupid fellow.) Equally, there would be no point in saying this *to* one's stupid interlocutor. Doing so would, for him, have the character of slanging. But either way, it would not be a fallacy.

A second case is taken from Johnson and Blair (2006), p. 99.

> An irrelevant attack on the person instead on the position, is the fallacy called ad hominem. Here's an example. In his 1989 book, *The Closing of the American Mind,* Allen Bloom attack rock music as an overtly sexual form of music which contributes to an overall climate of promiscuity .... In a review in *Rolling Stone,* William Grieder wrote:
>
> > Bloom's attack is inane. Still the professor is correct about one important distinction between the kids of the 50s and those of the 80s: in the 50s the kids talked endlessly about sex; today the young people actually do it.[418] This seems to drive the 56 year old Bloom – who is still a bachelor – crazy. Bloom denounces Jagger with such relish that one may wonder if the professor himself is turned on by Mick's pouty lips and wagging butt.

Johnson and Blair take this as typical of the abusive version of the fallacy. Grieder's attack, they say, is "largely personal", full of innuendo. Instead of attacking Bloom's dubious causal claim, Grieder snidely insinuates that Bloom is a repressed homosexual. "But", they say, "even if this were true, it would not have any relevance to an appraisal of Bloom's argument. Bloom's argument is about the effects of rock music on sexuality; his own sexuality is not at issue".

Well, let's see. First, where are Bloom's arguments? I myself can't find any. True, he does make the causal claim that rock promotes sexual license, which is certainly true. Bloom also implies that promiscuity is a bad thing. Grieder's responses are telling. Grieder does not attack Bloom's causal claim. He accepts it. In fact, he rejoices in it. What rankles Grieder is Bloom's denunciation of promiscuity. In his response, Grieder makes no attempt to undo Bloom's argument for his

---

[418] Clearly Grieder wasn't around in the 50s!

rejection of promiscuity; but this is because Bloom advances none, expressly or implicitly. What, then, is Grieder about? He thinks – certainly he insinuates – that Bloom is a repressed homosexual alarmist, terrified of being outed – as he would be a few years later by his buddy Saul Bellow. Grieder is suggesting that Bloom's opposition to promiscuity is insincere or unbalanced, a product of the troubled denial of his own sexuality. There is reason to think that this is so. Grieder, himself a promoter of promiscuity, is making the point that an insincere denunciation of it, even by a literary critic of Bloom's standing, is insufficient ground to give it up. Grieder is right. It is also true that Grieder's conduct is repulsive. Someone should teach Grieder some manners. But there's nothing wrong with his reasoning.

Here is a third case. Somewhere in *Technology and Empire* (1969) George Grant is critical of North American Jews for having squandered their large talents on the vulgarizations of mass culture. No one to my knowledge has ever accused Grant of anti-semitism, and I rather imagine that his remarks occasioned little notice in New York or Hollywood. But suppose now that you have the misfortune to be pitched to the same effect by a neo-Nazi skin-headed punk – tattoos, body piercings and all. If we were to heed the traditional wisdom, if I were prepared to give Grant's argument[419] a hearing, I couldn't in consistency not give young Adolph's argument a hearing. But what counts here is that arguing is nearly always discretionary. I am under no obligation to bother with Grant. I am under no obligation to engage with Adolph. Whether I do or not is up to me. It doesn't need a justification. While I might spend an hour or so going over this with Grant, I wouldn't give the punk the time of day. I happen to think that Grant is wrong in suggesting that North American Jews have dishonoured their talents. If that were also Adolph's view, it would also be wrong. But, while I might take it up with Grant, I wouldn't with Adolph.

A good part of the difference turns on bias – on bias in the preconception sense. I already know what else the punk would say, and I already reject it. I don't already know what else Grant is going to say. But I already know that what Grant is going to say might be worth hearing. So I wouldn't wait to see what Adolph's going to say. But I might wait to see what Grant's going to say. If I do, it might not be true, but it could be thought-provoking or in some way *surprising*.

---

[419] If argument there be, it is largely implicit.

Those who favour the "non-interactive" model of argument assessment will undoubtedly point out that the logician's task is to evaluate the argument irrespective of who advances it. I have three things to say about this. First, when this is in fact what's happening, it follows from the definition of "non-interactive", hence is trivial, that reference to Grant's anglophilia or to Adolph's thuggery is irrelevant. So it is also trivial that if offered as telling for or against the non-interactive argument it would be unavailing. A second point is that in non-interactive argument analysis – of the sort undertaken by academic philosophers – *ad hominem* remarks occur with nothing like the frequency that would qualify them as fallacies. How many philosophers do we know who mock Aquinas' Five Ways on grounds that he is a precommitted theist, that he's a Christian *saint*?[420] The third point is that where *ad hominem* exchanges are most frequent – and wholly natural – are interactive contexts, where they are subject to all the rich, and largely benign, variability that we have been at pains here to expose.

13.7 Ad hominem *inference?*

People who are in the post-1847 tradition claim that the *ad hominem* involves an error of inference. We may schematize the inference as follows:

1. Sarah makes her *ad hominem* retort.
2. She *concludes* from this that the adequacy of her opponent's case should be called into doubt.
3. She *concludes* from *this* that there is reason to think that her interlocutor's position is false.

No one denies that without the qualifications "in doubt" and "reason to think" this is a pretty suspect bit of reasoning, and a manifestly bad one if intended deductively.

It is at this point that we encounter an individuation problem. Some people are of the view that an *argumentum ad hominem* is constituted by all three components, the retort of (1) and the inferences of (2) and (3). So understood – and leaving to one side for now the question of the presence or absence of deductive intent – an *ad hominem* begins with (1) and ends with (3). Others are of the view that the *ad hominem* has a slighter constitution, one that begins with (1) and ends with (2). We have it, then, that in any dispute about whether the *ad hominem* is a

---

[420] Not even the abusive Christopher Hitchens, may he rest in peace.

fallacy, it is necessary that we solve this individuation problem. Complicating this issue is the question of deductive intent. Let us try to deal with it first. Even with the qualifications "in doubt" and "reason to think", there can be no serious doubt that the inferences at both (2) and (3) are invalid. If so, this takes the pressure off the individuation question, since whether the *ad hominem* ends at (2) or at (3), it is an inferential failure both times. But is it a fallacy?

If the question now before us is whether the mistake of inferring with deductive intent (2) alone, or (2) and (3) together, is a fallacy, the answer is that it is not. For, while drawing these conclusions with deductive intent is an error, there is not as I keep saying a jot of evidence in the empirical record to suggest that when these inferences are made in real life they are typically (or even frequently) drawn with deductive intent.

What if the intent is other than deductive? I have no solid answer to this question. In this I believe that I am not alone. Part of the difficulty is that, as the empirical record again will attest, typically our *ad hominem* behaviour on the ground carries with it an indeterminacy that underdetermines even the fairly simple schema that we are currently considering, and certainly does not in general help us in solving the individuation problem. It is true that the presence of the qualifiers "in doubt" and "reason to think" indicate the presence of some non-deductive generation relation, which might be thought to underlie some or other form of presumptive or (perhaps plausibilistic) reasoning, as we have already said. But the logic of presumption and plausibility is so far from having been worked out definitively, that citing it at this stage is more promissory than helpful.[421] In the circumstances, perhaps we can do no better than yield to the guidance of the principle of charity, putting it that if the inference of (2) (or (2) and (3)) is said to be an error, albeit a non-deductive one, the burden of proof rests with him who makes the accusation.

The burden of proof also suggests a presumption of innocence. If it falls to the charger of fallacy to show that *ad hominem* inferences actually embed some, as yet not theoretically well-understood, non-deductive generation relation which our *ad hominem* behaviour routinely mismanages, then, until that onus is met, the charge of fallacy is unfounded. I doubt whether this onus can be met at present. Doing so

---

[421] See again Rescher (1976) and Walton (1992), for criticisms, Gabbay and Woods (2005). Mathematical approaches are developed in Lehmann (1992a, 1992b) and Schlecta (2004), chapter 3.

would require the critic to show that there is *no* generation relation thanks to which these inferences are in *some* sense reasonable – that there is nothing whatever to be said for them. For all the strides made since Bolzano and Tarski, the general logic of consequence and third-way conclusionality is not in nearly good enough shape to make this a forseeably attainable target. Even if we allow that there is *an* inductive consequence relation that is well catered-for by conditional probability, most of the work on ampliative generation is still to be done.

Doubtless some readers will find this a trifle over-defensive and "legalistic". Surely something more forthcoming can be offered in support of the negative thesis. Let us see. Consider a class of cases. These are cases in which the point in contention is one of serious importance to the parties, and where there already exists a substantial record of advocacy and counter-advocacy surrounding it. This might even be something that has already been pretty much "argued to death". Let it be the case, in particular, that Lou is a well-known and rather fanatical spokesman for α, the Netherlands' policy of physician-assisted suicide. Harry now makes his *ad hominem* intervention: "Look, Lou, the fact that this is still an open question – after all, we're still arguing about it – tells us that you have not succeeded in discharging your burden by making the case for α? Doesn't this show that there is something wrong with that case?"

In effect, Harry has drawn a type-2 conclusion from the fact attributed in his *ad hominem* remark. Could he now with any plausibility go on to draw a type-3 conclusion? It would seem that he could. Granted that Lou's case is now in some doubt, Harry might reason autoepistemically, as follows.

1. If there were a sound case for α, then surely Lou, of all people, would have it in hand by now.
2. But, as we see, there is reason to think that he doesn't.
3. So it can be doubted that a successful case for α exists.
4. Moreover, if α were actually true, it is reasonable to suppose that by now a successful case for it would have presented itself.
5. But so far we know of no such case.
6. So there is some reason to think that α is not true.[422]

---

[422] I take it that since ⌐~α⌐ was long since the established position in the Netherlands (and everywhere else) – in fact, is a kind of taboo – the burden of proof lies with Lou. For an explanation of why Harry's position with regard to α

It is interesting to note that Harry's hypothetical autoepistemic reasoning embeds two of what Isaiah Berlin identifies as the defining conditions of the Enlightenment's conception of rationality.[423] One is that every question has a determinately correct answer. The other is that, with regard to all issues, sooner or later the truth will out. If we allow these assumptions to stand, Sarah's reasoning, tentative though it rightly is, has merit. In her dispute with Harry, the burden of proof rests with him. The merit of Sarah's argument is that it casts doubt on whether Harry can meet it.

We should not, of course, give these assumptions a free pass; nor, as a devoted critic of them, would Berlin wish us to do so. The kind of pluralism espoused (but not very well explained) by Berlin leaves it open that these assumptions are incorrect. In which case, it becomes a question of considerable importance as to which are the questions that lack determinate answers and concerning which the truth will *not* out. This accords to Harry's autoepistemic reasoning the necessity to amend its ultimate conclusion, (6). In its place, we write

> (6′) So there is some reason to think that either α is not true or that there is no determinate answer as to whether it is true as opposed to false, or false as opposed to true.

Either way, this is bad news. In the first instance, it is bad news for Lou. In the second instance, it is bad news for them both. In neither case, however, does "bad" mean *awful*. Even if there is reason to think these bad consequences might obtain, it is far from certain that they do. Optimists will want to press on, hoping for greater determinacy down the road. Pessimists, on the other hand, should probably call it a day. So let's make it official.

Proposition 13.7a
ANOTHER NON-FALLACY: *In neither its abusive, circumstantial or* tu quoque *variations is the* ad hominem *in the extension of the traditional concept of fallacy.*

---

is not subject to the same kind of type-2 and/or type-3 argument from Lou, see Woods (2000) on the dialectical frailty of taboos.
[423] Berlin (1999), pp. 21-22.

13.8 *Dialectically benign?*

So much for the negative thesis. What might be said about its positive vis-à-vis? *Ad hominem* remarks are challenges. By and large, they are not deal-breakers. They put the *ad hominem*'s interlocutor on notice, and they invite him to do what he can to remove or circumvent his challenger's doubts about his *bona fides* as a case-maker concerning the issue that presently divides them. When Harry and Lou have a stake in getting it right as to whether the Netherlands' policy is morally sound, they have a stake in jointly producing as objective an examination of reasons pro and con as lies within their means. They also have a stake in not quitting the issue prematurely. The *ad hominem* retort is an aid to both objectives. It warns the other party about possible difficulties with his case-making wherewithal, but it also keeps the discussion going. *In extremis*, it presents the other party with a (usually implicit) autoepistemic argument in the form (1) to (6′). It presents it as a challenge. It is a way of keeping the advocacy honest short of falling into irreconciliation or, as Locke would say, silence. If that is not a virtue, I don't know what is.

CHAPTER FOURTEEN

TAKING

"We hold these truths to be self-evident: ...". Thomas Jefferson

"... experience must be put through a conceptual grinder that in many cases is excessively coarse. Once the experience is passed through the grinder, often in the form of the quite fragmentary records of the complete experiment, the experimental data emerge in canonical form and constitute a model of the experiment. Patrick Suppes

"Intuitions are, surprisingly, not always obvious."
Alison Gopnik and Eric Schwitzgebel

14.1. *Respect for data*

The main objective of these chapters has been to advance our knowledge of mistaken reasoning, and to conduct the investigation by means of an empirically sensitive enquiry, a naturalized logic that is agent-centred, resource-based, task-sensitive and oriented to the individual reasoner. Success with the project would help advance the logical part of a positive, rather than privative, theory of error. A thick theory, not thin.

Theories decompose into constituent parts, which we could think of as their *targets*, their *programmes* and their *theses*. In the approach taken here, the target of a theory is the reasoning behaviour of human individuals on the ground. The programme is the specification of conditions under which such behaviour qualifies as mistakes. It is the theory's *modus operandi*. The theses are, nearly enough, those same conditions.

In chapter one and various places thereafter, I have urged the importance of the respect for data principle, which bids the theorist to be as clear as possible about what his theory's target phenomena actually are, and to give due consideration to contextual constraints and behavioural indeterminacies. Of equal importance is the care to be shown to a theory's confirmational data. The importance of getting the target and confirmational phenomena right can hardly be over-emphasized.

Target phenomena serve two crucial purposes. They are a large part of the theory's motivation, and they also function as its data. A datum is something *taken for granted, taken as given.* Data, we might say, are the objects of those takings-for. They impose significant constraints on a theory's programme and thesis-generation procedures. In a first approximation they require the programme to fit the target phenomena, and the truths generated by the theory's provisions to be true of them. Taking is essential to enquiry. It is essential to belief revision in general. It is essential to premiss-conclusion reasoning.

It is no easy thing to explain in any detail how these factors work, but clear-cut violations of them are sometimes readily discernible. In the first instance, if the theory's target were the behaviour of economic agents, then it would be a misconceived programme to seek a knowledge of error-commision conditions without some consideration of supply and demand and costs and benefits. In the second, a theory would have failed the data constraint if the sentences derivable in the theory are neither true of the target data nor in some appropriate way relevant to them. This is important – and a complication – as the example of normative theories makes clear. We don't want to rule it out in advance that the theory's theorems might be true normatively, even if not true on the ground. But when this happens, the gap between what the theory says is true and what is true empirically should not be tolerated in the absence of independent reason to think that the theory's norms have a binding legitimacy for target phenomena.

There is a second way in which data play a role in scientific theories. Not only are they a theory's target phenomena, but they are also implicated in the theory's explanatory or analytical apparatus. This is especially clear in the case of experimental theories of human behaviour – for example, a psychological theory of deductive reasoning. Here the target data cover reasoning behaviour on the ground, expressible by propositions such as "Individual reasoners hardly ever close their beliefs under deduction." Experimental data cover behaviour on the ground elicited by the theory's experimenters. Suppose, for example, that experimenters wanted to test the claim that there are ways to teach a deductive reasoner to significantly up his rate of belief-closure under deduction. Then an experimental datum would describe behavioural elements in the execution of such tests, along with the elicited responses to them. Data of both kinds are at risk for misinterpretation. In taking it as given that human beings are only very modest belief-closers in deductive contexts, care should be taken that the contexts in which closure is resisted are in fact contexts for deduction. Equally, data which report the outcomes of belief-closure enhancement trials should be examined for the subjects' deductive intent.

I want now to explore these ideas in a bit more detail. I have two particular objectives in mind. One is to develop an appreciation of the complexity of the relations between a theory and its data, and, between a theory of reasoning errors and the actual commission of them. The other is to use the discussion as background for an examination of some prominent examples of how psychologists and applied mathematicians deal with what they take to be errors of statistical and probabilistic reasoning. I will concentrate here on the three examples best known to philosophers, logicians and computer scientists. They are: the conjunction fallacy, the gambler's fallacy, and the base rate fallacy.

14.2. *Two kinds of model*

I begin with the first of these tasks, the connection between a theory and its data. One of the places in which the connection is most interesting and yet most difficult to get right is model based natural science. Consider that large class of the model-based theories meeting the following conditions.

1. They have as their targets the advancement of our knowledge of certain phenomena. Target phenomena are a theory's motivation.
2. Target phenomena are phenomena on the ground. They are empirical entities – as opposed to theoretical. They are creations of nature, not of the theorist.
3. The target phenomena furnish data for a theory.
4. Successful theories proceed by way of models. Since model-based theories account for their motivating data, we may say that they model those data.
5. Two of the most prominent respects in which scientific theories model their data are by way of idealization and abstraction.
6. Idealizations say what is false. Abstractions suppress what is true. Here "false" means false of the data, and "true" means true of the data. (Think, in the first case, of the infinitely

480

large populations of population biology[424] and mathematical sampling theory. Consider, in the second, the isosceles triangle of ordinary geometry.)

When model based theories meet these conditions, they are purposeful distortions of their target phenomena. This raises the obvious question of the extent, if any, to which a scientific theory in which models stand to data in fulfillment of these six conditions really does advance our knowledge of them. If it does, then the theory's distortions are *cognitively virtuous*. And if so, there must be a convincing answer to the obvious question: "How can the knowledge of those phenomena acquired thus be reconciled with the falsehood and the truth-suppressions visited upon them by the theory's models?"[425]

In tackling this question, philosophers of science sometimes impose simplifying assumptions of their own. Of particular interest is the assumption that data for a theory are – to put it metaphorically – *raw*. In fact, however, in the actual cut-and-thrust of model based science, this assumption is hardly ever true; and the respect for data principle bids us now to give this fact fair notice. I do so in the spirit of Patrick Suppes' "Models of data" (1962), a classic paper which emphasized – although not in these words – the sheer extent to which the data for experimental science are not raw.

The idea that the empirical phenomena of a scientific theory – the events and happenstances on the ground – don't present themselves in their (intuitively) raw states is at least as old as Bacon. It is embedded in the more recent rejection of the positivist distinction between observational and theoretical terms[426], and it plays a central role in Quine's theory of knowledge. Many more examples could be mentioned, but these will do for present purposes.

Discernible in this rather substantial literature are two dominant conceptions – neither of which is particularly well articulated or explained – of the factors that deny to scientific data their intuitive rawness. In the metaphorical spirit of lines above, let's for the present say that data that aren't raw are *cooked*. So, then, the thesis that one finds in

---

[424] Actually, it doesn't sound quite right to say that transfinite largeness *models* the actual largeness of large populations. Transfinitely large populations help us model natural selection on the ground. Still, let that pass.

[425] See here Woods and Rosales (2010a) and Woods (2013b).

[426] Consider for example, Sellars' myth of the given (1956) and Hanson's theory-ladenness of observational data (1958).

these writings is that for a considerable part of science the only data are cooked data; and the two conceptions of cooking that we find there are answers to the question, "How are data cooked?"

According to one answer, data are cooked by virtue of the need to give them a representational presence in the theory. Representation here can be perceptual or linguistic. I will say that someone subscribes to the "representational" view of data-cooking just in case he holds that data for theory are representations which in nontrivial ways are distortive. Consider, in the perceptual case, the representation of natural objects as coloured, and, in the linguistic case, the hard-edgedness of our discourse about a world that is generically blurry.

An important peculiarity of representational cooking is that its embedded distortiveness is a quite general feature of our cognitive and discursive encounters with the empirical, and not a peculiarity of our scientific engagements with it. It is in this observation that we find a major difference between the two ways of understanding data-cooking. For, on a second conception, cooking is indeed peculiar to science, or anyhow is not a general feature of empirical experience and reflection. I will say that someone holds the "massaging" view of data cooking just in case he takes it that data for a scientific theory must be interfered with in ways that *exceed* the distortive reach of perceptual or linguistic representation. Massaging, we might say, is data-distortion at second remove. It is the distortion of representational distortions.

What is the difference between representing and massaging? What is the difference between first and second remove? A not implausible suggestion is that representational distortion is spontaneous and that massaging is deliberate. Since representations are the product of how nature has put us together as perceiving and speaking beings, they are non-discretionary features of our phenomenal engagements. Massaging is different. It is a distortive treatment of phenomena that is "down to" the theorist rather than "down to" his experiential devices. When a theorist massages a theory's target data, he does so with a view to their suitability for theoretical engagement. He is doing something on purpose. He is doing something that he had to learn to do. He is imposing conditions which are not in general natural to impose.

The metaphor of cooking carries the suggestion of "cooking the books". It is not an entirely happy association, made so by our present distinction. Since representational cooking is not down to us, but rather is down to the way we are built as experiencing beings, it is not at all like book-cooking. And while massage-cooking is down to us and is distortion on purpose, it is by no means typically or even frequently for shady ends.

Mention of shadiness calls to mind that class of cookings which Gigerenzer calls data-bending, briefly discussed in chapter 2. Here the basic idea is that data on the ground (or as close to the ground as their representationality permits) are not only massaged, but are massaged in ways that facilitate their engagement by the theoretical contrivances of the theory that seeks a model for them; in other words, massaged to fit the theory's *modus operandi*. What Gigerenzer finds alarming about data-bending is that it allows theoretical machinery to wear the trousers in the very *identification* of data, rather than requiring it to deal with the data as they come, that is, as they lurk about as rawly as sober representation permits. One of Gigerenzer's bug-a-boos is the common assumption among decision theorists, belief-revision theorists and psychologists that the individual cognitive agent is an "intuitive statistician", and that the assumption is justified even against the grain of empirical fact precisely because it facilitates engagement with the heavy machinery of statistical probability in the production of the theory's theorems.

A theory charged with data-bending is said to acquire a theoretical grip on its data at the cost of missing the boat as to how target phenomena really are. There are three things one might want to say against data-bending. It is cynical; it is *ad hoc*; and it achieves its theorems only by changing the subject.

If cooking has a whiff of the shady about it, it might be the right idiom for data-bending, but it seems excessive for representing and massaging. So this would be a good point to stop talking of cooking and start talking of modelling. In this I follow Suppes. Given what Suppes means by it, models of data occupy an intermediate position between representation and data-bending. They also mark, or appear to mark, an intermediate position between represented data and theoretical models. Data-modelling bears some resemblance to data-bending. It distorts a theory's represented data in an effort to ready them for theoretical engagement; it primes them for a theory's programme; it sets them up for it; but it does not (it is supposed) bend them out of Gigerenzerian shape.[427]

So, then, the literature discloses two conceptions of model – theoretical models and data models, of which the former are inputs to the

---

[427] While the remarks of the past several paragraphs about the massaging and the bending of data have been framed for target data, it is easy to see their applicability to experimental and conformational data as well.

latter and constraints upon them. Relatedly, we see that the words "models of the data" are ambiguous as between these two conceptions. As an expository convenience, I will call data-models "D-models" and theoretical models "T-models". This leaves us with a pair of interesting questions:

    i.   Why aren't the distortions of D-models data-bending in Gigerenzer's disapproving sense?

    ii.   How do a theory's D-models differ from its T-models?

14.3. *Complications.*

I think that it would not be going too far to say that at a certain level of generality the present questions are well formed and that the assumptions on which they rest are true as far as they go. Whatever clarity attends this picture, its own debt to simplification is considerable. The issues before us are in fact a good deal more complex than these simplifications allow, a satisfactory sorting out of which would not be entirely dissimilar to the pacification of some cats in a sack. Various problems lie in wait. Here are two of them.

Controlling the removes. Let m be the massage of some perceptually or
linguistically represented state of affairs. The function of m is to introduce an element of distortion. How is this done? And, however it is done, is there an honest way of determining where the result of m ends up in the theory? Does it end up in its D-model or does it go straight to its T-model? Do we have a proper command of the distinction between distortions at first and second remove (a variation of question (ii) just above)? Consider, in this respect, the far from uncommon situations in which a theory's T-models are continuous and infinistic, but whose D-models must be discrete and finitistic.

Distortion and conceptual change. Suppose that there's not much science to speak of for a certain class of phenomena $e_1, \ldots e_n$. Suppose that a good part of the reason for this is that the mechanisms and procedures of systematic science have had difficulty in extending their reach to the e_i. The e_i are in no fit condition to connect with these methods. Even so, whatever else might be said of them, the e_i exhibit a certain *conceptual* coherence, on the whole we seem to have a workable informal consensus about what it is to be a thing of the e kind.

Suppose now that the following happens. A model-based scientist imposes on the e_i a description that

    a.   violates the informal meaning of "e", and yet
    b.   enables the machinery of his theory to be engaged in ways that generate deep theorems about the e_i under this conceptually distorting redescription.

On the face of it, this is a disaster. It is data-bending to no good Gigerenzerian end. But as the history of science amply attests, massaging of precisely this sort has achieved substantial – even medal-winning – approbation. When this happens, what began as tendentiously induced conceptual errors are now accepted as welcome occasion of *conceptual change*. This faces the theorist with some pressing options: Either stick with his understanding of what is to be a thing of the e-kind and forget about having a science of them. Or change the meaning of "e" and luxuriate in an ensuing scientific abundance. Neoclassical economists stipulate that the utilities sought for by beings like us are infinitely divisible. What would explain such an assumption? The answer is that it enables economic theory to appropriate the fire power of the calculus. Is this data-bending or is it honest data-massaging? And *where* is it the case that utilities are infinitely divisible? Certainly not on the ground. In D, then? Or in T?

    There is not room in this chapter to follow up these questions with requisite care and detail. But I think, even so, we may claim some slight progress with them, which could be summarized as follows.

Some protocols for D-modelling in a theory of reasoning

    1.   When a theory's programme is to produce an account of what makes for erroneous reasoning, proper care must be taken in specifying the events on the ground to which the theory's analysis is intended to apply. Call this "the target-phenomena identification requirement".
    2.   If in the course of target-phenomena identification data-massaging is required to facilitate engagement with some productive and well-understood theoretical machinery, two possibilities must be properly negotiated. Call this "the gap-negotiation requirement".
    3.   The first possibility is that the gap between the raw and massaged is such that the massaged data can plausibly be

taken as giving an improved understanding of what it is to be something of the kind in question. When this happens, I will say that the massaged data are a conceptual clarification of their raw counterparts. In the language of late in chapter 2, they provide an *explication* of the relevant concept.

4. The second possibility is that the gap between the raw and massaged is such as to make it implausible that the massaged data are a conceptual clarification of the raw, but plausible to say that they constitute a reconceptualization of the raw; that the concept under which the raw fall and the concept under which the massaged fall are different concepts. When this happens, I will say that data-massaging effects a conceptual change. In the language of chapter 2, the model provides a *rational reconstruction* of the target concept.

5. One negotiates the gaps contained in these possibilities only if satisfactory cases are made for the plausibility and implausibility respectively claimed for them here.

6. When a theory stipulates into being a new concept – or, in the language of chapter 2, synthesizes it – there is no gap between the theory's concept and any pre-existing concept. So the gap-negotiation problem doesn't arise here. Even so, see the section to follow

A theory of reasoning is one of a large family of theories about a particular kind of human behaviour. It is behaviour which is normatively assessable. Roughly speaking, these are theories that make up the social, and what used to be called the moral sciences. A quite large, and significantly simplifying, assumption of such theories is that their target phenomena are or inhere in the behaviour of human subjects. In the case of a theory of erroneous reasoning, the target phenomena are or inhere in reasoning's behavioural manifestations.

The respect for data principle takes some note of this when it instructs the theorist of erroneous reasoning to take pains to get his target phenomena right, but it gives little instruction on how this is to be done. Even so, a number of considerations offer useful, if loose, guidance. Operating in the background of a target-phenomena identification are what are taken as general facts about the human cognitive agent, including the familiar ones. Although human beings make lots of errors, we are good enough at knowing things to survive, flourish and occasionally fill up the Tate and the Prado. Accordingly, it is misbegotten to interpret people's behaviour on the ground in ways that reflect cognitive shortcomings on a scale which, if true, would strain against these background givens. Also important are the associated relativities: If you think that an agent's behaviour on a given occasion manifests some reasoning about something, take care to identify its informational inputs, its sought-for outcome, and its available resources. Take care to judge its success and failure in the light of these determinations.

There is lots of stupidity round about, and just about everybody is stupid about something or other. But we do not have it from this, nor is it true, that when it comes to conclusion spotting, conclusion drawing and premissory update, people are stupid. So some quite useful general advice for the would-be theorist of reasoning is to try to respect this fact in his target-phenomena identifications.

14.4 *Pretheoretic belief*

Data for theories aren't always flecks of human behaviour on the ground, needless to say. There are interesting cases in which data are beliefs of a certain kind. In matters nonempirical – a fuzzy notion, to be sure – data for theory are sometimes called *intuitions*. Think here of mathematics or analytic epistemology. Again in simplified form, let there be such a theory T whose subject matter is S. S might be the concept of set. It might be the concept of knowledge. It might be the concept of deductive consequence. Then an intuition I for T is a belief about S taken for true by the theorist independently of what his theory might eventually provide for it. I is something we already take ourselves as knowing about S going in; it is "pretheoretic". An important function of I is to help identify T's target. (If we had no beliefs about sets, it is hard to know what a theory of sets would be about. It is hard to know how a set theorist should proceed.)[428] A further function of I is to constrain the things that might admissibly be said about S in T. That is, intuitions constrain a

---

[428] This is the awkwardness in which Frege and Russell found themselves with respect to sets, after having become convinced that, owing to the paradox bearing Russell's name, there is nothing to be understood of them. How could there be if there is no concept of set, if there is nothing that counts as knowing what it is to be a set? In short order, Frege gave up on set theory. Russell invented (or tried to) a new concept of set, or more correctly of set*.. See here Woods (2003), chapter 4.

theory's analysis of sets. Two things are required of T. It must disclose facts about S which weren't previously known and it must do so in a way that honours what was known beforehand. T must try to produce new knowledge in ways that honour the old.

Even a nodding acquaintance with the literature reveals how rough a picture this is. For one thing, that T should preserve the truth of its pretheoretic intuitions was never considered an absolute requirement. It was allowed that T's new disclosures might rightly convince us to refine a founding intuition or perhaps to abandon it altogether. In extreme cases, theoretical attention to an I might show it up as seriously mistaken, if not outright inconsistent (as with Russell's fateful letter of 1902 to Frege). But again, even when T leaves its data in place, it is not at all uncommon for theorists to take the view that they are left in place by T in an improved state, improved by having been subject to T's clarifying or explicating or precisifying powers.

But here, too, we frequently have the suggestion of conceptual change. A good theory of S will sometimes produce a knowledge of S in ways that go beyond clarification all the way to reconceptualization. When this happens, T itself models its own pretheoretic data. At the point of input, they can be as raw as common belief allows. But at the point of output they possess properties different from and often incompatible with the properties possessed at the point of entry. Let's call this aspect of T's relation to its data, "retro-modelling". A theory that effects a reconceptualization of its input data constitutes a *retro-model* of them.

Retro-models are a significant thing to take note of. They remind us that not all D-models are prior to the theories they motivate. A key question for retro-models is the epistemic legitimacy of the conceptual deviations it sanctions. In plain terms, what is it about T that justifies our now supposing that a better understanding of S is achieved by an account in which S, the old concept, is displaced by S', the new concept? Isn't this just another form of data-bending? And is there any epistemically secure difference between "changing the subject" before getting started with T and "changing the subject" in the course of finishing up with T? I will say more about this in the book's closing chapter.

This chapter began by re-emphasizing the importance of the respect for data principle. At first, the principle will strike us as obviously right and straightforward to implement. The upshot of our discussion here is the naïvete of this assumption of implementational ease.

When someone has been judged to have made a mistake of reasoning and that the mistake is a mistake of kind k, the respect for data principle requires the accuser to have taken pains to determine that the behaviour in question instantiates or manifests reasoning of a kind for which k is a definable mistake and that it actually fails in the K-way. If you charge someone with a deductive error, you should have reason to believe that he was reasoning with deductive intent or in ways that make its deductive assessment an appropriate means of judgement. You should then identify that feature of his reasoning that violate the standards of deductive correctness. I have already said why I think that deductive validity is nearly always the wrong yardstick to apply to the reasoning in which real life agents are involved. If this is right, the respect for data principle has been violated by those who invoke it excessively (that is, more than hardly ever). I want now to consider three further such violations.

14.5 *The conjunction fallacy*

In a famous experiment, Amos Tversky and Daniel Kahneman presented their subjects with the following set of facts F.[429] A certain Linda is 31 years of age, single, outspoken, and extremely intelligent. As a student she majored in philosophy and was deeply involved in movements for anti-discrimination and social justice. She was also active in anti-war campaigns. Subjects were then asked to select the more likely claim from a list containing the following:

1. Linda is a bank teller.
2. Linda is a bank teller active in the feminist movement.

A large percentage of the responses favoured proposition (2) over (1). The experimenters took this as evidence that their subjects had committed the probabilistic conjunction fallacy. For the probability of a conjunction is never greater than the probability of any of its arguments.

Tversky and Kahneman are applying a theory – a theory which includes the probability calculus as a central part – to account for some experimental data. What are these data? To hear them tell it, the data inhere in their own instructions to the experiment's subjects, including the experiment's set-up information, and in the subject's response to

---

[429] Kahneman *et al.* (1982).

these instructions, including their reports of the requisite probability estimates. They also took it that in reporting the probabilities, subjects were putting into play the Pascalian or aleatory concept of probability.[430] This is a further datum for them. It is essential for the accuracy of how the TK theory assessed the subjects' response that these data are accurately ascribed. Let us see.

Subjects familiar with the rules of the probability calculus would know that for any α and β the probability of $^-\alpha \wedge \beta^-$ never exceeds the probability of α or the probability of β. Further, they would know that this held true no matter what the set-up information is. Part of the cause of the subjects' difficulty is the TK set-up. Its very introduction presupposes its relevance, when in fact it is of no relevance at all to the problem the experimenters took themselves to have posed.[431]

It is a fact that the experimenters asked their subjects to pick the likelier of (1) and (2). It is a fact that a goodly majority of them judged (2) as the more likely. It is also a fact that Tversky and Kahneman convicted the majority answer of the conjunction fallacy. The respect for data principle requires their negative assessment to have been preceded by a reasonable determination that the students did indeed interpret the instruction given them as calling for a reckoning of which of (1) and (2) had the greater Pascalian probability. Also required is that some care be taken to determine that the reasoning reported by the subject's voting behaviour was reasoning that was truly responsive to their presumed understanding of the instructions received. (In other words, they weren't disobeying the instructions or just larking about.) The principle also demands that the options provided by Tversky and Kahneman were taken by their subjects to be the options available to them precisely as stated. It is not hard to see that there is some question as to whether these conditions were actually met.

The experimenters asked for a judgement of greater likelihood. In asking for the likelihood of something, one is sometimes asking for Pascalian probability. But one might also be calling for what Jonathan Cohen calls Baconian probability.[432] In still other cases, the call is for plausibility. Recall the lexical underdetermination problem. It is possible to say "probable" and mean "plausible".[433] By and large, people aren't

---

[430] "Aleatory" derives from the Latin for games of chance.

[431] Similar reservations may be found, for example, in Gigerenzer (2000), especially pages 19, 169, 197, 242, 248-250.

[432] Cohen (1977).

[433] Mellers et al. (2001)

familiar with the notion of Pascalian probability. So why would we think that the students were putting the Pascalian concept into play?

Similarly, how likely is it that, given the implied relevance of the set-up information F, the students were *not* interpreting the wording of (2) as formulating the option

> (2′)   Given that F and that she is a bank teller, Linda is a bank teller active in the feminist movement

and were *not* interpreting (1) as saying

> (1′)   Linda is a bank teller and not active in the feminist movement.

How likely is it that the students were not interpreting their instructions as calling for a judgement of the greater plausibility as between (1′) and (2), or the greater plausibility as between (1) and (2′)?

Not much, as I say, is yet known in a formal way about the structure of plausibilistic inference Even so, two things stand out as fundamental.

1. If propositions are internally consistent but jointly incompatible, they can have the same plausibility values but not the same probability values.
2. Unlike aleatory probability, conjunctive plausibility is not multiplicative. In particular, it could easily be the case that "*Pl* (Linda is a bank teller and a feminist activist)" has a greater plausibility than the plausibility of its leftmost conjunct and lesser probability than the plausibility of its rightmost conjunct.

Certainly when plausibility is conferred by evidence, given that (1) and (2) are forced options – that is, imposed by the experimenters, the evidence contained in F offers stronger evidential support for (2) than for (1). The TK subjects had to choose (1) or (2). F is no evidence for (1) and a lot of evidence for (2). Indeed F is evidence *against* (2). (1) is not a choosable option. (2) is the necessary choice by elementary application of the disjunctive syllogism rule. Besides, isn't F ∧ (2) more *explanatorily coherent* than F ∧ (1)?[434]

---

[434] For the notion of explanatory coherence see Thagard (1992). See also Gabbay and Woods (2005), section 6.4. Note here that, unlike conditional probability,

Before quitting this section, it would be appropriate to acknowledge a tactical liability. As already noted, the mathematical theory of probability is a robust, highly successful theory. Its range is equally impressive. There is virtually no aspect of theories of ampliative reasoning that it has not influenced in some major way. By what in chapter 2 I called the Can Do Principle, and given its track record to date, there is ample reason to try to extend and deepen the influence of probability theory. Are we not all inclined to prefer successful mature theories to alternatives that are fledgling and callow? Nothing succeeds like success.

I am here pleading a plausible case for plausibility. In so doing, I encumber myself with the liability that, next to aleatory probability, hardly anything is known of plausibility in an appropriately deep way. This matters. We are bidden by Can Do to try to cover the behavioural and linguistic data which – I say – favour plausibilistic treatment, probabilistically. For if it is a *theory* we are after, probability offers us an available candidate and plausibility doesn't. So how can it be preferable to play the plausibilistic card?

Against this I have two cards of my own. One was played in chapter 1. Ordinary probabilistic reasoners do not in the general case achieve their ampliative ends by making the calculations required of them by the aleatory theory of probability. Probability theory has to be learned, and given the vastness of the human community at large, hardly anyone knows it. Of course this alone doesn't guarantee that probability theory has nothing to tell us about successful probabilistic reasoning of the everyday sort. For it is possible that whenever an ordinary probabilistic reasoner reasons successfully, he produces – *somehow* – results that would independently be sanctioned by a calculus that he himself is unable to implement. But I say again that, if true, this would be serendipity of the highest order, and quite plausibly a coincidence too good to be true.

A second card to play against Can Do in the present case is the quite general fact that any theory no matter its track record can be extended beyond its natural reach. In earlier chapters I said the same thing, pointing out that Can Do is always at risk for the degenerate case which we call Make Do or paradigm creep. There is reason to think that paradigm creep is precisely what the TK experiment exemplifies. Suppose that the TK subjects had been responding to the experimental

---

explanatory coherence is symmetric. That is, whereas Pr (H | E) ≠ Pr (E | H), H coheres with H if and only if H coheres with E.

inputs as if they were to solve a plausibility problem. Then the good news (I say) is that they did solve it. This is not just a plausible explanation. On the fairness version of the Charity Principle, it is the better explanation. The day is long off before plausibility claims the level of theoretical articulation presently claimed by probability. But we have to start somewhere.

The extent to which the plausibility interpretation is plausibly the right interpretation is the extent to which Tversky and Kahneman are guilty of experimental misdesign, occasioned by a bending of the data which a faithful application of the respect for data principle would have at least discouraged. For the conjunction fallacy to qualify as a fallacy in the logician's traditional sense, it must meet the traditional conditions. It would have to be the case that when beings like us have occasion to judge the probability of conjunctive events or states of affairs, two things happen with a notable frequency: First, that their estimates run foul of the conjunction rules of the probability calculus. Second, that in making these determinations, they are putting in play the Pascalian or aleatory conception of probability.

I am bound to say that while the first of these requirements might appear to be met, the second is not; and that the reason that the second is not met is the main reason that the first appears to be met.

Proposition 14.5a
NOT A FALLACY: *The conjunction fallacy is not a fallacy.*

If Proposition 14.5a is right, this is further confirmation of an extension of the negative thesis to a list that now contains the conjunction fallacy. Does it also lend weight to a like extension of the misalignment thesis? It does if, as understood by those who worry about it, it is understood with traditional intent. I think we must say that this is indeed the way the conjunction fallacy is understood by the TK community. It is an error. It happens with notable frequency. It is an attractive mistake. Rates of recidivism are high. Accordingly, given this reading,

Proposition 14.5b
NON-INSTANTIATION: *The conjunction fallacy falls into the cachment area of the extended misalignment thesis. Although not on the extended traditional list, it does not in fact instantiate the traditional concept of fallacy.*

Against the present suggestion, one might argue as follows. Tversky and Kahneman re-tested their original subjects. The experimental contest was now quite different. Subjects had been made aware of their original (alleged) errors, and appeared to have accepted that they were errors. It was explained to them that the majority choice in the first experiment violated the conjunction rules of the probability calculus. Subjects were then given new information and a new list of possible choices, in which was embedded the design of the original problem. Again they were asked to choose the most probable of a list of propositions. Although the group average was better than the first time out, there was a significant degree of recidivization. What makes the subsequent experiment important is that there was reason to believe that the subjects believed that giving $\ulcorner(\alpha \wedge \beta)\urcorner$ a higher probability than α alone is virtually always an error; yet numbers of them did so anyhow.

Tversky and Kahneman took this as even more compelling evidence of the probabilistic irrationality of even quite intelligent human subjects. Subjects knew they were being asked to play the aleatory game. They knew that, with those integral exceptions, $Pr(\alpha \wedge \beta)$ is always lower that $Pr(\alpha)$. But when it came to the decision, numbers of subjects were unable to stay on track. The diagnosis of Tversky and Kahneman cannot be rejected out of hand. It is possible that human subjects are probability-misfits. But another possibility is that human subjects find it unnatural to apply the aleatory concept in contexts in which information such as F is given a set-up role. It is possible that, when primed by F, the reasonable course is to operate with a different concept of probability. It is possible that in such circumstances it is more reasonable to operate plausibilistically. A key factor is the precise wording of the experimenters' instructions in the subsequent experiments. If subjects were told to *compute* or to *determine* the higher probability *value*, perhaps they could be found guilty of not paying attention. If, on the other hand, they were told to choose what *felt* like the more probable, it is very much harder to pin on them the rap of aleatory misconduct. Again, this would breech the fairness version of the Principle of Charity.

A final quick word about Linda's experimental misdesign. The comparative lack of rigorous attention to the design requirements of the experimental physical sciences is a serious problem not sufficiently appreciated by philosophers of science.[435] The same is true, but more so,

---

[435] Again Suppes is a significant exception. See his (1962), (1967), (1979) and (2002).

of the experimental social sciences, where various peculiarities of human subjects, including the need for ethical constraints, lead to difficulties in controlling relevant variables. Aside from the problems already mentioned, the Linda experiment leaves certain key questions inadequately dealt with. I will mention just two of them. One is the seriousness with which subjects are likely to take the Linda exercise, and the concomitant likelihood of effort and truthfulness. What is known about Linda-like cases that if the context is changed in certain ways, the majority for (2) goes down. If, for example, there is a nontrivial cash pay-off for the winning answer, this implicates that the answer is not obvious. What is the sense in paying people to tell you what's obvious to them? Why not just ask? Since the answer given in the original experiment is obvious, subjects have a reason to avoid it in the present circumstances, and to give their attention to what a good non-obvious answer might be.

A related difficulty is the element of trickery. Linda, and the legions of experiments like it, have the feel of trickiness about them. Here is why. On – I say – the preferred reading of the set-up information and experimental instructions, the answer to the Linda question is obvious. It is obvious that (2) is more likely than (1). This helps answer a puzzling question. Why, if the answer is obvious, did so many of the Linda subjects – not the majority, but still a goodly percentage – opt for (1)? Suppose these people asked themselves, "Why would TK be wasting our time and theirs with a question whose answer is so obvious?" A plausible reply is that the experimenters are *not* trifling with them, hence that the correct answer is not the obvious answer. This alone is grounds to opt for (1), since (1) is the only alternative to (2). In those circumstances, the experiment involves a trick. The trick is *not* to opt for (1) or for (2) on the basis of the set-up information, but rather on the basis of the subjects' expected inferences about the experimenters' real intentions. And if, in doing *that,* subjects opted for (1), it is not in the least necessary, or even very likely, that they were implementing the conjunction rule of the probability calculus. More plausible is the suggestion that they were implementing the much more intuitive rule of disjunctive syllogism: Either (1) or (2) holds. But we have reason to think that (2) doesn't hold. So there is reason to think that (1) holds.

14.6 *The gambler's fallacy*

Consider now the simple example of a coin-toss game. In its elementary form, the gambler's fallacy is committed when a subset of the history of tosses to date influences a player to believe that the probability of the coin's showing a head on the next toss is either higher or lower than 0.5. Suppose that the last sixteen tosses have nothing but heads. This gives two basic ways of committing the fallacy, "counterinductively" and inductively. The counterinductive solution is to switch to tails. The inductive solution is to stick with heads.

Let's deal first with the counterinductive option. Let us suppose that Harry knows full well that sixteen straight heads is a very rare occurrence.[436] Let us suppose that Harry is on to something when he thinks that the rarity of sixteen straight heads is reason to think that toss seventeen will turn up a tail. Suppose that Harry gives voice to this belief as follows:

> (1) It's not likely that seventeen will show a head because sixteen straight is already very unusual.

Of course, this is rather rough. What, we might ask, is Harry really saying? What is the logical structure of (1)? One possibility is that he read (1) as something in the form

$(1')$ $\Pr(\alpha/\beta) > 0.5$.

Another possibility is

$(1'')$ $\Pr(\alpha) > 0.5/\beta$.

If the utter rarity of sixteen straight (β) is indeed reason to think that seventeen will show a head (α), then (1) would be a reasonable thing to say. There would be some reading on which (1) would not be an erroneous way to think. $(1')$ furnishes such a reading, and $(1'')$ does not. $(1'')$ commits the gambler's fallacy. The question is whether we have a principled basis on which to attribute to Harry the defective reading $(1'')$ over the alternative reading $(1')$. If we judge Harry to have reasoned correctly that the rarity of the sixteen is reason to believe seventeen will show a tail, why in the world would we stick Harry with $(1'')$? Why would we not go with $(1')$?

Of course, we could always ask. We could ask Harry to tell us whether what he is saying has the form of $(1')$ or the form of $(1'')$. But there is a problem. Harry – and the rest of us – might have difficulty in answering. That would not be surprising, given that hardly anyone knows anything about the calculus of probability. By this I mean, hardly anyone

---

[436] Assuming that Harry believes the coin to be true and tosses fair.

knows how to articulate the details of the calculus of probability. Hardly anyone is able to describe the operation of his reasoning devices, even as they turn over in just the right ways. (This takes us straight to the visiting scientists' problem of reasoning-manifestation behaviour; and it reminds us that not knowing how correctly to describe one's reasoning processes is not necessarily evidence of bad reasoning).

This calls to mind the problem posed by Sleigh's fallacy. Suppose that Harry believes that α entails β and that α is true. Then he believes that

(2) Since α necessarily β.

Of course, (2) embodies an ambiguity of its own. It could mean

(2') □(since α, β)

or it could mean

(2″) Since α, □β.

If we attribute (2″) to Harry we accuse him of Sleigh's fallacy. If we attribute (1'), he is home-free. So why would we pin (2″) on Harry? What have we got against Harry?

Here too we could ask Harry what he means. Does he mean (1') or (1″)? Does he accept or does he not a Barcan formula for necessity? Here too Harry might not be able to tell us. But why should Harry's ignorance of the logical structure of what he's saying count against its correctness? Why are we being so tough on Harry? Sleigh's fallacy turns on the misplacement of the term "necessarily". The gambler's fallacy turns on the misplacement of the term "the probability of". Each is an error, make no mistake about it. But it is attributed, in the first instance on the basis of Harry's utterance of (1) and, in the second, on the basis of his utterance of (2), it is an attribution that runs foul of the charity principle, and it discomports with the respect for data principle. To qualify as a fallacy, the attribution must be fair and attentively respectful to utterances like (1) and (2) *in the general case.* There is no reasonable expectation of meeting this requirement.

Proposition 14.6a

NOT A FALLACY: *In its counterinductive form the gambler's fallacy is not a fallacy in the traditional sense.*
COROLLARY: *Add the gambler's fallacy to the misalignment thesis confirmation tally.*

It was said in Detroit by his then-colleagues at Wayne State University that Sleigh's Fallacy is so-named not because of the frequency of Robert Sleigh's commission of it, but the frequency of his attribution of it. It is a novel rationale for a baptism, but it is a mistaken baptism all the same. Sleigh's fallacy is the confusion of (2′) with (2″). Each of these in turn is a reading of (2). (2) is the sentence "Since α necessarily β". As we have it here, (2) is *punctuationally* underdetermined. It is underdetermined in the same way we saw earlier in sentences containing adverbial occurrences of the idioms of possibility, probability and plausibility. Punctuational underdetermination makes for ambiguities, and ambiguities are the occasion of errors of reasoning that arise from ambiguity-mismanagement. It is desirable that prior to their assessment such sentences be disambiguated. When the adverb is "necessarily" it is sometimes easy to do this. But given the particular wording of (2), it is not. It is easy enough to get (2″) out of (2), but for the everyday speaker of English, (2′) is a much less likely finding. Again, the reason for this is not that the everyday speaker doesn't understand (2). It is that he doesn't know how to punctuate what he understands. Whey should he? People have to be taught to punctuate (2) as (2′). They have to be taught not to punctuate it as (2″). People learn these things in First Year logic courses, which is also where they learn of Sleigh's Fallacy, though not necessarily under that name. Most everyday speakers don't go to university. A great many who do don't take First Year logic. Many who do aren't much interested.

It is a mistake to confuse (2) with (2′). If it happened with a notable occasioned frequency, it might justify the name of Sleight's Fallacy. One occasion to commit it would be an utterance of (2) or the holding of a belief expressible as (2). For the mistake to be committed on those occasions it would have to be the case that, in uttering or thinking that (2) the speaker means that (2′). What is the evidence that this condition is met? The evidence is that if we charged him with the task of removing (2)'s punctuational underdeterminedness, he would likely choose (2″) and not likely choose (2′). Why would this be so? Because he knows how to punctuate in the manner of (2″) and doesn't know how to do it in the manner of (2′). Yes, but doesn't his choice of (2″) show that what he means by (2) is (2″)? No, it means that he thinks that (2″) means what he means by (2).

The attribution to such speakers of Sleigh's fallacy is itself a fallacy. What are the occasions for its commission? They are occasions on which people seek punctuationally complete readings of (2). People so

498

inclined are mainly philosophers who have an interest in the alethic modalities. They are a fairly rarified bunch. How likely is it that any of them would seriously doubt that utterances of (2) are ambiguous as between (2') and (2")? How likely is it that any of them would fail to accuse a reading of (2) as (2") of a modal fallacy? How likely is it that any of them would not attribute this fallacy to Harry? The attribution of Sleigh's fallacy is a fallacy.

Proposition 14.6b
SLEIGH'S "FALLACY": *Although not on the traditional list, the* attribution *of Sleigh's fallacy instantiates the traditional concept of fallacy.*

This leaves us with the inductive form of the fallacy. We can deal with this more briefly. In heads-again case, there is no question of sixteen straight lending evidential support to another head on seventeen.[437] Harry is not reasoning that toss seventeen will show another head. It is possible that it will, and Harry knows it. What Harry is doing is *betting* that it will. He is paying the price of finding out for sure whether it will. In this, Harry is like the buyer of a ticket in a mega-lottery.

Of course, there is always the question of whether Harry's bet is a rational one to make. It depends on two factors. One is how much he can afford to satisfy his present degree of interest. The other is what he presently believes. If he actually believes that the next toss will show another head, that would affect the rationality of his bet if two conditions were met: He could afford the bet, and has reasons for believing what he does. Assuming true coins and fair tosses, the second condition is failed, leaving the rationality of the bet to considerations that lie elsewhere. In essence, it's all down to what turns Harry on and what Harry can afford to do about it.

It would be nonsense to ignore that places like Monte Carlo (after which the fallacy is sometimes named) and Las Vegas are fairly swimming with people who think themselves versed in the probability calculus, some of whom actually are. It would be silly to overlook all the probability calculations that go on when professional gamblers or professional statisticians set out to break the casino's bank. But the gambler's fallacy has nothing to do with these trappings. It is the simple

---

[437] Still assuming that Harry believes the coin to be true and tosses fair.

failure to recognize that track record is irrelevant to the pre-toss odds exhibited by the next toss when the coin is true and the toss is fair.

It is also worth emphasizing that it is no part of the received view of the gambler's fallacy that those who commit it are expert gamblers or expert mathematicians or expert anythings. The committer can perfectly well be a regular guy – Harry – who enjoys an occasional flutter or who frequents the Legion on Friday nights. It is a fallacy for Everyman.

Proposition 14.6c
STILL NOT A FALLACY: *In its inductive form, the gambler is not an error of reasoning, hence not a fallacy.*

14.7 The base-rate fallacy

Consider a standard example. Suppose that the rate of disease D in the homosexual population is three times its rate in the non-homosexual population. Imagine that you come to know the following fact about someone called Leslie and that you know nothing else in particular about him or her: Leslie has D. Suppose now that you are asked to determine the likelihood that Leslie is a homosexual.

There are problems with the design of this set-up. It is not made clear whether you are to answer the Leslie question based exclusively on the information given, or whether it is allowed (and perhaps expected) that your answer will also be shaped by other relevant information you might reasonably be expected to have. The first possibility is easily disposed of. It restricts the premisses of the inference you are asked to make to these:

1. The D-rate is three times greater for homosexuals than for non-homosexuals.
2. Leslie has D.

Concerning the likelihood of Leslie's own homosexuality there is nothing whatever to be learned just from (1) and (2). No one in his right mind would think otherwise.

What's missing here is the *base-rate* of homosexuality in the population at large. Suppose that it is 10%. Then we have a third premiss:

3. Homosexuals are 10% of the population at large.

Suppose the question is repeated. Given (1), (2) and (3), what is the probability that Leslie is a homosexual? Before answering this question, consider a somewhat different one: Given (1), (2) and (3), is Leslie *more* likely to be a homosexual than would be a randomly selected person from the population at large? Experimental evidence reveals a large majority for an affirmative answer. It is the right answer.

Let us now slightly re-phrase the original question. What, given (1), (2) and (3), is the *precise probability value* of Leslie's homosexuality? Here the experimental evidence clearly indicates a strong disposition to give the wrong answer – to assign a value that is significantly higher than the premisses allow. When this happens, it is routinely said that subjects have committed the base-rate fallacy.[438] That is, that in calculating the over-high probability of Leslie's homosexuality, they ignored premiss (3).

This is a surprising diagnosis. As we saw, no one would be inclined to draw any inference about the likelihood of Leslie's homosexuality from premisses (1) and (2) alone, that is, in the absence of the base rate premiss (3). Equally, however, once the base-rate premiss has been activated, people haven't the slightest hesitation in giving the qualitatively comparative – and correct – estimate that Leslie's being a homosexual is greater than that of a randomly related person. Why, then, when asked to be *more precise* about this probability, would these same people suddenly take the base-rate premiss out of play?

It is true that more often than not people get the precise value too high, higher in the present case than 25%, which happens to be the right answer. Why would this be so? That is, why would it be so, given the considerable unlikelihood that they have suddenly retired the base rate premiss, that they have decided to ignore as irrelevant what just before they realized was not irrelevant? The answer is that they don't in the general case know how to calculate the precise value. They don't know Bayes' Theorem. It is not that they aren't sensitive to base rate information. Rather it's their not knowing how to weave it into the computations demanded of them that counts here. They can't do it, so they guess. And they guess wrong.

Why would they guess? Why wouldn't they just down tools? The mere fact that the question has been put to them presumes a capacity for answering it. Given that subjects at large don't know how to filter base rate information into the needed calculations, they assume that the answer

---

[438] Tversky and Kahneman (1982).

which it is in their means to give is available from those elements of the case of which they have a better calculational grasp. So they ignore what they don't know how to make use of.

Who would seriously press upon a willing interlocutor a question whose answer requires the engagement of Bayes' Theorem, knowing that his addressee doesn't know the theorem, and likely has never heard of it? What is this supposed to show? The received answer is that it shows what a rotten probabilistic reasoners beings like us are. It does not show that. It shows that we can be tricked into error.

For most of human history, from the fateful excursion some fifty thousand years ago into the arts and technology no one had much of a clue about the precise probability of anything except some very simple cases. By the standards of the present day, the Leslie problem is pretty small beer, hardly a problem at all. But Aristotle would have been stymied by it and Spinoza defeated by it. No one much before Pascal and his colleagues could have handled it. Pascal and his epistolary intimates are great figures precisely because they finally discovered a way of getting this right. It would do us some good to call it to mind every now and then how difficult it was to suss out desired probabilities with mathematical precision.

Much the same could be said about our command of the gravitational laws before Einstein. Even the learned folk of the day didn't have the big picture. If presented with problems solvable only in the general theory, they could only get them wrong. How could they not? They were ignorant of the theory of general relativity. Nowhere in the history or philosophy of science is there the slightest suggestion of *the gravitational fallacy.* If those same questions were today put to Harry, the chances are that he wouldn't know how to answer them. Relativity theory is an utterly dominant part of physics, but in the population at large hardly anyone knows how it goes. I mean by this that hardly anyone knows how to solve its equations. If Harry were in dialectical circumstances similar to those in which he was pressed for the precise probability value of Leslie's homosexuality, he might guess at the answer instead of declaring his inability to answer it. The difference between the Leslie case and the relativity case is that Harry had reason enough in the former to give the right comparative estimate, and this is less likely to be true in the relativity case. But the similarities between the two situations are robust enough to make a useful point. It is that guessing wrong is not a fallacy. How could it be, given that by far the standard response to a request to do something you simply don't know how to do is to say that you don't. The standard response is not to guess. In the problem-space created by one's inability to do the thing asked, guessing at all – let alone guessing wrong – is not something that occurs with a notable frequency. So in its guessing-form, the base rate fallacy is not a fallacy.

Proposition 14.7a

NOT A FALLACY: *The base-rate fallacy lacks the occasioned frequency to qualify as a fallacy in the traditional sense.*

The nub of Gigerenzer's complaint about the paradigm creep of probability theoretic representations and assessments of the ampliative reasoning of individual agents is the uncritical assumptions that the human individual is an intuitive statistician. The weight of our present reflections is that it makes about as much sense to postulate the intuitive relativity theorist. But it makes no sense to postulate the intuitive relativity theorist. *Modus tollendo tollens.*

Proposition 14.7b

MISALIGNMENT UPDATE: *The following items are not instantiations of the traditional concept of fallacy or fail to meet the burden of proof to show otherwise: Hasty generalization, false cause, post hoc ergo propter hoc, ad verecundiam, ad ignorantiam, affirming the consequent, denying the antecedent, ad populum, begging the question, including circularity, many questions, ad hominem, conjunction, gamblers' and base rate. That's fourteen out of eighteen.*

COROLLARY: *I venture to say that this inflicts a mortal wound on the alignment thesis, the thesis that the eighteen instantiate the traditional concept of fallacy. In other words it craters it. This redeems a promissory note of chapter 1.*

CHAPTER FIFTEEN

THINKING IT OVER

15.1 *The traditional concept of fallacy*

I have already said that the concept-list misalignment thesis is a bold idea, one that many a fair-minded fallacy theorist would dismiss out of hand. If, as I proposed in chapter 1, the traditional list of fallacies is not to be found in the extension of the traditional concept of them, the fallacies project has been off-track since the birth of logic. If the misalignment thesis is true even in my cratered version of it, there is a serious sense in which those who have pursued the fallacies project in logic haven't known what they've been talking about. It is a hefty indictment, carrying a proportionately weighty burden of proof. Logic is one of those disciplines – unlike philosophy some would say – crowned with definitive successes from its start, not in an unbroken line to be sure, but to good gathering effect overall. In the long march from the syllogistic to linear logic, these advances could not have occurred without the talent to drive them forward. The history of logic brims with gifted people. How could they have missed the fallacies boat? (Hamblin's Question). A short possible answer is that they didn't miss it. It is not convincing response. Hamblin wouldn't have accepted it; nor do I. A more circumspect reply is that even if logic did miss the fallacies boat, it is not on account of concept-list misalignment that it did. Perhaps this is right. If so, the imputed mismatch of concept to list doesn't exist. If that in turn were so, the two traditions would have lined up in the traditionally expected way. The fallacies on the traditional list would have instantiated the traditional concept. True or not, this is a lot of weight for the concept of *tradition* to bear. We should give it some further attention, in particular the tradition's conception of fallacy.

The traditional way of thinking about fallacies is not without its critics. There are logicians and others who reject the claim that the fallacies – the gang of eighteen – are dominantly errors of inference. The dissenting views fall into two camps. If we wanted to be light-hearted about it, we could call them *heretics* and *pirates*. A heretic holds that what I have identified as the traditional concept is not indeed the traditional concept. A pirate holds that what I have called the traditional concept is in fact the traditional concept, but that the eighteen would be better served by having the word "fallacy" denote a non-traditional concept. Heretics are to fallacy theory as Luther was to the Pope. Pirates are to fallacy theory as same-sexers are to marriage. The difference between heretics and pirates is clearer on the page than in the historical record. But I think we may safely identify Jaakko Hintikka as a leading heretic. In the pirates' camp, one thinks of Frans van Eemeren and Rob Grootendorst.

If the heretics are right, then what I identify as the traditional list may be right, but what I identify as the traditional concept is clearly wrong. If so, the concept-list misalignment thesis, in the terms in which I have given it expression here, is also wrong. If the pirates are right, the traditional conception is either theory-resistant, or of insufficient interest to justify serious intellectual effort. In that case, the misalignment thesis is, in the terms in which I have given it expression here would either be paralyzed by the theoretical intractability of the concept or made not interesting enough to bother with.

15.2 *Heresy*

Are the heretics right? Do they disable the misalignment thesis? What might we learn about such things from Hintikka? Hintikka is a logician of impressive originality and range.[439] Given limitations of space, there is no reasonable prospect of identifying, still less giving analytical attention to, the many points at which his work intersects with my own interest in agent-centered logics. For our purposes we must make do with a good deal less of Hintikka than is actually relevant to these concerns.

For all his mastery of and contributions to mainstream techniques, Hintikka is a logical renegade.[440] He holds that those four

---

[439] Among his pathbreaking contributions are his early work in the semantics of modal logic (Hintikka, 1957, 1967), the adaption of S4 to epistemic and doxastic contexts (Hintikka, 1962), the logic of induction and probability (Hintikka, 1970, 1976a) and (Hintikka and Niiniluoto, 1976), pioneering work in information semantics (Hintikka, 1973), game-theoretic/interrogative logics (Hintikka, 1976b, 2007), alternative approaches to the ramified conception of sets (Hintikka, 1998, 1999), and independence-friendly logics (1995). There are several papers on Hintikka's logical contributions, with replies by him, in the Hintikka volume of the Library of Living Philosophers (Auxier and Hahn, 2006).
[440] This is not perhaps how Hintikka himself would see it. There is an old joke about a mother viewing the march-past of her son's regiment. "Why", she

staples of modern mathematical logic – set theory, model theory, proof theory and recursion theory – have strayed from a conception of logic adopted by its originator and, as a result, that logic has developed in ways less good than would have been the case had Aristotle's founding insight been retained. On Hintikka's telling, Aristotle's insight is that logic is a theory of question-and-answer games. Accepting this, Hintikka's own view is that the best general theoretical framework for logic is what we now know as game theory and that the best provision for logic's particular rules is what is now known as interrogative logic.

Hintikka's case against (my version of) the traditional conception of fallacy is set out in "The fallacy of fallacies" (1987). As Hintikka sees it, Aristotle's insight first emerges in his treatment of the fallacies in *Topics* and *On Sophistical Refutations.*[441] Except for fallacies such as consequent and accident (p. 212), Hintikka characterizes Aristotle's position as follows:

> Instead of being mistaken inference-types, the traditional "fallacies" were mistakes or breaches of rules in the knowledge-seeking games which were practiced in Plato's Academy and later in Aristotle's Lyceum. Accordingly, they must not be studied by reference to the codifications of deductive logic, inductive logic, or informal logic, for these are all usually thought of as codifications of inferences .... Instead, the so-called traditional fallacies are best studied by reference to the theory of information-seeking questioning processes .... (p. 211)

In this same spirit, Hintikka goes on to say that "all Aristotelian fallacies are *essentially* mistakes in questioning games, while some of them are *accidentally* mistakes in deductive (more generally, logical) reasoning". (p. 213; emphases added) In short, fallacies are necessarily I-theory mistakes and only contingently, if at all, mistakes of inference.

I-theory is Hintikka's logic of information-seeking questioning processes. I-theory is an agent-based, goal-directed logic. It regulates the interactions of a questioner and an oracle, that is, an information-seeker and an information-provider. Oracles can but needn't be people. An oracle can be a data base or tacit knowledge, or even nature herself. (p.

---

exclaims, "everyone's out of step but my Harry!" Like the mother, Hintikka would be drawn to the suggestion that it is mainstream logicians who are out of step, not he.

[441] Hintikka (1987, 1989).

217) At each stage of the inquiry, a questioner makes one of two moves. Either he addresses a question to the oracle or he derives a conclusion from answers already given. The correctness of derivations is governed by logical rules and the appropriateness of questions by interrogative rules. The rules of logic are "definatory"; this means that they determine whether a move is *permitted*. The rules of interrogation are "strategic"; this means that they determine when a permissible move is *smart* or as we might also say *apt*.[442]

Hintikka's inclination is to attribute to Aristotle the following claim:

Proposition 15.2a
FALLACIES AS FAILED STRATEGIES: *Aristotle's fallacies are, essentially and without exception, violations of interrogative rules.*

This is a view which, with slight variation, Hintikka himself is tempted to adopt as his own.

The problem is that Proposition 15.2.a simply can't be made to square with what Aristotle expressly asserts in *On Sophistical Refutations*. Hintikka writes: "[N]o perceptive reader of the *Topics* or *De Sophisticis Elenchis* can fail to realize that *elenchus* [refutation] compromises much more than logical inferences in any sense of inference". (p. 213) It is surprising that Hintikka doesn't take note of what Aristotle says about *elenchus* in the opening passages of *On Sophistical Refutation*. According to Aristotle, an *elenchus* "is a [syllogistic] deduction to the contradictory of the given conclusion [thesis]". (165[a] 3) There may be more to an *elenchus* than logical inference, but it is certainly also the case that the notion embeds essentially the logical concepts of deduction (*sullogismos*) and contradiction. Aristotle's reason for introducing this element at the beginning of his monograph is that he wants to make it clear that fallacies are errors of *elenchus*.

---

[442] In a rough and ready way, the definatory-strategic distinction is our distinction between conditions for consequence-having and rules for consequence-drawing. *Modus ponens* is a definatory rule. It provides that the rule's tail is a deductive consequence of the rule's head. But, like Harman, Hintikka doesn't think that every consequence a premiss has is one that should be drawn. It is strategic rules that give guidance in such matters.

Aristotle's own words force the abandonment of Proposition 15.2a and set the stage for the consideration of weaker versions of it, to wit:

Proposition 15.2b
MOSTLY STRATEGIC: Most *of Aristotle's fallacies are in essence violations of interrogation rules.*

Proposition 15.2c
STRATEGIC AT THEIR CORE: *Even if not most, the* hardcore *of Aristotle's fallacies are in essence violations of rules of interrogation.*

Let us note in passing that if Proposition 15.2c were true it would not necessarily define Hintikka's own view of what the fallaciousness of a fallacy consists in. I will return to this point below.

Proposition 15.2.1c presses us to ask what fallacies Hintikka sees at the core of Aristotle's treatment. In "The fallacy of fallacies", begging the question and many questions are clearly intended. But also discussed are the *ad hominem* and babbling, neither of which is in Aristotle's gang of thirteen.[443] It may be that Proposition 15.2c is Hintikka's way of riding two horses at once. If question-begging and many questions are Aristotle's core fallacies, then Proposition 15.2c is about question-begging and many question. But since "The fallacy of fallacies" makes much, as well, of the *ad hominem*, it may be that Hintikka thinks that something like Proposition 15.2c is true of it. So Proposition 15.2c is true of *Aristotle's* fallacies as regards question-begging and many questions, and is true of *Hintikka's* fallacies as regards these same two and the *ad hominem* as well. Since Hintikka recognizes the three as the traditional

---

[443] Hintikka doesn't attribute to Aristotle the classification of the *ad hominem* as a fallacy. Hintikka discusses it here (at pp. 226-227) because it has come to be regarded by the later tradition as a fallacy, stemming from some remarks in *On Sophistical Refutations* (see, for example, 177^a 12-19, 177^b 31-34, 178^b 16-23; see also *Metaphysics* 1062^a 2-3), and because it has a natural fit with the interrogative model. As for babbling, which is hardly an obvious candidate for deductive error, Aristotle denies that it is a fallacy. At 165^b 14-17, Aristotle sets out five distinct targets at which competitors may aim. They are: "*elenchus*, falsity (*fallacy*), paradox, solecism, and fifthly to reduce the opponent in the discussion to babbling." So babbling is distinguished from fallacy as fallacy in turn is distinguished from syllogism. Hintikka cannot have been unaware of this, and it might explain why he refers to it as a "pseudo-fallacy". (p. 225)

fallacies of the present day, his view is that the core traditional fallacies involve essentially the violation of interrogative rules.

Further inspection of *On Sophistical Refutations* reveals what Aristotle is up to. Aristotle is operating with a twofold conception of logic. On the one hand, he wants a wholly general theory of two-person argument. On the other he wants to develop an account of the syllogism. A syllogism is a context-free sequence of (on most accounts) three categorical propositions, of which the terminal member is the conclusion and the other two are premisses. It takes successful negotiation of a number of conditions before any such sequence qualifies as a syllogism. One is that the premisses necessitate (that is, entail) the conclusion. Another is that the conclusion may not come from any single premiss by immediate inference. A third is that, with respect to the necessitation by premisses of the conclusion, no premiss be redundant. Other conditions are that syllogisms not have multiple conclusions and that premiss-sets not contain a proposition and its contradictory.

Dialectical arguments are different. Where a syllogism is a sequence of propositions, a dialectical argument is a series of alternating speech acts between two opponents, one of whom plays the role of questioner and the other the role of respondent. Respondents also have the duty to post a thesis τ for the refutatory attention of the questioner. The questioner is required to put to the respondent – who is the proponent of τ – questions that occur one-at-a-time, and each of which is properly and completely answerable with a Yes or a No. Answers generate propositions. A yes-answer to the question whether virtue is teachable is that virtue is teachable. A no-answer to the same question generates the proposition that it is not the case that virtue is teachable. It is permitted to the questioner that any proposition generated by a yes-or no-answer be used at the questioner's sole discretion as a premiss in a syllogism. The overriding goal of the questioner is to elicit from the answer questions which generate propositions usable as premisses of a syllogism whose conclusion is ⌜~τ⌝, that is, the contradictory of the answerer's original thesis τ. When such a syllogism is produced it – the syllogism itself – is a refutation, of the respondent's thesis.

It is clear from Aristotle's remarks in *On Sophistical Refutations* that there are rules governing the asking and answering of questions (interrogative rules in Hintikka's sense). Questions must admit of complete answers yes-or-no, they must be asked one at a time, and so on. Violations of these rules are a serious matter for Aristotle. They spoil the integrity of *elenchi*. Hintikka thinks that Aristotle thinks that these violations are fallacies, and he is tempted himself to think that they are fallacies too. By this I mean, in particular, that these are fallacies as conceived of by Aristotle and passed onto in a tradition that hardly anyone has honoured. (Recall the Mum of footnote 2 who thinks that every soldier in the march-past is out of step except for her boy Harry). It is true, as we have seen, that Hintikka's thesis about Aristotle may well have to be boiled down to the core fallacies: begging the question and many questions. Certainly Aristotle would allow that these are interrogative rule-violations, but it is not this that makes them fallacies. Here is Aristotle on the point:

> It is altogether absurd to discuss refutation without first discussing syllogisms; for a refutation is a syllogism, so that one ought to discuss syllogisms before describing sophistical refutations; for a refutation of that kind is a merely apparent syllogism of the contradictory of a thesis. (*Soph. Ref.* 10, 171ᵃ, 1-5)

Concerning the gang of thirteen, there is one case in which Aristotle contemplates a variation on this theme. It is *ignoratio elenchi* or misconception of refutation.[444] The fallacy is committed in one or other of two ways. It is a fallacy to construct a good syllogism from properly conceded premisses whose conclusion is not the contradictory of the respondent's thesis τ. It is also a fallacy to assemble a τ-contradicting syllogism from premisses the answerer has not conceded or to which he is not committed by answers already given. In this case the fallacy is either one of misconceiving the contradictory of τ or it is a fallacy arising from illicit premiss-selection.

Refutations and syllogisms are tightly linked in Aristotle's logic. A dialectical argument can succeed only if the requisite syllogism is available from premisses the questioner concedes, and it will fail if the required syllogism is not forthcoming – that is, the questioner's purported syllogism is not a syllogism in fact, or it is a syllogism whose conclusion involves a contradictoriness error, or whose premisses involve a premiss selection error. Aristotle was very proud of his discovery of syllogisms. He thought that the logic of syllogisms would be the theoretical core of a wholly general theory of real life interpersonal argument, not only of dialectical arguments, but also of pedagogical arguments fashioned by what we know as the socratic method, and other modes of question-and-

---

[444] That is, misconception of what constitutes a good refutation.

answer enquiry, as well as – not so interpersonal in character – scientific demonstrations from first principles.[445]

Except for begging the question and many questions, there is nothing on Aristotle's list that hooks up in *Hintikka*'s way with question-and-answer games. It is true that, since question-and-answer games are competitions in reasoning, knowledge-seeking exchanges are a natural context for the confusion of non-syllogisms with syllogisms. But the question is whether – with the exceptions noted – that connection is anything but occasional and contingent, even if frequent. The answer is that it is not. Inspection of Aristotle's list makes this clear. In *On Sophistical Refutations, secundum quid* is the fallacy of concluding that Mr. Mandella is white from the premiss that he is white-haired. Ambiguity is the fallacy of giving to different occurrences of a common term in a syllogism different meanings, occasioning thereby the syllogistic error of four terms. Combination of words is the fallacy of confusing the proposition that, even while sitting Socrates has the power to start walking, with the proposition that Socrates has the power to walk and sit at the same time, that is to say, to walk while sitting. Non-cause as cause is (in the *Rhetoric*) the fallacy of confusing constant conjunction with causation (24, 1401^b, 30-31), something that crops up among the eighteen as the fallacy of *post hoc, ergo, propter hoc,* also sometimes called the fallacy of false cause. It takes no imagination to see that these, and the others, are committable errors entirely apart from the stimulus of questions and answers. Accordingly, if Aristotle's gang of thirteen is the right list of the Aristotelian fallacies, it is immediate that the great majority of them owe in their commission nothing to the interrogative rules. So, then, the nub of our present question is this: *Do the thirteen constitute Aristotle's list?* The answer is: Of course they do. Read *On Sophistical Refutations.*[446]

I have taken some time with Hintikka's interpretation of Aristotle not because of Aristotle's relevance to our project, but rather because of what it reveals of Hintikka's own conception of logic and of the place in it occupied by fallacy-making. For all his importance to contemporary

---

[445] For a more detailed treatment of Aristotle on fallacies see Hitchcock's excellent paper (2000). For a discussion of whether the logic of syllogisms is a game theoretic logic see my (2013).

[446] Even Aristotle's many questions fallacy has nothing inherently to do with mistaking a non-syllogism for a syllogism. See Woods (2009b).

logical theory, Hintikka has written very little about the fallacies.[447] Still, Hintikka is of rock-hard importance to my project. He is a heretic. He is drawn to the view that there is a traditional conception of logic descending from Aristotle, and that the logic community has either ignored or misdescribed it. So he disagrees with my conflation of the traditional concept and the modulo 1970 modern concept. He thinks that the modern concept is a misreading of the traditional concept. In this there is unmistakably clear advice for the would-be fallacy theorist. If he wants to get the logic of the eighteen right he'd better bring to bear on them not the received view of the traditional conception but the actual traditional conception which the present day theorist has lost sight of.

Readers may be unhappy with the purported distinction between what the tradition has actually bequeathed us and what the tradition is merely taken to have bequeathed us. As set out at the beginning of this chapter, there doesn't seem to be much space between traditions as they actually are and traditions on the received view of them. If this is right, Hintikka ceases being a heretic and becomes instead a pirate, albeit a rather reactionary one. Hintikka's advice is that the concept that the eighteen instantiate is not the modern concept of fallacy, but rather the ancient conception that Hintikka thinks he finds in Aristotle. If that were so, it would give us not my but rather *Hintikka*'s version of the thesis of concept-list misalignment:

Proposition 15.2d
HINTIKKA'S MISALIGNMENT THESIS: *On its received or modern interpretation – the EAUI interpretation – the eighteen aren't in the extension of the concept of fallacy.*

And, of course, to this we must add

Proposition 15.2e
HINTIKKA'S INSTANTIATION THESIS: *On Hintikka's own conception of fallacy, the gang of eighteen are indeed instantiations of* it.

---

[447] Even the three articles cited here are – rather shamefully – not standard reading for present day theorists of argumentation, and his other more voluminous writings on the game theoretic and interrogative approach to logic are given still scanter notice in the argument literature.

512

There is nothing in Proposition 15.2e for me to take issue with; Hintikka's misalignment thesis is just my own misalignment thesis. But we must part company over Proposition 15.2e. Hintikka's *re*instantiation thesis is false[448]

Hintikka is not by any means the only logician of importance to hold that the modern and traditional conceptions of fallacyhood fail to converge. Hamblin, who is the modern reinvigorator of the fallacies project in logic, also questions the convergence claim. In chapter 8 of *Fallacies,* devoted to formal dialectic, Hamblin can be read as holding that, apart from the strictly formal fallacies, the traditional concept of fallacies provides that the traditional list is made up of violations of question-and-answer games. There are interesting differences between Hamblin and Hintikka, and between still others and them, which are well-worth our formal notice. But it is also true that if my answer stands against Hintikka it also stands against the others. So in the interest of space, let Hintikka speak for them all.

15.3 *Piracy*

Heresy is the view that since the EUAI conception is not the traditional notion of fallacy, the concept-list misalignment thesis is a trivial failure. Piracy takes a different tack. Pirates grant that the EAUI conception and the traditional conception are one. Piracy's point of departure is that the traditional concept is not worth our theoretical bother or, worse, doesn't even admit of successful theoretical articulation. Its suggested alternative is that we search out a non-traditional conception of fallacy, a conception that is not only interesting but both theory-worth and theory-possible. Piracy is the business of the present section.

Pragma-dialectics is a normatively idealized theory of how to conduct "critical discussions".[449] Critical discussions are a subclass of dialectical disputes, made distinctive by the rules by which they are governed. In a critical debate, the contending parties seek to overcome

---

[448] Not that Hintikka is convinced of this. See his (1997), which is a rejoinder to Woods and Hansen (1997).

[449] Pragma-dialecticians are not and do not represent themselves as logicians, although some of their sympathizers are logicians. Even so, the original stimulus for the PD approach was Else Barth's work in the logic of the articles, and dialectically structured dialogue logics. See Barth (1981) and Barth and Krabbe (1982), and citations therein. Barth's work, in turn was stimulated by the dialogue logics of Lorenzen (1961) and Lorenzen and Lorenz (1978).

their disagreement and reach a common position. Sometimes this is achieved when one of the contenders gives up his position and yields to his opponent. Sometimes this is achieved when both parties amend their respective views in ways that converge on a mutually acceptable compromise. In other cases, agreement is not possible, and the rules provide for an orderly recognition of the fact, and correspondently for the parties' honourable retirement. Much of the nature of critical discussion is reflected in its rules of "rational conduct". They require the parties to be fair, truthful, open-minded, cooperative, equable and reasonable. As I said earlier, critical discussion so conceived of, exemplifies the Goody Two-Shoes model of argument. The rules that enforce these constraints owe their normative force not to empirical compliance on the ground – for often the rules are violated in actual practice – but rather to their status as idealizations imposed by the theory. Van Eemeren and his co-authors write as follows:

> In order to clarify what is involved in viewing argumentative discourse as aimed at resolving a difference of opinion, the theoretical notion of a critical discussion is given shape in an idea model specifying the various stages, in the resolution process and the verbal moves that are instrumental in each of these stages. (van Eemeren *et al.* (1996, p. 280)

However,

> Even a discourse which is clearly argumentative will in many respects not correspond to the ideal model of a critical discussion. (p. 299, nt. 49)

Accordingly,

> In this model, the rules and regularities of *actual* discourse are brought together with *normative* principles of goal-directed discourse. The model of critical discussion is an abstraction, a theoretically motivated system for ideal resolution-oriented discourse. (p. 311; emphases added)

Anyone who finds himself persuaded by the admonitions of chapter 2 will not be entirely at one with the normatively idealized cast of the PD approach. The failures it attributes to empirical discomportment with its prescriptions place PD in the crowded company of those who purvey the irrationality and approximate rationality theses. I have already had my say about the sheer presumptuousness of the normative idealizers, and I will say no more on that score beyond observing van Eemeren's and

Grootendorst's membership in it.

PD presents us with other questions of interest. Does it work as a general theory of conflict resolution? Does it deliver the right account of the fallacies? On their heels comes a third question, a question about method. Is it possible to answer the fallacies question without first having to settle the general theory of conflict resolution question? The answer is yes; so let's get on with it.[450]

Pragma-dialecticians are pirates about fallacies. They have appropriated the name of fallacy and applied it to a concept significantly different from the traditional concept. Writes Grootendorst in "Some fallacies about fallacies" (1987):

> Fallacies just are any violation of the rules of conduct for rational discussants [in critical discussions]. (p. 337; see also van Eemeren and Grootendorst, 1984, p. 189)

In contrast to this, Hintikka is a heretic. He allows that the word names the traditional concept but disagrees with modern commentators about what the traditional concept actually is. In the section above, I conceded that there might be occasion to push Hintikka from heresy to piracy, and I shall now say the same thing about van Eemeren and Grootendorst, in the opposite direction, so to speak. These are not matters that call for, or would greatly reward, the sifting of the textual minutiae required to settle them. So I shan't press the heresy-piracy duo beyond its usage as a loosely identifying convenience. As for van Eemeren and Grootendorst, suffice it to say that there is some reason to think that they themselves see themselves in the slipstream of Aristotle under a broadly Hintikkian characterization of it. If so, they are heretics if Hintikka is. On the other hand, there is a very clear story to tell of the PD turn in fallacy theory. I make bold to say that were I myself inclined to adopt it, the following is the course I would set for myself:

1. The EAUI conception is indeed the traditional concept of fallacy and the gang of eighteen the traditional list.

---

[450] Apart from our focus on fallacies, the question of whether PD is a successful general account of conflict resolution arguments, even of that species of them collected under the title "critical discussion", is not a matter of indifference to me, and is an important matter for argument theory. My views on this question can be found in Woods (1991), which is reprinted with changes in Woods (2004). See also Woods (2006). A valuable collection of retrospective essays on the PD approach is Houtlosser and van Rees (2006).

2. As Hamblin and others have noted, twenty-four hundred years of effort have brought forth nothing resembling a stable, comprehensive or deep theory of fallacies.

3. A plausible explanation of this omission – in the spirit of the rejection of folk psychological concepts by eliminative materialists – is that the EAUI-concept-eighteen list duo is theory-resistant.

4. This presents us with two options. One is to drop the fallacies project as undischargeable. The other is to retain the project, but give up on the EAUI-eighteen duo.

5. Modern mainstream logic has, to Hamblin's chagrin, exercised the first option.

6. This leaves contemporary argumentation theory in a difficult situation. It wants to retain the fallacies project, but it can't get on with it, owing to the theory-resistance of the EAUI-eighteen duo.

7. Accordingly, there are further options. One is to give up on the eighteen, and to replace it with a different list. Another is to retain the list but give up on the EAUI concept. A third is to give up on them both.

8. Suppose that we retained the fallacies project but abandoned the EAUI concept. Then we would have retained the name of fallacy but not the traditional concept of it. Correspondingly, we would have applied the name to a different concept, to something other than the traditional concept. We would now be talking not about fallacies but about, so to speak, fallacies*.

If it were my call, I would avail myself of the provisions of (8). There is nothing in it against which to take principled offence. It happens all the time. Words change their meanings, often on the heels of unplanned changes in usage, but sometimes as a matter of a theorist's creative intent. Before the mathematical theory of classes, no class obeyed the extensionality axiom. Before the actions of various Canadian courts, marriage lacked the conceptual reach for Beulah and Betty together. Before quantum theory, no particle could tolerate the distortions of complementarity and wave-packet collapse. It is a truism that science is there to tell us about how particles behave, but it is also there to tell us what we should now take "particle" to mean; and rather routinely the two endeavours are inextricably bound up with one another. Simply put, quantum theory couldn't tell us how particles behave without adjusting the meaning of "particle". It was an important example of the notion that theoretical progress often demands conceptual change.[451] There is reason to think that the founders of the pragma-dialectical approach take the same kind of position with respect to fallacies. In "Some fallacies about fallacies", Grootendorst emphasizes that

> [f]allacies are not buttercups. Fallacies do not lead a life of their own independently of a properly articulated theory. Something is a fallacy only within a theory of fallacies. (p. 335)

I find in the construction I've just now put on the PD approach the wherewithal for a serious objection to my own position. My position is that the eighteen don't instantiate the EAUI conception. If this is right, and if the eighteen nevertheless strike us as erroneous in some distinctive sense – as fallacious *somehow* – then it is the job of a logic of erroneous reasoning to explain what makes them that way. On my own insistence, the fallaciousness of the eighteen cannot be that they are EAUI instantiations, for it is precisely this that my misalignment thesis denies. So, granted that they are *somehow* fallacies, the defectiveness of the eighteen must consist in their falling under a conception of fallacy other than the EAUI conception. Further to that, if all I've got to say about the eighteen is that they're not fallacies in that sense of the term, then, in the absence of a demonstration that there is no other sense of the term that they do instantiate, my nay-saying has an unsatisfying air of nit-picking. If my interpretation of PD intentions is right, then pragma-dialecticians are in a glorious position to repair this omission. "Look", we can imagine them saying, "we both agree, albeit for different reasons, that the eighteen aren't fallacies in the EAUI sense. As Woods himself has pointed out, that something is not a EAUI fallacy is no automatic vindication of it. This leaves it open that the eighteen are fallacies in some other sense both worthy of and responsive to theoretical attention. Pragma-dialectics provides that attention, and in so doing it supplies what Woods demands and the misalignment thesis fails to deliver on. It provides for the eighteen a principled account."

I am not in the least doubt that this is a serious challenge. If I am to rise to it, I will need to examine the discursive rules whose violations PD theorists identify as fallacies.

---

[451] For this and like cases, see Torretti (1990) and Thagard (1992).

15.4 *PD-misalignment.*

The rules whose violations are pragma-dialectical fallacies are set out in van Eemeren *et al.* (2002), pp. 182-183:

1. *The freedom rule.* Parties must not prevent each other from putting forward standpoints or casting doubt on standpoints.
2. *The burden-of-proof rule.* A party who puts forward a standpoint is obliged to defend it if asked to do so.
3. *The standpoint rule.* A party's attack on a standpoint must relate to the standpoint that has indeed been advanced by the other party.
4. *The relevance rule.* A party may defend his or her standpoint only by advancing argumentation relevant to that standpoint.
5. *The unexpressed premise rule.* A party may not falsely present something as a premiss that has been left unexpressed by the other party or deny a premiss that he or she has left implicit.
6. *The starting-point rule.* No party may falsely present a premise as an accepted starting point, or deny a premise representing an accepted starting point.
7. *The argument scheme rule.* A standpoint may not be regarded as conclusively defended if the defence does not take place by means of an appropriate argument scheme that is correctly applied.
8. *The validity rule.* The reasoning in the argumentation must be logically valid or must be capable of being made valid by making explicit one or more unexpressed premises.
9. *The closure rule.* A failed defence of a standpoint must result in the protagonist retracting the standpoint, and a successful defence of a standpoint must result in the antagonist retracting his or her doubts.
10. *The usage rule.* Parties may not use any formulations that are insufficiently clear or confusingly ambiguous, and they must interpret the formulations of the other party as carefully and accurately as possible.[452]

---

[452] These same rules, slightly re-expressed and expanded, can also be found in van Eemeren *et al.* (2009), pp. 20-24.

Pragma-dialecticians are radical pluralists about fallacies. There is *no* pre-theoretical fact of the matter about whether something is or is not a fallacy. Something is a fallacy only if some theory pronounces it so. The ten rules are sound. That is, they are sound in PD. How they fare elsewhere is the business of elsewhere. This is a well-intended pluralism, as well-intended as, in a quite general way, it is a widespread presence in contemporary logical theory.

Pragma-dialecticians have their own concept of fallacy. The contrast with the EAUI conception could hardly be more pronounced. There is nothing in the PD framework that tells us that its rules carry any far flung inducement to violate them; so violating them does not as such satisfy the universality and attractiveness conditions for EAUI fallacies. Neither, for the same reason, does the PD conception satisfy the incorrigibility condition. So, while the PD conception is different from the EAUI conception, PD's aggressive pluralism cancels the obligation to determine which of the two, if either, has got the concept of fallacy *right*. For there is no prior truth of the matter about how fallacies are to be conceived of.

The striking thing about the PD approach to fallacies is that whereas it shows no reluctance to forgo the traditional concept of fallacy, it hangs on tightly to the traditional *list*. Even though all violations of the rules are incorrect moves in a critical discussion,

> [it is necessary so to interpret these infelicities that they] "correspond roughly to the various kinds of defects *traditionally referred to as fallacies.*" (van Eemeren and Grootendorst, 1987, p. 284; emphasis added)

This is puzzling. It may be that part of the explanation of it is that, in the absence of the eighteen, the PD account of fallacies is a RR-rule violation account. Putting the eighteen in creates the illusion of depth. But it's not a depth on which PD theorists have yet to deliver. There are still further grounds for puzzlement. With the possible exception of begging the question (see below), which PD theorists may see as a violation of the unexpressed premiss rule, none of the eighteen is in any obvious way the violation of a PD rule. What PD rule does many questions violate? What does hasty generalization violate? PD apologists tend to select the argument-scheme rule, which forbids the employment of argument schemes "not correctly applied". If hasty generalization *is* a fallacy, no doubt it violates that rule. But it does so, no more and no less, than any of the other eighteen, since all of them presumably are − or can be modelled as – argument schemes not correctly applied. What now counts is *why* they are not correctly applied. The intuitive answer – namely, that it is because they are fallacies – is not available here. For nothing is a fallacy except that it is made so by a theory; and the present theory makes the eighteen fallacies solely by virtue of their violation of a PD rule. When the rule is the argument scheme rule, that the eighteen are not correctly applied cannot be because they violate the rule that forbids the not-correctly-applied property. So in the absence of an independent specification of the not-correctly-applied property, saying that something is a fallacy if it violates the argument scheme rule, is more hand-waving than elucidating. It is in the relevant sense empty. It is sleep's dormative virtue. Much the same must also be said about the fallacy of equivocation in relation to the usage rule, which says, in part, "Don't be confusingly ambiguous" or, in a slightly more formal way, "Don't commit the fallacy of ambiguity". More emptiness still.

My more general reservation is this. If things of kind K necessarily instantiate things of kind K*, then that something is a K-thing *implies* that it is a K*-thing. But, as we saw with Hintikka, there is no PD rule and no traditional fallacy for which any way of committing the fallacy is, of necessity, a violation of the rule. Nor is question begging an exception to this. Suppose that in a dispute about whether α is the case, Harry defends by saying that β. Suppose that Sarah replies that since β, it follows that ~α. It would be a rare traditionalist who failed to see this as question-begging, but it is clearly no violation of the unexpressed premiss rule. B is a premiss each expressly accepts. So, contrary to what PD theorists themselves appear to think,

Proposition 15.4
PD CONCEPT-LIST MISALIGNMENT: *The traditional* list *of the fallacies don't instantiate the PD* conception *of them.*

If Proposition 15.4 can be made to stand, there would be reason to say that pragma-dialecticians have betrayed their own pluralism, having honoured it with regard to how fallacyhood should be conceived of, but having got cold feet with regard to how, so conceived, it is instantiated.

I regard this as a tactical mistake, but far from fatal to the PD enterprise as such. If I were a PD theorist, I'd drop the eighteen like a hot potato. Given the expansive latitude of the PD conception there are many more ways of being fallacious than is accounted for by the gang of eighteen. There is nothing in the PD approach that requires the slightest loyalty to the eighteen. Why, then, bother with them?

My own answer to this question is a twofold one. Whether or not they instantiate the EAUI conception, I find the eighteen brimming with interest and wholly deserving of the attention of a logic of argument. If, as I believe, they don't instantiate the EAUI concept of fallacy, don't instantiate Hintikka's conception of fallacy, and don't instantiate the PD conception of fallacy, the idea that there is no theoretically respectable conception of fallacy for which the eighteen are indeed fallacies takes on a certain appeal. Also shown in a favourable light is our distant friend Proposition 1.3c, according to which not only are the eighteen not fallacies, they are in the main *cognitively virtuous* reasoning strategies. Even so, it is prudent to emphasize that neither thesis – not the misalignment thesis and not the cognitive virtue thesis – will stand much of a chance in the absence of close theoretical examination of the misalignment relation's relata.

Suppose, then, that the following could be shown: The various ways in which the eighteen fail the EAUI conception are ways that lend support to the cognitive virtue thesis. If this were so, knowing it would make a positive contribution to a logic of reasoning. It would reveal to us heretofore unnoticed ways in which reasoning is good. It would do so in a fashion that would excuse our appropriation of Hintikka's title. For it would reveal that the common belief among logicians that these are not good ways to reason was a misconception. *It was the fallacy of fallacies.*

Douglas Walton is another influential proponent of the pragmatically oriented dialectical approach to the fallacies. Like the PD theorists, he positions the analysis of the fallacies within a more general theory of argument, although the reach of Walton's approach to argument is taxonomically more extensive than that of van Eemeren and Grootendorst. As might be expected, there are a number of respects in which Walton's work has advantages not present in the structurally slighter programme of pragma-dialectics. But *au fond* Walton is a dialectician about fallacies. With respect to how fallacy is to be conceptualized, this makes him a pirate in the same sense, and to the same degree, that the PD theorists are. So, for the purposes of this book, it is unnecessary to submit the Walton *oeuvre* to independent and detailed scrutiny. It suffices to have the PD approach stand in for them all. Douglas Walton's output is enormous – indeed it is amazing. For those who lack the time to read it all, perhaps the most informative one-volume example of his treatment of the fallacies is Walton (1995).

15.5 *Emptiness*

The time has come to take the pulse of the cratering thesis. The cratering thesis asserts that sufficiently many of the gang of eighteen fail to instantiate the traditional concept of fallacy to make the concept-list misalignment thesis all but unanswerable. The misalignment thesis contradicts the instantiation thesis. The instantiation thesis holds that all the eighteen are in the extension of the eighteen. I have argued that indeed hardly any are. This I say craters the instantiation thesis and in so doing lends ample support to the concept-list misalignment thesis. I have tried to show that the cratering thesis is confirmed by removal of the following: *ad hominem, ad populum, ad verecundiam, ad ignorantiam,* affirming the consequent, denying the antecedent, begging the question (and circularity), many questions, hasty generalization, equivocation, gambler's fallacy and *post hoc, ergo propter hoc* (including false cause). I've thrown in for good measure the conjunction and base-rate fallacies, and have done so in a way that reflects some of my dissatisfaction with the biased statistics fallacy. That gives us a cratering ratio of fourteen to eighteen. Is this enough to sustain the claim that traditionalists have blown their own fallacies project by stacking their accounts with fallacies that can't be fallacies as they themselves conceive them to be? This I leave for the reader to judge.

As for the others, the four that I've not discussed here, I would say that the manner in which I've treated the fourteen give ample guidance about how I would handle the four that remain.[453]

So, then, my position is that if the eighteen were the only candidates for membership in the extension of the traditional concept, then the traditional concept of fallacy would be empty. This raises a necessary question. How plausible is it to suppose that the eighteen are indeed the only errors eligible for consideration as fallacies in the traditional sense? How plausible is it to suppose that the very idea of fallacy is indeed empty?

I retain the view that both components of the EAUI-eighteen duo are interesting enough to be theory-worthy, and not so intractable as to be theory-impossible. If this is right, then the two traditions have potentially rosy futures – the traditional conception and the traditional list. I have been suggesting just now that some of the rosy promise of the list of

---

[453] Concerning composition, it is encouraging to see in recent treatments some overlap with the kind of approach I've been proposing here. See, in particular, Finochiaro (2013b) and (2013c).

eighteen is its constructive role in the advancement of our cognitive agendas. But what of the traditional conception? Where does its promise lie? We might think that one thing that would kill outright the very idea of its theoretical promise would be not only the failure of the eighteen to instantiate it, but the failure of *anything* to instantiate it. If the EAUI concept were completely empty, there would be nothing whatever for an EAUI theorist to theorize about.

This is too fast, and a long way from true. If the emptiness of the EAUI concept could be demonstrated, that would tell us something important about premiss-conclusion reasoning. It would tell us that either we don't make errors in conclusion-drawing, or that none of the errors we do make is typical for us to commit, habitually committed with a notable occasioned frequency, and with notably high postdiagnostic rates of recidivism. And if *this* were true, at least some of the following would also be true.

1. In contexts of conclusion-drawing, error types are not uniformly distributed over the human species.
2. Such errors aren't committed with a notable frequency.
3. There is nothing type-habitual about error making. That is, all such errors once diagnosed are easy to avoid.

I am not much drawn to the romantic conceit that human reasoning is error-free, and that conclusion-drawing is pretty much tickety-boo, but I am not prepared simply to dismiss the possibility without ado. The other consequences of EAUI emptiness are not so easily dismissed either. They require some reflection. Let's say something about them now.

An empirically sensitive logic of conclusion-drawing is one that pays attention to the empirical record. It is one that eschews automatic or over-hasty condemnation of empirical disconformity with a theory's putative norms. It pays attention to the lawlike disclosures of the relevant empirical sciences. The first of these features bids the logician to take note of what actually happens and to avoid data-bending. The second bids him (within reason) to seek normative guidance from what typically happens on the ground. The third permits us idealizing assumptions – and to that extent distortions of the empirical record – but not to do so beyond the reach of what is required for an empirical theory's descriptive adequacy. There are tensions between and among these requirements, needless to say. A good logic of erroneous reasoning will be one that manages their harmonization in the best way. In particular, there will be some facts which can't be idealized away even for the sake of descriptive adequacy. Here, from chapter 3, is one of them: Beings like us make errors, lots of them.

Isn't the error abundance thesis a rather large cat among some rather timid pigeons? How can it be true that how we actually reason is by and large how we should reason? Even if errors of reasoning lag behind errors of fact, isn't it perfectly plain that we make errors of reasoning in hefty multiplicities? Or in Quine's words about another thing: Doesn't there exist a teeming prosperity of error-making? And if part of what my negative thesis rests on is that the errors of reasoning associated with the traditional fallacies aren't frequent enough to fulfill the universality condition, doesn't this conflict with Error Abundance?

Fallacy theorists have always seen fallacious reasoning as bunching into kinds. If our conclusion-drawing errors do indeed bunch into kinds, it is necessary to ask, "What kinds?" The Gang of Eighteen has been advanced by the tradition as an answer to that question. The question at issue is whether Error Abundance lends support to the eighteen in ways that undermine the negative thesis and, with it, the concept-list misalignment thesis. The answer is that it does not.

Here is why. The error abundance thesis could be true, the bunching, attractiveness and universality assumptions could be true, and the negative thesis could also be true. A standard way for the negative thesis to be true *and* to appear to be false is that, for every item on the list of eighteen there are occasions on which it is an error. But, as we have been trying to show, the occasions on which they *are* errors they are errors whose frequency doesn't begin to satisfy the universality condition. Even so, what Error Abundance may be telling us is, in effect, that it is implausible to suppose that the traditional concept of fallacy has a null extension. That is, unless we are prepared to accept that the errors humans commit with a frequency that amounts to abundance are random both with respect to types and commitors, then it must be the case that there are errors that it is characteristic of us all to commit in numbers that fulfill the frequency condition of Error Abundance. And that is tantamount to saying that for Error Abundance to be true, the EAUI conception must have instantiations. This being so, it won't do to throw up our hands on grounds that the eighteen aren't EAUI instantiations. If Error Abundance is true EAUI has instantiations, *never mind* that the eighteen aren't among them. Clearly it would be going too far to insist that it is logic's job to track them all down. But if a logic of error could establish the non-emptiness of the EAUI concept as a matter of principle, that would be a contribution of some significance to the errors of reasoning research programme, wherever it is in force.

15.6 *Beyond the eighteen*

To be theory-worthy the traditional EAUI conception requires a non-traditional list of instantiations. It is a rather striking thing that logic's mainstream literature shows little interest in tracking such instantiations down. Logicians should try to do better. Here is an example. It is, as we see, Powers Paradox.

1. If $\alpha \supset \beta$ then if α then β.
2. If, if α then β, then $\alpha \supset \beta$.
3. Therefore, $\alpha \supset \beta$ if and only if if α then β.

This is rather striking. (3) provides that implication and material implication are one and the same relation. This is problematic if true. If true, implication would be stricken to the paradox of material implication. Every falsehood really would imply every proposition whatever, and every truth really would be implied by them. This is implicational generosity on a scale that makes *ex falso* comparatively small beer. *Ex falso* is a thesis about *necessary* falsehood and *necessary* truth. It is a controversial thesis and may well be false. But when re-framed is a thesis about any kind of falsity and any kind of truth, there isn't the remotest chance of its being true. So (3) is a *reductio* of its premisses; at least one of them is untrue.

The conclusion of the paradox is unwelcome enough to make itself a *reductio* of its premisses. No one doubts the truth of (2). No conditional "if … then"-statement can withstand the joint truth of its "if"-part and falsity of its "then"-part. This makes (1) our culprit, and so it is. It looks right and isn't. It looks right because it looks like the propositional variant of *modus ponens*:

(1') If $\alpha \supset \beta$ and α, then β.

But what it looks like it isn't and what it is is wrong.

(1) passes the attractiveness test for EAUI fallacies. It also passes the universality test. Of course, most people won't have had a prior acquaintance with the connective $\supset$, and won't therefore have had occasion to assent to (1). Logic students in large culturally diverse universities are a different story. They have plenty of occasion to assent to (1) and do assent to it with a striking frequency when presented with Powers' Paradox (although not as so-named), and to do so in ways

unaffected by differences of sex or ethnicity or age.

Line (1) of the paradox has had an interesting career in the recent history of non-classical logics. It plays a pivotal role in an attempt by Stephen Read to convict the proof of *ex false quodlibet,* briefly entertained late in our chapter 1, of the fallacy of equivocation.[454] For years I have given my logic students[455] the occasion to trip over (1), and they do. After a decent interval, I ask them to assess Read's argument. With a striking frequency they fail to catch its error, the same error as we find at premiss (1) of the paradox. So the incorrigibility condition appears to have been met as well.

The EAUI conception is not empty. Powers' Paradox is a convincing instantiation of it at line (1). The Powers' Paradox fallacy is one of a type that it lies within the province of logic to detect. No doubt the detection of some of the others will fall rather more naturally to the empirical sciences of cognition. When it becomes aware of them, they too are cases for logic to take note of. But we should not desist from searching out other examples of EAUI's non-emptiness with the tools and perspectives with which logicians are more at home.

I have already had my say about Sleigh's fallacy, the conjunction fallacy, the gambler's fallacy and the base rate fallacy. I have said that they aren't fallacies in the traditional or EAUI sense. Still, they too may be further evidence of the non-emptiness of the traditional concept of fallacy, the very concept that virtually none of the traditional eighteen falls under.[456]

15.7 *Hamblin's Question*

We were barely into chapter 1 before the necessity arose to ask Hamblin's Question. Having noted logic's persistent failure to achieve the appropriately deep and reasonably stable theoretical grasp of the fallacies Hamblin asks for, what would explain this embarrassing omission? Hamblin responds to his own question with an heretical explanation. He thinks that modern logicians have misconceived the nature of logic. He thinks in the manner of Hintikka and Barth that they have lost sight of its intrinsically dialectical character. Since Hamblin

---

[454] For details, see Read (1988), pp. *22* ff.

[455] In Canada, Britain, the Netherlands and China.

[456] Although let's not forget that the cases which draw Sleigh's ire might well be rescued by the punctuational undetermination of utterances in the form "P and if P then Q, so necessarily Q."

also thinks that the traditional fallacies are themselves intrinsically dialectical errors, he concludes that they are destined to elude the theoretical grasp of the logics of the non-dialectical mainstream. So, lacking the theoretical means to plumb the eighteen's dialectical structures, the non-dialectical mainstream abandoned the fallacies project without note. They simply picked up their ball and left the field.

Mine is a different answer to Hamblin's Question. My answer is the concept-list misalignment thesis. It asserts that the modern orthodoxies have become fixated on the wrong list. It asserts that what makes the eighteen the wrong list is its failure to instantiate the traditional concept of fallacy, that is to say, the EAUI concept. In chapter 1 I proposed a standard of proof against which to assess the merits of my claim. I proposed that, in effect, that it would stand as a convincing answer to Hamblin's Question, if it could be shown that at least most of the eighteen as they typically occur they fail to instantiate the traditional concept. In that event, I said that although the traditional concept-list instantiation thesis might not be exhaustively confuted, it would at a minimum be a "cratered" thesis. For it would show at best the traditional eighteen fall mainly outside the extension of the concept of fallacy on its traditional interpretation.

In some significant degree, this book has been a contextually attentive survey of the eighteen with a view to showing, one by one, that they suffer the fate the misalignment thesis attributes to them, and that such cases are sufficiently numerous and sufficiently typical of the eighteen to sustain the cratered version of the thesis. Whether I have succeeded in this effort I leave to the reader to judge. But let's at least make it official:

Proposition 15.7
ANSWERING HAMBLIN: *The truth of the concept-list misalignment thesis in either its strong or cratering forms, suffices to answer Hamblin's Question.*

15.8 *Finocchiaro's question*

There was also a second question that made an early appearance. It is Finocchiaro's Question. Why, it asks, have mainstream logicians been unable to see this failure of fit between concept and its purported extension? I want now to suggest a three-part answer.

The first part pivots on what in earlier work I called the Fallacies Dilemma.[457] If fallacies are bit(s) of bad reasoning whose badness is inapparent, how is it possible to *exemplify* them? Successful exemplication requires that the person to whom the example is directed must recognize the reasoning as bad and yet not see it as bad. So, on the face of it at least, to see something as a fallacy is to see it as bad and not to see it as bad. But doesn't this make the fallacies impossible to recognize? Perhaps this is not a very strict impossibility, but it is not a matter for light-hearted dismissal. The problem occasioned by a fallacy's badness is at least as deep as the hiddeness of error in general. It is true that lots of errors are subject to easy detection after the fact. Errors of simple misinformation are a notable example of this. The error of fallacies is different. Error is an intrinsic feature of them. Fallacies are errors as such, and the error of a fallacy is hidden as such; that is to say, the inapparency of error is not removed or erased by subsequent detection. Nor is the persistent inapparency of error the only problem for fallacy recognition. If anything, the postdiagnostic recidivism condition poses an even tougher recognition challenge.

Suppose that a piece of reasoning is correctly seen to be erroneous. This satisfies only one of the five conditions required to identify something as a fallacy. An error of reasoning is not fallacious unless it is reasoning of a type that is attended by a notable frequency of postdiagnostic recurrence. To make this determination it is necessary to arrive at an accurate opinion of the postdiagnostic behaviour of those parts of the population at large for whom there is occasion to re-offend. However this is to be done, it is implausible to suppose that it can be brought about simply by dwelling on the sample at hand or even on the kind of reasoning it is. Consider again the fallacious line (1) of Powers' Paradox. It took me awhile to discover its frequency of recurrence. I discovered it because I was in a position to do so. I gave my students occasion to re-offend, and they did so with a notable frequency. This is important. It tells us that attributions of high levels of postdiagnostic recidivism require a certain investigative effort and, in at least some cases, an experimental twist. The same applies to the occasioned universality of fallacies. Whether an error of reasoning is one that the population at large commits or is disposed to commit with a notable

---

[457] Woods (2004), chapter 1. It first appeared in 1992 in the Proceedings of the 1990 International Conference of the International Society for the Study of Argument, held at the University of Amsterdam.

occasioned frequency is something that has to be looked into. It won't tell us what we want to know of it without some probing about.

Problematic as the inapparency of error might be for fallacy detection, the difficulties occasioned by the universality and postdiagnostic recidivism conditions are worse. We can now see why. They are conditions encumbered by a wide confirmational reach. Determinations of fulfillment carry an investigative onus that extends to occasions to commit in the population at large. Sometimes quite a lot is known to us individuals at a time about the behaviour of populations at large. When this is so, it is often a matter of, as we say, common knowledge. But it would be a large mistake to suppose that the reach of our knowledge here and now of the frequencies of reasoning errors and of their recomission in the population at large comes anywhere close to the *actual* frequencies of commission and recomission. As far as I am able to tell, vastly more logicians are susceptible to the Powers' Paradox fallacy than are alert to it, and hardly any at all have any idea of its postdiagnostic recidivistic frequencies. But make no mistake. Powers' Paradox embodies a fallacy, and the fallacy it embodies instantiates the EAUI concept of it.

Proposition 15.8a
DISCOVERED NOT FOUND: *Given the wide confirmation reach of their defining conditions, fallacies are not objects trouvées. Fallacies have to be discovered, not found.*

The second third of my answer to Finocchiaro has been with us since chapter 1. It need not detain us long. It is true that most of the eighteen fail to hit correctness standards traditionally imposed by the logical orthodoxies. With the possible exception of *petitio*, the eighteen are neither deductively valid nor inductively strong. They aren't truth preserving and they aren't probabilistically clinching. There are contexts in which these are the right standards to impose. Mathematical proofs require a validity standard, and the inductions of sample-to-population theoretical enquiry and the other forms of statistico-experimental science require the standard of inductive strength. But most human reasoning occurs in different contexts. Beings like us are situated in cognitive ecologies in which our cognitive agendas answer to our interests and circumstances, and our reasonings are proportionate to our wherewithal for successful outcomes with a frequency that signals an attendant adaptive advantage. Most reasoning is third way reasoning, and third way reasoning is reasoning for which neither validity nor inductive strength is an appropriate criterion of correctness.

Proposition 15.8b

WRONG STANDARDS: *Traditional logicians heavily over-invest in the validity and inductive strength standards, making the error rates in reasoning artificially high, to the point of a blanket rejection of third way reasoning.*[458]

The last third of my answer to Finocchiaro flows directly from these reflections and can be written as a corollary to Proposition 15.8b. Proposition 15.8b tells us that traditional logicians have conceptualized error in ungenerous ways. It shows their readiness to treat reasoning in the way that traditional epistemologists treat knowledge. Traditional epistemologists are content to conceptualize most of knowledge out of existence. Traditional logicians are content to conceptualize most correct reasoning out of existence. Traditional epistemologists violate the maxim of epistemological generosity. Traditional logicians are likewise disposed.

COROLLARY: *Traditional logicians are sceptics about good reasoning. They violate what we might call the maxim of logical generosity toward reasoning.*

Thus

Proposition 15.8c

ANSWERING FINOCCHIARO: *Propositions 15.8a and 15.8b, and the latter's corollary, are my answer to Finocchiaro's Question.*

Whether or not this is a convincing answer I leave to the tender mercies of Finocchiaro to judge, as well as those others of like curiosity.

15.9 *Woods' Question*

The answer to Finocchiaro's Question serves as part of my answer to Woods's Question. One of the reasons the post-Hamblin fallacies project has not achieved a deep and settled theoretical maturity

---

[458] Again, many questions and certain forms of *petitio* might be exceptions. In their modern forms this appears to be so. The trouble they occasion is different from invalidity and inductive weakness. But it is interesting to note that all five of Aristotle's ways of committing the *petitio* are indeed (syllogistically) invalid.

is over-attachment to the gang of eighteen. Whether the traditional or a more dialogically-oriented conception is in play, the gang of eighteen occasions a misalignment problem for it. A second part of the answer helps explain the first. What explains this lingering attachment to the traditional list? The reason is that fallacy theorists, I among them, have under-appreciated the importance of the Fallacies Dilemma. A piece of reasoning whose erroneousness is easy to spot is not likely to meet the conditions on fallacyhood, notably, the attractiveness condition. Fallacies are attractive, they invite wide-spread commission, because they are hard to spot. That being so, fallacy theorists tend to stick, *faute de mieux*, with the canonical list. What is more, notwithstanding some occasional disavowals, contemporary fallacy theorists oversubscribe to the validity and inductive strength standards. A fourth part of the answer is that contemporary theories tend to claim a normative authority which they fail to establish. The fifth part of the answer is a variation of a point advanced years ago in the Woods-Walton Approach. It was our view that the appropriate response to Hamblin's challenge would be, one by one, to appropriate a logic suitable to the particular fallacy up for analysis. That, I think, was the right sort of thing to propose. If you want to get the fallacies right, take care to select the right logics. My present view is that the various logics we invoked for this purpose – deductive logic, inductive logic, intuitionist logic, relatedness logic, epistemic logic, and dialogue logic, are not sufficiently refined for the total job at hand. The injunction, "Pick the right logic" is right. But by and large, the logics we chose, one by one, aren't quite right enough. So the last part of my answer to Woods's Question is that it has yet to occur to the theoretical mainstream that the best instrument for handling the fallacies is a *naturalized* logic. If I am not mistaken, we can sum up these five points as follows:

Proposition 15.9
ANSWERING WOODS: *We owe the thinness and theoretical unsettledness of contemporary work on fallacies to a general failure to heed the respect for data principle, and to appreciate the impact of the decision to naturalize the logic of thought.*

15.10 *What now?*

I said at the beginning that this would be an exploratory book, that it would raise many more questions than it would answer. It won't have taken readers long to appreciate how much of an understatement this was. How could it have been otherwise? The modern form of naturalized epistemology has been with us for just over forty years, and it too leaves more questions asked than answered. The naturalization of logic has barely got underway, and might well not last long. The jury is still out on epistemological naturalism. The jury on logical naturalism has yet to be empanelled. Perhaps no one will agree to serve.

Not only are fewer questions answered here than raised, most of my answers make negative rather than positive points. I am fairly certain that the eighteen aren't in the extension of the traditional concept. I am quite sure that validity and inductive strength are comparatively often the wrong standards to apply. I am satisfied that the modelling of premiss-conclusion reasoning as argument is the wrong way to go. I am as firm in my resolve to resist the normative presumptuousness of ideal modelling, as I am in resisting justificationism in epistemology.

On the positive side, I am more than satisfied that we owe our cognitive well-being to the smooth functioning of our cognitive devices, and that smooth functioning is that state that they are in as a matter of course. But I don't have much of an idea about what constitutes smooth functioning or of the connections between cognitive device and cognitive output. I take as given our proficiency with premiss-management, our facility in separating the relevant from irrelevant, the plausible from the implausible, and the projectible from the not. But I haven't much of an idea about the natural go of these things; nor am I alone in this. I have no doubt that we have more of a nose for good reasoning than for the screening out upon arrival of misinformation. But here too the causal and explanatory particulars escape me. Most of the work in the logic of reasoning has yet to be done. Most of what needs doing has yet to be formulated, never mind achieved.

The naturalistic turn in logic is a radical development. Its predecessor was a radical departure in epistemology. No less was required of it than the dethroning of its methods of conceptual analysis, most especially their reliance on the intuitions of practitioners. This has not proved an easy thing to do. This derives in part from the fact that conceptual analysis is what most of reformers were themselves trained for. A related consideration is that comparatively few of them had any first hand acquaintance with the empirical sciences they demanded philosophy's sensitivity to. These alienations have lightened of late. Whole squadrons of younger philosophers flood into alliances with their counterparts in cognitive and experimental psychology and the brain sciences.

It will prove a harder slog in logic, predicated on a key and pre-existing difference between epistemology and logic. Epistemology naturalized was analytic philosophy's first excursion into interdisciplinarity. Modern logic has had two successful such alliances. The first was the transformation of logic into a branch of theoretical mathematics. The second was a like make-over in theoretical computer science. These are the targets of logic's would be naturalizers. The toughness of the task speaks for itself. The problem here is not logic's unfamiliarity with cross-disciplinary retrofitting. The problem is that mathematics and theoretical computer science have long since achieved a methodological footing more deeply dug in than the semantic intuitions of old school epistemologists. It is true that in its heyday, the intuitions crowd was firmly seized of the irrelevance of empirical science. The same is true of mathematics and theoretical computer science. The difference is that the second of these claims, even if untrue, has a plausibility that gives to the first the appearance of smoke-blowing.

I have already had my say about the place (and value) of mathematical virtuosity in the theoretical treatment of normatively assessable kinds of human performance. Mathematized models are distortions on purpose. These distortions have to be paid for. There are three ways of doing this. The theory that contains the modelling must do well at the empirical checkout counter or it must pay its dues at the normative authority board. Or it must create new concepts with experimental intent, in the hopes that the time might come when thinking in these new ways influences further thinking that actually does pay off either empirically or normatively. Think again of Riemann's geometry. Anything short of this is the way of paradigm creep.

BIBLIOGRAPHY

Peter Aczel, *Non-Well-Founded Sets,* Stanford: CSLI Publications, 1988.

Jonathan Adler, "Transmitting knowledge", *Noûs,* 30 (1996), 99-111.

Jonathan Adler, *Belief's Own Ethics,* Cambridge MA: MIT Press, 2002.

Jonathan Adler, "Withdrawal and conteptualism", *Analysis,* 66 (2006), 280-285.

Jonathan Adler, "Introduction [to *Reasoning*]." In Jonathan E. Adler and Lance J. Rips, editors, *Reasoning: Studies of Human Inference and its Foundations,* pages 1-34. Cambridge: Cambridge University Press 2008.

R. Adolphs, D. Tranel, H. Damásio and A. Domásio, "Impaired recognition in facial expressions following bilateral damage to the amygdale", *Nature,* 372 (1994), 669-672.

Peter Adriaans and Johan van Benthem, editors, *Philosophy of Information.* A volume in *The Handbook of the Philosophy of Science,* editors, Dov Gabbay, Paul Thagard and John Woods. Amsterdam: North-Holland, 2008.

Carlos E. Alchousson, Peter Gärdenfors and David Makinson, "On the logic of theory change: Partial meet contraction and revision functions", *Journal of Symbolic Logic*, 50 (1985) 510-530.

Atocha Aliseda, *Abductive Reasoning: Logical Investigation into the Processes of Discovery and Evaluation.* Amsterdam: Springer, 2006.

William P. Alston, *Perceiving God,* Ithaca: Cornell University Press, 1991.

N. Ambady and R. Rosenthal, "Half a minute: Predicting teacher evaluations from thin slices of behaviour and physical attractiveness", *Journal of Personality and Social Psychology,* 64 (1993), 431-441.

A.R. Anderson and N.D. Belnap, Jr. *Entailment: The Logic of Relevance and Necessity,* volume 1. Princeton: Princeton University Press, 1975.

A.R. Anderson, and N.D. Belnap, Jr. and J.M. Dunn. *Entailment: The Logic of Relevance and Necessity,* volume 2. Princeton: Princeton University Press, 1992.

G.A. Antonelli, "Defeasible inheritance over cyclic networks", *Artificial Intelligence,* 92 (1997) 1-23.

The Complete Works of Aristotle, Jonathan Barnes, editor. Princeton: Princeton University Press, 1985.

D.M. Armstrong, *Belief, Truth and Knowledge.* Cambridge: Cambridge University Press,
1973.

D.M. Armstrong, *A World of States of Affairs.* Cambridge: Cambridge University Press,
1997.

Kenneth J. Arrow, *Social Choice and Individual Values,* 2nd edition,. New York: Wiley,
1963.

Sergei Artemov, "The logic of justification", *Review of Symbolic Logic,* 1 (2008), 477-513.

Nicholas Asher and Michael Morreau, "Commonsense entailment: A modal theory of nonmonotonic reasoning". In J. Molopoulos and R. Reiter, editors, *Proceedings of the Twelfth International Joint Conference on Artificial Intelligence,* pages 387-392. Los Altos: Morgan Kaufman, 1991.

Nicholas Asher, "Commonsense entailment: A conditional logic for some generics". In Gabriella Crocco, Luis Farinas del Cerro and A. Herzig, editors, *Conditionals: From Philosophy to Computer Science,* pages 103-145. Oxford: Oxford University press, 1995.

Nicholas Asher and Johan Kamp, "The knower's paradox and representational theories of attitudes", *Proceedings of the 1986 conference on theoretical aspects of reasoning about knowledge,* pages 131-147, San Francisco: Morgan Kaufman, 2006.

Fred Attneave, *Applications of Information Theory to Psychology: A Summary of Basic Concepts*. New York: Holt, 1959.

R. Audi, "The place of testimony in the fabric of justification and knowledge", *American Philosophical Quarterly*, 34 (1997), 405-422.

Randall E. Auxier and Lewis Edwin Hahn, editors, *The Philosophy of Jaakko Hintikka*, Library of Living Philosophers volume XXX. Chicago and La Salle: Open Court, 2006.

A.J. Ayer, *The Problem of Knowledge*, London: Macmillan, 1958.

Roberta Ballarin, "The interpretation of necessity and the necessity of interpretation", *Journal of Philosophy*, 101 (2004), 609-638.

Alexandric Baltag, Lawrence Moss and Slawomir Solecki, "The logic of common knowledge, public announcements and private suspicions" In *Proceedings of the Seventh Conference on Theoretical Aspects of Rationality and Knowledge*, pages 43-46. Morgan Kaufmann, 1998.

C. Baral, *Knowledge Representation, Reasoning and Declarative Problem Solving*. New York: Cambridge University Press, 2003.

J.A. Bargh and M.L. Ferguson, "Beyond behaviorism: On the automaticity of higher mental processes", *Psychological Bulletin*, 126 (2000), 925-945.

Yehoshua Bar-Hillel, *Language and Information: Selected Essays on Their Theory and Application*, Reading, MA: Addison-Wesley, 1964.

John A. Barker, "The fallacy of begging the question", *Dialogue*, 15 (1976), 241-255.

E.M. Barth, *The Logic of the Articles in Traditional Philosophy: A Contribution to the Study of Conceptual Structures*, second edition, Dordrecht: Kluwer, 1981.

Else M. Barth and Erik C.W. Krabbe, *From Axiom to Dialogue: A Philosophical Study of Logics and Argumentation*, Berlin and New York: de Gruyter, 1982.

Jon Barwise and Robin Cooper, "Generalized quantifiers and natural languages, *Linguistics and Philosophy,* 4 (1981), 159-219.

Jon Barwise and John Perry, *Situations and Attitudes,* Cambridge, MA: MIT Press, 1985.

Jon Barwise and L. Moss, *Vicious Circles,* Stanford: CSLI Publications, 1996.

Jon Barwise and Jerry Seligman, *Infiormation Flow: The Logic of Distributed Systems,* Cambridge: Cambridge University Press, 1997.

Robert Batterman, "Idealization and modelling", *Synthese,* 169 (2009), 9-63.

Mark E. Battersby, "Critical thinking as applied epistemology: Relocating critical thinking in the philosophical landscape", *Informal Logic,* 11 (1989), 91-100.

JC Beall & Greg Restall, *Logical Pluralism,* Oxford: Clarendon Press, 2006.

Isaiah Berlin, *The Roots of Romanticism: The A.W. Mellon Lectures in the Fine Arts,* 1965, edited by Henry Hardy, London: Chatto and Winders, 1999.

Francesco Berto, *How to Sell a Contradiction: The Logic and Metaphysics of Inconsistency,* London: College Publications, 2008.

Francesco Berto, *There's Something About Gödel,* Malden, MA: Wiley-Blackwell, 2009.

Leopoldo E. Bertossi, Anthony Hunter and Torsten Schaub, editors, *Inconsistency Tolerance,* Berlin/Heidelberg: Springer 2004.

Jean-Yves Beziau, Walter Carnielli and Dov Gabbay, editors, *Handbook of Paraconsistency,* London: College Publications, 2007.

Ken Binmore, *Rational Decisions,* Princeton: Princeton University Press, 2009.

538

M.H. Birbaum, "Base rates in Bayesian inference: Signal detection analysis of the Cab problem", *American Journal of Psychology*, 96 (1983), 85-94.

J. Biro and H. Siegel, "Epistemic normativity, argument and fallacies", *Argumentation*, 11, (1997), 277-292.

Michael A. Bishop and J.D. Trout, *Epistemology and the Psychology of Human Judgement*, Oxford: Oxford University Press, 2005.

Patrick Blackburn and Johan van Benthem, "Modal logic: A semantic perspective". In Patrick Blackburn, Johan van Benthem and Frank Wolter, editors, *Handbook of Modal Logic*, pages 1-84. Amsterdam: Elsevier, 2007.

J. Anthony Blair, *Groundwork in the Theory of Argumentation: Selected Papers of J. Anthony Blair*, Dordrecht: Springer, 2012.

Alexander Bochman, "A causal approach to nonmonotonic reasoning", *Artificial Intelligence*, 160, (2004) 105-143.

Alexander Bochman, "Nonmotonic reasoning". In Dov M. Gabbay and John Woods, editors, *The Many Valued and Nonmotonic Turn in Logic*, pages 557-632, volume 8 of the *Handbook of the History of Logic*. Amsterdam: North-Holland, 2007.

Gary Botterill and Peter Carruthers. *The Philosophy of Psychology*, Cambridge: Cambridge University press, 1999.

M.D.S. Braine and D.P. O'Brian, "A theory of if: A lexical entry, reasoning program, and pragmatic principles", *Psychological Review*, 98 (1991), 182-203.

Valentino Braitenberg, *Vehicles: Experiments in Synthetic Psychology*, Cambridge MA: MIT Press, 1984.

Robert Brandom, *Articulating Reason: An Introduction to Inferentialism*, Cambridge, MA: Harvard University Press, 2000.

M.E. Bratman, *Intention Plans and Practical Reason,* Cambridge, MA: Harvard University Press, 1987.

G. Brewka and K. Konolige, "An abductive framework for general logic programs and other nonmonotonic systems". In *Proceedings IJCAI,* pages 9-17, 1993.

Bryson Brown, "Preservationism: A short history". In *The Many Valued and Nonmonotonic Turn in Logic* volume 8 of the *Handbook of the History of Logic,* editors Dov M. Gabbay and John Woods, pages 95-128. Amsterdam: North-Holland, 2007.

P.D. Bruza, D.W. Song and R.M. McArthur, "Abduction in semantic space: Towards a logic of discovery", *Logic Journal of the IGPL,* 12, (2004), 97-110.

Peter D. Bruza and Richard Cole, "Quantum logic of semantic space: An exploratory investigation of contexteffects in practical reasoning". In Sergei Artemov, Howard Barringer, Artur d'Avila Garcez, Luis C. Lamb and John Woods, editors, *We Will Show Them! Essays in Honour of Dov Gabbay on his 60th Birthday,* volume 1, pages 339-361, London: College Publications, 2005.

Peter Bruza, Dominic Widdows and John Woods, "A quantum logic of down below". In Dov M. Gabbay and Kurt Engesser, editors, *Handbook of Quantum Logic*, pages 625-660. Amsterdam: Elsevier, 2007.

Tyler Burge, "Content preservation", *The Philosohical Review,* 102 (1993), 457-488.

Guido Calabresi, *The Cost of Accidents: A Legal and Economic Analysis.* New Haven: Yale University Press, 1970.

A.J. Calder, J. Keane, J. Manes, N. Antoun and A. W. Young, "Impaired recognition and experience of disjust following brain injury", *Nature Revies Neuroscience*, 3 (2000) 1077-1078.

Gregory N. Carlson and Francis Jeffry Pelletier, editors, *The Generic Book,* Chicago: Chicago University Press, 1995.

Rudolf Carnap, *The Logical Syntax of Language,* London: Routledge & Kegan Paul, 1939.

S. Carrere and J. M. Gottman, "Predicting divorce among newlyweds from the first three minutes of a marital conflict discussion", *Family Process,* 38 (1999) 293-301.

Nancy Cartwright, *How the Laws of Physics Lie,* New York: Clarendon, 1983.

G.J. Chaitin, "Gödel's theorem and information", *International Journal of Theoretical Physics,* 22 (1982), 941-954.

G.J. Chaitin, *Algorithmic Information Theory.* New York: Cambridge University Press, 1987.

N. Chater and M. Oaksford, "The probability heuristics model of syllogistic reasoning", *Cognitive Psychology,* 39 (1999) 191-258.

Brian Chellas, "Basic conditional logic", *Journal of Philosophical Logic,* 4 (1975), 133-154.

Chi-Lun Cheng and John W. Van Ness, *Statistical Regression with Measurement Error.* New York: Oxford University Press, 1999.

Christoher Cherniak, *Minimal Rationality,* Cambridge, MA: MIT Press, 1986.

Roderick M. Chisholm, *Perceiving: A Philosophical Study,* Ithaca: Cornell University Press, 1957.

David Christensen, *Putting Logic in Its Place: Formal Constraints on Rational Belief.* Oxford: Oxford University Press, 2004.

Alonzo Church, "An unsolvable problem of elementary number theory", *Annals of Mathematics,* 33 (1936), 346-366.

Paul Churchland, "Cognitive neurobiology: A computational hypothesis for laminar cortex", *Biology and Philosophy,* 1, (1985), 25-51.

Paul M. Churchland, *The Engine of Reason, The Seat of the Soul,* Cambridge, MA: MIT Press, 1995.

A. Clark, *Being There: Putting Brain, Body and World Together Again,* Cambridge, MA: MIT Press/Radford Books, 1997.

C.A. J. Coady, *Testimony: A Philosophical Study,* Oxford: Oxford University Press, 1992.

Ronald Coase, "The problem of social cost", *Journal of Law and Economics,* 3 (1960), 1-44.

Lorraine Code, *Epistemic Responsibility,* Cambridge, MA: Harvard University Press, 1987.

L. Jonathan Cohen, *The Probable and the Provable,* Oxford: Clarendon Press, 1977.

L. Jonathan Cohen, "Can human irrationality be experimentally demonstrated?", *The Behavioral and Brain Sciences,* 4 (1981), 317-331.

L. Jonathan Cohen "Are people programmed to commit fallacies: Further thoughts about the interpretation of experimental data and probability judgement", *Journal of Theory and Social Behavior,* 12 (1982), 251-274.

L. Jonathan Cohen, *The Dialogue of Reason: An Analysis of Analytical Philosophy.* Oxford: Clarendon Press, 1986.

Stewart Cohen, "Basic knowledge and the problem of easy knowledge", *Philosophy and Phenomenological Research,* 65 (2002), 309-329.

William S. Cooper, *The Evolution of Reason: Logic as a Branch of Biology,* Cambridge: Cambridge University Press, 2001.

John Corcoran, editor, *Ancient Logic and Its Modern Interpretation,* Dordrecht: Reidel, 1974.

Gary Cornell, J.H. Silverman and G. Stevens, editors, *Modular Forms and Fermat's Last Theorem,* New York: Springer, 1997.

Robert Cummins, "Reflections on reflective equilibrium". In Michael R. DePaul and William Ramsay, editors, *Rethinking Intuition: The Psychology of Intuition and its Role in Philosophical Inquiry*, pages 113-127. Lanham: Rowman and Littlefield, 1998.

Newton Da Costa and Steven French, *Science and Partial Truth: A Unitary Approach to Models and Scientific Reasoning,* New York: Oxford Univrsity press, 2003.

Robertson Davies, "The relevance and importance of the humanities in the present day". In John Woods and Harold Coward, editors, *Humanities in the Present Day,* pages 1-10, Waterloo, ON: Wilfred Laurier University Press, 1979.

A.S. d'Avila Garcez, K. Broda and Dov M. Gabbay, *Neural-Symbolic Learning Systems: Foundations and Applications.* Berlin: Springer-Verlag, 2002.

A.S. d'Avila Garcez and L.C. Lamb, "Reasoning about time and knowledge in neural symbolic learning systems". In S. Thrum and B. Schoelkoph, editors, *Advances in Neural Information Processing Systems 16: Proceedings of the NIPS 2003 Conference Vancouver BC.* Cambridge MA: MIT Press, 2004.

Artur d'Avila Garcez, Dov M. Gabbay, Oliver Ray and John Woods "Abductive reasoning in neural-symbolic systems", *Topoi,* 26, (2007), 37-49.

Artur d'Avila Garcez, Howard Barringer, Dov M. Gabbay and John Woods, *Neuro-fuzzy Argumentation Networks,* to appear.

Donald Davidson, "Theories of meaning and learnable languages". In Yehoshua Bar-Hillel, editor, *Proceedings of the Congress for Logic, Methodology and Philosophy of Science,* Amsterdam: North-Holland, 1965.

Augustus DeMorgan, *Formal Logic,* London: Taylor, 1847.

John Dewey, *The Later Works,* 17 volumes, Carbvadate: Southern Illinois Press, 1981-1991.

Zoltan Dienes and Josef Perner, "A theory of implicit and explicit knowledge", *Behavioral and Brain Sciences*, 22 (1999), 735-808.

Jon Doyle, "A truth maintenance system", *Artificial Intelligence,* 12 (1979), 231-272.

Fred Dretske, *Knowledge and the Flow of Information.* Cambridge, MA: MIT Press, 1981.

Fred Dretske, "Epistemology and information". In Pieter Adriaans and Johan van Benthem editors, *Philosophy of Information,* pages 29-47. A volume in the *Handbook of the Philosophy of Science,* edited by Dov M. Gabbay, Paul Thagard and John Woods. Amsterdam: North-Holland 2008.

R. Dukas, editor, *Cognitive Ecology: The Evolutionary Ecology of Information Processing and Decision Making.* Chicago: University of Chicago Press, 1998.

R. Dukas and J.M. Ratcliffe, editors, *Cognitive Ecology II.* Chicago: University of Chicago Press, 2009.

P.M. Dung, "An argumentation-theoretic foundation for logic programming", *Journal of Logic Programming,* 22, (1995), 151-177.

Graham Dunn, *Statistical Evaluation of Measurement Errors: Design and Analysis of Reliability Studies,* 2nd edition, New York: Oxford University Press, 2004.

J. Michael Dunn, "Intuitive semantics for first-degree entailments and 'coupled trees'", *Philosophical Studies,* 29 (1976), 149-168.

A. Edalat, "Dynamical systems, measures and fractals via domain theory", *Information and Computation,* 120 (1995), 32-48.

Gerald Edelman, *Second Nature: Brain Science and Human Knowledge,* New Haven: Yale University Press, 2006.

Ward Edwards, "The theory of decision making", *Psychological Bulletin,* 41 (1954), 380-417.

Ward Edwards, "Behavioral decision theory", *Annual Review of Psychology,* 67 (1961), 441-452.

W. Edwards, "Conservatism in human information processing". In *Formal Representation of Human Judgement,* edited by B. Kleinmuntz. New York: Wiley, 1968.

A.E. Eiben and J.E. Smith, *Introduction to Evolutionary Computing,* Berlin: Springer, 2010.

Brian Ellis, *Rational Belief Systems,* Totawa, NJ: Rowman and Littlefield, 1979.

Reneé Elio and Francis Jeffry Pelletier, "Belief change as a propositional attitude". In Jonathan E. Adler and Lance J. Riips, editors, *Reasoning: Studies of Human Inference and its Foundations,* pages 566-596. New York: Cambridge University Press, 2008.

D.W. Etherington, and R. Reiter, "On inheritance hierarchies and exceptions". In *Proceedings of the National Conference on Artificial Intelligence.* Los Altos, CA: Morgan Kaufmann, 1983.

J. St. B.T. Evans and D.E. Over. *Rationality and Reasoning,* Hove, UK: Psychology Press, 1996.

J.St. B.T. Evans, "Logic and human reasoning: An assessment of the deduction paradigm", *Psychological Bulletin,* 128 (2002), 978-996.

J. St B.T. Evans, *Hypothetical Thinking: Dual Processes in Reasoning and Judgement,* Hove and New York: Psychology Press, 2007.

S. Evnine, *Donald Davidson,* Stanford: Stanford University Press, 1991.

R. Fagin, J.D. Ullman and M.Y. Vardi, "On the semantics of updates in databases", *Proceedings of Second ACM SIG ACT-SIGMOND,* 1983, 352-365.

T.S. Ferguson, "Who solved the secretary problem?", *Statistical Science,* 4, pp. 282-296, 1989.

R.E. Fikes and N.J. Nilsson, "STRIPS: a new approach to the application of theorem proving to problem solving", *Artificial Intelligence*, 2, (1971), 189-208.

S. Fillenbaum, "Mind your P's and Q's: The role of content and context in some uses of and, or, and if". *Psychology of Learning and Motivation*, 11 (1977), 41-100.

Kit Fine, *Semantic Relationism*, Oxford: Blackwell, 2007.

Maurice Finocchiaro, *Galileo and the Art of Reasoning: Rhetorical Foundations of Logic and Scientific Method,* (Boston Studies in the Philosophy of Science, Vol. 61) Boston: D. Reidel, 1980.

Maurice Finocchiaro, "Fallacies and the evaluation of reasoning", *American Philosophical Quarterly,* 18, (1981), 13-22. Reprinted in Finocchiaro (2005) as chapter 6.

Maurice Finocchiaro, "Six types of fallaciouness: Toward a realistic theory of logical criticism", *Argumentation,* 1 (1987), 263-282.

Maurice Finocchiaro, *Arguments About Arguments: Systematic, Critical and Historical Essays in Logical Theory,* Cambridge: Cambridge University Press, 2005.

Maurice AFinocchiaro, *Meta-argumentation: An Approach to Logic and Argumentation Theory,* London: College Publications, 2013a.

Maurice Finocchiaro, "Debts, oligarchies and holisms: Deconstructing the fallacy of composition", *Informal Logic*, 33 (2013b), 143-174.

Maurice Finocchiaro, "The fallacy of composition and meta-argumentation", Proceedings of the OSSA Conference, Windsor, 2013c.

R.A. Fischer, "Statistical methods and scientific induction", *Journal of the Royal Statistical Society,* (1955) 69-78.

Peter A. Flach and Antonis C.Kakas, editors, *Abduction and Induction: Essays on Their Relation and Interpretation.* Dordrecht and Boston: Kluwer, 2000.

546

D. Floreano and C. Mattiussi, *Bio-inspired Artificial Intelligence: Theories, Methods and Technologies,* Cambridge, MA: MIT Press, 2008.

L. Floridi, editor, *The Blackwell Guide to the Philosophy of Computing and Information,* Oxford: Blackwell 2003.

Luciano Floridi, *Philosophy of Information and Information Ethics: Critical Reflections and the State of the Art.* A special issue of *Ethics and Information Technology,* Charles Ess, editor, 10 (September 2008).

Jerry Fodor and Zenon Pylyshyn, "Connectivism and cognitive architecture: A critical analysis", *Cognition,* 28 (1988), 1-2, 3-71.

Robert J. Fogelin, "The logic of deep disagreement", *Informal Logic,* 7 (1985), 1-8.

Richard Foley, "Egoism in epistemology". In Frederick F. Schmitt, editor, *Socializing Epistemology,* pages 55-73, Lanham, MD: Rowman and Littlefield, 1994.

Dagfinn Follesdal and Douglas Quine, editors, *Quine in Dialogue,* Cambridge, MA: Harvard University Press, 2008.

M. Forti and F. Honsell, "Set theory with free construction principles", *Annali Scuolo Normale Superiore di Pisa, Classe di Scienze,* 10 (1983), 493-522.

James Freeman, *Acceptable Premises: An Epistemic Approach to Informal Logic,* Cambridge: Cambridge University Press, 2005.

Gottlob Frege, *Begriffsschrift, eine der arithmetischen nechgebeldete Formelsprache des reinen Denkens.* Translation in Jean van Heijenoort, editor, *From Frege to Gödel,* Cambridge MA: Harvard University Press, 1967.

Tim French and Hans van Ditmarsch, "Undecidability for arbitrary public announcement logic", *Advances in Modal Logic,* 7 (2008), 23-42.

N. Friedman and J.Y. Halpern, "Plausibility measures: A user's guide", *Proc of the Eleventh Conference on Uncertainty in AI,* 1995, pp. 175-184.

N. Friedman and J.Y. Halpern, "Plausibility measures and default reasoning", *Journal of the ACM,* 48 (2001), 648-685.

André Fuhrmann, "Theories of belief change", in Sven Bernecker and Duncan Pritchard, editors, *The Routledge Companion to Epistemology,* pages 621-638, New York: Routledge, 2011.

Dov M. Gabbay "Theoretical foundations for non-monotonic reasoning in expert systems." In K.R. Apt, editor, *Logics and Models of Concurrent Systems,* pages 439-459. Berlin: Springer Verlag, 1985.

Dov M. Gabbay, "Conditional implications and non-monotonic question". In Gabriella Crocco, Luis Farinas del Cerro and A. Herzig, editors, *Conditionals: From Philosophy to Computer Science,* pages 337-359. Oxford: Oxford University Press, 1995.

Dov M. Gabbay, *Logic for AI and Information Technology,* London: College Publications, 2007.

Dov M. Gabbay, J. Christopher and J.A. Robinson, editors, *Handbook of Logic in Artificial Intelligence and Logic Programming,* volume 3, *Nonmonotonic Reasoning and Uncertain Reasoning,* Oxford: Oxford University Press, 1994.

D.M. Gabbay, R.H. Johnson, H.J. Ohlbach and J. Woods, *Handbook of the Logic of Argument and Inference,* Amsterdam: North-Holland, 2002.

Dov M. Gabbay, Odinaldo Rodrigues and John Woods, "Belief contraction, anti-formulae and resource overdraft: Part I Deletion in resource bound logics", *Logic Journal of IGPL* 10 (2002), 601-652.

Dov M. Gabbay, Odinaldo Rodrigues and John Woods "Belief contraction, anti formulae and resource overdraft: Part II". In Shahid Rahman, J.M. Torres, J.P. van Bendegem and Dov M. Gabbay, editors, *Logic, Epistemology and the Unity of Science*, pages 291-326, Dordrecht and Boston: Kluwer, 2004.

Dov M. Gabbay and John Woods, "The new logic", *Logic Journal of the IGPL,* 9 (2001a), 157-190.

Dov M. Gabbay and John Woods, "Non-cooperation in dialogue logic: Getting beyond the Goody Two-Shoes model of argument", *Synthese,* 127 (2001b), 161-186.

Dov M. Gabbay and John Woods, *Agenda Relevance: A Study in Formal Pragmatics,* vol. 1 of *A Practical Logic of Cognitive Systems,* Amsterdam: North Holland, 2003

Dov M. Gabbay and John Woods, *The Reach of Abduction: Insight and Trial,* volume 2 of *A Practical logic of Cognitive Systems,* Amsterdam: North Holland, 2005.

David Gale and F.M. Stewart, "Infinite games with perfect information", in H.E. Kuhn and A.W. Tucker, editors, *Contributions to the Theory of Games II,* pages 245-286, Princeton: Princeton University Press, 1953.

Peter Gärdenfors, "Epistemic importance and minimal changes of belief", *Australasian Journal of Philosophy,* 62 (1984), 137-157.

Peter Gärdenfors, *Knowledge in Flux: Modeling the Dynamics of Epistemic States.* Cambridge, MA: MIT Press, 1988.

Peter Gärdenfors and David Makinson, "Revisions of knowledge systems using epistemic entrenchment." In *Proceedings of the Second Conference on Theoretical Aspects of Reasoning about Knowledge,* pages 83-95. Los Altos: Morgan Kaufman, 1988.

Peter Gärdenfors and David Makinson, "Nonmonotonic inferences, based on expectations", *Artificial Intelligence,* 65 (1994), 197-245.

Jelle Gebrandy, *Bismulations on Planet Kripke,* PhD thesis, ILLC, University of Amsterdam, 1999.

H. Geffner, *Default Reasoning: Causal and Conditional Theories.* Cambridge, MA: MIT Press, 1992.

Edmund L. Gettier, "Is justified true belief knowledge?", *Analysis,* 23 (1963), 121-123.

G. Gigerenzer, "From tools to theories". In Carl Graumann and Kenneth J. Gergen, editors, *Historical Dimensions of Psychological Discourse,* pages. 336-359, Cambridge: Cambridge University Press 1996.

G. Gigerenzer, P.M. Todd & the ABC Group, *Simple Heuristics That Makes Us Smart,* New York: Oxford University Press, 1999.

G. Gigerenzer, *Adaptive Thinking: Rationality in a Real World,* Oxford: Oxford University Press, 2000.

G. Gigerenzer and R. Selten, *Bounded Rationality: The Adaptive Toolbox,* Cambridge, MA: MIT Press, 2001.

Gerd Gigerenzer, *Reckoning with Risk: Learning to Live with Uncertainty,* London: Penguin Books, 2002.

Gerd Gigerenzer, "I think, therefore I err", *Social Research,* 1 (2005), 195-217.

Gerd Gigerenzer, *Gut Feeling,* New York: Viking USA, 2007.

J.P. Gilbert and F. Mosteller, "Recognizing the maximum of a sequence", *American Statistical Association Journal,* 61 (1966), 35-73.

M. Ginsberg, *Essentials of Artificial Intelligence,* Waltham, MA: Morgan Kaufmann, 1993.

Jean-Yves Girard, "Linear logic", *Theoretical Computer Science,* 50 (1987), 1-102.

Clark Glymour and David Danks, "Reasons as causes in Bayesian epistemology". *Journal of Philosophy,* 104 (2007), 464-474.

Kurt Gödel, "On formally undecidable propositions of *Principia Mathematica* and related systems "I", *Montashefte für Mathematik und Physik,* 38 (1931), 173-198.

Kurt Gödel, "Russell's mathematical logic", in Paul A. Schilpp, editor, *The Philosophy of Bertrand Russell,* pages 123-153, Evanston, IL: Northwestern University Press, 1944.

Kurt Gödel, "What is Cantor's continuum problem?", *American Mathematical Monthly,* 54 (1947), 515-525.

Peter Godfrey-Smith, "Environmental complexecity and the evolution of cognition". In R. Sternberg and J. Kaufman, editors, *The Evolution of Intelligence,* pages 233-249. Mahwah: Lawrence Erlbaum, 2002.

Peter Godfrey-Smith, "Untangling the evolution of mental representation". In António Zihâo, editor, *Evolution, Rationality and Cognition: A Cognitive Science for the Twenty-First Century,* pages 85-102. London: Routledge, 2005.

Alvin I. Goldman, "A causal theory of knowing", *The Journal of Philosophy,* 64, 12, (1967), 357-372.

Alvin I. Goldman, "What is justified belief?" in George Pappas, editor, *Justification and Knowledge,* pages 1-23, Boston: D. Reidel, 1979

Alvin I. Goldman, *Epistemology and Cognition.* Cambridge, MA: Harvard UniversityPress, 1986.

Alvin I. Goldman, "Epistemic folkways and scientific epistemology". In *Goldman, Liasons: Philosophy Meets the Cognitive and Social Sciences.* Cambridge, MA: MIT Press, 1992.

Alvin I. Goldman and C.S. Sripada, "Simulationist models of face-based emotion recognition", *Cognition,* 94, (2005),193-213.

Alvin I. Goldman, S*imulating Minds: The Philosophy, Psychology and Neuroscience of Mindreading,* Oxford: Oxford University Press, 2006.

Nelson Goodman, *Fact, Fiction and Forecast,* Cambridge, MA: Harvard University Press 1983/first published in 1954.

Trudy Govier, "Trust and testimony: Nine arguments on testimonial knowledge", *International Journal of Moral and Social Studies,* 8 (1993), 21-39.

John Greco, "Agent reliabilism". In J. Tomberlin, editor, *Philosophical Perspectives 13: Epistemology.* Atascadero: Ridgeview, 1999.

John Greco, *Achieving Virtue: A Virtue-Theoretic Account of Epistemic Normativity,* New York: Cambridge University Press, 2010.

J.D. Greene, *The Moral Brain and How to Use It,* New York: Penguin, 2011.

Nicholas Griffin, *Russell's Idealist Apprenticeship,* Oxford: Clarendon Press, 1991.

R.A. Griggs and J.R. Cox, "The elusive thematic-materials effect in Wason's selection task", *British Journal of Experimental Psychology,* 73 (1982), 407-420.

Z. Griliches, "Errors in variables and other unobservables", *Econometrics,* 42, pp. 971-998, 1974.

P.D. Grünwald and P.M.B. Vitányi, *Complexity and Information Theory,* F. Emmert-Streib and M. Dehwer, eds. Springer-Verlag, 2008.

Marcello Guarini, "A defence of connectionism against the SYNTACTIC argument", *Synthese,* 128, (2001), 287-317.

Jürgen Habermas, *Knowledge and Human Interests,* J.J. Shapiro, translator, Boston: Beacon, 1971. First issued in German in 1968.

Ulrike Hahn, M. Oaksford and H. Bayinder, "How convinced should we be by negative evidence?". In B. Bara, L. Barsalou and M. Bucciarelli, editors, *Proceedings of the 27th Annual Conference of the Cognitive Science Society,* pages 887-892. Mahwah, NJ: Erlbaum, 2005.

Ulrike Hahn and Mike Oaksford, "A Bayesian approach to informal argument fallacies", *Synthese,* 152 (2006a), 207-236.

Ulrike Hahn and Mike Oaksford, "A normative theory of argument strength: Why might one want one and why might one want to be Bayesian?" *Informal Logic,* 26 (2006b), 1-24.

Ulrike Hahn and Mike Oaksford, "The rationality of information, argumentation: A Bayesian approach to reasoning fallacies", *Psychological Review,* 114 (2007a), 704-732.

Ulrike Hahn and Mike Oaksford, "The burden of proof and its role in argumentation", *Argumentation*, 21 (2007b), 39-61.

Charles L. Hamblin, *Fallacies,* London: Methuen, 1970.

Robert Hanna, *Rationality and Logic.* Cambridge, MA: MIT Press, 2006.

Hans V. Hansen and Robert Pinto, *Fallacies: Classical and Contemporary Readings,* University Park, PA: The Pennsylvania State University Press, 1995.

Hans V. Hansen, "The straw thing of fallacy theory", *Argumentation,* (2002), 133-155.

N.R. Hanson, *Patterns of Discovery,* Cambridge: Cambridge University Press, 1958.

Sven Ove Hansson, "Economic (is) rationality in risk analysis", *Economics and Philosophy,* 22 (2006), 231-241.

John Hardwig, "Epistemic dependence", *Journal of Philosophy,* 82 (1985), 335-349.

John Hardwig, "The role of trust in knowledge", *Journal of Philosophy* 88 (1991), 693-708.

D. Harel, "Dynamic logic". In D.M. Gabbay and F. Guenthner, editors, *Handbook of Philosophical Logic, Volume 11: Extensions of Classical Logic,* pages. 497-604. Dordrecht: Reidel, 1984.

Gilbert H. Harman, "The inference to the best explanation", *Philosophical Review,* 74 (1965), 88-95.

Gilbert Harman "Induction". In Marshall Swain, editor, *Induction, Acceptance and Rational Belief,* Dordrecht: Reidel, 1970.

Gilbert Harman, *Thought.* Princeton: Princeton University Press, 1973.

Gilbert Harman, *Change in View: Principles of Reasoning,* Cambridge, MA: MIT Press, 1986.

Gilbert Harman and Sanjeev Kulkarni, *Reliable Reasoning: Induction and Statistical Learning Theory*, Cambridge, MA: MIT Press, 2007.

William Harper, "Rational conceptual change", *PSA 1976*, 462-494.

J.C. Harsanyi, "Games with incomplete information played by 'Bayesian' players I. The basic model", *Management Sciences,* 14 (1967), 159-182.

H.L.A. Hart, "The ascription of responsibility and rights", *Proceedings of the Aristotelian Society,* 49 (1949), 171-194.

R.V.L. Hartley, "Transmission of information", *Bell Technical Journal,* 7 (1928), 535-563.

R. Hassin, J. Uleman and J. Bargh, editors, *The New Unvonscious*, New York: Oxford University Press, 2005.

Vincent F. Hendricks, *Mainstream and Formal Epistemology*, New York: Cambridge University Press, 2006.

Leon Henkin, "Some remarks on infinitely long formulas", in *Infinistic Methods,* editor not named, pages 167-183, Oxford: Pergamon Press, 1961.

David Hilbert and Paul Bernays, *Grundlagen der Arithmetik*, Berlin: Springer, 1939.

David Hilbert, "On the foundations of logic and arithmetic", in Jean van Heijenoort, editor, *From Frege to Gödel: A Source Book in Mathematical Logic,* pages 129-138, Cambridge, MA: Harvard University Press, 1967. First published in 1904.

Jaakko Hintikka, "Modality as referential multiplicity", *Ajatus,* 20 (1957), 27-47.

Jaakko Hintikka, *Knowledge and Belief,* Ithaca, NY: Cornell University Press, 1962.

Hintikka, Jaakko. 1968. "Language games for quantifiers". In Nicholas Rescher, ed, *Studies in Logical Theory,* pages 46-72, Oxford: Blackwell.

Jaakko Hintikka, "Two studies on probability". In *Reports from the Institute of Philosophy,* University of Helsinki, 1970.

Jaakko Hintikka, *Logic, Language-games and Information,* Oxford: Clarendon Press, 1973.

Jaakko Hintikka, "Quantifiers in logic and quantifiers in natural language". In Stephan Körner, editor, *Philosophy of Logic Proceedings of the 1974 Bristol Colloquium,* pages 208-232, Oxford: Blackwell, 1976.

Jaakko Hintikka and Ilkka Niiniluoto, "An axiomatic foundation for the logic of inductive generalization." In M. Prztecki *et al.,* editors, *Formal Methods in the Methodology of Empirical Sciences,* pages 57-81. Dordrecht: Reidel, 1976.

Jaakko Hintikka, "The fallacy of fallacies", *Argumentation,* 1, (1987), 211-238.

Jaakko Hintikka, "The role of logic in argumentation", *Monist,* 72, (1989), 3-24.

Jaakko Hintikka, "What is elementary logic? Independence-Friendly logic as the true core area of logic". In Kostas Gavroglu, John Stachel and Mary W. Wartofsky, editors, *Physics, Philosophy and the Scientific Community,* pages 301-326, Dordrecht: Kluwer, 1995.

Jaakko Hintikka, *The Principles of Mathematics Revisited,* Cambridge: Cambridge University Press, 1996.

Jaakko Hintikka, "What was Aristotle doing in his early logic, anyway? A reply to Woods and Hansen", *Synthese,* 113, (1997), 241-249.

Jaakko Hintikka, "Truth definitions, Skolem functions and axiomatic set theory", *Bulletin of Symbolic Logic,* 4 (1998), 303-337.

Jaakko Hintikka, "Is the axiom of choice a logical or set-theoretical principle?" *Dialectica* 53 (1999), 283-290.

Jaakko Hintikka, *Socratic Epistemology: Explorations of Knowledge-Seeking by Questioning,* Cambridge: Cambridge University Press, 2007.

M. Hirt and W. Pithers, "Selective attention and levels of coding in schizophrenia", *British Journal of Clinical Psychology,* 30 (1991), 139-149.

David Hitchcock, "Fallacies and formal logic in Aristotle", *History and Philosophy of Logic,* 21 (2000), 207-221.

David Hitchcock, "Is ther an *argumentum ad hominem* fallacy?" In Hans V. Hansen and Robert C. Pinto, editors, *Reason Reclaimed,* pages 187-199. Newport News, VA: Vale Press, 2007.

Thomas Hobbes, *Leviathan,* edited by John Plamenatz, Glasgow: Fontana, 1962. First published in 1651.

T. Horgan and J. Tienson, "Settling into a new paradigm", *Southern Journal of Philosophy,* 26, (1988), 97-113.

J.F. Horty, "Skepticism and floating conclusions", *Artificial Intelligence Journal,* 135 (2002), 55-72.

J.F. Horty, R.H. Thomason and D.S. Touretsky, "A sceptical theory of inheritance in nonmonotonic semantic networks", *Artificial Intelligence,* 42 (1990), 311-348.

J.F. Horty, "Some direct theories of nonmonotonic inheritance". In D. Gabbay, C. Hogger and J. Robinson, editors *Handbook of Logic in Artificial Intelligence and Logic Programming,* volume 3, pages 111-187. Oxford and New York: Oxford University Press, 1994.

J.F. Horty, "Nonmonotonic logic". In L. Goble, editor, *The Blackwell Guide to Philosophical Logic*, pages 336-361. Oxford: Blackwell, 2001.

J.F. Horty, *Reasons as Defaults,* New York: Oxford University Press, 2012.

Peter Houtlosser and Agnes van Rees, editors, *Considering Pragmadialects: A Festschrift for Frans H. van Eemeren on the Occasion of his 60th Birthday,* Mahwah, NJ: Erlbaum, 2006.

W.A. Howard, "The formulae-as-type notion of construction in 1969". In Jonathan Selden and Roger Hindley, editors, *To H.B. Curry: Essays in Combinatory Logic, Lambda Calculus and Formalism,* pages 479-490. New York: Academic Press, 1980.

P. Husbands and J.A. Meyer, *Volumtary Robotics: Proceedings of the First European Workshop, Evo Robot 98.* Berlin: Springer, 1998.

Edwin Hutchins, *Cognition in the Wild,* Cambridge, MA: MIT Press, 1995.

Daniel D. Hutto, *Folk Psychological Narratives: The Sociological Basis of Understanding Reasons,* Cambridge, MA: MIT Press, 2008.

Ray Jackendoff, "Toward an explanatory semantic interpretation", *Linguistic Inquiry,* 7 (1976), 89-150.

Dale Jacquette, "Review of Dov Gabbay and John Woods *Agenda Relevance: A Study in Formal Pragmatics* (A Practical Logic of Cognitive Systems Volume 1)" *Studia Logica* 77 (2004) 133-139.

Dale Jacquette, editor, *Philosophy, Psychology and Psychologism,* paperback edition, Berlin: Springer, 2010a.

Dale Jacquette, "Circularity or lacunae in Tarski's truth-schemata", *Journal of Logic, Language and Information,* 19 (2010b), 315-326.

Mark Jary, *Assertion,* London: Palgrave Macmillan, 2010.

S. Jaśkowski, "On the rules of suppositions in formal logic". In Storrs McColl, editor, *Polish Logic: 1920-1939,* Oxford: Oxford University Press, 1967. First published in *Studia Logica,* 1 (1934).

L.R. Jauch and W.F. Glueck, *Business Policy and Strategic Management.* New York: Macmillan 5th edition, 1982.

Richard Jeffrey, "Bayesianism with a human face", in John Earman, editor, *Testing Scientific Theories,* pages 133-156, Minneapolis: University of Minnesota Press, 1984.

M.K. Johnson and J.A. Reeder, "Consciousness as meta-processing". In J.D. Cohen and J.W. Schooler, editors, *Scientific Approaches to Consciousness,* pages 261-293. Mahwah, NJ: Erlbaum, 1997.

Ralph H. Johnson and J.A. Blair, "The recent development of informal logic". In J.A. Blair and Ralph H. Johnson, editors, *Informal Logic,* pages 3-28, Inverness, CA: Edgepress, 1980.

Ralph H. Johnson, "Logic naturalized: Recovering a tradition". In F. H. van Eemeren, R. Grootendorst, J.A. Blair and C. A. Willard, eds, *Argumentation: Across the Lines of Discipline,* pp. 47-56, Foris, 1987.

Ralph H. Johnson, *The Rise of Informal Logic,* Newport News, VA: Vale Press, 1996.

Ralph H. Johnson, *Manifest Rationality,* Mahwah, NJ: Erlbaum, 2000.

P.N. Johnson-Laird and R.M. J. Byrne, *Deduction: Essays in Cognitive Psychology.* Mahwah, NJ: Erlbaum, 1991.

P.N. Johnson-Laird, R.M.J. Byrne and W. Schaeken, "Propositional reasoning by Model", *Psychological Review,* 99 (1992), 418-439.

Henry W. Johnstone Jr. "Persons and selves", *Philosophy and Phenomenological Research* 28 (1967), 205-212.

H.W.B. Joseph, *An Introduction to Logic,* 2nd revised edition, Oxford: Clarendon Press, 1916. First edition 1906.

Michael Jubien, *Possibility,* New York: Oxford University Press, 2009.

D. Kahneman and A. Tversky, "Prospect theory: An analysis of decisions under risk", *Econometrica,* 4 (1979), 263-291.

D. Kahneman, P. Slovic and A. Tversky, *Judgment Under Uncertainty: Heuristics and Biases,* Cambridge: Cambridge University Press, 1982.

D. Kahneman and A. Tversky, "On the reality of cognitive illusions: A reply to Gigerenzer's critique", *Psychological Review,* 103 (1996), 582-591.

D. Kahneman and A. Tversky, editors, *Choices, Values and Frames*, New York: Cambridge, 2000.

A. Kakas, R.A. Kowalski and F. Toni, "Abductive logic programming", *Journal of Logic and Computation*, 2 (1995), 719-770.

Kant, Immanuel. *Inquiry Concerning the Distinctness of Principles of Natural Theology and Morality, and Logic,* Indianapolis, Ind: Bobbs-Merrill, 1974. First published in 1764.

Kant, Immanuel, *Logic,* Indianapolis, Ind: Bobbs-Merrill, 1974. First published in 1800.

Jerrold Katz, *Semantic Theory,* New York: Harper Row, 1972.

H.H. Kelley, "Attribution theory in social psychology". In D. Levine, editor, *Nebraska Symposium on Motivation,* Lincoln: University of Nebraska Press, 1967.

Ruth Kempson, Greg Restall and John Woods, *Relevance*, a guest-edited number of the *Journal of Logic and Computation*, 22 (2012).

Philip Kitcher, "The division of cognitive labor", *Journal of Philosophy*, 87 (1990), 5-22.

Philip Kitcher, *The Advancement of Science,* New York: Oxford University Press, 1993.

Joshua Knobe and Shaun Nichols, editors, *Experimental Philosophy,* New York: Oxford University Press 2008.

Hanna Kokko, *Modelling for Field Biologists (And Other Interesting People),* Cambridge: Cambridge University Press, 2007.

Barteld Kooi, "Expressivity and completeness for public update logic via reduction axioms", *Journal of Applied Non-Classical Logics,* 17 (2007), 231-253.

Hilary Kornblith, *Naturalizing Epistemology, 2nd edition.* Cambridge, MA: MIT Press, 1994.

Robert A. Kowalski, *Logic for Problem Solving,* Amsterdam and New York: Elsevier, 1979.

Erik C.W. Krabbe, Renee Jose Dalitz and Pier A. Smit, editors, *Empirical Logic and Public Debate: Essays in Honour of Else M. Barth,* Amsterdam: Editions Rodopi, 1993.

Erik C.W. Krabbe, "Metadialogues". In Frans H. van Eemeren, J. Anthony Blair, Charles A. Willard and A. Francisca Snoeck Henkemans, editors, *Anyone Who Has a View: Theoretical Contributions to the Study of Argumentation,* pages 83-90. Dordrecht and Boston: Kluwer 2003.

S. Kraus, D. Lehmann and M. Magidor, "Nonmonotonic reasoning, preferential models and cumulative logics", *Artificial Intelligence,* 44 (1990), 167-207.

A. Krautz, "The logic of Persistence", *Proceedings of the Fifth National Conference on Artificial Intelligence,* pages 401-405, Philadelphia, PA: AAAI-86, 1986.

Deanna Kuhn, *The Skills of Argument,* Cambridge: Cambridge University Press, 1991.

Manfred Krifka, Francis Jeffry Pelletier, Gergory N. Carlson, Alice ter Meulen, Godehard Link and Germano Chierchia, "Genericity: An introduction". In Gregory N. Carlson and Francis Jeffry Pelletier, editors, *The Generic Book,* pages 1-124, Chicago, IL: The University of Chicago Press, 1995.

Theo Kuipers, "Abduction aiming at empirical progress or event truth approximation leading to a challenge for computational modeling". *Foundations of Science,* 4 (1999) 307-323.

A. Kurz, "Coalgebras and their logics", *ACM SIGAT News,* 37 (2006), 57-77.

G. Lakemeyer and H. Levesque, "Evaluation-based reasoning with disjunctive information in first-order knowledge bases", *Proceedings of the KR-2002 Conference,* Toulouse, France, 2002.

George Lakoff, "On generative semantics". In J.D. McCawley, editor, *Semantics: An Interdisciplinary Reader in Philosophy, Linguistics and Psychology*, pages 232-296, Cambridge: Cambridge University Press, 1971.

C.H. Langford, "The notion of analysis in Moore's philosophy. In Paul Schilpp, editor, *The Philosophy of G.E. Moore*, pages 321-342, Evanston: Northwestern University Press, 1942.

Jan Lauwereyns, *The Anatomy of Bias: How Neural Circuits Weigh the Options*, Cambridge, MA: MIT Press, 2010.

A.D. Lawrence, A.J. Calder, S.M. McGowan and P.M. Grasby, "Selective disruption of the recognition of facial expressions of anger", *NeuroReport,* 13 (2002), 881-884.

A.D. Lawrence and A.J. Calder, "Homologizing human emotions". In D. Evand and P. Cruse, editors, *Emotions, Evolution and Rationality,* pages 15-47. New York: Oxford University Press, 2004.

D. Lehmann, "Plausibility logic", Proceedings of CSLI91, 1992a.

D. Lehmann, "Plausibility logic", Tech. Rept. TR-92-3, Feb. 1992, Hebrew University, Jerusalem 91904, Israel, 1992b.

D. Lehmann and M. Magidor, "What does a conditional knowledge base entail?", *Artificial Intelligence,* 55, (1992) 1-60.

Ernest Lepore and Kirk Ludwig, Donald Davidson: *Meaning, Truth, Language and Reality,* New York: Oxford University Press, 2007.

Hector J. Levesque, "All I know: An abridged report, *AAAI 1987,* 426-431.

Isaac Levi, "Subjunctives, dispositions and chances", *Synthese,* 34 (1977), 423-455.

Isaac Levi, *The Enterprise of Knowledge,* Cambridge MA: MIT Press, 1980.

C.I. Lewis, "Implication and the algebra of logic", *Mind* 21 (1912), 522-531.

C.I. Lewis and C.H. Langford *Symbolic Logic,* New York: Century-Croft, 1932.

M. Li and P. Vitányi, *An Introduction to Kolmogorov Complexity and its Applications,* Berlin: Springer-Verlag, second edition, 1997.

V. Lifschitz, "Formal theories of action: Preliminary report". In John McDermott, editor, *Proceedings of the Tenth International Joint Conference on Artificial Intelligence,* Los Altos, CA: Morgan Kaufmann, 1987.

V. Lifschitz, "Circumscriptive theories: A logic-based framework for knowledge representation", *Journal of Philosophical Logic,* 17, (1988), 391-441.

V. Lifschitz, "Benchmark problems for formal nonmonotonic reasoning:, in *Proceedings of the Second International Workshop on Non-Monotonic Reasoning,* 1989, 202-219.

V. Lifschitz, "Circumscription". In D.M. Gabbay, C. Hogger and J.A. Robinson, editors, *Handbook of Logic in Artificial Intelligence and Logic Programming,* volume 3, *Nonmotonic Reasoning and Uncertain Reasoning,* pages 297-352. Oxford: Oxford University Press, 1994.

F. Lin and Y. Shoham, "Argument systems: A uniform basis for nonmonotonic reasoning". In *Proceedings of 1ˢᵗ International Conference on Principles of Knowledge Representation and Reasoning,* pages 245-255, Stanford, CA, 1989.

Peter Lipton, *Inference to the Best Explanation,* London: Routledge, 1991. Second edition 2004.

John Locke, *An Essay Concerning Human Understanding,* edited and with an introduction by John W. Yolton, in two volumes. London: Dent 1961. First published in 1690.

G. Loomes and C. Taylor, "Non-transitive preferences over gains and losses", *Economic Journal*, 102 (1992), 357-365.

L.L. Lopes, "Doing the impossible: A note on induction and the experience of randomness, *Journal of Experimental Psychology: Learning, Memory and Cognition*, 8, (1982), 626-636.

Paul Lorenzen, "Ein dialogischesches Konstrucktivitätskriterium", *Infinistic Methods. Proceedings of the Sumposium on the Foundations of Mathematics*, pages 193-200, Oxford: Pergamon and Warsaw: PWN, 1961.

Paul Lorenzen and Kuno Lorenz, *Dialogische Logik*, Darnstadt: Wissenschaftliche Buchzesellscraft, 1978.

W. Łukaszewicz, "Nonmonotonic logic for default theories", *Proceedings of the European Conference on Artificial Intelligence – '84*, Pisa, 305-314.

Hugh MacColl, "If and Imply", *Mind* 17 (1908), 151-152.

John Macnamara, *A Border Dispute: The Place of Logic in Psychology.* Cambridge, MA: MIT Press, 1986.

N. McCain and H. Turner, "Causal theories of action and change", *Proceedings AAAI 97*, pp. 460-465, 1997.

J. McCarthy and P.J. Hayes, "Some philosophical problems from the standpoint of artificial intelligence". In B.M.D. Mickie, editor, *Machine Intelligence 4*, pages 463-502. University of Edinburgh Press, 1969.

John McCarthy, "Circumspection – A form of non-monotonic reasoning", *Artificial Intelligence*, 13 (1980), 27-39.

John McCarthy, "Applications of circumscription to formalizing common sense knowledge", *Artificial Intelligence*, 26 (1986), 89-116.

D. McDermott and J. Doyle, "Nonmonotonic logic I", In D.G. Bobrow, editor, Special Issue on Nonmonotonic Logics, *Artificial Intelligence*, 13 (1980), 41-72.

Drew McDermott, "A temporal logic for reasoning about processes and plans", *Cognitive Science*, 61, (1982), 101-155.

J.D. Mackenzie, "Question-begging in non-cumulative systems", *Journal of Philosophical Logic,* 8 (1978), 117-133.

Lorenzo Magnani, *Abduction, Reason and Science: Processes of Discovery and Explanation.* Dordrecht: Kluwer and New York: Plenum, 2001.

Lorenzo Magnani, *Morality in a Technological World: Knowledge as a Duty*, Cambridge, Cambridge University Press, 2007.

Lorenzo Magnani, *Abduction Cognition: The Epistemological and Eco-Cognitive Dimensions of Hypothetical Reasoning.* Heidelberg: Springer, 2009.

Lorenzo Magnani, "The non-fictional nature of scientific models. Model-based science as epistemic warfare", to appear in the not yet named Proceedings of MBR-2012, Heidelberg: Springer, 2013.

D.C. Makinson, "The paradox of the preface", *Analysis,* 25 (1965), 205-207.

D. Makinson, "General patterns in nonmonotonic reasoning". In D. Gabbay, C. Hogger and J. Robinson, editors, *Handbook of Logic in Artificial Intelligence and Logic Programming,* volume 3, pages. 35-110. Oxford and New York: Oxford University Press, 1994.

David Makinson, *Bridges From Classical to Nonmonotonic Logic,* London: College Publications, 2005.

Ruth Barcan Marcus, *Modalities: Philosophical Essays.* Oxford and New York: Oxford University Press, 1993.

S.L. Marcus and L.J. Rips, "Conditional reasoning", *Journal of Verbal Learning and Verbal Behavior,* 18 (1979), 199-224.

Edward Mares, "Relevant logic and the theory of information, *Synthese* 109 (1997) 1-19.

Marion, Mathieu. 2013. "Game semantics and the history of logic: The case of Greek dialectics". Forthcoming.

Marion, Mathieu & Rückert, Helge, "Aristotle on universal quantification". Forthcoming.

Gerald J. Massey, "Are there any good arguments that bad arguments are bad?", *Philosophy in Context*, 4, (1975a), 61-77.

Gerald J. Massey, "In defense of the asymmetry", *Philosophy in Context*, 4 (1975b), 44-45, supplementary volume.

Gerald J. Massey, "The fallacy behind fallacies", *Midwest Studies in Philosophy*, 6 (1981), 489-500.

Deborah Mayo and Rachelle Hollander, editors, *Acceptable Evidence: Science and Values in Risk Management*, New York: Oxford University Press, 1991.

Deborah Mayo, *Error and the Growth of Experimental Knowledge*, Chicago: University of Chicago Press 1996.

Deborah Mayo and Aris Spanos, *Error and Inference: Recent Exchanges on Experimental Reasoning, Reliability and the Objectivity and Rationality of Science*, Cambridge: Cambridge University Press, 2010

Joke Meheus, Liza Verhoeven, Maarten van Dyck and Dagmar Provijn, "Ampliative adaptive logics and the foundation of logic-based approaches to abduction". In *Logical and Computational Aspects of Model-based Reasoning*, editors, Lorenzo Magnani, Nancy J. Nersessian and Claudio Puzzi. Dordrecht and Boston: Kluwer, pages 39-77, 2002.

B.R. Mellers, R. Hertwig and D. Kahneman, "Do frequency representations eliminate conjunction effect? An exercise in adversarial collaboration", *Psychological Science*, 12 (2001), 269-275.

Hugo Mercier and Dan Sperber, "Why do human beings reason? Arguments for an argumentative theory", *Behavioural and Brain Sciences,* 34 (2011), 57-111.

John-Jules Meyer and Wiebe van der Hoek, *Epistemic Logic for AI and Computer Science,* Cambridge: Cambridge University Press, 1995.

John Stuart Mill, *A System of Logic,* London: Longman's Green, 1959. Originally published in 1843.

John Stuart Mill, *On Liberty,* E. Rapaport, editor. Indianapolis: Hackett, 1978. Originally published in 1859.

Ruth Millikan, *Language, Thought and Other Biological Categories,* Cambridge, MA: MIT Press, 1984.

Jack Minker "Logic and databases: Past, present and future", *AI Magazine* 18 (1997), 21-47

Marvin Minsky "A framework for representing knowledge", *Tech Report* 306, Artificial Intelligence Laboratory, MIT, 1974.

Marvin Minsky, "A framework for representing knowledge". In P. Winston, editor, *The Psychology of Computer Vision,* pages 211-277. New York: McGraw-Hill, 1975.

E. Mishan, *Cost Benefit Analysis: An Introduction.* New York: Praeger, 1971.

Philippe Mongin, "Consistent Bayesian aggregation", *Journal of Economic Theory,* 66 (1995), 313-351.

Richard Montague, *Formal Philosophy: Selected Papers of Richard Montague.* Richmond H. Thomason, editor. New Haven and London: Yale University Press, 1974.

James Montmarquet, *Epistemic Virtue and Doxastic Responsibility,* Lanham, MD: Rowman and Littlefield, 1993.

Robert C. Moore, "Semantical consideration on non monotonic logic". In *Proceedings of IJCAI* (1983) pages 272-279, Karlruhe: Morgan Kaufmann.

Robert C. Moore, "A formal theory of knowledge and action". In Jerry R. Hobbs and Robert C. Moore, editors, *Formal Theories of the Commonsense World*, pages 319-358. Norward, NJ: Ablex, 1985.

Robert C. Moore, "Autoepistemic logic". In P. Smets, E.H. Mamdani, D. Dubois and H. Prade, editors, *Non-Standard Logics for Automated Reasoning*, pages 105-136, New York: Academic Press, 1988.

Robert C. Moore, "Autoepistemic logic revisited", *Artificial Intelligence*, 59 (1993), 27-30.

Robert C. Moore "Visualizing the group homomorphism theorem", *College Mathematics Journal*, 26 (1995),143.

Charles Morgan "The nature of nonmonotonic reasoning", *Minds and Machines*, 10 (2000), 321-360.

John Morgan and Gang Tian, *Ricci Flow and the Poincaré Conjecture*, Cambridge, MA: Clay Mathematics Institute, 2007.

Adam Morton, "Knowing what to think about: When epistemology meets the theory of choice". In Stephen Hetherington, editor, *Epistemology Futures*, pages 111-130, New York: Oxford University Press, 2006.

Adam Morton, *Bounded Thinking: Epistemic Virtues for Limited Agents*, New York: Oxford University Press, 2013.

L. Moss and I. Viglizzo, "Final coalgebras for functors on measurable spaces", *Information and Computation*, 204 (2006), 610-636.

B.B. Murdock, Jr., "A theory for the storage and retrieval of item and associative information", *Psychological Review*, 89 (1982), 609-625.

B. Nebel, "Belief revision and default reasoning: Syntax-based approaches." In *Proceedings of the Second Conference on Knowledge Representation*, pages 417-428, San Mateo: Morgan Kaufman, 1991.

Y. Neuman, "Go ahead, prove that God doesn't exist!", *Learning and Instruction*, 13 (2003), 367-380.

Y. Neuman and E. Weitzman, "The role of text representation in students' ability to identify fallacious arguments", *Quarterly Journal of Experimental Psychology: Human Experimental Psychology*, 56A (2003), 849-864.

Y. Neuman, M.P. Weinstock and A. Glasner, "The effect of contextual factors on the judgment of informal reasoning fallacies", *Quarterly Journal of Experimental Psychology: Human Experimental Psychology*, 59A (2006), 411-425.

Shaun Nichols, Stephen Stich and Jonathan M. Weinberg, "Meta-skepticism: Meditations in ethno-epistemology". In S. Luper, editor, *The Skeptics,* pages 227-247, Aldershot: Ashgate Publishing, 2003.

Friedrich Nietzsche, *Beyond Good and Evil,* translated by Judith Norman and edited by Rolf-Peter Norstmann, Cambridge: Cambridge University Press, 2002. First published in 1886.

Friedrich Nietzsche, *The Gay Science: With a Prelude in Rhymes and an Appendix of Songs,* translated with commentary by Walter Kaufmann, New York: Vintage Books, 1974. First published in 1882.

R.E. Nisbett and L. Ross, *Human Inference: Strategies and Shortcomings of Social Judgment,* Englewood Cliffs, NJ: Prentice-Hall, 1980.

R. Nozick, *Philosophical Explanations,* Oxford University Press, 1978.

G. Nuchelmans, "On the fourfold root of the *argumentum ad hominem"*. In E.C.W. Krabbe, R.J. Dalitz and P.A. Smit, editors,. *Empirical Logic and Public Debate: Essays in Honour of Else M. Barth* pages 37-47, Amsterdam: Rodopi, 1993.

M.R, Oaksford, N. Chater, B. Grainger and J. Larkin, "Optimal data selection in the reduced array selection task (rast.) *Journal of Experimental Psychology: Learning, Memory and Cognition,* 23 (1997), 441-458.

Mike Oaksford and Ulrike Hahn, "A Bayesian approach to the argument from ignorance", *Canadian Journal of Experimental Psychology,* 58 (2004), 78-85.

James W. Oliver, "Formal fallacies and other invalid arguments", *Mind*, 76 (1967), 463-478.

Eric Pacuit, Oliver Roy and Johan van Benthem, "Toward a theory of play: A logical perspective on games and interaction", *Games* 2 (2011), 52-86.

Fabio Paglieri and John Woods "Enthymematic parsimony", *Synthese*, 178 (2011), 461-501.

F. Paoli, *Substructural Logics: A Primer*, Kluwer Academic, 2002.

Woosuk Park, "On classifying abduction", *Foundations of Science*, forthcoming.

Marc Pauly and Rohit Parikh, editors, *Game Logic*, a special number of *Studia Logica*, 72 (2003).

L. Paulson, "Final coalgebras as greatest fixed points in ZF set theory", *Mathematical Structures in Computer Science*, 9 (1999), 545-567.

Pauly, Marc & Parikh, Rohit, (eds.). 2003. *Game Logic*, a special issue of *Studia Logica*, 72, 163-256.

Giuseppe Peano, *Formulario Mathematico*, volume 5. Turin: Bocca, 1908.

Judea Pearl, *Heuristics: Intelligent Search Strategies for Computer Problem Solving*, Reading, MA: Addison-Wesley, 1984.

Judea Pearl, *Probabilistic Reasoning in Intelligent Systems: Networks of Plausible Inference*. San Mateo, CA: Morgan Kaufmann, 1988.

Judea Pearl, "System Z: A Natural Ordering of Defaults with Tractable Applications to Default Reasoning". In Rohit Parikh, editor, *Proceedings of the Third Conference on Theoretical Aspects of Reasoning about Knowledge*. San Mateo, CA: Morgan Kaufmann, 1990.

C.S. Peirce, *Collected Works,* Cambridge, MA: Harvard University Press, 1931-1958. A series of volumes, the first appearing in 1931.

C.S. Peirce, *Reasoning and the logic of things: The Cambridge Conference lectures of 1898*, Cambridge, MA: Harvard University Press, 1992. Kenneth Laine Ketner, editor, with an introduction by Kenneth Laine Ketner and Hilary Putnam.

Francis Jeffry Pelletier, editor, *Kinds, Things and Stuff: Mass Terms and Generics*, New York: Oxford University Press, 2010.

F.J. Pelletier and R. Elio, "The case for psychologism in default and inheritance reasoning", *Synthese*, 46 (2005), 7-35.

Francis Jeffry Pelletier and Nicholas Asher, "Generics and defaults". In John van Benthem and Alice ter Meulen, editors, *Handbook of Logic and Language*, pages 1125-1177. Amsterdam: North-Holland, 1997.

Alex (Sandy) Pentland, *Honest Signals: How They Shape Our World*, Cambridge, MA: MIT Press, 2008.

Andy Perkins, *The Mangle of Practice: Time, Agency and Science*, Chicago: University of Chicago Press, 1985.

Robert Peston, *Brown's Britain*. London: Short Books, 2006.

Stanley Peters and Dag Westerståhl, *Quantifiers in Logic and Language*, Oxford: Oxford University Press, 2006.

Robert C. Pinto, *Argument, Inference and Dialectic*, Dordrecht and Boston: Kluwer, 2001.

Alvin Plantinga, *Warrant and Proper Function*. New York: Oxford University Press, 1993.

Jan Playa, "Logics of public communications", in M.L. Emrich, M. Hadzikadic, M.S. Pfeifer and Z.W. Ras, editors, *Proceedings of the Fourth International Symposium on Methodologies for Intelligent Systems*, 1989, 201-216.

John Pollock, *Cognitive Carpentry*. Cambridge, MA: MIT Press, 1995.

John L. Pollock, *Thinking About Acting: Logical Foundations for Rational Decision Making,* Oxford: Oxford University Press, 2006.

John Pollock, "Defeasible reasoning". In Jonathan Adler and Lance Rips, editors, *Reasoning: Studies in Human Influence and its Foundations,* pages 451-471. Cambridge and New York: Cambridge University Press, 2008.

Karl Popper, *Logic of Scientific Discovery.* London: Hutchinson, 1959. German original published in 1934.

K. R. Popper, *Conjecture and Refutation,* Routledge & Kegan Paul, 1963.

Richard Posner, *Economic Analysis of Law,* Boston: Little Brown, 1973.

Sandeep Prasada, "Conceptual representation and some forms of genericity". In Francis Jeffry Pelletier, editor, *Kinds, Things and Stuff: Mass Terms and Generics,* pages 36-59. New York: Oxford, 2010.

Graham Priest, "The logic of paradox", *Journal of Philosophical Logic,* 8 (1979), 219-241.

Graham Priest, Richard Routley and Jean Norman, editors, *Paraconsistent Logic,* Philosophia Verlag, 1989.

A. Prince and P. Smolensky, *Optimality Theory: Constraint Interactions in Generative Grammar,* Cambridge, MA: MIT Press, 1993.

Hilary Putnam, "Is logic empirical?". In Robert S. Cohen and Marx W. Wartofsky, editors, *Boston Studies in the Philosophy of Science,* col. 5, pp. 216-241. Dordrecht: D. Reidel, 1968.

W.V. Quine, "Natural kinds". In *Ontological Relativity and Other Essays,* pages 114-138. New York: Columbia University Press, 1969.

W.V. Quine, *Methods of Logic,* New York: Rinehart & Winston, 1950. Revised edition by Harvard University Press, 1982.

W.V. Quine, *Philosophy of Logic.* Englewood Cliffs, NJ: Prentice-Hall, 1970, 2nd edition 1986.

W.V. Quine, *Theories and Things.* Cambridge, MA: Harvard University Press, 1981.

W.V. Quine, *Pursuit of Truth.* Cambridge, MA: Harvard University Press, 1990.

Rahman, Shahid & Rückert, H., editors, *New Perspectives in Dialogical Logic,* a special issue of *Synthese,* 78 (2001).

H. Raiffa, *Decision Analysis,* Reading, MA: Addison-Wesley, 1968.

F.P. Ramsey, *The Foundations of Mathematics and Other Logical Essays,* editor R.B. Braithwaite. Routledge and Keegan Paul, 1931.

Petrus Ramus, *Aristotelica Animadversiones,* Paris, with an introduction by W. Risse, Stuttgart-Bad Cannstatt: Fromann, 1964.

John Rawls, *A Theory of Justice,* Harvard University Press, 1971.

Stephen Read, *Relevant Logic,* Oxford: Blackwell, 1988.

Hans Reichenbach, *Experience and Prediction,* Chicago: University of Chicago Press, 1952, originally published in 1938.

Raymond Reiter, "On closed world data bases", *Journal of Logic and Data Bases* (1978) 55-76.

Raymond Reiter, "A logic for default reasoning", *Artificial Intelligence,* 12, (1980), 81-132.

Raymond Reiter, "Non-monotonic reasoning", *Annual Reviews of Computer Science,* 2 (1987), 147-186.

R. Reiter and J. de Kleer, "Formal foundations for assumption based truth maintenance systems: preliminary report" In *Proceedings of AAAI-87* pages 183-188, Seattle Washington, 1987.

Bryan Renne, *Dynamic Epistemic Logic with Justification,* PhD dissertation, Computer Science Department, City University of New York Graduate Center, 2008.

Bryan Renne, "Evidence-elimination in multi-agent justification logic". In A. Heifetz, editor, *Theoretical Aspects of Rationality and Knowledge, Proceedings of the Twelve Conference (TARK 2009)*, ACM Publications, pages 227-236, 2009.

Nicholas Rescher, *Studies in Logical Theory,* Oxford: Blackwell, 1968.

Nicholas Rescher, *Plausible Reasoning,* Amsterdam: van Gorcum, 1976.

Nicholas Rescher, *The Limits of Science,* Berkeley and Los Angeles: University of California Press, 1984.

Nicholas Rescher, *Error: Our Predicament When Things Go Wrong,* Pittsburgh: University of Pittsburgh Press, 2007.

Greg Restall, "Information flow and relevant logic", *Logic, Language and /computation: The 1994 Morago Proceedings,* 1995.

Greg Restall, "Negation in relevant logic (How I stopped worrying and learned to love the Routley star".) In Dov M. Gabbay and Hans Wansing, editors, *What is Negation*, pages 53-76, Dordrecht: Kluwer, 1999.

Greg Restall, "Relevant and substructural logic". In Dov Gabbay and John Woods, editors *Logic and the Modalities in the Twentieth Century*, pages 289-446, volume 7 of *The Handbook of the History of Logic.* Amsterdam: North-Holland, 2006.

R.B. Ricco, "The macrostructure of informal arguments: A proposed model and alalysis", *Quarterly Journal of Experimental Psychology: Human Experimental Psychology,* 56A (2003), 1021-1051.

Lance Rips, "Cognitive processes in propositional reasoning", *Psychological Review* (1983), 38-71.

Lance Rips, *The Psychology of Proof: Deduction in Human Thinking,* Cambridge, MA: MIT Press, 1994.

L. J. Rips, "Circular reasoning", *Cognitive Science,* 26 (2002), 767-795.

E.T. Rolls and T.R. Scott, "Central taste anatomy and neurophysiology". In I.R.L. Doty, editor, *Handbook of Olfaction and Gustation*, New York: Dekker, 1994.

Eleanor Rosch, "Principles of categorization". In Eleanor Rosch and B.B. Lloyd, editors, *Cognition and Categorization*, pages 27-48. Mahwah, NJ: Erlbaum, 1978.

S.J. Rosenschein, "Plan synthesis: A logical perspective". In P.J. Hayes, editor, *Proceedings of the International Joint Conference on Artificial Intelligence,* pages 331-337. Vancouver, BC: Morgan-Kaufmann Publishers, 1981.

A. Ross, "Why do we believe what we are told?", *Ratio*, 26 (1986), 69-88.

William W. Rozeboom, "Why I know so much more than you do", *American Philosophical Quarterly*, 4 (1967) 281-290.

Bertrand Russell, "Mathematical knowledge as based on the theory of types", *American Journal of Mathematics,* 30 (1908) 222-262.

S. Russell and P. Norvig, *Artificial Intelligence: A Modern Approach,* Upper Saddle River, NJ: Prentice-Hall, 1995.

J.M. Saguillo, "Methodological practice and complementary concepts of logical consequence: Tarski's model-theoretic consequence and Corcoran's information-theoretic consequence", *History and Philosophy of Logic,* 30 (2009), 21-48.

Gabriel Salomon, *Distributed Cognitions: Psychological and Educational Considerations,* New York: Cambridge University Press, 1997.

Paul Samuelson, "A note on the pure theory of consumer's behaviour", *Economics,* 5 (1938), 61-71.

Erik Sandewell, "An approach to the frame problem and its implementation." In B. Metzer and D. Michie, editors, *Machine Intelligence 7*, pages 195-204. Edinburgh: Edinburgh University Press, 1972.

Crispin Sartwell, "Knowledge is merely true belief", *American Philosophical Quarterly*, 28 April (1991), 157-165.

Crispin Sartwell, "Why knowledge is merely true belief", *The Journal of Philosophy*, 89, April (1992), 167-180.

Robert Schank and Robert Abelson, *Scripts, Plans, Goals and Understanding: An Inquiry into Human Knowledge Structures*, Hillsdale, NJ: Erlbaum, 1977.

Karl Schlecta, *Coherent Systems*, volume 2 of *Studies in Logic and Practical Reasoning*, Amsterdam: Elsevier, 2004.

Peter Schotch, Bryson Brown and Raymon Jennings, editors, *On Preserving: Essays on Preservationism and Paraconsistent Logic*, Toronto: University of Toronto Press, 2009.

John Searle, *Rationality in Action*, Cambridge, MA: MIT Press, 2001.

Teddy Seidenfelt and Joseph Kadane, "On the shared preferences of two Bayesian decision makers", *Journal of Philosophy*, 86 (1989), 225-244.

S.A. Selesnick and G.S. Owen, "The logic of schizophrenia", *Journal of Applied Logic*, 10 (2012), 115-126.

Wilfrid Sellars, *Empiricism and the Philosophy of Mind*, Minneapolis: University of Minnesota Press, 1956.

Sextus Empiricus, *Works*, with a translation by R.G. Bury, London: Loeb Classical Library 1933-1949. 4 volumes.

Glenn Shafer, *A Mathematical Theory of Evidence*, Princeton: Princeton University Press, 1976.

C. E. Shannon, "A mathematical theory of communication", *Bell System Technical Journal*, 27 (1948), p. 379-423, 623-656.

C.E. Shannon and W. Weaver, *The Mathematical Theory of Communication*, Urbana: University of Illinois Press, 1949.

Gila Sher, *The Bounds of Logic: A Generalized Viewpoint*. Cambridge, MA: MIT Press, 1991.

Richard M. Shiffrin, "Automatism and consciousness". In Jonathan D. Cohen and Jonathan W. Schooler, editors, *Scientific Approaches to Consciousness*, pages 49-64. Mahwah, NJ: Erlbaum, 1997.

Yoav Shoham, *A Semantical Approach to Nonmonotonic Logics*. San Mateo, CA: Morgan Kaufmann, 1987.

Yoav Shoham and Kevin Layton-Brown, *Multiagent Systems: Algorithmic, Game-Theoretic and Logical Foundations*, Cambridge: Cambridge University press, 2009.

R. Shope, *The Analysis of Knowing*, Princeton: Princeton University Press, 1983.

P. Shor, "Algorithms for quantum computation: Discrete logarithms and factorizing", *Proceedings of the 35th Annual Symposium on the Foundations of Computer Science*, 1994, 124-134.

Harvey Siegel, *Educating Reason: Rationality, Critical Thinking, and Education*, London: Routledge, 1988.

Ori Simchen, *Necessary Intentionality: A Study in the Metaphysics of Aboutness*. Oxford: Oxford Univrsity Press, 2012.

Herbert Simon, *Models of Man*, New York: John Wiley, 1957.

Herbert Simon, *Models of Thought*, volumes one and two, New Haven: Yale University Press, 1979.

Herbert Simon, *Reason in Human Affairs*, Stanford: Stanford University Press 1983.

Herbert Simon, *The Sciences of the Artificial*, third edition, Cambridge, MA: MIT Press, 1996.

Simon Singh, *Fermat's Enigma*, New York: Anchor, 1998.

Hartley Slater, *Logic is Not Mathematical*, London: College Publications, 2011.

D.M. Small, D.H. Zald, M. Jones-Gotman, R.J. Zatorre, J.V. Pardo, S. Frey and M. Petrides, "Brain imaging: human cortical gustatory areas: a review of functional neuroimaging data", *NeuroReport*, 10, (1999), 7-14.

Edward E. Smith and Douglas L. Medin, *Categories and Concepts*, Cambridge, MA: Harvard University Press, 1981.

Elliott Sober, "The adaptive advantage of learning and *à priori* prejudice". In Elliott Sober, *From a Biological Point of View: Essays in Evolutionary Philosophy*. Cambridge: Cambridge University Press 1994.

R.J. Solomonoff, "The discovery of algorithmic probability", *Journal of Computer and System Science*, 55 (1997), 73-88.

Roy A. Sorensen, *Blindspots*, Oxford: Clarendon Press, 1988.

Ernest Sosa, *Knowledge in Perspective*, Cambridge: Cambridge University Press, 1991.

Ernest Sosa, *A Virtue Epistemology: Apt Belief and Reflective Knowledge*, volume I, Oxford: Oxford University Press, 2007.

Dan Sperber and Deirdre Wilson, *Relevance: Communication and Cognition*, Oxford: Blackwell, 1986.

Dan Sperber and Deirdre Wilson, *Relevance: Communication and Cognition*. Second edition. Oxford: Basil Blackwell, 1995.

Wolfgang Spohn "A brief comparison of Pollock's defeasible reasoning and ranking functions", *Synthese*, 131 (2002), 39-56.

R. Sprengelmeyer, A.W. Young, U. Schroeder, P.G. Grossenbacher, J. Federlein, T. Buttnes and H. Przuntek, "Knowing no fear", *Proceedings of the Royal Society (Series B: Biology)*, 266, pages. 2451-2456, 1999.

Chandra Sripada and Alvin Goldman, "Simulationist models of face-based emotion", *Cognition*, 94 (2005), 193-213.

Robert Stalnaker, "Nonmonotonic consequence relations", *Fundamental Information*, 21 (1994), 7-21.

David Stanford, "Begging the question", *Analysis*, 32 (1972), 197-199.

Edward Stein, *Without Good Reason: The Rationality Debate in Philosophy and Cognitive Science,* Oxford: Clarendon Press, 1996

Keith Stenning, *Seeing Reason: Image and Language in Learning to Think,* Oxford: Oxford University Press, 2002.

K. Sterelny, *Thought in a Hostile World,* Oxford: Blackwell, 2004.

Stephen Stich and Richard Nisbett, "Justification and the psychology of human reasoning", *Philosophy of Science*, 47 (1980), 188-202.

Stephen Stich, *The Fragmentation of Reason,* Cambridge, MA: MIT Press, 1990.

George Joseph Stigler, "The economics of information", *Journal of Political Economy,* LXIX, volume 3, pp. 213-224, 1961.

Michael Strevens, *Depth* Harvard University Press, 2008.

Barry Stroud, *Engagement and Metaphysical Dissatisfaction: Modality and Value,* Oxford: Oxford University Press, 2012.

Patrick Suppes, "Models of data" in Ernest Nagel, Patrick Suppes and Alfred Tarski, editors, *Logic, Methodology and the Philosophy of Science: Proceedings of the 1960 International Congress,* pages 252-261, Stanford: Stanford University Press, 1962.

Patrick Suppes, "What is a scientific theory?" In Sidney Morgenbesser, editor, *Philosophy of Science Today,* pages 55-67. New York: Basic Books, 1967.

Patrick Suppes, "The role of formal methods in the philosophy of science". In P.D. Asquith and H.E. Kyburg, editors, *Current Research in the Philosophy of Science*, pages 16-27. East Lansing: Philosophy of Science Association, 1979.

Patrick Suppes, *Representation and Invariance of Scientific Structures,* Stanford: CSLI Publications, 2002.

Patrick Suppes, "Where do Bayesian Priors come from?" *Synthese,* 156 (2007), 441-471.

E.E. Sweetser, *From Etymology to Pragmatics: Metaphorical and Cultural Aspects of Semantic Structure,* Cambridge: Cambridge University Press, 1990.

John Symons, "Logical formal semantics for epistemology. In Sven Bernecker and Duncan Pritchard, editors, *The Routledge Companion to Epistemology,* pages 571-585. New York: Routledge, 2011.

W.P. Tanner and J.A. Swets, "A decision-making theory of visual detection", *Psychological Review,* 61 (1954), 401-409.

A. Tettamanzi, M. Thomassine and J. Jansen, *Soft Computing: Integrating Evolutionary, Neural and Fuzzy Systems,* Berlin: Springer, 2010.

Paul Thagard, *Conceptual Revolutions,* Princeton: Princeton University Press, 1992.

Paul Thagard, "Abductive inference: From Philosophical analysis to neural mechanisms". In A. Fieney and E. Heit, editors, *Inductive Reasoning: Experimental, Developmental, and Computational Approaches,* pages 226-247, Cambridge: Cambridge University Press, 2007.

Richmond H. Thomason, "Some limitations to the psychological orientation in semantic theory", *Journal of Philosophical Logic,* 40 (2011), 1-14; original ms 1983.

Christopher W. Tindale, *Fallacies and Argument Appraisal,* Cambridge and New York: Cambridge University Press, 2007.

Roberto Torretti, *Creative Understanding,* Chicago: University of Chicago Press, 1990.

Stephen Toulmin, *The Uses of Argument,* Cambridge: Cambridge University Press, 1958.

Stephen Toulmin and D.E. Leary, "The cult of empiricism in psychology, and beyond". In Sigmund Koch and David E. Leary, editors, *A century of Psychology as a Science,* pages 594-617, New York: McGraw-Hill, 1985.

D.S. Touretsky, *The Mathematics of Inheritance Systems,* Los Altos: Morgan Kaufmann, 1986.

J.D. Trout and Michael Bishop, *Epistemology and the Psychology of Human Judgment,* Oxford: Oxford University press, 2005.

Alan Turing, "On computable numbers, with an application to the *Entscheidungsproblem*", *Proceedings of the London Mathematical Society,* 42 (1936-1937), 230-265.

Johan van Benthem, "Resetting the bounds of logic", *European Review of Philosophy,* 4 (1999) 21-44.

Johan van Benthem, "Dynamic logic for belief revision", *Journal of Applied Non-Classical Logics,* 14 (2004), 1-26.

Johan van Benthem, Jan van Eijck and Barteld Kooi, "Logics of communication and change", *Information and Computation,* 204 (2006), 1660-1662.

Johan van Benthem, *Logical Dynamics of Information and Interaction,* Cambridge: Cambridge University Press, 2011.

P. Vanderschrsaf and G. Sillari, "Common knowledge", in Edward Zalta, editor, *The Stanford Encylopia of Philosophy,* 2007, http://plato.Stanford.edu/archives/spr2009/entries/common-knowledge/

Hans van Ditmarsch, Wiebe van der Loek and Bartold Kooi, "Dynamic epistemic logic with assignment". In F. Dignum, V. Dignum, S. Koening, S. Kraus, M. Singh and M. Wooldridge, editors, *Proceedings of the Fourth International joint Conference on Autonomous Agents and Multiagent Systems,* ACM, 2005, pages 141-148.

Frans van Eemeren and Rob Grootendorst, *Argumentation, Communication and Fallacies*, Hillsdale, NJ: Erlbaum, 1992.

Frans H. van Eemeren and Rob Grootendorst, *A Systematic Theory of Argumentation*, New York: Cambridge University Press, 2004.

Frans van Eemeren, Bart Garssen and Bert Meuffels, *Fallacies and Judgments of Reasonableness*, Dordrecht: Springer, 2009.

Bas C. van Fraassen, "Belief and the will", *Journal of Philosophy*, 81 (1984), 235-256.

Bas C. van Fraassen, "Figures in a probability landscape", in J. Michael Dunn and Anil Gupta, editors, *Truth or Consequences,* pages 345-356, Dordrecht: Kluwer, 1990.

Bas C. van Fraassen, "The day of the dolphins: Puzzling over epistemic partnership". In Kent A. Peacock and Andrew Irvine, editors, *Mistakes of Reason: Essays in Honour of John Woods,* pp. 111-133, Toronto: University of Toronto Press, 2005.

Bas C. van Fraassen, "Thomason's paradox for belief, and two consequence relations", *Journal of Philosophical Logic*, 40 (2011) 15-32.

Tim van Gelder, "Enhancing and augmenting human reasoning". In António Zilhão, editor, *Evolution, Rationality and Cognition: A Cognitive Science for the Twenty-First Century,* pages 162-181, London and New York: Routledge, 2005.

Lev Vygotsky, *Thought and Language*, MIT Press, 1962.

Jonathan Vogel, "Reliabilism levelled", *Journal of Philosophy,* XCVII (2000), 602-23.

Jonathan Vogel, "Epistemic bootstrapping", *Journal of Philosophy,* CV (2008), 518-539.

John R. Vokey and Philip A. Higham, "Opposition logic and neural network models in artificial grammar learning", *Consciousness and Cognition,* 13, (2004), 565-578.

John von Neumann and Oscar Morgenstern, *The Theory of games and Economic Behavior*. Princeton: Princeton University Press, 1947.

Peter P. Wakker, *Prospect Theory For Risk and Ambiguity*, New York: Cambridge University Press, 2010.

Douglas Walton, "What is reasoning? What is argument?", *Journal of Philosophy*, 87 (1990), 399-419.

Douglas Walton, *Begging the Question*, New York: Greenwood Press, 1991.

Douglas Walton, *Plausible Argument in Everyday Conversation*, Albany: State University of New York Press, 1992.

Douglas Walton, *A Pragmatic Theory of Fallacy*. Tuscaloosa: University of Alabama Press, 1995.

Douglas Walton, *Scare Tactics*, Dordrecht: Kluwer, 2000.

Hans Wansing, "Informational interpretation of substructural logics, *Journal of Logic, Language and Information*, 2 (1993a), 285-308.

Hans Wansing, *The Logic of Information Structures*, Berlin: Springer-Verlag, 1993b.

P.C. Wason, "Reasoning". In B.M. Foss, editor, *New Horizons in Psychology*, Hammondsworth, UK: Penguin, 1966.

Daniel Wegner, *The Illusion of Conscious Will*, Cambridge, MA: The MIT Press, 2002.

Jonathan M. Weinberg, Shaun Nichols and Stephen P. Stich, "Normativity and epistemic intuitions". In Joshua Knobe and Shaun Nichols, editors, *Experimental Philosophy*, pages 17-45. New York: Oxford University Press 2008.

Mark Weinstein, "Informal logic and applied epistemology". In Ralph H. Johnson and J. Anthony Blair, editors, *New Essays in Informal Logic*, pages 140-161. Windsor: Informal Logic Press 1994.

Mark Weinstein, *Logic, Truth and Inquiry,* London: College Publications, 2013.

P. Weirich, *Realistic Decision Theory,* New York: Oxford University Press, 2004.

Max Weiss, "A closer look at manifold consequence", *Journal of Philosophical Logic,* to appear in 2013.

M. Welbourne, "The transmission of knowledge", *The Philosophical Quarterly,* 29 (1979), 1-9.

M. Welbourne, "The community of knowledge", *The Philosophical Quarterly,* 3 (1981), 302-314.

M. Welbourne, *The Community of Knowledge,* Aldershot: Gregg Revivals, 1993.

Dag Westerståhl, "Decomposing generalized quantifiers", *Review of Symbolic Logic,* 1 (2008), 355-371.

Richard Whately, *Elements of Logic.* New York: William Jackson, 1836.

M. Wheeler and A. Clark, "Genie representation: Reconciling content and causal complexity". *British Journal for the Philosophy of Science,* 50, (1999), 103-135.

B. Wicker, C. Keysers, J. Plailly, J.-P. Royer, V. Gallese and G. Rizzelatti, "Both of us disgusted in *my* insula: the common neural basis of seeing and feeling disgust", *Neuron,* 40, (2003), 655-664.

W.A. Wicklegreen and D.A. Norman, "Strength models and serial position in short-term recognition memory", *Journal of Mathematical Psychology,* 3 (1966), 316-347.

Dallas Willard, "The case against Quine's case for psychologism". In Mark A. Notterno, editor, *Perspectives on Psychologism,* pages 286-295. Leiden: E.J. Brill, 1989.

L. Willard and L. Yuan, "The revised Gärdenfors postulates and update semantics". In, S. Abitebasel and P. Konellakis, editors, *Proceedings of the International Conference in Database Theory,* volume 470 of *Lecture Notes in Computer Science,* pages 409-421. Berlin: Springer Verlag, 1990.

E.O. Wilson, *Sociobiology: The New Synthesis,* Cambridge, MA: Harvard University Press, 1975.

Mark Wilson, "Generality and nomological form", *Philosophy of Science,* 46 (1979), 161-164.

W. Wimsatt, "Forms of aggregativity". In A. Donagan, N. Perovich and M. Wedin, editors, *Human Nature and Natural Knowledge,* pages 259-293, 1986.

Ludwig Wittgenstein *Philosophische Bemerkunzen,* Blackwell: Oxford University Press, 1956.

Ludwig Wittgenstein, *Lectures on the Foundations of Mathematics,* Ithaca: Cornell University Press, 1956.

John Woods, "Ideals of rationality in dialogic", *Argumentation, 2* (1988a), 395-408.

John Woods, "Rationality ideals and mentality", *Argumentation, 2* (1988b), 419-424.

John Woods, "Pragma-dialectics: A radical departure in fallacy theory", *Communication and Cognition,* 24 (1991) 43-53. Reprinted as chapter 9 of Woods (2004).

John Woods, "Dialectical Blindspots", *Philosophy and Rhetoric,* 26 (1993), 251-265.

John Woods "Deep disagreements and public demoralization". In Dov M. Gabbay and Hans Jurgen Ohlbach, editors, *Practical Reasoning: Springer Notes on Artificial Intelligence.* Berlin: Springer-Verlag, 650-662, 1996. Reprinted in Woods (2004).

John Woods "Aristotle", *Argumentation* 13 (1999b), 203-220.

John Woods, "Slippery slopes and collapsing taboos", *Argumentation,* 4 (2000), 107-134.

John Woods, *Paradox and Paraconsistency: Conflict Resolution in the Abstract Sciences*, Cambridge: Cambridge University Press, 2003.

John Woods, *The Death of Argument: Fallacies in Agent-Based Reasoning,* Dordrecht: Kluwer 2004.

John Woods, "Pragma-dialectics reconsidered". In M.A. van Rees and P. Houtlosser, editors, *Considering Pragma Dialectics,* pp. 301-311. Mahwah, NJ: Erlbaum, 2006.

John Woods "Lightening up on the *ad hominem",* *Informal Logic,* 27 (2007a) 109-134.

John Woods, "Should we legalize Bayes' theorem?" In Hans V. Hansen and Robert C. Pinto, editors, *Reason Reclaimed: Essays in Honor of J. Anthony Blair and Ralph H. Johnson,* pages 257-267. Newport News, VA: Vale Press, 2007b.

John Woods, "Agendas, relevance and dialogic ascent", *Argumentation* 3 (2007c), 209-221.

John Woods, "Beyond reasonable doubt: An abductive dilemma in criminal law", *Informal Logic,* 28 (2008a), 60-70.

John Woods "Rescher on aporetics and consistency". In Robert Almeder, editor, *Rescher Studies: A Collection of Essays on the Philosophical Work of Nicholas Rescher,* pages 493-512, Frankfurt: Ontos Verlag, 2008b.

John Woods, "SE 176a 10-12: Many questions for Julius Moravcsik". In Dagfinn Follesdall and John Woods, editors, *Logos and Language: Essays in Honour of Julius Moravcsik,* pages. 211- 220, London: College Publications, 2009b.

John Woods, "Making too much of worlds". In *Possible Worlds,* edited by Guido Imaguire and Dale Jacquette, pages 172-217, Munich: Philosophia Verlag, 2010d.

John Woods "Abduction and proof: A criminal paradox". In Shahid Rahman *et al.,* editors *Approaches to Legal Rationality,* pages 217-238. Dordrecht: Springer 2010e.

John Woods "MacColl's elusive pluralism". In Amirouche Moktefi and Stephen Read, editors, *Hugh MacColl After One Hundred Years,* a special number of *Philosophia Scientiae* 15 (2011), 163-189.

John Woods, "A history of the fallacies in western logic". In Dov M. Gabbay, Jeffry Pelletier and John Woods, editors, *Logic: A History of its Central Concepts* pages 513-610. Volume 11 of the *Handbook of the History of Logic.* Amsterdam: North-Holland, 2012.

John Woods, "Ancestor worship in the logic of games: How foundational were Aristotle's contributions?", to appear in Proceedings of the Games, Game Theory and Game Semantics Conference, Riga, Latvia, 2013a.

John Woods, "Against fictionalism", to appear in Proceedings of MBR-2012, Heidelberg: Springer, 2013b

John Woods and Hans V. Hanson, "Hintikka on Aristotle's fallacies", *Synthese*, 13 (1997), 217-239.

John Woods and Douglas Walton, "On fallacies", *Journal of Critical Analysis* 5 (1972), 103-111. Reprinted in Woods and Walton (1989/2007).

John Woods and Douglas Walton, *"Argumentum ad verecundiam"*, *Philosophy and Rhetoric*, 7 (1974) 135-153. Reprinted in Woods and Walton (1989/2007).

John Woods and Douglas Walton, "Petitio principii", *Synthese*, 31 (1975), 107-128.

John Woods and Douglas Walton, *"Petitio* and relevant many-premissed arguments", *Logique et Analyse,* 20 (1977), 97-110. Reprinted in Woods and Walton (1989/2007).

John Woods and Douglas Walton, "Arresting circles in formal dialogues", *Journal of Philosophical Logic,* 7 (1978), 73-90. Reprinted in Woods and Walton (1989/2007).

John Woods and Douglas Walton, "Question begging and cumulativeness in dialectical games", *Noûs,* 16 (1982), 585-605. Reprinted in Woods and Walton (1989/2007).

John Woods and Douglas Walton, *Argument: The Logic of the Fallacies,* Toronto: McGraw-Hill Ryerson, 1982.

John Woods and Douglas Walton, *Fallacies: Selected Papers 1972-1982,* 2nd edition, with a Foreword by Dale Jacquette, xvi, 322. London: College Publications, 2007. First published in 1989.

John Woods, Dov M. Gabbay, Ralph H. Johnson and Hans Jürgen Ohlbach, "Logic and the practical turn". In D.M. Gabbay, R.H. Johnson, H.J. Ohlbach and J. Woods, editors, *Handbook of the Logic of Argument and Inference: The Turn Toward the Practical,* pages 1-39. Amsterdam: North-Holland, 2002.

John Woods and Alirio Rosales, "Virtuous distortion in model-based science". In Lorenzo Magnani and Walter Carnielli, editors, *Model-Based Reasoning in Science and Technology: Abduction, Logic and Computational Discovery,* pages 3-30, Berlin: Springer, 2010a.

John Woods and Alirio Rosales, "Unifying the fictional". In John Woods, editor, *Fictions and Models: New Essays,* pp. 345-388, Munich: Philosophia Verlag, 2010b.

Michael Wooldridge and Nicholas R. Jennings, "Intelligent agents: Theory and practice", *The Knowledge Engineering Review,* 10 (1995), 115-152.

J.C. Wylie, *Military Strategy: A General Theory of Power Control.* New Brunswick, NJ: Rutgers University Press, 1967.

Linda Zagzebski, "Virtue in eithics and epistemology", *American*

Catholic Philosophical Quarterly, 71 Supp (1997), 1-17.

Linda Zagzebski, "Virtue epistemology". In the *Routledge Encyclopedia of Philosophy.* New York: Routledge, 2005.

Jésus Zamora-Bonilla, "Science studies and the theory of games", *Perspectives on Science,* 14 (2006), 525-557.

Manfred Zimmerman, "The nervous system and the context of information theory". In R.F. Schmidt and G. Thurs, editors, *Human Physiology,* 2nd edition, pages 166-175, translated by Marguerite A. Biederman-Thorson. Berlin: Springer-Verlag, 1989.

INDEX

591

filter, 207, 214, 208

Moore's paradox, 161
Moorean blindspots, 123
Morgan, C., 275
multiagency, 316, 317, 318

N

naturalistic turn in logic, 532
naturalized logic, 11, 15, 125
  of reasoning, 65
necessitation relation, 264
negation as failure, 348, 354, 403
negative "character" campaign, 454
neglect fallacy, 9
Nilsson, N. J., 172
NN-convergence, 54, 55, 57, 59,
  61, 68, 86, 195
  thesis, 214, 221, 313
no telling-no fact principle, 352,
  353
nomological force, 252
non sequitur, 392
non-cause as cause, 137
noncircularity condition, 425, 426
nonmonotonic consequence, 69,
  232, 239, 265
nonmonotonic inheritance
  reasoning, 234
nonmonotonic logics, 230
nonmonotonic reasoning, 70
non-universal generalization, 219
non-universal quantifications, 372
    plural, 251
non-wellfounded sets, 438
normal and normative, 53, 55
normative adequacy, 42
normative authority, 533
normative legitimacy, 51
normative models, 68
normative self-insulation, 106
normativity and normalcy, 52
notable frequency, 142

O

occasion, 423, 424
occasioned frequency, 144
occasioned universality, 423
occurrent belief, 360
Oliver, J., 3
one-shot generalization, 149
opportunity costs, 197
optimality, 195
ordinary language philosophers, 80
orthodoxy, 70, 71, 72
OSC orientation, 195
OSCAR, 242
overconfidence bias, 9

P

paraclassical consequence, 239
paraconsistency, 34
paraconsistent logic, 34, 37
paraconsistentists, 37
paradigm creep, 533
paradox of analysis, 433
paradox of validity, 433
partial abductions, 381
Pascalian probability, 490
PD conception, 519
PD-misalignment, 518
Peirce, C. S., 98, 164, 184, 297,
  357, 374, 376, 387
Peirce's plausibilities, 298, 299
Peirce's schema, 375
Peirce's Triad, 383
Peircean surprise, 357, 349, 368,
  369, 376, 382, 398
Pelletier, F. J., 157, 218, 257
perceptual errors, 193
Perelman, G., 320
Perner, J., 192
Perry, R.B., 165
phenomenal inapparency, 164, 178
philosophy's most difficult
  problem, 89
pirates, 493, 504, 515, 513

quasi-showing, 421
quasi-success, 421
question-begging, 400, 403, 405,
    409, 411, 412, 415, 418, 419
    430, 431, 425
Quine, W. V., 17, 19, 23, 79, 99

R

Rabelais, F., 446
Raiffa, H., 49, 54
Ramus, P., 4
rational monotony, 237
rational reconstruction, 78, 486
Read, S., 526
reasoning down below, 125
recursive definition, 437
red herring, 141
referentialism, 31
reflective equilibria, 57
reflexivity, 235, 239, 248
refutation, 416, 417
regularity, 356
Reichenbach, H., 184
Reiter, R., 223, 229, 233
relevance, 29, 182
representational distortions, 482
repugnancy-to-reason, 194, 189
Rescher, N., 4, 29, 38, 39, 85
resolving Nixon, 255
resolving Tweety, 252
resource allocation, 197
respect for data principle, 14, 478,
    486
retro-model, 488
Riemann, B., 81
Riemannian geometry, 81
right reasoning, 2
ring of truth, 212, 297, 298
Rips, L., 158
Routley, R., 410
Rozeboom, W., 168
rule-breaking, 195
rules, 201
rule-breaking, 201

rupture, 228
Russell, B., 36, 79, 276, 488

S

satisficing, 189, 196, 197
saturated end state, 160, 163
sayso, 314, 322
    -chains, 332
    -elimination, 338
    -filters, 337, 339
    -manifold, 321, 322, 405
    -Schema, 326
s-belief, 161
scant resources, 196, 382, 383
scantness, 204
scarce information, 204
scarcity, 204
schizophrenia, 182
scripts, 220
Scriven, M., 139
secundum quid, 5, 137, 511
self-referential definition, 438
SEU orientation, 195
Sextus Empiricus, 427
short-selling, 106
showing false, 415
shrinkage, 301
shrinking violet fallibilism, 183,
    185
simulation theory, 468
slanging, 446
Sleigh, R., 498
Sleigh's Fallacy, 268, 281, 497,
    498, 526
small tasks, 15
small-surface, 340
smartening up, 204
Sperber, D., 29
S-proof, 419, 422, 424
star semantics, 22
start up costs, 197
Stich, S., 170
stipulation, 78
straw man, 5, 134

Lightning Source UK Ltd.
Milton Keynes UK
UKHW021402220223
417454UK00016B/336